Laser Techniques Applied to Fluid Mechanics

Springer

Berlin
Heidelberg
New York
Barcelona
Hong Kong
London
Milan
Paris
Singapore
Tokyo

R.J. Adrian · D.F.G. Durão · F. Durst · M.V. Heitor
M. Maeda · J.H. Whitelaw (Eds.)

Laser Techniques Applied to Fluid Mechanics

Selected Papers from the 9th International Symposium
Lisbon, Portugal, July 13-16, 1998

With 442 Figures

 Springer

Editors

R.J.Adrian

Dept. of Theoretical
and Applied Mechanism
University of Illinois
at Urbana - Champaign
Urbana, Illinois 61801, USA
e-mail: rja@curly.tam.uiuc.edu

D.F.G. Durão

Dept. of Mechanical Engineering
Instituto Superior Técnico
Av. Rovisco Pais
1049-001 Lisbao, Portugal
e-mail: ddurão@dem-ist.utl.pt

F. Durst

Lehrstuhl für Strömungstechnik
University of Erlangen - Nürnberg
Cauerstraße 4
91058 Erlangen, Germany
e-mail: fdurst@lstm.uni-erlangen.de

M.V. Heitor

Dept. of Mechanical Engineering
Instituto Superior Técnico
Av. Rovisco Pais
1049-001 Lisbao, Portugal
e-mail: mheitor@dem.ist.utl.pt

M. Maeda

Dept. of Mechanical Engineering
Keio University
1-14-1 Hiyoshi, Kohuku
Yokohama 223, Japan
e-mail: maeda@sd.keio.ac..jp

J.H. Whitelaw

Imperial College of Science
Technology and Medicine
Dept. of Mechanical Engineering
Exhibition Road
London SW7 2BX, England
e-mail: 100773.2135@compuserve.com

ISBN 3-540-66738-5 Springer-Verlag Berlin Heidelberg New York

CIP-data applied for

Die Deutsche Bibliothek - CIP-Einheitsaufnahme
Laser techniques applied to fluid mechanics : selected papers from the 9th international symposium, Lisbon, Portugal, July 13 - 16, 1998 / R. J. Adrian ... (ed.). - Berlin ; Heidelberg ; New York ; Barcelona ; Hong Kong ; London ; Milan ; Paris ; Singapore ; Tokyo : Springer, 2000
ISBN 3-540-66738-5

Springer-Verlag is a company in the specialist publishing group Bertelsmann/Springer
© Springer-Verlag Berlin Heidelberg 2000
Printed in Germany

The use of general descriptive names, registered names, trademarks, etc. in this publication does not imply, even in the absence of a specific statement, that such names are exempt from the relevant protective laws and regulations and therefore free for general use.

Typesetting: Camera-ready by editors
Cover layout: Struve & Partner, Heidelberg
SPIN: 10721860   Printed on acid-free paper   61 / 3020 hu - 5 4 3 2 1 0

Preface

This volume includes revised versions of selected papers presented at the **Ninth International Symposium on Applications of Laser Techniques to Fluid Mechanics** held at *the Calouste Gulbenkian Foundation* in Lisbon, during the period of July 13 to 16, 1998.

The papers describe *Instrumentation Developments* and results of measurements of *Turbulent Shear Flows, Aerodynamics Flows, Mixers and Rotating Flows, Combustion and Engines,* and *Two-Phase Flows*. The papers demonstrate the continuing and healthy interest in the development of understanding of new methodologies and implementation in terms of new instrumentation.

The prime objective of the Ninth Symposium was to provide a forum for the presentation of the most advanced research on laser techniques for flow measurements, and communicate significant results to fluid mechanics. The applications of laser techniques to scientific and engineering fluid flow research was emphasized, but contributions to the theory and practice of laser methods were also considered where they facilitate new improved fluid mechanic research. Attention was placed on laser-Doppler ancmometry, particle sizing and other methods for the measurement of velocity and scalars, such as particle image velocimetry and laser induced fluorescence.

We would like to take this opportunity to express our thanks to those who contributed to the success of the conference. The assistance provided by the Advisory Committee, by assessing abstracts, was highly appreciated. The companies who participated in the exhibition of equipment are also acknowledged. In addition, thanks go to the participants who contributed actively in discussions, learned from the presentations and were essential to the success of the Symposium. And last, but not least, we are highly indebted for the financial support provided by the Sponsoring Organizations that made the Symposium possible.

Sponsoring Organizations:

European Research Office: United States Army, Navy and Air Force; Luso-American Foundation for Development, FLAD; Calouste Gulbenkian Foundation; Portuguese Science and Technology Foundation, FCT; Instituto Superior Técnico, IST.

The Editors

Table of Contents

CHAPTER I. INSTRUMENTATION DEVELOPMENTS

CHAPTER II.   TURBULENT SHEAR FLOWS

CHAPTER III.   AERODYNAMICS

CHAPTER IV.   MIXERS AND ROTATING FLOWS

CHAPTER V.   COMBUSTION AND ENGINES

CHAPTER VI.   TWO-PHASE FLOWS

CHAPTER I.   INSTRUMENTATION DEVELOPMENTS

I.1. Wavenumber Spectrum Estimation from Irregularly Spaced Data: Application to PIV Data Processing
W. Hübner[‡], C. Tropea[†] and J. Volkert[‡]

[†] Fachgebiet Strömungslehre und Aerodynamik, TU Darmstadt,
Petersenstr. 30, 64287 Darmstadt, Germany
[‡] Lehrstuhl für Strömungsmechanik, University of Erlangen-Nürnberg,
Cauerstr. 4, 91058 Erlangen, Germany

Abstract. The problem of 2D wavenumber spectrum estimation from randomly sampled data is examined. Two new estimators are proposed, one based on a spatial sample and hold procedure and one based on the projection slice theorem, commonly used in tomography. Each estimator is tested using simulated PTV data. Initial results indicate that these estimators have the potential to extend spectral estimates beyond the convential Nyquist limits imposed when using regularly sampled data.

Keywords. 2D Wavenumber Spectrum, PIV

1. Introduction

When describing turbulent motions in a statistical sense much of our fundamental theoretical knowledge exists in the form of either spatial correlations or wavenumber spectrum (Tennekes & Lumley (1972)). Examining flows experimentally using point measurement techniques such as Hot Wire Anemometry or Laser Doppler Anemometry, requires therefore a transformation of the obtained velocity time series into spatial information, either through the simultaneous use of multiple probes or through application of the Taylor hypothesis. The first possibility is tedious at best and the second is only applicable for f-type correlations. In this respect Particle Image Velocimetry (PIV) provides a unique opportunity to estimate spatial correlations of velocity fluctuations, even for non-stationary and possibly also for non-homogeneous turbulence fields.

The term PIV is understood here to encompass also Particle Tracking Velocimetry (PTV), the usage being generally chosen according to particle density but in essence residing in the data processing applied. A good overview of current PIV and PTV techniques is given in (Meas. Sci. Technol. **8** (1997) Raffel et al. (1998)), whereby the increased use of 3D PTV is noteworthy. One main distinction between PIV and PTV is the fact that PIV velocity vectors are equally distributed in space, corresponding to a regular pattern of interrogation spots. PTV on the other hand provides a velocity vector for every identifiable tracer particle pair and as such, is a spatially irregular sampling of the velocity field. While most commercial systems offer only a PIV processing, methods do exist which allow, even for high particle densities, an individual particle tracking to be accomplished. The most well known of these is the Super-Resolution PIV presented in (Keane et al. (1995)).

Addressing the problem of wavenumber spectrum estimation from PIV or PTV data has many analogies with frequency spectrum estimation from velocity time series obtained using an LDA. In this analogy the roles of time and space are almost perfect counterparts. In the more familiar case of frequency spectrum estimation, it has long been recognized that the irregular (or near random) sampling of the velocity in time can be both advantageous and disadvantageous. The advantage lies in the principle ability to estimate the spectrum far beyond the average Nyquist frequency, while the disadvantage lies in having to find an estimator suitable for randomly sampled data. This topic has been exhaustively studied in recent years and an appropriate estimator for frequency spectra can be found in Nobach et al. (1998 & 1996). The main subject of the present paper is to investigate similar estimators applied to spatially irregular sampled data, to yield wavenumber spectra. Emphasis will be placed on 2D PTV data, although no fundamental exclusion of 3D PTV data can yet be identified.

Returning to the space/time analogy, it can be seen that PIV velocity vectors, which are local spatial averages over one interrogation spot, would correspond to short time averages of LDA data. Such time averaging of LDA data prior to further processing is never performed, as the high time resolution of LDA is sacrificed. Similarly, except for the aesthetics of a regular pattern of velocity vectors, there also doesn't appear to be any fundamental advantage of locally averaging the spatial data from a PIV system. Certainly other techniques have been proposed which dispense with this practice altogether and retain the spatial resolution offered by the chosen particle density (Ruck (1995)). One exception is the use of PIV vectors in the pre-estimation procedure of super-resolution PIV, which indeed will constitute its main justification in the present paper.

A final remark concerning the analogy is that stationarity in time will correspond to homogeniety in space and for the following discussion these states will be assumed.

Therefore, the starting point of the present paper is that PTV data is available and that an estimate of wavenumber spectra is to be made from the spatially irregularly sampled data. Much can be learned from previous experience of spectral estimation in time domain before proceeding. One of the first suggestions to deal with the irregular sampling in time, was to perform a direct Fourier transform (DFT) on the data, utilizing the arrival time information in the estimate but with the computational penalty of not being able to apply FFT algorithms (Gaster & Roberts (1977 & 1975)). This is not a viable approach due to the extremely high variability of the estimate, which was recognized by its authors at the time and since confirmed. Therefore, there is no reason to pursue a 2D DFT as a procedure to apply in the present case.

A more widely used technique is the slot correlation, in which resolution and variability are traded off through the choice of slot (or bin) width in lag time τ (Mayo et al. (1974) & Scott(1974)). This is no doubt a possible route to also obtain a 2D spatial correlation from PTV data. However it has been shown that refined reconstruction methods can achieve a lower estimator variance at high frequencies than the slot correlation (Nobach et al. (1998)).

In time series analysis, the slot correlation also displays a bias if the data set mean was poorly estimated or if there was a velocity/sample rate correlation (the origin of statistical velocity bias in LDA data processing). With PTV data there is no reason to believe that there will exist any correlation between particle spatial density and velocity, at least if all the particles are following the flow with no slip. Therefore this is one aspect of time series analysis , which, in the analogy, does not appear in the analysis of spatially distributed velocity information.

Two procedures for wavenumber spectrum estimation from irregularly spaced data will therefore be pursued. The first is a direct extension of the technique given by (Nobach et al. (1998 & 1996)) from one dimension (time) to two dimensions (space).

The second technique has its origins in tomography and it involves the application of the projection slice theorem (PST) (Nobach (1997)). The following two sections describe each of these techniques respectively. First results are obtained using simulated data, derived from direct numerical simulations (DNS), for which the wavenumber spectrum was known. The influence of noise has not yet been considered. These results and their discussion are given in the final section.

2. Wavenumber Estimation Using a 2D Sample and Hold with Refinement

To review briefly the approach of Nobach et al. (1998), the LDA time series data is resampled at regular time intervals $(1/f_r)$ using a sample and hold (S+H) reconstruction. This procedure is illustrated schematically in Fig.1. The velocity data is available from the LDA at the particle arrival times t_i. A reconstruction is obtained by resampling at a frequency f_r, along t using the last valid velocity value. The expected autocorrelation of the resampled data is derived similar to the procedure given in Adrian & Yao (1987) and is denoted by $R_{uu}^{(r)}(k\Delta\tau_r)$. This function is related to the true autocorrelation function through a linear filter matrix $\mathbf{F}$,

$$R_{uu}^{(r)} = \mathbf{F} R_{uu} \tag{1}$$

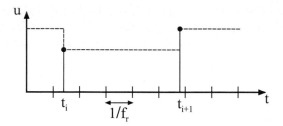

Figure 1: Sample and Hold reconstruction from LDA velocity data.

which when inverted, yields a non-biased and consistent estimator of the true autocorrelation function

$$\hat{R}_{uu} = \mathbf{F}^{-1}\hat{R}_{uu}^{(r)} \tag{2}$$

The frequency spectrum can be obtained via a Fourier transformation. The only parameter involved in the filter $\mathbf{F}$ is the mean data rate $\dot{n}$ (particles/s). All details of this procedure can be found in Nobach et al. (1998).

Applying this technique to two spatial dimensions first requires an equivalent spatial sample and hold, which can be envisioned graphically as illustrated in Fig. 2. For a given grid point lying on a regular mesh (ξ_i, ψ_j), a veloc-

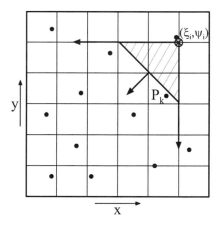

Figure 2: Spatial Sample and Hold

ity value will be assigned which corresponds to the nearest sample lying in the third quadrant, P_k. (The choice of quadrant is presumably immaterial). Using the next particle in a direction aligned with the mesh, would result not only in using particles far removed from the mesh point, but many mesh points would get assigned the same value. Thus the choice of a 45° direction. It should also be remarked, that the value to be assigned may just be a single velocity component or the velocity magnitude, depending on what wavenumber spectrum is to be estimated.

The next step is to estimate the probability that the velocity of a given particle will be used. There are two cases to be considered, as shown graphically in Fig. 3.

Case 1: In this case a particle occurs at (x_1, y_1) which is to the lower left of

(ξ_1, ψ_1) and no further particle is found in the triangle

$$x = \xi_1 \quad y = \psi_1 \quad y = (x_1 + y_1) - x \tag{3}$$

Furthermore at least one particle occurs in the triangle

$$x = \xi_2 \quad y = \psi_2 \quad y = (x_1 + y_1) - x \tag{4}$$

excluding the triangle given in Eq. (3). No further particle is found in the triangle

$$x = \xi_2 \quad y = \psi_2 \quad y = (x_2 + y_2) - x \tag{5}$$

The probability of this case occurring can be evaluated by assuming the particle distribution follows a Poisson distribution, characterized by a mean data density of $\dot{n}$ particles/area. The probability of finding at least one particle in an area $\Delta\xi\Delta\psi$ is then

$$P(\Delta\xi, \Delta\psi) = 1 - e^{-\dot{n}\Delta\xi\Delta\psi}, \tag{6}$$

where $\Delta\xi\Delta\psi$ represents the resolution of the resampling grid. The probability for the above case becomes

$$P_1(x_1, y_1, x_2, y_2) = \left(1 - e^{-\dot{n}\Delta\xi\Delta\psi}\right)^2$$
$$\cdot e^{-\frac{1}{2}\dot{n}(\xi_1 - x_1 + \psi_1 - y_1)^2} \cdot e^{-\frac{1}{2}\dot{n}(\xi_2 - x_2 + \psi_2 - y_2)^2} \tag{7}$$

Case 2: In this case a particle is found at a position (x_1, y_1) to the lower left of (ξ_1, ψ_1) and no further particle is found in the triangle (shaded area in Fig. 3b)

$$x = \xi_2 \quad y = \psi_2 \quad y = (x_1 + y_1) - x \tag{8}$$

The probability for this case occurring is

$$P_2(x_1, y_1) = \left(1 - e^{-\dot{n}\Delta\xi\Delta\psi}\right) \cdot e^{-\frac{1}{2}\dot{n}(\xi_2 - x_1 + \psi_2 - y_1)^2} \tag{9}$$

These two cases can now be used to relate the measured spatial correlation on the ξ/ψ mesh to the spatial correlation of the particles.

An equation results in the form of Eq. (1), in which the filter matrix $\mathbf{F}$ is now only a function of the particle area density $\dot{n}$. This equation has been derived, however due to its length, it is not shown. Reference is made to [4]. The next step is to find the inverse function $\mathbf{F}^{-1}$. This has not yet been accomplished.

8

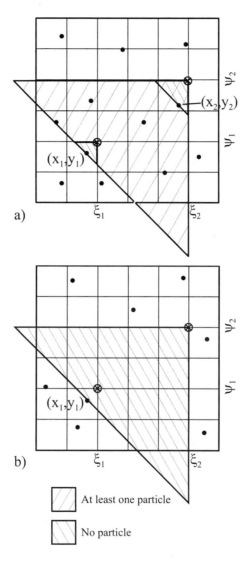

Figure 3: Possible cases for S+H resampling: a) Case 1; b) Case 2.

3. Wavenumber Estimation Using the Projection Slice Theorem (PST)

The projection slice theorem used in tomography will be used to derive an estimator for wavenumber spectra. The projection slice theorem (PST) is shown pictorially in Fig. 4. Basically it says that the one-dimensional Fourier transform of a projection of the function $f(x, y)$, is a slice of the two-dimensional Fourier transform $F(\omega_x, \omega_y)$ of the original function. The

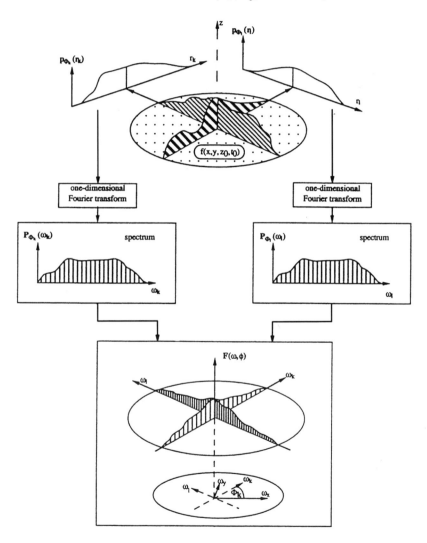

Figure 4: Pictorial representation of the projection slice theorem.

Fourier transform of the projection of the field $f(x, y)$ onto the axis r rotated an angle ϕ with respect to x is given by

$$P_\phi(\omega) = \int\limits_{-\infty}^{\infty} \int\limits_{-\infty}^{\infty} f(r, s)\, ds \quad e^{-jr\omega}\, dr \tag{10}$$

where the integral over s is the projection. The PST states that

$$P_\phi(\omega) \quad = \quad F(\omega, \phi) = F(\omega_x, \omega_y) \tag{11}$$
$$\text{for} \quad \omega_x = \omega \cos\phi \tag{12}$$
$$\omega_y = \omega \sin\phi$$

It can also been extended to apply to the power spectral density. This means that the power spectrum of the projected field is related to the power spectrum of the two-dimensional field through

$$P_\phi^*(\omega)\, P_\phi(\omega) = F^*(\omega_x, \omega_y)\, F(\omega_x, \omega_y) \tag{13}$$

A further consideration of applying the PST to PTV data leads to the necessity of introducing small slices of width dr to carry out the integration in the s direction, as illustrated in Fig. 5.

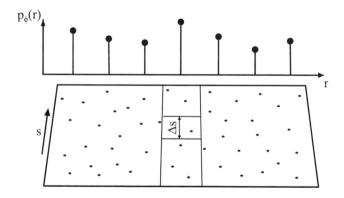

Figure 5: Slice of width dr to integrate in s-direction.

Introducing a slice of width dr will obviously be equivalent to a a spatial filter, the question to be investigated is whether dr can be made substantially smaller than a typical PIV interrogation spot, in which case some advantage in resolution can be gained using the PST.

4. Application to Simulated Data

The estimation algorithms described in the previous sections have been applied to simulated PTV data. With simulated data the correct wavenumber spectrum is known and thus both the extension and the the variance of the estimator can be evaluated. Furthermore, the data can be noise free, which for the present purpose is advantageous.

Data was generated using homogeneous isotropic turbulence fields com-

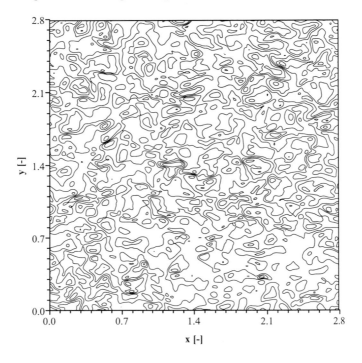

Figure 6: Turbulence field simulated with DNS

puted by direct numerical simulation (DNS) of the Navier Stokes equations. By DNS homogeneous isotropic turbulence fields are computed according to the following procedure.

First, a fluctuating velocity field is generated on an isotropic mesh. Both the integral length scale and the turbulent kinetic energy are introduced by scaling this velocity field with a prescribed spectrum. As usual in DNS, the mesh size implicitly defines the Kolmogorov length scale. The viscosity is then evaluated to satisfy that scale. The computation was performed in a box, using periodic boundary conditions in all directions. To get a turbulence field with higher moment statistical properties, the fluctuating field must freely evolve satisfying the Navier Stokes equations.

A typical input turbulence field is shown in Fig. 6, in which lines of constant velocity are shown. Note that there is no mean flow. This input field is a slice of the DNS data available on a mesh of 256x256 grid points. The autocorrelation of this field is shown in Fig. 7, where the integral length scale is determined to be $I = 0.093$. The total field dimensions are 2.8x2.8

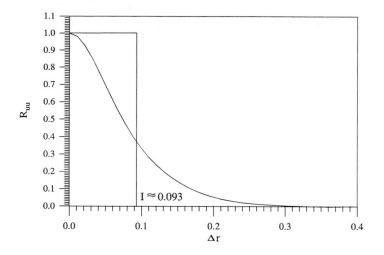

Figure 7: Integral length scale

or 30x30 length scales. There were approximately 9 DNS grid points per integral length scale.

Velocity samples were randomly distributed in the field according to a Poisson distribution with density parameter $\dot{n}$. The velocity at any given point was interpolated using weighted mean of the neighboring four DNS grid points. The resolution of the input DNS fields and also the linear interpolation used for assigning velocity values will also limit the validity of the following comparisons, however some first parameter dependencies can still be identified.

Sample and Hold Estimate

In Fig. 8 and 9 the dependence of the sample and hold (S+H) spectral estimate, without the refinement given in Eq. (2), on the particle density and the resampling rate is given, respectively. In Fig. 8 a resample mesh of 1024x1024 is used and the particle density is varied between $1/I^2$ and $60/I^2$. The wavenumber spectrum obtained directly from the DNS field is shown for comparison. To simplify the comparison only a slice of the spectrum at $90°$ is shown for each of the estimates.

The DNS spectrum corresponds to the prescribed spectrum, showing small fluctuations which correspond to its variability, since only one realization has been used. At low particle rates the S+H wavenumber spectrum exhibits a filter like behavior, with a slope of approximately -1.7 and a cut-off at approximately 50. Note, that the wavenumber is defined as $k = 2\pi/\lambda$. The smallest resolvable wavenumber is therefore 2.24 $(2\pi/2.8)$ and the maximum wavenumber is 1149 $\left(\frac{2\pi}{2(2.8/1024)}\right)$. The wavenumber corresponding to the integral length scale is 67 $(2\pi/0.093)$. Clearly the low particle density acts

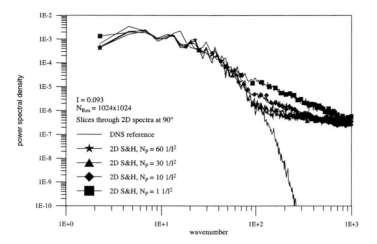

Figure 8: Power spectral density from 2D S&H at different particle densities

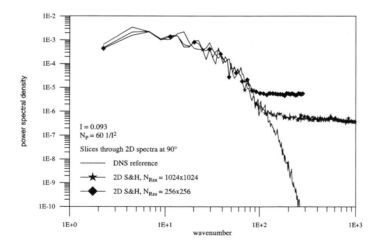

Figure 9: Power spectral density from 2D S&H with different resampling rates

as a filter, through which high wavenumber fluctuations can no longer be resolved. A similar behavior is seen in the time domain, whereby the slope is -2 and the cut-off frequency is given by $\dot{n}/2\pi$ [1]. However in time series analysis, a S+H resampling also leads to so called "step noise", which raises the entire spectrum uniformly across all frequencies. An equivalent to this step noise is not observed in the wavenumber spectrum.

Increasing the particle density improves the estimate, however at a value of $60\ P/I^2$, an asymptotic distribution is obtained. The asymptotically reached spectrum follows the flow spectrum until about $k = 100$, after which it flattens out. In time series analysis a continual improvement with increasing particle density could be expected. In the present case this is not observed, probably due to the interpolation step necessary in generating the simulation data. The wavenumber spectrum obtained for $60\ P/I^2$ presumably corresponds to the true spectrum of linearly interpolated data from the DNS data set.

In Fig. 9 the influence of the resample rate is investigated. In time series analysis the resample rate does not influence the estimation (Nyquist frequency). Fig. 9 indicates that a higher resample rate in space does however improve the wavenumber spectra estimate.

Estimate from PST

[h] Estimates of the wavenumber spectrum using PST are shown in Fig. 10

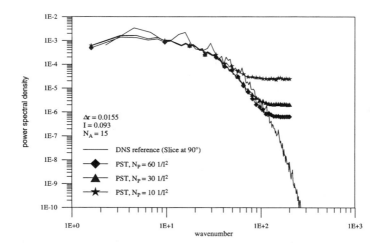

Figure 10: Power spectral density from PST at different particle densities

for a bin width of $\Delta r = 0.0155$. This leads to a maximum wavenumber of $202\ (\frac{2\pi}{2\Delta r})$. In Fig. 10 the particle density has been varied between 10 and $60\ P/I^2$. As with the S+H estimate, the spectrum improves with increasing particle density, as can be expected. However at all particle densities the

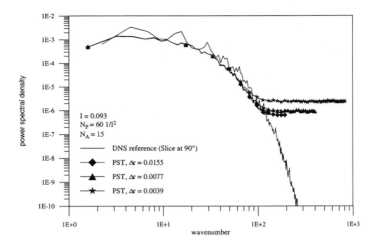

Figure 11: Power spectral density from PST with different sizes of the bins (Δr)

high wavenumber behavior is uniformly flat, indicating that this part of the spectrum corresponds to an interpolation noise present already in the simulation data. At 60 P/I^2 there is on average approximately 1 particle/DNS grid cell.

In Fig. 11 the influence of the bin size Δr on the spectrum is shown. In fact the estimate improves with broader bins. For the narrowest bin of $\Delta r = 0.0039$, there are in average about 80 particles per integration slice at a particle density of 60 P/I^2 and an angle of 0° or 90°. This increases to 330 particles for $\Delta r = 0.0155$.

Fig. 12 shows the influence of the number of slices (N_A) being averaged to estimate the mean spectrum. Note, that for the slice under 90° and the mean over 8 angles the curves are shifted up/down. As expected, the variance of the estimate decreases with increasing the number of angles. Investigating isotropic turbulent fields, this represents an important advantage of the PST procedure compared to the 2D S+H method, since the number of velocity fields to be measured can be reduced.

In Fig. 13 a comparison of the result from both the PST and the 2D S+H procedure and an estimate of the PSD as derived from PIV measurements is presented. The chosen number of of interrogation arrays (IA) corresponds to a number of 14 particle pairs per IA. While the maximum resolvable wavenumber according to the PIV resolution is 71, the spectra obtained by the presented estimators both follow the theoretical spectrum until $k = 100$. The main difference between the two proposed estimators for irregularly sampled data is expressed by a smaller variance of the PST results through averaging over several angles.

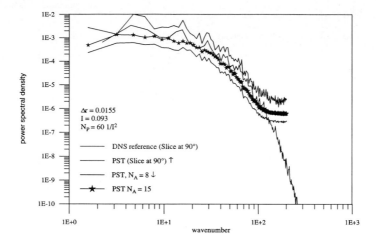

Figure 12: Power spectral density from PST, means over a different number of angles

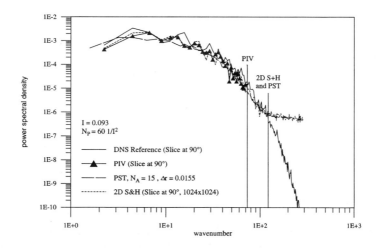

Figure 13: Comparison between the estimated power spectral density from 2D S&H, PST and PIV

5. Summary and Outlook

The present paper is a first presentation of two estimators for wavenumber spectra from PTV data. The first of these estimators, the sample and hold with refinement has not yet been fully formulated since a closed form for the refinement step must still be derived. However a sample and hold procedure for two dimensions could be presented. The procedure for the second estimator, based on the projection slice theorem is complete.

The performance of each of these estimators could not yet systematically evaluated. The main obstacle is the generation of homogeneous isotropic turbulence fields with known spectral content. The technique presently used, taking velocity field data from DNS computation, is rather unsatisfactory because the resolution of the primary data is quite coarse. Much finer grids become very demanding on computational time. Alternative methods of signal simulation must be sought.

A restriction which has not yet been discussed at length is that of homogeniety. In the analogy with time series analysis, that corresponds to statistical stationarity. If a signal is transient in time, a spectrum estimated over a longer period of the signal will be some weighted average of the changing spectral content. Similar averaging can be expected over space with the proposed estimators, if the turbulence field is non-homogeneous. At best, a smaller observation window can be applied and traversed across the field.

Although the present estimators are far from being complete, already other applications can be recognized. For instance wall arrays of hot-film sensors are used to monitor the passage of velocity fluctuations. The minimum spacing between sensors is physically limited, thus also the maximum resolvable wavenumber of disturbances. A random placement of arrays together with one of the above wavenumber estimators may prove to be advantageous in this case.

Acknowledgements

The close consultation with Holger Nobach in deriving the spatial S+H scheme is gratefully acknowledged. W. Hübner was supported by the PRO-COPE program during his stay at École Centrale de Lyon, where Marc Michard was very helpful in assisting with organizing suitable PIV experiments. Thanks are also due to Henry Pascal for providing the DNS fields used to simulate PTV data and to Lance Benedict for several useful discussions.

References

Adrian, R.J. and Yao, C.S. 1987 *Power Spectra of Fluid Velocities Measured by Laser Doppler Velocimetry* Exp. in Fluids **5** 17-28

Gaster, M. and Roberts, J.B. 1977 *The Spectral Analysis of Randomly Sampled Records by a Direct Transform* Proc. R. Soc. Lond. A. 27-58

18

Gaster, M. and Roberts, J.B. 1975 *Spectral Analysis of Randomly Sampled Signals* J. Inst. Maths. Applics. **15** 195-216

Hübner, W. 1998 *Abschätzung von zweidimensionalen Leistungsdichtespektren aus nicht-äquidistant abgetaste- ten Skalarfeldern: Anwendung auf Particle Tracking Velocimetry* Diplomarbeit Universität Erlangen-Nürn- berg, Germany

Keane, R.D. Adrian, R.J. and Zhang, Y. 1995 *Super Resolution Particle Image Velocimetry* Meas. Sci. Technol. **6** 754-768

Mayo, W.T. Jr. Shay, M.T. and Riter, S. 1974 *Digital Estimation of turbulence power spectra from burst counter LDV data* Proc. of the Second Int. Workshop on Laser Velocimetry, Purdue Univ., West Lafayette, Indiana 16-26

Measurement Science and Technology 1997 Special Issue: Particle Image Velocimetry, vol. **8**, no. 12, 1379-1583

Mersereau, R.M. and Oppenheim, A.V. 1974 *Digital Reconstruction of Multidimensional Signals from their Projections*, Proc. of the IEEE, vol. **62**, no. 10, 1319-1338

Nobach, H. Müller, E. and Tropea, C. 1998 *Efficient Estimation of Power Spectral Density from Laser Doppler Anemometer Data* Exp. Fluids **24** 5/6 499-509

Nobach, H. 1997 *Verarbeitung stochastisch abgetasteter Signale*, Dissertation Universität Rostock, Germany

Nobach, H. Müller, E. and Tropea, C. 1996 *Refined Reconstruction Technique for LDA Data Analysis*, Proc. 8th Int. Symp. of Appl. of Laser Techn. to Fluid Mechanics, Lisabon, Portugal **23.2**

Raffel, M. Willert, C. and Kompenhans, J. 1998 **Particle Image Velocimetry: A Practical Guide** Berlin, Springer

Ruck B. 1995 *A new Laser-Optical Method for Visualization and Real-Time Vectorization of Flow Fields* Proc. 6th Int. Conf. Laser Anemometry, Hilton Heat, South Carolina, FED vol. **229** 179-194

Scott, P.F. 1974 *Random Sampling Theory and its Application to Laser Velocimeter Turbulence Spectral Measurements* Report 74CRD216, General Electric Co, Cooperate Research and Development

Tennekes, H. and Lumley, J.L. 1972 **A First Course in Turbulence**, The MIT Press, Cambridge, Massachusetts

I.2. PIV Error Correction
D.P. Hart

Department of Mechanical Engineering, Massachusetts Institute of Technology,
Cambridge, MA 02139-4307, USA

Abstract. *A non-post-interrogation method of reducing subpixel errors and eliminating spurious vectors from particle image velocimetry (PIV) results is presented. Unlike methods that rely on the accuracy or similarity of neighboring vectors, errors are eliminated before correlation information is discarded using available spatial and/or temporal data. Anomalies are removed from the data set through direct element-by-element comparison of the correlation tables calculated from adjacent regions. The result is a correction method that improves subpixel accuracy and effective spatial resolution and is highly robust to out-of-boundary particle motion, particle overlap, inter-particle correlations, and electronic and optical imaging noise.*

Keywords. PIV, correlation, error correction, super-resolution, recursive-correlation

1 Introduction

Because Particle Image Velocimetry (PIV) is based on the statistical correlation of imaged subregions to determine local flow velocities, it is subject to inherent errors that arise from finite tracer particle numbers, sample volume size, and image resolution. These errors, in extreme cases, are relatively easy to detect as they tend to vary substantially from neighboring vectors in both magnitude and direction. Despite this, correcting these errors is often difficult as present computer algorithms lack the innate pattern recognition ability of humans. Furthermore, such errors need not present themselves in obvious manners. Velocity vectors determined by correlating finite subregions of tracer particle images are often biased to varying degrees by; out-of-boundary particle motion, correlations occurring between unmatched particle pairs, particle overlap, non-uniform particle distribution, and variations in image intensity [Keane et. al. 1992; Prasad et. al. 1992; Westerweel et. al. 1993, 1994, 1997; Raffel et. al. 1994; Lourenco et. al. 1995; Fincham et. al. 1997]. Such errors along with errors associated with excessive velocity gradients and the finite sample volume size necessary to image a statistically meaningful number of tracer particles, limit accuracy and resolution and thus, limit the usefulness of PIV.

Currently, the most widely used and accepted technique to eliminate correlation errors is to compare vectors with their neighbors to determine if they are in some statistical or physical sense inconsistent. This technique, analyzed in detail by Westerweel (1994), is based on the assumption that vectors resulting from correlation errors are far removed in magnitude and/or direction from neighboring vectors. It assumes that the resolution of PIV data is high enough and the flow features benign enough that apparent discontinuities in the flow will not present themselves and be eliminated. It is a method of detecting errors and not a method of resolving tracer particle displacement. Detailed correlation information is discarded before interrogation. Consequently, errors can only be eliminated from the results and replaced by interpolated values. Furthermore, this error correction method addresses only the most obvious of correlation errors and does not address the more subtle problems that severely limit subpixel accuracy and resolution. Although extremely useful, post-interrogation error correction is not ideal.

Presented herein is a robust and computationally efficient method of removing errors from PIV results and resolving vectors from regions where noise in the correlation table obscures tracer particle displacement. This method, based on an element-by-element comparison of the correlation tables taken from adjacent regions, does not rely on the accuracy or similarity of neighboring vectors, as does post-interrogation correction. Errors are directly eliminated from the correlation data improving spatial resolution and subpixel accuracy.

2   Correlation Errors

PIV images are typically processed by subdivision into a regular grid of overlapping windows that bound regions of similar flow velocity; a velocity vector is then found for each window by autocorrelation or cross-correlation. Autocorrelation and cross-correlation produces a table of correlation values over a range of displacements, and the overall displacement of particles in the window is represented by a peak in this correlation table. Errors occur primarily from insufficient data whether from a lack of imaged flow tracers or poor image quality, and/or from correlation anomalies generally resulting from unmatched tracer images within the correlated sample volume. Currently, errors are held to a minimum by using high resolution imaging equipment and carefully controlling seeding density and interrogation size.

Seeding Density

As the number of tracer particles within a sample volume increases, the probability of any finite region existing with a particle set of similar intensity and pattern decreases. Thus, the probability of obtaining an accurate measure of the displacement of a set of particles using correlation increases as the number of particles increases. Keane and Adrian (1992) demonstrated this by showing that the number of spurious vectors that appear in PIV data drop dramatically as particle numbers within correlated subregions are increased to an average of about ten particle images per region. Very high seeding densities, however, can alter the characteristics of the flow being measured and make it difficult if not impossible to adequately illuminate and image tracer particles within a specific region of interest. Consequently, there generally exists a limit to the density that a flow can be seeded.

Interrogation Size

An alternative is to increase the size of the sample volume. This increases the number of tracer particles in the interrogation region without increasing the seeding density. Furthermore, it reduces correlation anomalies associated with particles entering and exiting the sample volume in the time between exposures. As the sample volume is increased, a smaller fraction of particles enter and exit relative to the total number of particles that remain within the sample volume between exposures. Consequently, there is a reduction in correlation values from unmatched particle images between exposures. Keane and Adrian (1992) conjectured and Westerweel, Dabiri and Gharib (1997) demonstrated that significant errors due to an asymmetry in the peak correlation within the correlation table are associated with this phenomenon. Westerweel *et. al.* (1997) went on to show that these *bias errors* could be significantly reduced by correlating with an interrogation window offset by the integer value of local particle pixel displacement. This effectively eliminates out-of-boundary particle motion and the correlation anomalies associated with them. There is a significant reduction in error as long as the local particle displacement is approximately an integer value. It is less effective when the residual of the local displacement is around a half-pixel. Unfortunately, this technique addresses only errors due to translational flow within the interrogation plane and does not account for asymmetries in the correlation table caused by other sources such as out-of-plane particle motion or velocity gradients.

In particular, large velocity gradients are troublesome, not only because it introduces bias errors, but also because it limits the maximum interrogation size. A large local velocity gradient can result in unequal particle displacements causing one part of an interrogation region to correlate at a significantly different location than another part. The correlation table, rather than having one prominent peak representing the average particle displacement in a region, has multiple peaks or one shorter wider peak representing individual particle correlations from different areas in the sample volume. Increasing the sample volume exacerbates the problem as the relative separation in individual particle correlations from one region to another due to gradients in the flow increases as well. Consequently, there is an upper limit to both the seeding density and the interrogation size used to process PIV data. These limits severely constrain the use of PIV and force an often-unsatisfactory compromise between processing accuracy, resolution, and robustness.

3   Correlation Error Correction

Both errors resulting from insufficient data and errors caused by correlation anomalies can be eliminated during processing, regardless of the method of correlation, simply by multiplying the correlation table generated during processing by the correlation table generated from one or more adjacent regions.

This *correlation error correction* technique is illustrated in Fig.1. Here, the correlation table calculated during processing of one region (Fig. 1A) is multiplied, element-by-element, by the correlation table calculated from an adjacent region that overlaps the first region by fifty-percent (Fig. 1B). Neither of the correlation tables in this example (Fig. 1A or 1B) has a discernable peak representing tracer particle displacement. The resulting correlation table (Fig. 1C), however, has very few correlation anomalies and has a very prominent correlation peak in the lower right hand corner.

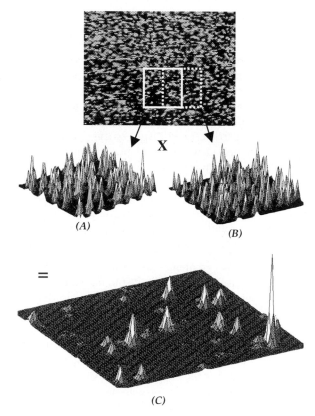

(A)

(B)

(C)

Figure 1. – *Elimination of correlation anomalies by multiplying the correlation tables from adjacent regions. Correlation values that do not appear in both tables are eliminated allowing tracer particle displacement to be resolved.*

Correlation error correction is effectively a correlation of two or more correlation tables. It is not an averaging technique. Any correlation value that does not appear in each of the combined correlation tables is eliminated from the resulting table. As the probability of exactly the same anomalies appearing in different regions is very small, correlation anomalies, regardless of their source, are eliminated from the data. Conversely, correlation values that are identical in location and magnitude in each of the combined tables are amplified. Thus, even if tracer particles displacement is not discernable in any of the combined correlation tables, multiplied together, the peak is either easily resolved or it becomes evident that at least one of the combined tables does not contain sufficient information to resolve particle displacement.

Consider the two non-overlapping NxN adjacent interrogation regions shown in Figure 2A. If the velocity within these interrogation regions is linear such that $\vec{V} = \left(v_x \cdot \dfrac{x}{N} + v_o \right) \hat{j}$ then the peaks resulting from the correlation of these regions falls within the y-displacement envelope defined by

$$\phi_y = \int_{-N/2}^{N/2} e^{-\left(\frac{y - v_x \frac{x}{N} - v_o}{r} \right)^2} dx$$

where r is the characteristic tracer particle radius. This envelope, graphed in Fig. 2B and 2C, has a Gaussian profile that is elongated in the direction of the velocity gradient and centered on the average velocity within the interrogation region. If the correlation tables from the two regions are multiplied element-by-element together, the resulting correlation peak is forced into the axisymmetric Gaussian profile shown in Fig. 2D. The symmetry of this envelope allows accurate subpixel interpolation. Furthermore, correlation errors from non-uniform illumination and from non-uniform tracer particle distribution within the interrogation regions that result in asymmetric correlation profiles are minimized as they have little influence on the combined profile, Fig. 3. This forced symmetry can be accomplished in the y-direction as well simply by combining correlation tables from more than two adjacent interrogation regions located along both the x and y directions.

Correlation error correction is not equivalent to correlating a larger region equal to the sum of the combined regions. Such a correlation would not eliminate correlation anomalies. It would, assuming no local velocity gradient, only strengthen the correlation peak representing the average particle displacement in the combined regions. This is not true of correlation error correction. The correlation peak found in the table resulting from correlation error correction is weighted to the displacement of the tracer particles within the overlap of the combined regions. Information within the overlapping regions identically effect the values in all of the correlation tables equally and are, therefore, not removed during processing. Particle displacements in regions outside the overlap influence the calculated displacement but to an extent that depends on the similarity in displacement. Thus, rather than a reduction in resolution, there is an improvement that depends on the size of the overlap and the gradient of the velocity relative to the size of the sample volume.

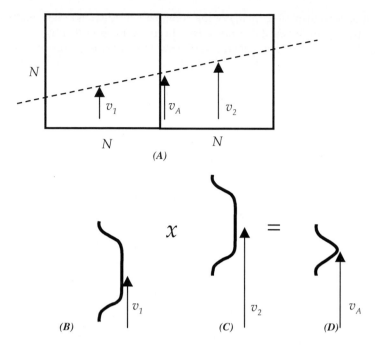

Figure 2. – *Forced correlation symmetry by multiplying the correlation tables from adjacent regions. Correlation values that fall outside one particle radius are eliminated forcing the correlation profile into a near axisymmetric shape centered on the tracer particle displacement of the adjoining region. The symmetry of this correlation profile aids in estimating subpixel displacement.*

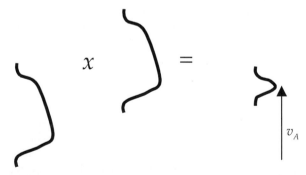

Figure 3. – *Correction of asymmetry in correlation profiles occurring due to non-uniform illumination and non-uniform seeding within interrogation regions. Because correlation error correction forces the correlation peak into a symmetric profile, velocity and illumination gradient errors that occur from correlating finite subregions are minimized.*

The level of effectiveness of this error correction technique at reducing correlation anomalies increases as the size of the overlapped region decreases. This is due to a reduction in the level of shared information. Correlation anomalies from image data within the overlapped region, appear equally in the correlation tables of each of the combined regions and are thus, not removed when the tables are multiplied together. Hence, it is desirable to maintain as small an overlap between combined regions as possible. Valid correlations, however, can be eliminated from regions with high velocity gradients if the flow results in a relative particle displacement greater than about one particle diameter between the combined regions. This is illustrated in Fig. 4 where correlation tables from two non-overlapping regions are multiplied together element-by-element. The resulting table has a much reduced correlation peak compared to Fig.1. The peak, found in the lower right hand corner of this table, represents tracer particle displacements that are roughly one particle image diameter apart.

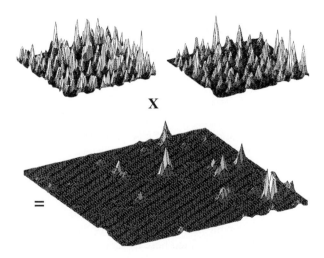

Figure 4. – *Illustration of the effect of an excessive particle displacement between regions using correlation error correction. Signal strength is significantly reduced when tables are multiplied from regions where particle displacement differs by more than one particle diameter.*

The optimum overlap between combined regions depends, largely, on the characteristics of the flow and the seeding density. In order to improve spatial resolution, most correlation algorithms currently use a fifty-percent overlap in adjacent regions [Willert et. al. 1991]. Therefore, fifty-percent overlapping regions were selected to evaluate the performance of this error correction method.

Although perhaps not optimum, this overlap serves as a baseline for comparison and it has been demonstrated to efficiently remove spurious vectors (Fig. 5).

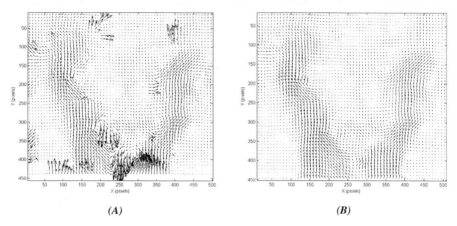

(A)   (B)

Figure 5. – Elimination of processing errors using correlation error correction. 3(A) represents uncorrected PIV results of a high Reynolds number swirling flow undergoing sudden expansion. 3(B) illustrates the same data processed using correlation error correction. The corrected vectors are true tracer particle displacements and not interpolated values.

4   Performance

Using Monte Carlo simulations, the effects of large velocity gradients, out-of-boundary motion, and seeding density on the performance of correlation error correction were investigated.   Pairs of synthetic images were generated with randomly distributed 256-grayscale Gaussian particle images 4 pixels in diameter. Processing was done using FFT spectral correlation on 64x64 pixel interrogation regions with 10 particle images per region.   Results using correlation error correction were compared with non-spurious results from uncorrected data.   (For simplicity, spurious vectors are defined here as vectors that have a x or y component greater than 0.5 pixels from the imposed mean displacement within the interrogation region.)

Correlation Signal-to-Noise Ratio

By eliminating correlation anomalies, correlation error correction significantly enhances the correlation signal-to-noise ratio.   This allows particle displacement to be resolved from regions that would otherwise be obscured by correlated noise from unmatched tracer particle images, inter-particle correlations, and electronic and optical imaging noise.   To illustrate this, synthetic images representing 16 *pixel* translational displacement were created with the equivalent of 32 *pixel* out-of-plane motion (50% of the equivalent interrogation plane thickness).   Using 64x64 *pixel* interrogation regions that overlapped 50%, 12% of the calculated vectors were determined to be spurious when processed *without* correlation error correction. When processed *with* correlation error correction, 70% of these spurious vectors were resolved and the remaining 30% were removed from the data set.

Fig. 6 illustrates the improved performance of correlation error correction. Here the correlation signal-to-noise ratio (SNR) is defined as the ratio between the second highest and highest correlation peaks. The percentage of the spurious and valid vectors with correlation signal strength (1-1/SNR) greater than indicated by the *x* axis are plotted for data processed with and without correlation error correction. When processed with correlation error correction, 90% of the valid vectors and none of the spurious vectors have a signal strength greater than 50%. In contrast, when the same data is processed without correlation error correction, only 30% of the valid vectors (shown as a dashed line) have a signal strength greater than 50%.

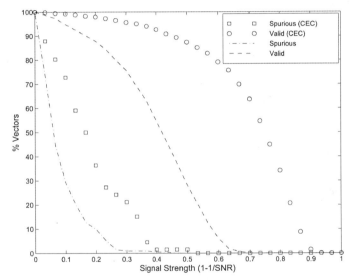

Figure 6. *– Improved signal-to-noise ratio resulting from correlation error correction (CEC). The percentage of spurious and valid vectors with correlation signal strengths greater than indicated by the x axis are compared for data processed with and without correlation error correction.*

Spurious Vector Elimination

The improved correlation signal-to-noise ratio resulting from correlation error correction, not only increases the number of valid vectors calculated, but also allows spurious vectors to be easily detected and removed. Two methods exist for detecting spurious vectors; (1) a threshold can be set and all vectors with a correlation signal-to-noise ratio less than the threshold discarded, or (2) the correlation peak resulting from the combined correlation tables can be compared with the peaks from each of the tables. If the peak in the combined table exists as a peak in at least one of the other tables, it is likely to represent tracer particle displacement.

The threshold method, although effective at removing spurious vectors, can result in the loss of significant numbers of valid vectors. In the previous example, illustrated by Fig. 6, if a threshold level of 50% is used for the correlation signal-to-noise ratio, all of the spurious vectors are removed but at a cost of about 10% of the valid vectors. This is not as excessive as it first appears considering that the valid vectors removed are the ones with the lowest correlation signal-to-noise ratio and are thus, the ones most likely to contribute to subpixel displacement inaccuracies in the results. Furthermore, correlation error correction in this example results in an 8% increase in the number of valid vectors calculated. Thus, a 10% loss in the number of valid vectors while eliminating spurious vectors results in only 2% fewer valid vectors than would be calculated without correlation error correction – a small price for improved accuracy.

Nonetheless, the second method of detection, peak comparison, provides a more robust and direct way of eliminating spurious vectors. This method of detection is similar to the *median* method suggested by Westerweel (1994) except that it is based on data that exists only in the interim of processing. The probability that the peak correlation value from a spurious vector found after multiplying correlation tables together is the same as the peak correlation value in one of the combined tables is remote. Furthermore, valid vectors are removed only when tracer particle displacement is obscured in all of the combined correlation tables. Consequently, this method of spurious vector detection is highly effective and results in little data loss.

When the data from the previous example is processed using peak comparison, 100% of the spurious vectors are removed and less than 2% of the valid vectors are removed. The result is a 6% increase in the number of valid vectors calculated using correlation error correction even after 100% of the spurious vectors are removed.

Spatial Resolution

As a way of illustrating the improvement in spatial resolution resulting from correlation error correction through velocity gradient enhancement, consider the two interrogation regions that overlap by 25% shown in Fig. 7. A velocity field can be artificially imposed such that particle displacement is one particle image diameter in the negative direction in the left region, $\Delta v_L = -d$, and one particle image diameter in the positive direction in the right region, $\Delta v_R = +d$. As illustrated in Fig. 8, calculated particle displacement of the left and right regions when processed without correlation error correction is almost unaffected by any velocity imposed on the overlapping region (Δv_o). When processed with correlation error correction, however, the resulting calculated displacement closely matches the imposed velocity in the overlapped region and does not equal the average of the imposed displacements on the left and right regions. Since the overlapped region is only

25% of the area of either the left or right regions, spatial resolution is effectively improved by 75%.

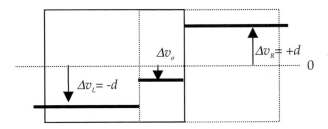

Figure 7. – *Schematic of imposed velocity on two interrogation regions that overlap 25%. This imposed velocity profile is used to illustrate the enhanced spatial resolution resulting from correlation error correction (Fig. 8).*

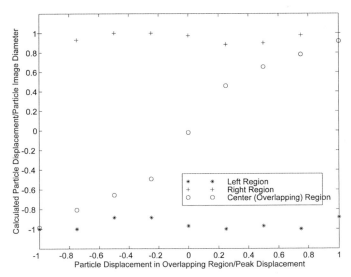

Figure 8. – *Illustration of improved effective spatial resolution by velocity gradient enhancement. The calculated particle displacement using correlation error correction (shown as 'o') follows the imposed velocity in the overlapped region in Fig. 7 and is not an average of the velocities imposed on the left and right regions (shown as '*' and '+' respectively).*

Out-of-Boundary Particle Motion

Fig. 9 illustrates the effect of translation on subpixel accuracy by comparing uncorrected results with results corrected from fifty-percent overlapping regions.

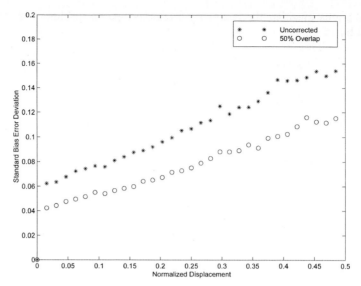

Figure 9. – *Effect of translational displacement on subpixel accuracy. Correlation error correction improves subpixel accuracy by eliminating anomalies in the correlation table and strengthening the peak correlation signal.*

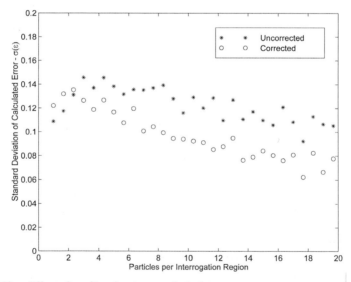

Figure 10. – *Effect of seeding density on subpixel accuracy.*

Here the standard deviation of the displacement error from one thousand interrogation regions are plotted as a function of imposed particle translational displacement normalized by the interrogation size (64 *px*). The correlation error corrected results show a twenty-five percent improvement in resolving subpixel displacement. Note that the bias error increases linearly with translational displacement. This is consistent with errors conjectured by Adrian (1991) and analyzed in detail by Westerweel, Dabiri, and Gharib (1997). It is, to some extent, the result of fewer matching particle image pairs between correlated regions. In translational flow, particles are convected out of and into the interrogation region. These convected particles do not appear in both exposures and thus, not all of the particle images in a region contribute to the peak correlation value. Consequently, translational flow has a similar effect to a reduction in seeding density, Fig. 10.

Errors associated with translational motion exist for out-of-plane particle displacement as well. Out-of-plane motion is particularly limiting to PIV as it restricts measurements to regions of nearly planer flow and restricts the minimum width of the sample volume, thereby restricting spatial resolution.

As shown in Fig. 11, correlation error correction is highly effective at resolving particle displacements from regions with significant out-of-plane flow. Here the percentage of valid vectors is plotted as a function of normalized out-of-plane translational particle displacement. The uncorrected data begins to produce excessive spurious results at fifty-percent displacement. Using fifty-percent overlapped regions, correlation error corrected data produces few spurious results until almost 90% displacement. At 100% displacement, flow information is lost as all of the particles within the interrogation region are transported out between exposures.

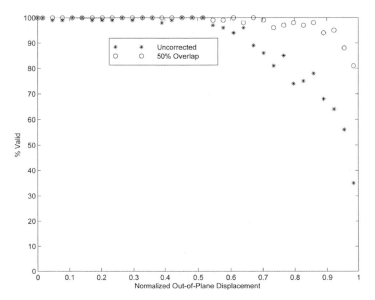

Figure 11. – Effect of out-of-plane flow on PIV processing. Correlation error correction using fifty-percent overlapped regions significantly improve the ability to resolve particle displacements.

Velocity Gradient Effects

While highly effective at resolving displacements from regions with significant out-of-plane motion, for a fixed interrogation size, correlation error correction is somewhat less effective at resolving regions of high velocity gradients. Fig. 12 shows a plot of the percentage of valid vectors as a function of normalized velocity gradient in the *ij* interrogation plane;

$$G_\nabla = \frac{m|\nabla\bar{v}_{ij}|N\Delta t}{d}$$ where *N* is the interrogation size (64 *px* in this case), *m* is the image

magnification, *Δt* is the time between exposures, and *d* is the average imaged particle diameter. As shown, correlation error correction increases the number of valid vectors calculated from regions of high velocity gradients. Above a gradient level of about $G_\nabla=1.2$, however, the probability of acquiring a valid vector quickly diminishes. A distinct advantage of correlation error correction, however, is that the improvement of the correlation signal strength allows smaller interrogation regions to be used. These smaller regions can resolve higher velocity gradients. Consequently, the simple analysis presented here does not accurately reflect the improvement in resolving velocity gradients that results from the use of correlation error correction. Such improvement depends largely on the seeding density of the flow.

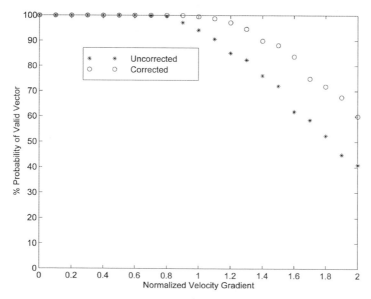

Figure 12. – *Effect of velocity gradient on PIV processing. Correlation error correction improves the ability to resolve tracer particle displacements in regions that have high velocity gradients.*

5 Experimental Demonstration: *High Accuracy/ High Resolution Processing*

The robustness of correlation error correction in removing spurious vectors and correlation anomalies allows the use of recursive correlation processing to iteratively arrive at local particle displacement without generating spurious results; a region is correlated, the interrogation window size is then reduced and offset by the previous result before re-correlating with the new window over a reduced region. Each correlation depends on the accuracy of the proceeding result. Without a highly robust and efficient means of detection, errors quickly propagate. With the aid of correlation error correction, the result of this correlation process can be quite dramatic as is illustrated in Fig. 13. Here recursive correlation along with correlation error correction is used to iteratively resolve the velocity of a swirling flow undergoing sudden expansion (backward step at bottom of figure).

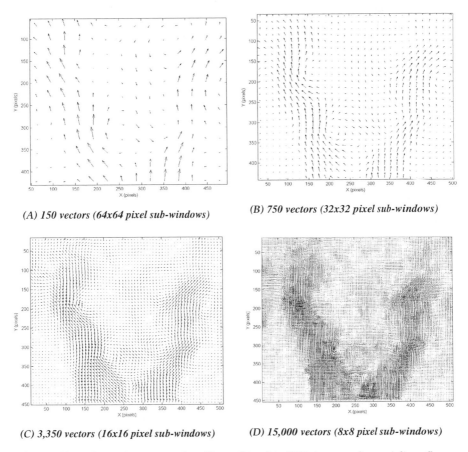

(A) 150 vectors (64x64 pixel sub-windows)

(B) 750 vectors (32x32 pixel sub-windows)

(C) 3,350 vectors (16x16 pixel sub-windows)

(D) 15,000 vectors (8x8 pixel sub-windows)

Figure 13. – *Recursive processing (A to D) of a PIV image of a swirling flow undergoing sudden expansion (Re=35,000, Ω=2.4).*

34

Figure 14 shows the same data resolved almost to the level of single particles (4x4 *pixel*). 60,000 vectors are displayed each representing a 0.8mm³ region of the flow. Using sparse array image correlation, results from the entire image are processed in less than one minute (>1,000 vectors/sec) [Hart, 1996]. Although it is unlikely that the images used to process this data contain information to the resolution implied by this plot, it illustrate the possibility of processing PIV to the limits of optical resolution without generating excessive errors even, as in this case, when significant out of plane flow exists.

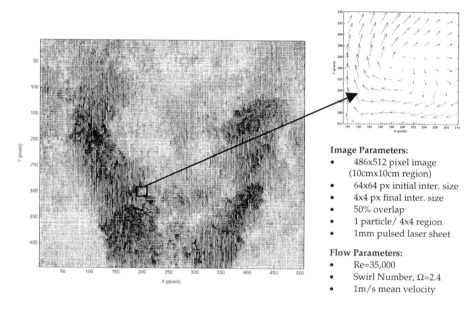

Image Parameters:
- 486x512 pixel image (10cmx10cm region)
- 64x64 px initial inter. size
- 4x4 px final inter. size
- 50% overlap
- 1 particle/ 4x4 region
- 1mm pulsed laser sheet

Flow Parameters:
- Re=35,000
- Swirl Number, Ω=2.4
- 1m/s mean velocity

Figure 14. – 60,000 vectors calculated from a 486x512 image of a swirling flow (Re=35,000, Ω=2.4) undergoing sudden expansion using recursive correlation. Each vector on average represents a single tracer particle in the flow and is the result of as many as five sub-window correlations. These results are processed at a rate of over 1,000 vectors/sec using sparse array image correlation with correlation error correction.

6 Conclusions

Stochastic correlation values can be eliminated during PIV processing by element-by-element multiplication of correlation tables calculated from adjacent regions. This correction method, which is essentially a correlation of two or more adjacent correlation tables, improves subpixel accuracy and eliminates spurious vectors resulting from unmatched particle pairs, out-of-boundary motion, particle overlap, inter-particle correlations, and electronic and optical imaging noise. Particle displacement information that does not correlate equally in the combined regions is minimized. The resulting calculated displacement is weighted to the adjoining area improving the effective spatial resolution in regions of high velocity gradients.

Correlation error correction, unlike post-interrogation methods, uses all available correlation information calculated from interrogation regions and does not rely on the accuracy or similarity of neighboring vectors. Consequently, it resolves tracer particle displacement from regions that are otherwise obscured by correlation anomalies. Thus, the correction method presented herein reduces bias errors and eliminates spurious vectors while improving spatial resolution and vector yields.

References

Adrian, R. J. (1991) "Particle imaging techniques for experimental fluid mechanics," *Annual Review of Fluid Mechanics* Vol. 23, pp. 261-304.

Fincham, A. M.; Spedding, G. R. (1997) "Low Cost, High Resolution DPIV for Measurement of Turbulent Flow," *Experiments in Fluids.*, Vol. 23, pp. 449-462.

Fincham, A. M.; Spedding, G. R.; Blackwelder, R. F. (1991) "Current Constraints of Digital Particle Tracking Techniques in Fluid Flows," *Bull Am Phys Soc.* Vol. 36, 2692.

Hart, D.P., (1996) "Sparse Array Image Correlation," 8th International Symposium on Applications of Laser Techniques to Fluid Mechanics, Lisbon, Portugal.

Keane, R. D.; Adrian, R. J., (1990) "Optimization of Particle Image Velocimeters," *Measurement Science and Technology*, Vol. 2, pp. 1202-1215.

Keane R.D.; Adrian, R.J., (1992) "Theory of cross-correlation of PIV images," *Applied Scientific Research*, Vol. 49, pp. 191-215.

Landreth, C. C.; Adrian, R. J., (1990) "Measurement and Refinement of Velocity Data Using High Image Density Analysis in Particle Image Velocimetry," Applications of Laser Anemometry to Fluid Mechanics, Springer-Verlag, Berlin, pp. 484-497.

Okamoto, K; Hassan, Y. A.; and Schmidl, W. D., (1995) "New Tracking Algorithm for Particle Image Velocimetry," *Experiments in Fluids.*, Vol. 19, pp. 342-347.

Prasad, A.K.; Adrian, R.J.; Landreth, C.C.; Offutt, P.W., (1992) "Effect of Resolution on the Speed and Accuracy of Particle Image Velocimetry Interrogation," *Experiments in Fluids*, Vol. 13, pp. 105-116.

Raffel, M.; Kompenhans, J., (1994) "Error Analysis for PIV recording Utilizing Image Shifting," Proc. 7th International Symposium on Applications of Laser Techniques to Fluid Mechanics, Lisbon, July, p. 35.5.

Westerweel, J. (1994) "Efficient Detection of Spurious Vectors in Particle Image Velocimetry Data," *Experiments in Fluids.*, Vol. 16, pp. 236-247.

Westerweel, J.; Dabiri, D.; Gharib, M., (1997) "The Effect of a Discrete Window Offset on the Accuracy of Cross-Correlation Analysis of Digital PIV Recordings," *Experiments in Fluids*, Vol. 23, pp. 20-28.

Willert, C. E.; Gharib, M. (1991) "Digital Particle Image Velocimetry," *Experiments in Fluids*, Vol. 10, pp. 181-193.

I.3. Effect of Sensor Geometry on the Performance of PIV Interrogation

J. Westerweel

Laboratory for Aero and Hydrodynamics, Delft University of Technology, The Netherlands

Abstract. This paper describes the mathematical investigation of the effect of the sensor geometry, i.e. the pixel size and pixel fill ratio, on the performance of PIV interrogation. Two sub-pixel estimators are investigated: the particle-image centroid, and the Gaussian peak fit. It is found that no bias errors occur when the particle-image diameter is at least two pixels, and the measurement error is determined by random errors only. When particle images are much smaller than one pixel, an irrecoverable signal loss deteriorates the measurement performance. For intermediate resolutions the bias errors are of the same magnitude as the random errors. It is demonstrated that image blurring by de-focussing reduces the bias error, but increases the random error. The analysis shows that sensors with a high fill ratio have a better performance.

Keywords. particle image velocimetry, optimization, CCD fill ratio, defocussing, pixel locking.

1 Introduction

In recent years the performance of PIV interrogation methods has improved considerably, and the resolution of the pixelization with respect to the particle-image diameter has been reduced substantially.

Originally, the pixelization of PIV images was done with a high resolution. When the discretization of the image matches the minimum sampling rate prescribed by the sampling theorem, the original continuous image can be reconstructed perfectly from the discrete samples. In that case the details of the sensor geometry are completely negligible, and typically occurs for the analysis of highly resolved PIV photographs. However, nowadays it is more common to record PIV images directly on electronic image sensors (viz., CCD arrays). The demand for PIV measurements with high spatial resolution often implies that the discretization no longer matches the Nyquist sampling criterion, and consequently, the measurement precision depends on the detailed geometry of the image sensor.

For example, an effect known as 'pixel locking' (Fincham & Spedding 1997) may interfere with the measurements. This pixel locking is described as a biasing of the measured displacement towards integer values of the displacement in pixel units, as is illustrated in Figure 1. The effect is usually

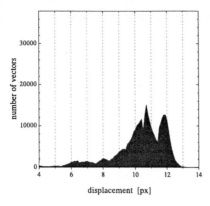 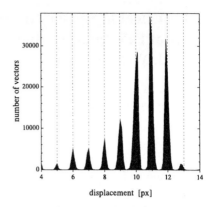

Figure 1: Histograms of the measured displacement for PIV in a turbulent pipe flow. An 'ideal' measurement (left), and a measurement that suffers from pixel locking (right).

ascribed to the fact that the particle images become too small with respect to the size of the pixels. Also, it is conjectured that CCD sensor arrays with low fill ratio are more susceptible to pixel locking.

The pixel locking effect leads to peculiar measurement results. This is illustrated in Figure 2, which shows how a vortex structure transforms into a polygon shape as a result of pixel locking.

In this paper the error for PIV interrogation is investigated mathemati-

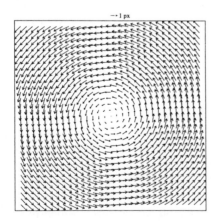 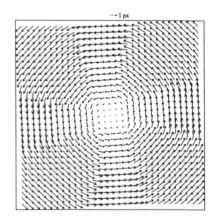

Figure 2: Vector plots of a Burgers vortex. The left plot is the actual displacement field, whereas the right plot suffers from pixel locking. The small arrow at the top of the plots corresponds to a displacement of one pixel.

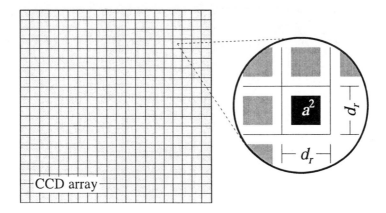

Figure 3: Schematic of the geometry of a CCD sensor array.

cally as a function of the sensor geometry, which is characterized by the size d_r of the pixels, and the relative fraction a^2 of the total pixel area that is light sensitive; see Figure 3. Generally the measurement error is split into a random error component and a bias error component. The random error is proportional to the width of the particle images, whereas the bias error is associated with the finite size of the pixels. Other effects that may influence the interrogation performance are left out of consideration.

The next section consists of a concise review of the main characteristics of particle images, spatial correlation and pixelization. In Section 3 two estimators are investigated: (1) the particle-image centroid (which is commonly applied at low image density), and (2) the so-called *Gaussian peak fit* (which is applied at high image density). The analysis for the particle-image centroid is based on the work by Alexander & Ng (1991), and is extended to include the effect of blur. Altough the analysis is done mathematically, the results are incidentally compared with results obtained from the interrogation of synthetic PIV images, generated by means of Monte-Carlo methods. Section 4 contains the conclusions and a discussion on the valid range for PIV measurements that are restricted by the sensor characteristics.

2 characteristics of particle images

2.1 continuous domain

The image of a small tracer particle is given by the convolution of its geometrical image and the diffraction-limited spot of the imaging optics (Goodman 1968). The Fourier transform of the diffraction-limited spot is commonly referred to as the optical transfer function (OTF). The OTF for a lens with a square aperture is given by

$$F(u,v) = \Lambda(u/u_0)\Lambda(v/u_0), \tag{1}$$

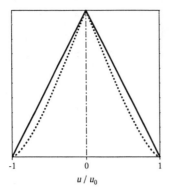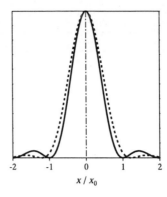

u / u_0 x / x_0

Figure 4: The optical transfer function (left) and diffraction-limited spot (right) for square (——) and circular (– – –) apertures.

with

$$\Lambda(u) = \begin{cases} 1 - |u| & \text{for } |u| < 1, \\ 0 & \text{elsewhere,} \end{cases} \tag{2}$$

and where u_0 is the cut-off spatial frequency, given by

$$u_0 = D/f\lambda(M + 1), \tag{3}$$

where D is the aperture diameter, f the lens focal length, λ the light wave-length, and M the image magnification. The corresponding diameter d_s of the diffraction-limited spot is given by

$$d_s = 2(M + 1)f\lambda/D = 2/u_0. \tag{4}$$

The OTF of a square aperture is separable in u and v, which makes it very convenient for mathematical analysis. It is also a good approximation of the OTF of a lens with a circular aperture, which has the same cut-off frequency, but is circularly symmetric. The diameter d_s of the diffraction-limited spot for a lens with a circular aperture is given by

$$d_s \cong 2.44(M + 1)f\lambda/D. \tag{5}$$

The optical transfer function and diffraction-limited spot for a lens with a circular aperture and for a lens with a rectangular aperture are shown in Figure 4.

The diffraction-limited spot and geometric particle image are commonly approximated by two-dimensional Gaussian functions, so that the particle-image diameter d_τ is approximately given by (Adrian 1984):

$$d_\tau \cong \left(M^2 d_p^2 + d_s^2\right)^{\frac{1}{2}}, \tag{6}$$

where d_p is the diameter of the tracer particle. Hence, for small M the size of the particle images is determined by the diffraction-limited spot, i.e. $d_\tau \cong d_s$.

2.2 spatial correlation

At high image density the particle-image displacement is determined from the estimated location of the displacement-correlation peak in the spatial correlation of two image fields I_1 and I_2, taken with a small time delay. It was shown by Keane & Adrian (1992) that the spatial cross-correlation can be separated into three terms: $R_D + R_C + R_F$. The term R_D represents the spatial correlation of the image intensity fluctuations, and holds all information with regard to the motion of the particle images. The terms R_C and R_F vanish when the (local) mean image intensity is subtracted from the instantaneous image fields. The ensemble mean of the term R_D over all possible realizations of I_1 and I_2 for a given uniform velocity field $\boldsymbol{u}$ is given by (Keane & Adrian 1992)

$$\langle R_D(\boldsymbol{s})|\boldsymbol{u}\rangle \propto N_I F_I F_O F_\tau(\boldsymbol{s} - \boldsymbol{s}_D), \tag{7}$$

where N_I is the image density (i.e., the mean number of particle images in the interrogation window), F_I and F_O are the loss-of-correlation due to in-plane and out-of-plane motion respectively of the tracer particles, F_τ is the self-correlation of the particle images, and $\boldsymbol{s}_D$ is the (in-plane) displacement of the particle-images. For Gaussian particle images with an e^{-2}-diameter d_τ, the displacement-correlation peak is also Gaussian, but with an e^{-2}-diameter of $\sqrt{2}d_\tau$.

2.3 discrete domain

The sensor geometry that was described in the Introduction can be represented by the spatial pixel sensitivity $p(x, y)$. For square pixels with a uniform sensitivity over the light sensitive area $p(x, y)$ is given by:

$$p(x, y) = \frac{1}{(ad_r)^2} \begin{cases} 1 & \text{for: } |x|, |y| \leqslant \frac{1}{2}ad_r, \\ 0 & \text{elsewhere.} \end{cases} \tag{8}$$

The discrete image field is given by the convolution of the continuous image intensity field and the pixel sensitivity, sampled at intervals d_r, i.e.

$$I[i, j] = \iint p(x - x_i, y - y_j)I(x, y)dxdy, \tag{9}$$

with: $x_i = id_r$, $y_j = jd_r$.

The pixelization of an image $f(x, y)$ can be expressed in the two-dimensional spatial frequency domain as a double infinite series of the Fourier transform $F(u, v)$ of $f(x, y)$, multiplied by the Fourier transform of the spatial pixel sensitivity, denoted by $P(u, v)$, i.e.:

$$G(u, v) = \sum_{n=-\infty}^{\infty} \sum_{m=-\infty}^{\infty} F\left(u - \frac{n}{d_r}, v - \frac{m}{d_r}\right) P\left(u - \frac{n}{d_r}, v - \frac{m}{d_r}\right) \tag{10}$$

(Alexander & Ng 1991; Oppenheim *et al.* 1983). For $p(x, y)$ given in (8), $P(u, v)$ is equal to

$$P(u, v) = \text{sinc}(ua/d_r) \, \text{sinc}(va/d_r), \qquad (11)$$

with $\text{sinc}(u) = \sin(\pi u)/\pi u$.

The discrete spatial correlation is given by convolution of continuous spatial correlation and self-correlation of pixel sensitivity, sampled at intervals d_r:

$$R_D[i, j] = \iint \Phi_{pp}(s - s_i, t - t_j) R_D(s, t) ds dt, \qquad (12)$$

with

$$\Phi_{pp}(s, t) = \iint p(x, y) p(x + s, y + t) dx dy \qquad (13)$$

(Westerweel 1993). For the pixel sensitivity defined in (8), the self-correlation $\Phi_{pp}(s, t)$ is given by

$$\Phi_{pp}(s, t) = \Lambda(s/ad_r) \Lambda(t/ad_r), \qquad (14)$$

where $\Lambda(u)$ is defined in (2). Note that $\Phi_{pp}(s, t)$ is separable in s and t.

2.4 optimal sampling

The sampling theorem states that a bandlimited signal can be reconstructed from its discrete samples when the sampling rate is at least twice the bandwidth (Oppenheim *et al.* 1983). Optical systems are essentially bandlimited, for which the bandwidth is given by the OTF cut-off spatial frequency (Goodman 1968).

In the case of an aberration-free lens with a square aperture, the OTF cut-off frequency is given by (1). This means that an exact reconstruction of the continuous image intensity field from the discrete samples (viz., discrete image) could be achieved when the size of the pixels is less than $1/2u_0$. Given that d_s is equal to $2/u_0$, the Nyquist sampling criterion is satisfied when the pixel size d_r is less than one quarter of the diffraction-limited spot diameter, i.e.

$$d_s/d_r \geqslant 4. \qquad (15)$$

Hence, errors that are associated with the sensor geometry may be expected when the size of a pixel is larger than about one-quarter of the diffraction-limited spot diameter.

3 Analysis

Consider a particle image or a displacement-correlation peak located at a position (d_x, d_y). The measurement error is the difference between the estimated position (m_x, m_y) returned by the interrogation, and the actual position. Consider only the error in the x-coordinate:

$$\varepsilon_x = m_x - d_x. \tag{16}$$

It is convenient to separate ε_x into mean and fluctuating parts, and average over many identical measurements:

$$\langle \varepsilon_x \rangle = \delta_x, \quad \text{and:} \quad \langle \varepsilon_x^2 \rangle = \delta_x^2 + \sigma_x^2, \tag{17}$$

where δ_x is denoted as the *bias error*, and σ_x as the r.m.s. *random error*. The random error is generally associated with variations of individual particle images with respect to the average shape of the particle image, whereas the bias error is associated with the variation of the average shape with respect to ideal shape of the particle images. Typically, the bias error arises due to the change in shape of the particle image or displacement-correlation peak as a result of image pixelization.

3.1 random error

In the continuous domain the uncertainty to determine the location of a particle image or displacement-correlation peak is proportional to the width of the particle image (Adrian 1991), i.e.

$$\sigma_x \sim c\, d_\tau, \tag{18}$$

where c is a proportionality constant. This constant is about 0.05 to 0.1, as obtained from experimental data (Prasad *et al.* 1992). It was shown by Wernet & Pline (1993) that this relationship also holds for discrete images, and that the value for c is inversely proportional to the signal-to-noise ratio for the image intensity field. For low light levels the signal strength is determined by the photon count rate, so that the signal-to-noise ratio is proportional to the square root of the intensity (viz., photon count). Hence, the signal-to-noise ratio for a given light level will be proportional to the square root of the active pixel area, so that a reduction of the pixel fill ratio at a fixed light level implies a proportional increase of the random error. This can be compensated for by properly increasing the total illumination of the tracer particles. However, note that by this argument a sensor with 20% fill ratio requires a 5 times more powerfull illumination source in comparison with a 100% fill-ratio sensor.

3.2 particle-image centroid

Consider a single particle image, with an intensity distribution $\tau(x, y) = f(x - d_x, y - d_y)$, where (d_x, d_y) is the location of the particle image. The

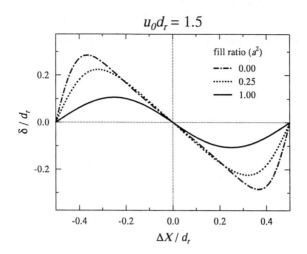

Figure 5: The bias error for the centroid of a diffraction-limited particle image as a function of the displacement, for $d_s/d_r = \frac{4}{3}$ and $a = 1.0$, 0.5 and 0.0.

centroid m_x of the particle image in the direction of the x-coordinate is then given by

$$m_x = \frac{\iint x\tau(x,y)\,dx\,dy}{\iint \tau(x,y)\,dx\,dy},\tag{19}$$

and *vice versa* for the centroid m_y of the particle image in the direction of the y-coordinate. The centroid bias error δ_x for the particle-image centroid in (19) can be written as

$$\delta_x = \frac{\iint x f(x,y)\,dx\,dy}{\iint f(x,y)\,dx\,dy} = \frac{\frac{dF}{du}(0,0)}{2\pi i F(0,0)},\tag{20}$$

(Alexander & Ng 1991) where $F(u,v)$ is the Fourier transform of $f(x,y)$. It is easily proven that the centroid (m_x, m_y) yields the *exact* location of the particle image when $f(x,y)$ is symmetric (Alexander & Ng 1991).

The image pixelization is accounted for by substitution of $G(u,v)$ in (10) for $F(u,v)$ in (20). In order to keep the mathematics simple, it is assumed that $G(u,v)$ is separable, i.e.:

$$G(u,v) = G_u(u)G_v(v), \quad \text{and:} \quad \frac{dG}{du}(u,v) = G'_u(u)G_v(v)\tag{21}$$

This reduces the subsequent analysis to a one-dimensional problem.

Following the analysis of Alexander & Ng (1991) yields the following exact expression for the bias error δ_x:

$$\delta_x = \frac{\sum\limits_{n=1}^{\infty} G'_u(n/d_r) \sin(2\pi d_x n/d_r)}{\pi \left[G_u(0) + \sum\limits_{n=1}^{\infty} G_u(n/d_r) 2 \cos(2\pi d_x n/d_r) \right]}, \tag{22}$$

and *vice versa* for δ_y. Note that the error is zero when the cut-off frequency is smaller than $1/d_r$. This implies that it is not necessary to satisfy the Nyquist sampling rate—which is equal to 2 times the cut-off frequency— to obtain error-free estimates of the particle-image centroid. So, the pixel resolution d_r should satisfy

$$d_s/d_r \geqslant 2. \tag{23}$$

This is a factor two smaller than the minimum sampling rate prescribed by (15).

3.2.1 *slight under-sampling*
If the OTF cut-off frequency is between $1/d_r$ and $2/d_r$ then (22) reduces to:

$$\delta_x = \frac{G'_u(1/d_r) \sin(2\pi d_x/d_r)}{\pi \left[G_u(0) + G_u(1/d_r) 2 \cos(2\pi d_x/d_r) \right]}. \tag{24}$$

So, the error is periodic with a period that is equal to the sample spacing d_r.

Let us evaluate this result for a diffraction-limited particle image. Given that $G_u(u) = F_u(u) P_u(u)$, we have:

$$\begin{cases} G_u(0) = 1, \quad G_u(1/d_r) = (1 - 1/u_0 d_r) \operatorname{sinc}(a) \\ G'_u(1/d_r) = d_r(1 - 1/u_0 d_r) \cos(\pi a) - d_r \operatorname{sinc}(a) \end{cases} \tag{25}$$

Substitution in (24) yields the following expression for δ_x at $a = 0$:

$$\delta_x/d_r = -\frac{(1/u_0 d_r) \sin(2\pi d_x/d_r)}{\pi \left[1 + (1 - 1/u_0 d_r) 2 \cos(2\pi d_x/d_r) \right]} \tag{26}$$

and for δ_x at $a = 1$:

$$\delta_x/d_r = -\frac{1}{\pi}(1 - 1/u_0 d_r) \sin(2\pi d_x/d_r). \tag{27}$$

In Figure 5 the error δ_x is plotted for $a^2 = 0$, $\frac{1}{4}$ and 1, for a sampling rate $u_0 d_r = \frac{3}{2}$. This graph shows that the error for contiguous pixels (i.e., $a = 1$) is smaller than for infinitesimal pixels (i.e., $a = 0$).

For the special case where $1/d_r$ exactly matches the cut-off frequency, the amplitude of δ_x is zero for $a^2 = 1$, whereas values of $a^2 < 1$ lead to a finite error amplitude. This demonstrates the importance of using CCD arrays with 100% fill ratio.

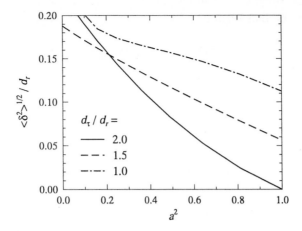

Figure 6: The root-mean-square amplitude $\langle \delta^2 \rangle^{1/2}$ of the bias error for the diffraction-limited particle-image centroid as a function of the pixel fill ratio a^2, for $d_s/d_r = 1.0$, 1.5 and 2.0.

Note that the curves in Figure 5 have negative gradients at $d_x = 0$. This implies that the measured location for a particle image is biased towards the center of a pixel. Consequently, measurements of the location and displacement of a particle image favor integer pixel values. This effect is commonly referred to as *pixel locking*. The analysis presented here has demonstrated that this effect can be associated with an undersampling in the pixelization of the image.

3.2.2 the effect of blur

In the literature it is often reported that a slight blurring of the image improves the precision of the estimated location. In this section this will be investigated by considering the tracking error for slightly de-focussed particle images. Since the tracking error is identical to zero for fully resolved pixelization, we consider the effect of blur for the case of slightly under-sampled pixelization. To keep the analysis concise only the case for contiguous pixels (i.e., $a^2 = 1$) will be considered.

The analysis in Section 3.2.1 showed that the shape and amplitude of the tracking error is determined by $G_u(1/d_r)$ and $G'_u(1/d_r)$, where $G_u(u)$ is the product of the optical transfer function (OTF) and the spatial pixel sensitivity. For an optical configuration that is not exactly in focus the OTF is given by (Goodman 1968):

$$G_u(u) = \Lambda(u/u_0) \, \mathrm{sinc}\left[8 \frac{w}{\lambda} \frac{u}{u_0} \left(1 - \frac{|u|}{u_0} \right) \right] \tag{28}$$

where w is the *maximum path-length error*, defined as:

$$w = \left(\frac{1}{d_i} + \frac{1}{d_o} - \frac{1}{f} \right) \frac{D^2}{8} \tag{29}$$

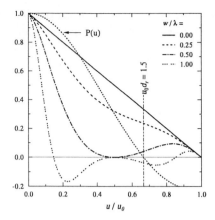

Figure 7: The optical transfer function for a (square aperture) optical system with different degrees of out-of-focus aberration. The dash-dotted line represents the transfer function of contiguous pixels ($a^2 = 1$) at a sampling rate of $1/d_r = \frac{2}{3}u_0$.

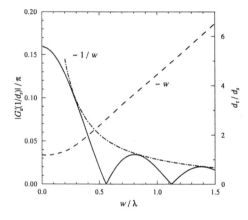

Figure 8: The amplitude of the bias error and e^{-2} width of the diffraction-limited spot, as a function the out-of-focus path length error w. The dashed dotted line represents the envelope of the curve for the bias-error amplitude.

where d_i and d_o are the image and object distances respectively. The maximum path-length error can be written as: $w/\lambda \approx -\frac{1}{2}h/\delta_z$, where h is the distance between the object plane and the plane of focus, and δ_z the object focal depth (Adrian 1991).

Figure 7 shows the optical transfer function (OTF) for an optical system with square aperture for different values of w, corresponding to an object plane that is located between zero ($w/\lambda = 0$) and twice ($w/\lambda = 1$) the focal depth from the true focal plane. Note that where the object plane is out of focus, the values of $G(u)$ and $G'(u)$ for $1 \leq 2u/u_0 \leq 2$ are generally *smaller* than for the case of perfect focus ($w/\lambda = 0$); this already indicates that one

48

may expect a smaller tracking error for blurred particle images.

For the case of contiguous pixels, the pixelization transfer function is zero for $u = 1/d_r$; see (11). Consequently, the tracking bias error is a sinus function with an amplitude given by:

$$\frac{1}{\pi} G'(1/d_r) = -\frac{1}{\pi} \Lambda \left(\frac{1}{u_0 d_r} \right) \mathrm{sinc} \left[8 \frac{w}{\lambda} \frac{1}{u_0 d_r} \left(1 - \frac{1}{u_0 d_r} \right) \right]. \qquad (30)$$

The absolute amplitude of the tracking error for pixelization with contiguous pixels with a sample spacing of $d_r = \frac{3}{2} u_0^{-1}$ is shown in Figure 8. This result indeed shows a considerable reduction for the tracking error for an increasing out-of-focus location of the object plane. The amplitude even becomes zero at $w/\lambda \cong 0.65$ and 1.25.

So, a slight blur can reduce the tracking error considerably, and for this particular example a properly chosen blur may even completely compensate for the tracking error. However, it should noted that blur *increases* the particle-image diameter, which also increases the random error and may cancel out the favorable effect of blur. This is illustrated by Figure 8, in which is also plotted the e^{-2}-diameter of the particle image as a function of w/λ; note that the particle-image diameter increases linearly for $w/\lambda > 0.5$. The envelope of the bias error amplitude (denoted by the dash-dotted line in Figure 8) is inversely proportional to w, and consequently the product of the envelope and the particle-image diameter is constant. Hence, for this particular case the *reduction* of the tracking bias error is replaced by a proportional *increase* of the random error.

3.3 Gaussian peak-fit

The observation that the shape of particle images and consequently the shape of the displacement-correlation peak is well approximated by a Gaussian curve has promoted the use of a Gaussian curve fit for the estimation of the

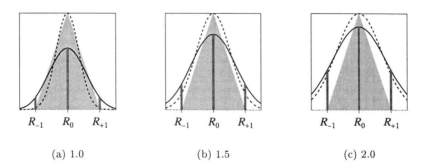

(a) 1.0	(b) 1.5	(c) 2.0

Figure 9: The continuous (– – –) and discrete (——) correlation peaks for Gaussian particle images with $d_\tau/d_r = 1.0$, 1.5 and 2.0, at zero fractional displacement with $a = 1$. The shaded triangle represents Φ_{pp}.

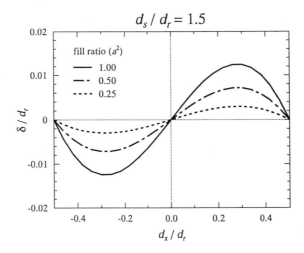

Figure 10: The bias error for the Gaussian peak fit for Gaussian particle images as a function of the displacement, for $d_\tau/d_r = 1.5$ and fill ratio $a^2 = 1.00$, 0.50 and 0.25.

particle-image or peak location (Willert & Gharib 1991; Westerweel 1993). For correlation peaks in the discrete domain it was shown that for small particle images, i.e. $d_\tau/d_r \sim 1-2$, only the correlation maximum and it adjacent correlation values are significant with respect to the background random correlations (Westerweel 1993). Hence, each component of the location is estimated by fitting a Gaussian curve to the correlation maximum R_0, and its two neighbors, denoted by R_{-1} and R_{+1} respectively. The x-component of the location is then given by:

$$m_x = i_0 d_r + \frac{1}{2}\frac{\ln R_{-1} - \ln R_{+1}}{\ln R_{-1} + \ln R_{+1} - 2\ln R_0}d_r, \tag{31}$$

with: $R_i = R_D[i_0 + i, j_0]$, where $[i_0, j_0]$ are the indices of the maximum correlation. For matter of simplicity it is assumed that $i_0 = j_0 = 0$. When the correlation peak has an exact Gaussian shape, then the estimate (31) would be exact. However, for digital image fields, the shape of the discrete correlation peak is given by (12); see also Figure 9. So, even for perfectly Gaussian particle images, the estimate (31) will deviate from the true location.

To determine the bias error as a function of the particle-image size and pixel fill ratio, the expression in (12) for the discrete correlation was substituted in (31). It was assumed that the continuous displacement-correlation peak was Gaussian with an e^{-2}-diameter of $\sqrt{2}d_\tau$. The result was determined by numerical integration of the substituted equations.

3.3.1 slight under-sampling
First, the bias error is considered for the range of particle-image sizes that can be considered as 'slightly under-sampled,' i.e. $1 \leqslant d_\tau/d_r \leqslant 2$.

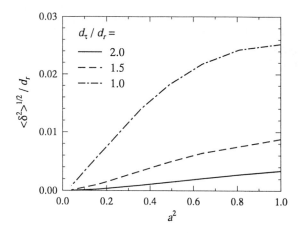

Figure 11: The root-mean-square amplitude $\langle \delta^2 \rangle^{1/2}$ of the bias error for the Gaussian peak fit as a function of the fill ratio a^2 for $d_\tau/d_r = 1.0$, 1.5 and 2.0.

In Figure 10 the bias error for the Gaussian peak fit is shown as a function of the displacement for $d_\tau/d_r = \frac{3}{2}$. Note that the derivative of the bias error at $d_x = 0$ is *positive*. Hence, in contrast with the bias error for the centroid, the Gaussian peak-fit estimate for the location of the displacement-correlation peak is biased *against* integer pixel values. This is opposite to what is observed in many practical situations, in which the Gaussian peak fit also suffers from a bias towards integer pixel values, rather than a bias against integer pixel values (Fincham & Spedding 1997). This is further discussed in the next section.

The bias error amplitude is plotted in Figure 11 as a function of the fill ratio a^2 for different values of d_τ/d_r. Note that the bias error for the centroid estimate was increased when the fill ratio was decreased, whereas the result in Figure 11 for the Gaussian peak fit estimate shows that the bias error *decreases* for decreasing fill ratio. This is not very surprising, given that a Gaussian shape for the continuous displacement-correlation peak was chosen originally. In a practical situation the shape of the particle images is not exactly Gaussian, so it is expected that bias error does not decrease that rapidly. In addition, a reduction of the fill ratio will also increase the random error (unless the total illumination is increased proportionally; see Section 3.1). The error reduction only sets in when a^2 is substantially smaller than unity, so that the total illumination needs to be increased by a considerable amount before one can expect any benefit from such a low fill ratio.

The bias error for the Gaussian peak fit is plotted in Figure 12 as a function of d_τ/d_r for $a = 0.5$ and 1.0. The bias error is inversely proportional to d_τ/d_r. It increases very rapidly for $d_\tau/d_r < 1$, and it practically vanishes for $d_\tau/d_r > 2$. This complies with the result obtained for the centroid estimate. Note that the difference between the curves for different fill ratio only differ slightly, indicating that the fill ratio only marginally affects the bias error.

In Figure 12 are also plotted the total root-mean-square error obtained from the analysis of synthetic PIV images. The straight dashed line in Figure 12 corresponds to the empirical relationship in (18) with $c \cong 0.02$. This rather low value for c can be explained by the fact that these synthetic PIV images do not include any noise source, other than the local variation of the number of tracer particles, so that the SNR is likely to be substantially higher than in most practical situations. Nonetheless, the minimum value for the total error occurs at about $d_\tau/d_r \sim 2$ with a value of about $0.04d_r$, which complies with simulation studies carried out by Willert (1996), and experimental data obtained by Prasad *et al.* (1992) and Westerweel (1997).

3.3.2 serious under-sampling
The preceding analysis was done for values of the pixel size that correspond to a slight under-sampling of the image field. When the size of the particle images with respect to the pixels is further reduced, then another effect sets in that will be explained here.

Consider the expression (12) in the limit $d_\tau/d_r \to 0$. The shape of the discrete correlation peak is dominated by the self-correlation of the pixel sensitivity Φ_{pp}. Evidently, the correlation amplitude is proportional to $\Phi_{pp}(d_x, d_y)$. So, the detectability of the displacement-correlation peak does not only depend on the image density, and the in-plane and out-of-plane loss-of-pairs, but also on the fractional part of the displacement. The lowest value of the detectability as a function of the fractional displacement occurs for $d_x/d_r = \pm\frac{1}{2}$. So, what happens is that the peak amplitude for non-zero fractional displacements is reduced, but the random correlations maintain the same amplitude.

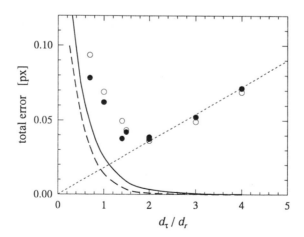

Figure 12: The total r.m.s. error for the Gaussian peak fit as a function of d_τ/d_r for $a = 1.0$ (—— and ●) and $a = 0.50$ (– – – and ○). The symbols represent results obtained from synthetic PIV images.

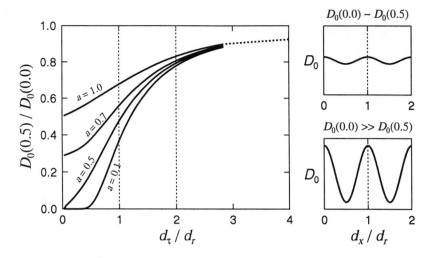

Figure 13: The ratio of the detectability D_0 for 0.5 and 0.0 fractional displacement as a function of d_τ/d_r. For $D_0(0.5)/D_0(0.0){\sim}1$ the detectability is practically independent of the displacement (top right), whereas for $D_0(0.5)/D_0(0.0){\ll}1$ the detectability strongly depends on the sub-pixel fraction of the displacement (bottom right).

Consequently, the probability that a random correlation peak is higher than the displacement-correlation peak is enhanced.

In Figure 13 the ratio of the peak height at a fractional displacement of one-half and the peak height at zero fractional displacement is plotted as a function of d_τ/d_r for different fill ratios. Assuming that the amplitude for the random correlations is the same for both values of the fractional displacement, this ratio is also a measure of the relative change in peak detectability.

For contiguous pixels (i.e., $a = 1$), the amplitude of the correlation peak drops by a factor 2 in the limit $d_\tau/d_r \to 0$ (i.e., $\Lambda(d_x/d_r) = \frac{1}{2}$ for $d_x/d_r = \frac{1}{2}$). However, when the pixel fill ratio is less than 0.25, the displacement-correlation peak vanishes completely (i.e., $\Lambda(d_x/ad_r) = 0$ for $a < d_x/d_r = \frac{1}{2}$). If one would plot a displacement histogram for such a situation, only displacements with a fractional displacement near zero yield valid measurements, whereas displacements with a fractional displacement near one-half result in a strong signal loss. This would explain why measurements obtained with the Gaussian peak fit display pixel locking at zero fractional displacements.

4   Discussion and Conclusions

The preceding analysis has shown that the effects of sensor geometry on the performance of PIV interrogation are negligible when the particle-image diameter is at least two pixels.

When the particle-images are slightly under-sampled, i.e. $1 \leqslant d_\tau/d_r \leqslant 2$, the bias errors are of the same order of magnitude as the random errors, so that measurements in this range of d_τ/d_r are still feasible. However, the performance depends strongly on the pixel fill ratio, and it deteriorates substantially for both the centroid error and Gaussian peak fit when the fill ratio is much smaller than unity.

For slightly under-sampled images a blurring of the image by means of a slight de-focussing can reduce the bias error amplitude. However, this also increases the random error amplitude, so there is no real net improvement. For a substantial improvement of the measurement performance it is necessary to increase the signal bandwidth, whereas de-focussing only changes the shape of the OTF, but not its bandwidth.

When the particle images are less than one pixel in diameter, the bias errors dominate. For the Gaussian peak fit this implies an un-recoverable signal loss, that is proportional to the magnitude of the fractional part of the displacement. This leads to a higher probability to detect displacements that lie near an integer value of in pixel units. Hence, the pixel locking effect for the Gaussian peak fit is more likely related to a signal loss, than a variation in peak shape as a function of the sensor parameters.

So, in what way does this affect how one can apply PIV with digital image sensors to measurements in flows? Given the experimental restriction $d_\tau/d_r \sim 1$–2, one may use (6) to determine a relation between the minimum f-number and the image magnification for given tracer size d_p and pixel size d_r. For a PIV measurement in air the specific gravity of the tracer particles (or droplets) is much higher than that of the fluid, so that one is forced to use very small tracer particles. The typical diameter of oil droplets is 1–2 μm. Hence, for measurements in air the particle-image diameter is usually determined by the diffraction-limited spot of the imaging optics; at small image magnification one has to use high f-number lenses to match the particle-image diameter to the size of the pixels (and use correspondingly powerfull light sources to obtain a significant signal).

For experiments in water the situation is somewhat different. It is usually possible to match the specific gravity of the tracer particles with that of the fluid, so that one may use considerably larger particles in comparison with PIV measurements in air. Typical diameters for tracer particles in measurements in water lie in the range of 10–20 μm. So, for measurements in water it is more likely that the particle-image diameter is determined by the diameter of the geometric particle image, i.e. Md_p, when M is large.

These aspects are summarized in Figure 14, which shows the minimum required f-number for the imaging optics that correspond to $d_\tau/d_r = 1$ and 2 as a function of the image magnification M for $d_p = 2$ μm and 20 μm (representing PIV measurements in air and water respectively), and for a given pixel size of 12 μm (which is a typical value for CCD sensor arrays). The symbols correspond to the conditions for actual PIV measurements, conducted both in air and water. The data that were used were part of the EuroPIV

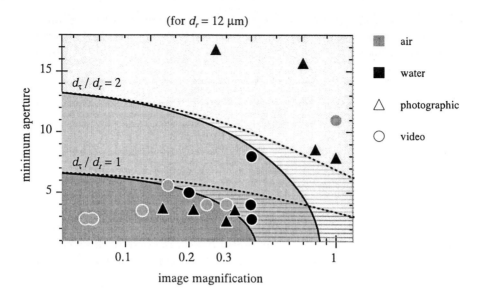

Figure 14: The minimum numerical aperture $f^{\#}$ $(= f/D)$ as a function of the image magnification M for 20 μm (——) and 2 μm (– – –) particles. The symbols represent PIV measurements conducted in air (light symbols) and water (dark symbols), using photographic (triangles) or digital (circles) recording.

project (Stanislas *et al.* 1999). It should be noted that for the measurements with photographic recording the f-number was rescaled to match the effective spatial resolution of $d_r = 12$ μm.

The image magnification for the typical small-scale laboratory experiment lies in the range between 0.25 and 1, which shows that low f-numbers can be used. However, for large-scale experiments, such as measurements in large wind tunnels, the image magnification may have a much lower value; imaging a 1 m^2 area on a 0.5 inch CCD array corresponds to $M = 0.012$, whereas imaging the same area on a large format 4''×5'' film corresponds to $M = 0.1$. Hence, for such situations PIV measurements with CCD arrays at low f-numbers would typically suffer from pixel locking effects.

PIV measurements that lie near the corresponding curve for $d_\tau/d_r = 2$ are 'ideal' in the sense that the total error is minimal, whereas measurements that lie below the $d_\tau/d_r = 1$ curve suffer from an irreversible form of 'pixel locking'; between the two curves, the image field is slightly under-sampled, and pixel locking only affects the measurements marginally.

The symbols in Figure 14 demonstrate a trend in which PIV measurements with digital recording are taken for decreasing values of the image magnification, i.e. indicating an increase in the scale of the measurements. To gain an adequate amount of light often low f-numbers are used, but this holds the danger of having too small particle images with respect to the image spatial resolution, so that these experiments are more prone to peak locking effects.

Acknowledgments

The research of dr.ir. J. Westerweel has been made possible by a fellowship of the Royal Netherlands Academy of Arts and Sciences.

References

ADRIAN, R. J. 1984. Scattering particle characteristics and their effect on pulsed laser measurements of fluid flow: speckle velocimetry vs. particle image velocimetry. *Appl. Opt.*, **23**, 1690–1691.

ADRIAN, R. J. 1991. Particle-imaging techniques for experimental fluid mechanics. *Ann. Rev. Fluid Mech.*, **23**, 261–304.

ALEXANDER, B. F., & NG, K. C. 1991. Elimination of systematic error in subpixel accuracy centroid estimation. *Opt. Engr.*, **30**, 1320–1331.

FINCHAM, A. M., & SPEDDING, G. R. 1997. Low cost, high resolution DPIV for measurement of turbulent fluid flow. *Exp. Fluids*, **23**, 449–462.

GOODMAN, J. W. 1968. *Introduction to Fourier Optics*. New York: McGraw-Hill.

KEANE, R. D., & ADRIAN, R. J. 1992. Theory of cross-correlation analysis of PIV images. *Appl. Sci. Res.*, **49**, 191–215.

OPPENHEIM, A. V., WILLSKY, A. S., & YOUNG, I. T. 1983. *Signals and Systems*. Englewood Cliffs (NJ): Prentice-Hall.

PRASAD, A. K., ADRIAN, R. J., LANDRETH, C. C., & OFFUTT, P. W. 1992. Effect of resolution on the speed and accuracy of particle image velocimetry interrogation. *Exp. Fluids*, **13**, 105–116.

STANISLAS, M., KOMPENHANS, J., & WESTERWEEL, J. (EDS.). 1999. *EUROPIV: A cooperative action to apply Particle Image Velocimetry to problems of industrial interest* in preparation. Dordrecht: Kluwer.

WERNET, M., & PLINE, A. 1993. Particle displacement tracking technique and Cramer-Rao lower bound error in centroid estimates from CCD imagery. *Exp. Fluids*, **15**, 295–307.

WESTERWEEL, J. 1993. *Digital Particle Image Velocimetry*. Ph.D. thesis, Delft University of Technology, The Netherlands.

WESTERWEEL, J., DABIRI, D., & GHARIB, M. 1997. The effect of a discrete window offset on the accuracy of cross-correlation analysis of digital PIV recordings. *Exp. Fluids*, **23**, 20–28.

WILLERT, C. E. 1996. The fully digital evaluation of photographic PIV recordings. *Appl. Sci. Res.*, **56**, 79.

WILLERT, C. E., & GHARIB, M. 1991. Digital particle image velocimetry. *Exp. Fluids*, **10**, 181–193.

I.4. Micron-Resolution Velocimetry Techniques

C.D. Meinhart[1], S.T. Wereley[1], and J.G. Santiago[2]

[1]Department of Mechanical and Environmental Engineering,
University of California, Santa Barbara, CA 93106

[2]Department of Mechanical Engineering,
Stanford University, Stanford, CA 94305

Abstract. This paper examines various techniques that have been developed to measure flow fields in microfluidic devices with $10^0 - 10^2$ micron-scale spatial resolution. These techniques include Scalar Image Velocimetry (SIV), Laser Doppler Velocimetry (LDV), and Particle Image Velocimetry (PIV). Advantages and disadvantages of these techniques are presented.

PIV results are presented for a liquid flow through a nominally 30 μm high × 300 μm wide microchannel. These velocity measurements have spatial resolutions of 5.0 μm × 1.3 μm × 2.8 μm. Using 50% overlap, the velocity vector spacing is 680 nm in the spanwise direction of the microchannel.

1. Introduction

There are several areas in science and engineering where it is important to determine the flow field at the micron scale. In many situations the flow is complicated and numerical models are not sufficient to describe the flow field. For example, the flow may be non-Newtonian, surface interactions may be important, or the boundary conditions may not be well understood. In these situations, it is desirable to apply non-intrusive diagnostic techniques to experimentally analyze the flow field. Examples of several microfluidic devices in the computer, aerospace, and biomedical industries are given below to illustrate the need for accurate diagnostic techniques at the microscale.

In the aerospace industry, micron-scale supersonic nozzles play an important role in the development of small-scale aircraft and spacecraft. Figure 1 is a Scanning Electron Micrograph (SEM) of a supersonic nozzle being developed at MIT. The throat of this nozzle is approximately 35 μm wide. The thickness of the nozzle is 500 μm. These nozzles are being designed for JPL/NASA to be used as microthrusters on micro-satellites (Bayt et al., 1997).

In the computer industry, inkjet printers account for 65% of the computer printer market (Kamisuki, 1998). Inkjets basically consist of an array of nozzles with exit orifices on the order of tens of microns in diameter. During expulsion, ink is forced out of a micro-reservoir by either a mechanical actuator or by a heating element

58

that boils a small volume of ink. This process can be repeated at a rate of 1-3 kHz. One of the main goals of the computer printer industry is to make inkjets smaller, to increase printer resolution, and to increase the repetition rate of inkjets for faster printing speed.

The biomedical industry is currently developing and using microfluidic devices for patient diagnosis, patient monitoring, drug delivery, and drug discovery. Many of these processes involve chemical reactions and the transport of macromolecules or cells through sub-millimeter channels. The details of the fluid motion in these small channels, coupled with interactions between macromolecules, cells, and the surface-dominated physics of the channels create complex phenomena, which can be difficult to simulate numerically.

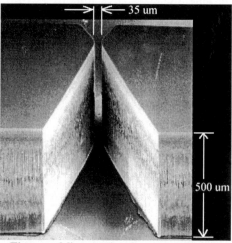

Figure 1. Scanning Electron Micrograph of a microfabricated supersonic nozzle (courtesy, Robert Bayt and Kenny Breuer, MIT).

The I-STAT device is the first microfabricated fluidic device that has been widely used in the medical community for blood analysis. All the chemical analysis is performed in a disposable microfluidic device at the point-of-care, providing doctors with immediate blood-work results without sending samples to a laboratory. Since the analysis chambers are completely disposable, the potential for cross contamination is eliminated.

Other examples of microfluidic devices for biomedical research include microscale flow cytometers for cancer cell detection (Krulevitch, 1997), micromachined electrophoretic channels for DNA fractionation, and polymerase chain reaction (PCR) chambers for DNA amplification (Northrup et al., 1995).

Mixing two fluids, or distributing chemical species and macromolecules in microchannels is a major challenge facing microfluidics researchers. The low

Reynolds numbers present in microfluidic devices preclude turbulence as a mechanism for mixing. In many situations, pure molecular diffusion is too slow to achieve adequate mixing. Therefore researchers are developing micromixing devices that promote chaotic advection to enhance mixing (Evans et al., 1997).

A wide range of diagnostic techniques has been developed for microfluidics research. Some of these techniques have been designed to obtain the highest spatial resolution and velocity resolution possible, while other techniques have been designed for application in non-ideal situations where optical access is limited (Lanzillotto et al., 1996), or in the presence of highly scattering media (Chen et al., 1997). Details of these techniques are discussed in the following sections.

2. Scalar Image Velocimetry (SIV)

Scalar image velocimetry (SIV) refers to the determination of velocity-vector fields by recording images of a passive scalar quantity, and inverting the transport equation for a passive scalar

$$\frac{\partial c}{\partial t} + (\boldsymbol{u} \cdot \nabla)\, c = D\, \nabla^2 c. \qquad (1)$$

Here, c is the concentration of the passive scalar, and D is the coefficient of molecular diffusion. Dahm et al. (1992) originally developed Scalar Image Velocimetry (SIV) for measuring turbulent jets. Successful velocity measurements depend on having sufficient spatial variations in the passive-scalar field and relatively high Schmidt numbers.

Since SIV uses molecular tracers to follow the flow, it has several advantages at the microscale over techniques such as PIV or LDV, which use discrete flow-tracing particles. Typically, molecular tracers have much higher diffusion coefficients than discrete particles, which can significantly lower the spatial resolution and velocity resolution of the measurements.

Paul et al. (1997) analyzed the motion of an uncaged fluorescent dye, using SIV to estimate velocity fields for pressure- and electrokinetically-driven flows in 75 μm diameter capillary tubes. A 20 μm × 500 μm sheet of light from a λ = 355 nm frequency-tripled Nd:YAG laser was used to uncage a 20 μm thick cross-sectional plane of dye in the capillary tube. In this technique, only the uncaged dye is excited when the test section is illuminated with a shuttered beam from a continuous wave Nd:YVO$_4$ laser. The excited fluorescent dye is imaged using a 10x, NA = 0.3 objective lens onto a CCD camera at two known time exposures. The velocity field is then inferred from the motion of the passive scalar. We approximate the spatial resolution of this experiment to be on the order of 100 μm

× 20 μm × 20 μm, based on the displacement of the fluorescent dye between exposures, and the thickness of the light sheet used to uncage the fluorescent dye.

Molecular Tagging Velocimetry (MTV) is a related technique that could, in principle, be applied to microfluidics research. In this technique, flow-tracing molecules phosphoresce after being excited by a grid of UV light. Two CCD cameras image the phosphorescent grid lines with a short time delay between the two images. Local velocity vectors are estimated by correlating the grid lines between the two images (Koochesfahani et al., 1996). In order to achieve spatial resolutions on the order of one micron, the molecules would have to be excited with grid lines which are about 500 nm thick (in the illumination plane). While it is possible to focus UV light down to about 500 nm using diffraction-limited low *f-number* optics, it is impractical to generate a grid of several lines that are 500 nm wide over an appreciable length. In principle, MTV could be used for micro-fluidics research, with spatial resolutions on the order on tens of microns.

Optical Flow refers to a class of velocimetry algorithms originally developed by the machine vision community to determine the motion of rigid objects. The technique can be extended to fluid flows by assuming the effect of molecular diffusion is negligible, and requiring that the velocity field be sufficiently smooth. Assuming that the image intensity of a passive scalar is proportional to the concentration of the scalar, one can form the constraint equation for optical flow

$$\frac{\partial I}{\partial t} + (\boldsymbol{u} \cdot \nabla)I = 0 , \tag{2}$$

where I is the image intensity when projected back into the flow field. Combining the above constraint with the smoothness criterion that $\|\nabla \boldsymbol{u}\|$ is small, yields the equation for determining optical flow. It is customary to solve optical flow by minimizing the integral

$$\iint_R \left[\lambda \left(\frac{DI}{Dt} \right)^2 + \|\nabla \boldsymbol{u}\|^2 \right] d^2\boldsymbol{x} , \tag{3}$$

where R is the region of interest, and λ is a parameter that specifies the importance of the constraint compared to the smoothness criteria.

Since the velocity field is computed from temporal and spatial derivatives of the image field, the accuracy and reliability of the velocity measurements is strongly influenced by noise in the image field. This technique imposes a smoothness criterion on the velocity field that effectively low-pass filters the data, and can lower the spatial resolution of the velocity measurements (Wildes et al., 1997).

Lanzillotto et al. (1996) applied optical-flow algorithms to infer velocity fields in 500 – 1000 μm diameter micro-tube. This technique indirectly images 1 – 20 μm diameter X-ray-scattering emulsion droplets in a liquid flow. A synchrotron is used to generate high-intensity X-rays that scatter off the emulsion droplets onto a phosphorous screen. A CCD camera imaging the phosphorous screen detects variations in the scattered X-ray field. Lanzillotto et al. (1996) report a mean velocity field, measured in an 840 μm diameter tube, with a velocity-vector spacing of about 40 microns and axial bulk velocities of 7–14 μm/s. The primary advantage of X-ray imaging technique is that one can obtain structural information about the flow field without having optical access.

Hitt, Lowe & Newcomer (1996) applied the optical flow algorithm to *in vivo* blood flow in microvascular networks, with seed particle diameters on the order of 100 μm. The algorithm spectrally decomposes sub-images into discrete spatial frequencies, and then correlates the different spatial frequencies to obtain flow field information. The advantage of this technique is that it does not require discrete particle images to obtain velocity information. Hitt et al. (1995) obtained *in vivo* flow measurements of blood cells flowing through a microvascular network using a 20x water-immersion lens. We estimate the spatial resolution of this technique to be on the order of 20 μm in all three dimensions.

3. Laser Doppler Velocimetry (LDV)

Laser Doppler Velocimetry (LDV) has been a standard technique in fluid mechanics for more than 25 years. In the case of a dual-beam LDV system, the intersection of the two laser beams defines the measurement volume. The measurement volumes of standard LDV systems have characteristic dimensions on the order of a few millimeters.

Turbulence researchers recognize the importance of obtaining spatial resolutions that are high enough to resolve small-scale turbulent motions. Compton & Eaton (1996) used short focal length optics to obtain a measurement volume of 35 μm × 66 μm. The short focal length optics allowed them to obtain measurements as close as 0.1 mm from the wall, which resolves their $Re_\theta = 1400$ boundary layer down to $y^+ = 3$.

Using very short focal length lenses, Tieu, Machenzie, & Li (1995) built a dual-beam solid-state LDA system that has a measurement volume of approximately 5 μm × 10 μm. The length of the probe volume is determined by

$$l = \frac{8f^2\lambda}{Dd\pi},\qquad(4)$$

where f is the focal length of the lens, λ is the wavelength of light, D is the beam diameter, and d is the beam spacing. For a given wavelength of light, a small probe volume can be achieved by using a short focal length, f, and a large beam diameter,

D, and a large beam spacing, d. Tieu et al. (1995) used $f = 12$ mm, $D = 3.3$ mm, $d = 10$ mm, and $\lambda = 685$ nm, to obtain a probe length of $l = 10$ µm.

Clearly, a micron-scale probe volume is desirable for microfluidic measurements. However, this introduces significant constraints on the system that can limit overall performance. The probability of a scattering particle entering the probe volume during a given time interval decreases with the probe volume. Since the uncertainty in determining the Doppler frequency depends on the number of fringes in the probe volume, N_{fr}, it is desirable to have a large number of fringes.

A micron-scale probe volume significantly limits the number of fringes, and subsequently limits the accuracy of the velocity measurements. Tieu et al. (1995) applied their micro LDV system to measure the flow through a 175 µm thick channel. The resulting time-averaged measurements compare well to the parabolic velocity profile, except within 18 µm of the wall, where the measurements fail.

Optical Doppler Tomography (ODT) has been developed to measure microscale flows embedded in highly scattering media. In the medical community, the ability to measure *in vivo* blood flow under the skin allows doctors to determine the location and depth of burns (Chen et al., 1997). ODT combines single-beam Doppler velocimetry with heterodyne mixing from a low-coherence Michelson interferometer. The lateral spatial resolution of the probe volume is determined by the diffraction spot size. The Michelson interferometer is used to limit the effective longitudinal length of the measurement volume to that of the laser coherence length. The ODT system developed by Chen et al. (1997) has a lateral and longitudinal spatial resolution of 5 µm and 15 µm, respectively. The system was applied to measure flow through a 580 µm diameter conduit. Here, 1.7 µm diameter particles were used to trace the flow.

The micro-sensor group at Caltech has developed a Miniature Laser Doppler Anemometer (MLDA) that measures 50 mm in length and 25 mm in diameter. The device uses an inexpensive commercially available diode laser, and is fiber optically connected to an external Photo Multiplier Tube (PMT). The probe can be designed to have a measurement diameter as small as 10 microns (Gharib, Modares & Taugwalder, 1998).

4. Particle Image Velocimetry (PIV)

Particle Image Velocimetry (PIV) can be used to obtain high-resolution 2-D velocity fields. Urushihara, Meinhart & Adrian (1993) used Particle Image Velocimetry (PIV) to make velocity measurements in turbulent flows with spatial resolutions of up to 280 µm × 220 µm × 200 µm. This high spatial resolution was obtained using conventional PIV interrogation and data processing algorithms (Meinhart et al., 1993). Keane, Adrian & Zhang (1995) developed *super-resolution PIV* and applied it to Urushihara's data to achieve spatial resolutions of approximately 50 µm × 50 µm × 200 µm. The super-resolution PIV algorithm consists of two parts: First, standard-resolution velocity fields are obtained from the image data using standard correlation-based algorithms. Second, high-resolution velocity fields are obtained by tracking individual particles, using the information from the standard-resolution measurements as a search guide.

The spatial resolution of a PIV system is limited by many factors such as particle size, particle seeding density, and image quality. Ultimately, PIV techniques are limited by the diffraction-limited resolution of the optical recording system. Diffraction-limited resolution can be increased by decreasing the *f-number* of the optical system, where the *f-number* is the ratio of the focal length, *f*, to the aperture diameter of the optical system, *D*. As *f-number* is decreased, geometric aberrations become increasingly important. Most optical systems have an optimal *f-number*, where resolution limits resulting from geometric aberrations are balanced by resolution limits resulting from diffraction. Microscopic objectives have extremely high resolving powers, because they operate with very low *f-numbers*, and they are corrected to be nearly aberration free. Microscope lenses are characterized by their numerical aperture, $NA = n \sin \theta$, where *n* is the index of refraction of the imaging medium, and θ is the half angle of the light collecting aperture. Oil-immersion objective lenses are available commercially with a $NA = 1.4$ and a magnification, $M = 100$. The diameter of the diffraction-limited point spread function, d_s, when projected onto the CCD camera is given by

$$d_s = 2.44M \frac{\lambda}{2NA}, \qquad (5)$$

where *M* is the geometric magnification of the objective lens. Recording images with $\lambda = 560$ nm wavelength light produces a diffraction-limited point spread function of diameter 488 nm, when projected back into the fluid. Therefore, an infinitesimally small particle will appear to have a 488 nm diameter. Assuming the ratio of the particle-image diameter to the pixel width, (d_τ / d_{pix}) is about three, the centroid of the correlation peak can be located to within $1/10^{th}$ of the particle-image diameter (Prasad et al., 1992). When projected back into the fluid, the particle appears to be 488 nm in diameter, implying that its position can be determined to within 48 nm (i.e. $1/10^{th}$ of the apparent particle-image diameter). Since we know *a priori* the shape of the particle, and since we are over-sampling the particle image by a factor of three, we can determine the particle's position to within $\sim \lambda/12$ (i.e. 48 nm $\sim$ 560 nm / 12).

Santiago et al. (1998a) developed a PIV system that is capable of measuring instantaneously 1000 velocity vectors with a spatial resolution of $6.9 \times 6.9 \times 1.5$ microns. The system uses an epi-fluorescent microscope and an intensified CCD camera to record 300 nm diameter polystyrene flow-tracing particles. The particles are illuminated using a continuous Hg-arc lamp. The continuous Hg-arc lamp is chosen for situations that require low levels of illumination (for example, flows containing living biological specimens). The images are obtained using an intensified frame-transfer CCD camera. The velocity-vector field shown in Figure 2 contains over 900 velocity vectors, with a spatial resolution of $6.9 \times 6.9 \times 1.5$ μm for each vector (Santiago, et al. 1998a).

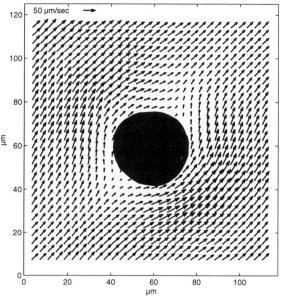

Figure 2. Micro PIV velocity vector map of a Hele-Shaw flow. The vector field contains over 900 velocity vectors, each measured with a spatial resolution of $6.9 \times 6.9 \times 1.5$ μm. (Reproduced from Santiago et al., 1998a.)

Meinhart et al. (1999) applied micro PIV to measure the flow field in a nominally 30 μm high × 300 μm wide rectangular channel, with a flow rate of 200 μl/hr. Figure 3 is a schematic of their micro PIV. The system utilizes a pulsed Nd:YAG laser to illuminate fluorescent particles and a cooled frame-transfer camera to record particle images. The pulsed Nd:YAG laser can be used in situations where low levels of light are not critical, and must be used in situations where the velocity is large enough that only a pulsed light source can freeze the particle motion. The flow is imaged by a 60x, NA = 1.4, oil-immersion lens. The 200 nm diameter polystyrene flow-tracing particles were chosen small enough so that they faithfully followed the flow and were 150 times smaller than the smallest channel dimension. The particles were chosen large enough so that errors due to Brownian motion were not significant. Reliable velocity data was obtained using an interrogation spot size of 5.0 μm × 1.3 μm × 2.8 μm, which was defined by the size of the first interrogation window. By overlapping the interrogation spot size by 50%, the velocity-vector spacing is 680 nm in the wall normal position. Figure 4 compares the velocity profile obtained by ensemble averaging the resulting velocity-vector fields to the analytical solution for Stokes' flow in a rectangular duct. The results show good agreement between PIV data and the analytical solution.

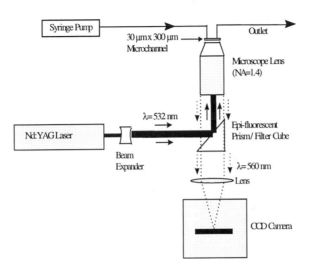

Figure 3. Schematic of a micro PIV system. A pulsed Nd:YAG laser is used to illuminate fluorescent 200 nm dia. flow-tracing particles, and a cooled CCD camera is used to record the particle images (Meinhart et al., 1999).

The first interrogation window was 44 pixels in the streamwise direction and 12 pixels in the spanwise direction, while the second interrogation region was 48 pixels × 16 pixels, respectively. The spanwise dimension of the interrogation region was reduced as much as possible to resolve the shear near the wall of the channel while the streamwise dimension of the interrogation region was elongated so that enough particle images were present in each interrogation region to obtain reliable velocity information.

High-resolution measurements are obtained by using an ensemble-average correlation technique. Twenty pairs of single exposure particle-image fields are cross correlated at each interrogation spot. The ten correlation functions at each interrogation spot are then ensemble averaged to produce an ensemble-averaged correlation function. The signal peak detected in the ensemble-averaged correlation function is an estimate of the ensemble-averaged velocity vector at that position in space. The ensemble-averaged correlation function has a much higher signal-to-noise ratio than an instantaneous correlation function. This allows one to obtain reliable velocity measurements with higher spatial resolution than is possible using traditional correlation algorithms.

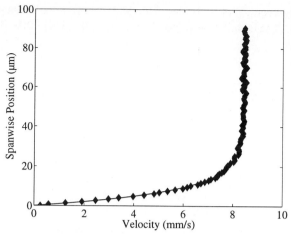

Figure 4. Ensemble-averaged velocity profile measured in a 30 μm × 300 μm channel. The symbols represent ensemble-averaged PIV data, and the line is the analytical solution of Stokes' equation for a rectangular channel. The velocity resolution is approximately 5 μm × 1.3 μm × 2.8 μm. With 50% overlap, the spacing between velocity vectors is 680 nm in the spanwise direction.

5. Spatial Resolution Issues

In order to compare the spatial resolution of different diagnostic techniques, one must have a consistent definition of spatial resolution that can be extended to all techniques. Here, we define the spatial resolution of an instrument as the volume in physical space that is averaged to obtain a sample measurement. Similarly, we define the temporal resolution of an instrument as the duration of time that is averaged to obtain a specific measurement. Many aspects of a measurement technique must be considered before one can properly estimate spatial resolution. For example, the spatial resolution of a specific technique cannot be better than the size of flow-tracing particle used to make the measurement. Technically, the dimension of the averaging volume must be at least several particle diameters large, because the motion of the fluid surrounding the particle affects the motion of the particle. Therefore, the averaging volume of a 1 μm diameter particle is at least several microns in length, in each direction. With this in mind, particles must be chosen significantly smaller than the smallest length scale that is to be resolved.

In PIV, the motion of a group of particle images is determined using correlation functions to spatially average the group of particles. In the case where particle images from the first window are correlated with particle images from second window, it is customary to assume that the size of the spatial average resulting from correlation is the size of the smallest interrogation window. This is correct in the limit of zero offset between interrogation windows, as is the case of autocorrelation. However, this is not correct in the case of cross correlation, where the window offset can become a significant fraction of the window size.

The spatial average resulting from a 2-D correlation may be defined more accurately as the size of the smallest correlation window plus the relative offset between the two windows (assuming that the offset between the windows closely approximates the measured displacement). PIV researchers commonly choose Window 1 as the smaller of the two windows. In the case where flow-tracing particles do not diffuse appreciably between successive exposures, one can choose Window 2 to be the smaller of the two windows without loss of generality (Westerweel, 1997). In micro PIV, sub-micron particles are often used to trace the flow. Since sub-micron particles can diffuse due to Brownian motion, it is important to choose Window 1 to be smaller than Window 2.

In addition to the averaging effects of spatial correlation, imposing a smoothness constraint on the flow field (as is the case for *optical flow*), or spatially low-pass filtering the velocity data further reduces spatial resolution.

Using the above criteria for defining spatial resolution, we have estimated the spatial resolution of several velocimetry techniques in Table 1. In many situations, the authors do not present estimates of spatial resolution, due to the difficulty of obtaining an accurate value. In these cases, we have given our best estimate of actual spatial resolution.

Conclusions

A significant number of researchers have developed diagnostic techniques for microfluidics. A large number of these techniques can obtain spatial resolutions on the order of, say, 10 μm. All the techniques presented here are restricted to liquid flows. Micro PIV has been used to obtain spatial resolutions of 5.0 μm × 1.3 μm × 2.8 μm in a nominally 30 μm thick × 300 μm wide microchannel (Meinhart et al., 1999). Using 50% overlap, the velocity-vector spacing in the spanwise direction is 680 nm. The PIV data agrees well with the analytical solution for Stokes' flow in a rectangular channel.

Acknowledgement

This work is supported at UCSB by a grant from AFOSR/DARPA number F49620-97-1-0515 under the direction of Dr. Mark Glauser, by JPL/NASA under the direction of Dr. Bill Tang, and by the College of Engineering at UCSB. The Beckman Institute for Advanced Science and Technology and a Ford Foundation Post-Doctoral Fellowship supported this work at the University of Illinois.

Table 1. Comparison of high resolution velocimetry techniques.

Technique	Author	Flow Tracer	Spatial Resolution (μm)	Observation
LDA	Tieu et al. (1995)	———	$5 \times 5 \times 10$	4 – 8 Fringes limits velocity resolution
Optical Doppler Tomography (ODT)	Chen et al. (1997)	1.7 μm Polystyrene Beads	5×15	Can image through highly scattering media
Optical Flow using Video Microscopy	Hitt et al. (1996)	5 μm Blood Cells	$20 \times 20 \times 20$	In vivo study of blood flow
Optical Flow using X-ray imaging	Lanzillotto et al. (1996)	1 – 20 μm Emulsion Droplets	~ 20 - 40	Can image without optical access
Uncaged-fluorescent dyes	Paul et al. (1997)	Molecular Dye	$100 \times 20 \times 20$	Resolution limited by molecular diffusion
Particle Streak Velocimetry	Brody et al. (1996)	0.9 μm Polystyrene Beads	~ 10	Particle Streak Velocimetry
Particle Image Velocimetry (PIV)	Urushihara et al. (1993)	1 μm oil droplets	$280 \times 280 \times 200$	Turbulent flows
Super-resolution PIV	Keane et al. (1995)	1 μm oil droplets	$50 \times 50 \times 200$	Particle Tracking Velocimetry
Micro PIV	Santiago et al. (1998a)	300 nm polystyrene particles	$6.9 \times 6.9 \times 1.5$	Hele-Shaw Flow
Micro PIV	Santiago et al. (1998b)	300 nm polystyrene particles	$6.9 \times 6.9 \times 1.5$	Silicon Microchannel Flow
Micro PIV	Meinhart et al. (1999)	200 nm polystyrene particles	$5.0 \times 1.3 \times 2.8$	Microchannel Flow

References

Adrian, R. J. 1983. Laser Velocimetry. Chapter in *Fluid Mechanics Measurements*, (ed. Goldstein, R. J), Taylor&Francis Publishing, Washington, DC, pp. 155-244.

Bayt, R. L., Ayon, A. A., Breuer, K. S. 1997. A performance evaluation of MEMS-bases micronozzles. *AIAA Paper 97-3169, 33rd AIAA/ASME/SAE/ASEE Joint Propulsion Conference & Exhibit*, July 7-9, Seattle, WA.

Brody J.P.; Yager P.; Goldstein R.E.; Austin R.H. 1996. Biotechnology at low Reynolds numbers. *Biophys J.*, **Vol. 71**, pp. 3430-3441

Chen Z; Milner T.E.; Dave D. & Nelson J.S. 1997. Optical Doppler tomographic imaging of fluid flow velocity in highly scattering media. *Optics Letters,* **Vol. 22**: pp. 64-66.

Compton, D. A. & Eaton, J. K. 1996. A high-resolution laser Doppler anemometer for three-dimensional turbulent boundary layers. *Exp. Fluids*, **Vol. 22**, pp. 111-117.

Dahm, W.J.A.; Su, L.K.; Southerland, K.B. 1992. A scalar imaging velocimetry technique for fully resolved four-dimensional vector velocity field measurements in turbulent flows. *Physics of Fluids A (Fluid Dynamics)*, **Vol. 4**, No. 10, pp. 2191 – 2206.

Evans, J. D., Liepmann, D. & Pisano, A. P. 1997. Planar laminar mixer*, MEMS 97, TheTenth Annual International Workshop on Micro Electro Mechanical Systems*, January 26 – 30.

Gharib, M., Modares, D. & Taugwalder, F. 1998. Development of a miniature and micro laser Doppler anemometer, *Personal Communication*.

Hitt, D. L., Lowe, M. L., Tincher, J. R., Watters, J. M. 1996. A new method for blood velocimetry in the microcirculation. *Microcirculation*, **Vol. 3** No. 3, pp. 259-263.

Hitt, D. L., Lowe, M. L., Newcomer, R. 1995. Application of optical flow techniques to flow velocimetry. *Phys. Fluids.*, **Vol. 7**, No. 1, pp. 6–8.

Kamisuki, S., Hagata, T., Tezuka, C., Nose, Y., Fujii, M., Atobe, M. 1998. A low power, small, electrostatically-driven commercial inkjet head. *Proceedings of MEMS'98*.

Keane, R. D., Adrian, R. J. & Zhang, Y. 1995. Super-resolution particle imaging velocimetry. *Meas. Sci. Tech.* **Vol. 6**, pp. 754-768.

Koochesfahani, M.M., Cohn, R. K., Gendrich, C. P., and Nocera, D. G. 1996. Molecular tagging diagnostics for the study of kinematics and mixing in liquid phase flows. *Proceedings in the Eighth International Symposium on Applications of Laser Techniques to Fluid Mechanics*, Lisbon, Portugal, July 8-11.

Krulevitch, P. 1997. Personal Communication.

Lanzillotto, A. M., Leu, T. S., Amabile, M., Wildes, R. and Dunsmuir, J. 1996. Applications of x-ray micro-imaging, visualization and motion analysis techniques to fluidic microsystems. *Proceedings from Solid-state sensors and actuators workshop*, Hilton Head, SC, June 13-16.

Meinhart, C. D, Prasad, A. K. and Adrian, R. J. 1993. A parallel digital processor for particle image velocimetry. *Measurement Science Technology*, **Vol. 4**, pp. 619-626.

Meinhart, C. D., Wereley, S. T. & Santiago, J. G. 1999. PIV Measurements of a Microchannel Flow, *Submitted to Exp. Fluids*.

Northup, M. A., Hills, R. F., Landre, R., Lehew, H. D., Watson, R. A. 1995. A MEMS-based DNA analysis system, *Transducers'95, Eighth International Conference on Solid State Sensors and Actuators*, Stockholm, Sweden, June, pp. 764-767.

Paul, P. H., Garguilo, M. G., Rakestraw, D. J. 1997. Imaging of pressure- and electrokinetically-driven flows through open capillaries. Submitted to *Anal. Chem.*

Santiago, J. G., Wereley, S. T., Meinhart, C. D., Beebe, R. & Adrian, R. J. 1998a. A Particle Image Velocimetry System for Microfluidics, *Exp. Fluids*, **Vol. 25** No.4, pp 316-319.

Santiago, J. G., Wereley, S. T., Meinhart, C. D., Beebe, R. & Adrian, R. J. 1998b. A Micron-Resolution Particle Image Velocimetry System, *8th International Symposium on Flow Visualization*, Sorrento, Italy.

Tieu, A. K., Mackenzie, M. R., Li, E. B. 1995. Measurements in microscopic flow with a solid-state LDA. *Exp. Fluid*, **Vol. 19**, pp. 293 – 294.

Urushihara, T., Meinhart, C. D. & Adrian, R. J. 1993. Investigation of the logarithmic layer in pipe flow using particle image velocimetry. In *Near-Wall Turbulent Flows*, R. So, et al. (Eds.), New York: Elsevier, pp. 433-46.

Westerweel, J. 1997. Fundamentals of digital particle image velocimetry. *Meas. Sci. Technol.*, **Vol. 8**, pp. 1379-1392.

Wildes, R., Amabile, M., Lanzillotto, A. M., Leu, T. S. 1997. Physically based fluid flow recovery from image sequences. In *Proc. IEEE CVPR*.

I.5. PIV in Two-Phase Flows: Simultaneous Bubble Sizing and Liquid Velocity Measurements

I. Dias and M.L. Reithmuller

Department of Environmental and Applied Fluid Dynamics, von Kármán Institute for Fluid Dynamics, Chaussèe de Waterloo 72, 1640-B Rhode-St-Genèse, Belgium

Abstract. A two-phase measurement technique based on the simultaneous sizing of the dispersed phase and PIV measurement of the continuous phase is presented. The referred technique has been applied to study the formation of bubbles, when air is injected into quiescent water through a single needle. The efficient application of the PIV technique to measure the liquid velocity field in bubbly flows is based on the use of fluorescent tracers combined with optical filters. Strong optical effects at the air/water interface limit the use of the PIV image for the accurate determination of the bubble contour. Therefore, a supplementary CCD-camera acquires simultaneously the bubble shadow images. The results show the evolution of the liquid velocity field around a growing bubble during its formation cycle, as well as a comparison between the PIV measurements in the vicinity of the bubble interface and the theoretical predictions using a potential flow solution.

Keywords. PIV, LIF, Shadow detection method, Two-phase flow, Bubbles

1   Introduction

The injection of gases into liquids in the form of bubbles is a common industrial operation in the casting of metals and chemical or nuclear reactors. Recently, strong effort has been conducted to develop theoretical and mathematical models able to predict the complex behavior of bubbly flows. Nevertheless, the details of the interaction between the liquid flow field and the forming bubbles are still not well established. Therefore, detailed experimental data is required to improve the understanding of the physical phenomena involved, as pointed out by Murai & Matsumoto (1995) and Hassan & Canaan (1991).

The present work concerns the experimental investigation of the bubble formation process. Individual bubbles are created when air is injected, at low flow rate, through a single needle into stagnant water. The upward movement of the detached bubbles to the free surface follows a helicoidal path, creating a strong asymmetric wake that might influence the formation of the subsequent bubble.

The first objective of this study is the development of a two-phase visualization and measurement technique to provide simultaneously the liquid velocity and the size of the forming bubble. The second objective is the application of the referred measurement technique to investigate the importance of the bubble wake effect on the bubble formation process.

PIV is a non-invasive measurement technique yielding quantitative, full-field, instantaneous velocity maps of the flow (Meynart & Lourenço, 1984). Several studies have been reported, where the velocity fields in a bubbly flow system are determined using PIV for the continuous phase and PTV - particle tracking velocimetry - for the dispersed phase [Sridhar *et al.* (1991); Liu & Adrian (1993); Philip *et al.* (1994); Oakley *et al.* (1995); Hilgers *et al.* (1995)], or simply PTV for both phases as reported by Hassan *et al.* (1993). Recently, Gui & Merzkirch (1996) developed a digital mask technique to separate the phases in a two-phase flow.

Fluorescent seeding of the liquid phase combined with optical filters allows the discrimination of the particles signal from the light reflected by the dispersed phase [Hassan *et al.* (1993); Philip *et al.* (1994); Hilgers *et al.* (1995)]. This aspect is important when a large fraction of the PIV image is occupied by reflections from the interface.

The present technique has the advantage of yielding accurate sizing of the dispersed phase structures and high-density velocity vectors of the continuous phase, which is particularly important in the analysis of the bubble formation process.

2 Experimental Investigation

2.1 Set-up

The experimental study of the bubble formation process is performed in the facility represented in Fig. 1. The measurement technique couples the liquid PIV measurement with the sizing of the bubbles using the shadow detection method.

The bubbles are generated by the injection of air into a stagnant water tank, through a needle mounted on its base. This container is constituted by a Plexiglas square base of 97 mm side and 4 mm thickness glass walls 154 mm high. It is filled with tap water up to a height of 100 mm, and the top is open to the atmosphere. The stainless steel needle has internal and external diameters, respectively of 3 and 4 mm, and is 30 mm long. Constant gas flow rate regime is ensured by the use of a sonic hole at the base of the needle in the air feeding system. Air flow is provided by a 50 l pressurized gas bottle. A needle valve allows a fine control of the flow rate that enters the air reservoir mounted underneath the water tank and connected to the needle. This stainless steel cylindrical reservoir (60 mm diameter and 40 mm height), is equipped with a Validyne transducer monitoring the pressure. The air issued from this reservoir passes through a 25 μm diameter sonic hole, ensuring a constant mean air flow rate entering at the needle base. "Constant gas flow rate" regime is guaranteed when sonic working conditions are attained, i.e. pressure in the reservoir must be at least 1.9 times larger than the pressure at the needle entry; and the pressure in the reservoir is kept constant. During the time duration of one experiment, this last requirement is easily satisfied due to the use of a pressurized gas bottle as feeding system. The absolute pressure in the gas reservoir is 1.92 bar, corresponding to an air flow rate of 0.1485 ml/s and a bubbling frequency around 2 Hz.

The PIV image is obtained illuminating the flow with an Argon-Ion laser, Spectra-Physics model 2016-05S. The laser is working on "single wavelength mode", having as primary output wavelength 514.6 nm and a power of 2 W. A system of lenses, pin-holes and a Bragg cell is employed to expand the initial 1 mm circular Gaussian laser beam into a flashing converging light sheet of approximately 1 mm thickness and 2 cm in width, at the camera field of view (FOV). The acousto-optic modulator (Bragg cell) used is from Isomet, model 1201E-1 and the Digital drive, model 221A-2 (series 200), works with a center frequency of 40 MHz. The laser sheet is aligned with the axis of the needle and the viewing area of the camera is 14x10 mm^2. A system composed by a normal lens is not suitable to visualize such a small observation area. Instead, a microscope Technical 2 from Jenoptik Jena, equipped with a 164 mm lens and an ocular (MF-Meßprojektif K 4:1), yielding a magnification of 1.6 is used and directly connected to the acquisition camera. The bubble obstructs the passage of the light to the opposite side of the laser. As a consequence a significant part of the FOV is in the dark. A circular flat mirror (46 mm diameter), forming an angle of 15° with the tank wall is used to reflect the upper part of the laser sheet. Since the needle or bubble does not block it, it illuminates the dark side of the bubble. The lack of symmetry of the flow surrounding the bubble justifies the need to illuminate both sides of the bubble. A red filter is placed in front of the camera which captures the images of the particles to block the intense green light reflected by the bubbles and allow the passage only of the light emitted by the fluorescent particles. The high pass filter (wavelength) used is opaque below 515 nm. The red filter, microscope and the CCD-camera are aligned and positioned at 90° with the laser sheet and parallel to the viewing area.

The bubble contour images are acquired by another CCD-camera equipped with a 105 mm Nikkor-P photographic lens (aperture 1:2.5) and an extension tube, aligned with another red filter and positioned with an angle of 60° with the side wall of the tank. The red filter is similar to the one used for the PIV camera, because its main objective is also to suppress the strong reflections of the bubble interface. The background illuminating source is a diffused white light (100 W) located on the opposite side of the tank, facing the acquisition camera which collects light in forward scattering mode. The image obtained with this camera is the shadow of the bubble. Although the light emitted by the fluorescent particles is not intercepted before reception by the acquisition camera, its intensity is low in comparison with the background white light and it does not reduce the contrast quality of the bubble contour images.

Both CCD-cameras used in this experiment are from i2S France, model IEC 800BC with 756(H)x581(V) pixels of resolution. The PIV camera works in Standard European video CCIR mode (40 ms). Laser flashes duration is equal in both fields of the interlaced image (2.4 ms), and the pulse separation is 16 ms. The bubble contour camera works in shutter mode (integration time of 1 ms), while the background white light is continuous. Fig. 2 shows the timing diagram of the CCD-cameras working modes and respective illumination systems.

74

The two CCD-cameras and their respective acquisition systems are synchronized in line mode by connecting the video output of the PIV camera (master) to the input of the contour camera (slave). The image acquisition system is composed by a Matrox PIP-1024 acquisition board and controlled by PC-Scope software. This system acquires four frames of 512x512 pixels each. The synchronization precision is found to be better than 1 ms.

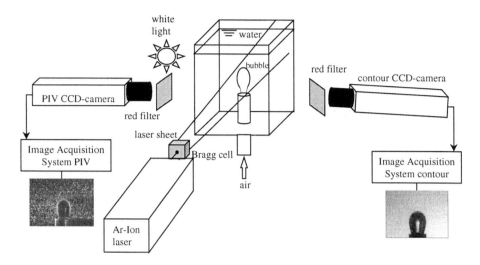

Fig. 1. Schematic representation of the experimental set-up

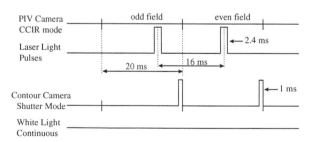

Fig. 2. Timing diagram of the cameras working modes and illumination systems

2.2 Fluorescent Particles

Fluorescent seeding of the liquid phase combined with optical filters allows a clear discrimination of the signal emitted by the particles and the intense neutral reflections occurring at the bubble interface. This aspect is particularly important in the analysis of the bubble formation process, as in this case the interface reflections might occupy a large part of the FOV. Fig. 3 presents the emission spectrum of the incident light, an Argon-Ion laser emitting at 514.6 nm (green), and the signal emitted by the fluorescent particles around 590 nm (red). The red filter used blocks the passage of light below 515 nm.

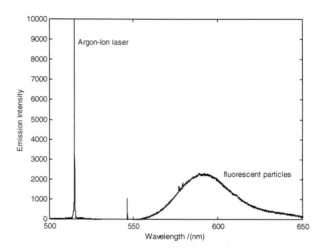

Fig. 3. Emission spectra of the laser (λ=514.6 nm) and fluorescent particles (λ~590 nm)

The fluorescent particles are a light orange vinyl pigment, from Lefranc & Bourgeois, soluble in water. They are neutrally buoyant in water, and the diameter is about 10 μm.

3　Data Processing

3.1　Shadow Detection Method

The interlaced images of the bubble shadows are separated into the odd and even fields. The contour of the bubble in each of the fields is then determined.

The algorithm developed to determine the bubble interface co-ordinates in an image is a new combination of different basic schemes. This method can be summarized by the following sequence of operations:

  i.   image gradient determination (G)
  ii. image gradient histogram equalization (G^*)
 iii. images subtraction (D)
  iv. binarization according to an appropriate threshold
  v.   contour determination.

Given a two-dimensional image function $F(x,y)$, the magnitude of the vector gradient $G[F(x,y)]$ is computed as follows:

$$G = \left[\left(\frac{\partial F}{\partial x} \right)^2 + \left(\frac{\partial F}{\partial y} \right)^2 \right]^{\frac{1}{2}} \tag{1}$$

Typically, the histogram of the field gradient has a narrow distribution band. To improve the contrast of the image, its histogram is flattened by performing a linear contrast stretching operation (*i.e.* dark becomes darker and bright becomes brighter).

As proposed by Bow (1992), the subtraction of the equalized image gradient (G^*) to the original noisy image (F) is the image difference (D):

$$D = F - G^* \tag{2}$$

that denotes a high pixel level transition at the edges and a low fine noise level. In this case, a single threshold value is suitable for the whole image binarization.

The proposed method has the advantage of eliminating fine noise effect and operates with an almost insensitive threshold level for the binarization. The bubble interface co-ordinates are obtained by detecting line-by-line the left and right transitions of the binarized image. An example of the sequence of operations previously described, starting from the original noisy image until obtaining the bubble contour, is shown in Fig. 4. The uncertainty in the size (equivalent diameter) determination is estimated to be 3%.

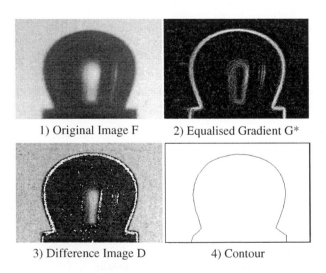

| 1) Original Image F | 2) Equalised Gradient G* |
| 3) Difference Image D | 4) Contour |

Fig. 4. Sequence of operations performed in the shadow image for the bubble contour detection

3.2   Particle Image Velocimetry

The digital PIV images are interrogated using the cross-correlation algorithm WIDIM, developed by Scarano (1997). This procedure performs the displacement of the second exposure interrogation area to compensate for the loss-of-pairs due to the in-plane motion and optimizes it with an iterative procedure.

The processing of the acquired images uses a Gaussian three-point-fit for the sub-pixel interpolation peak determination. The original interlaced PIV image of 512x512 pixels is separated into the odd and even fields of 512x256 pixels resolution. The initial window size is 64x64 pixels, one step refinement is performed yielding a final window size of 32x32 pixels. A 75% overlapping factor is used leading to a final grid spacing of 8x8 pixels, and a total number of 1769 vectors. The percentage of non-validated vectors keeps below 5%.

The velocity uncertainty is mainly associated with the uncertainty in determining the particles images displacement, which is less than a tenth of a pixel. The large dynamic range of the present case leads to an uncertainty of 2.5% within a confidence level of 75%.

4   Results and Discussion

4.1   Optical Effects

Important optical effects occur at the air/water interface, preventing the 'extraction' of the bubble contour from the PIV image.

A first result concerns this preliminary study, done in order to determine the nature of these optical effects. This test consists on the immersion of a hollow glass bulb (17 mm external diameter and 0.5 mm glass thickness) in the tank containing stagnant water and fluorescent particles. The bulb and surrounding water are illuminated with a green light sheet and an image is acquired (Fig. 5). A red filter is used to block the green light before reaching the camera. The glass bulb is used to simulate a static bubble, avoiding the complexity of a moving interface.

The external surface of the bulb (zone 1) can be clearly distinguished, because a fine layer of fluorescent particles is sticking on this static surface. A virtual image of the bubble surface (zone 2), that in the camera image plane is seen inside the real surface (zone 1) is yielded as a result of the internal refractions of the image of particles laying on the bubble surface. Area 3 is the region delimited by the real external surface of the bubble (zone 1) and the virtual image of the same surface (zone 2). It is important to understand that it does not correspond to the real thickness of the glass bulb, as it appears six times larger.

A simplified ray tracing model of a spherical air bubble, with a thin layer of fluorescent particles immersed in stagnant water and cross illuminated by a laser sheet, is schematized in Fig. 6.

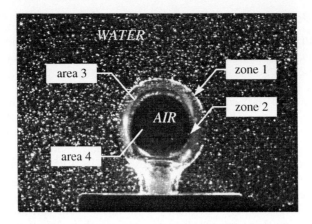

Fig. 5. Image of an hollow glass bulb and surrounding stagnant water with fluorescent particles

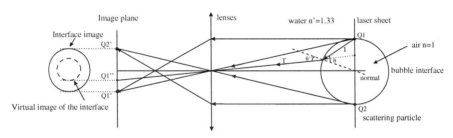

I - Incident Ray   T - Refracted Ray   θ - Angle of Incidence   θ'' - Angle of Refraction

Fig. 6. Simplified optical ray tracing diagram of the beams scattered from particles laying on the surface of a spherical bubble.

When a light ray excites the surface static particles (Q1 and Q2), the latter emit fluorescent light in all the directions. Following the optical path of one of the light rays inside the bubble (I), when it reaches the air/water interface part of the light is reflected inside the bubble and other part refracts through the water (T). The light ray is travelling from air to water, *i.e.* increasing refractive index, according to Snell's law the ray is deviated towards the normal ($\theta''<\theta$). In the image plane, this (internal) refraction results in a virtual image of the particle (Q1'' and Q2'') positioned closer to the axis center than the image of the particles (Q1' and Q2') resulting from the (external) reflections. The "integration" of this effect over all the bubble surface yields as result an internal concentric ring, called the virtual image of the interface, corresponding to the zone 2 in Fig. 5.

In area 3 (Fig. 5), external reflections of the particles in the surrounding liquid can be observed. The intensity of those reflections undergoes a rapid decay with respect to zone 1. The images of the particles in both sides of the interface, between the two phases, within a small region are identical (mirror effect), leading to the erroneous idea that this region (area 3) might correspond to the continuous phase. In the black core of the bulb image (area 4), low intensity image of the particles are still existing but can hardly be seen. The optical effects observed in the hollow bulb, such as the virtual image of the surface and external reflections of the seeding particles are expected also for a bubble interface.

In Fig. 7, an example of a bubble contour image and the corresponding PIV image of the liquid around the bubble can be observed. In the PIV image, the bubble core is identified as the dark central region apparently without particles images (area 4 in the bulb test).

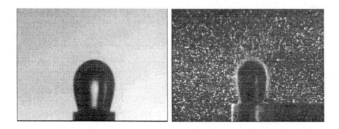

Fig. 7. Original bubble contour image (left) and corresponding PIV image of the surrounding liquid (right)

The co-ordinates of the bubble surface are obtained after processing the bubble image. This information is then superimposed to the corresponding PIV image (Fig. 8). The comparison between the bubble co-ordinates, obtained by the contour image, and the borders of the bubble core in the PIV image reveals an important underestimation of the bubble size given by the PIV image.

Since the interface of a growing bubble is continuously moving, the external surface of the bubble cannot be clearly distinguished in the PIV image as it is possible in the bulb test. Referring to the bulb test, this means that the external limit of area 3 (zone1) cannot be defined in the PIV image. The determination of the bubble interface co-ordinates using the limits of the bubble core region is therefore incorrect. A correction scheme, allowing the determination of the bubble surface (zone 1) knowing the borders of the bubble core region (zone 2), would yield inaccurate results because the local curvature of a growing bubble and therefore the thickness of the ring (area 2) is continuously changing. This fact emphasizes the need for a supplementary method to accurately determine the bubble contour. In the present work, the shadow detection method is used, requiring another CCD-camera.

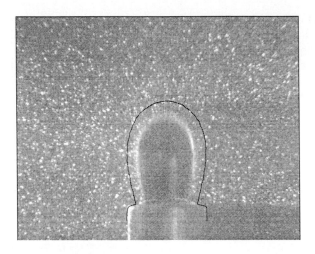

Fig. 8. Bubble contour superposed to the PIV image

4.2 Velocity Comparison

The liquid velocity field evolution around a growing and detaching bubble in stagnant water can be observed in Fig. 9. The reference velocity vector (12 mm/s) is presented in the plot for t=147.5 ms, and it is equal for all the vectors plots.

The first velocity field (for t=0) corresponds to the initial burst out of the air flow from the needle. This stage is one of the fastest of the whole formation cycle. Nevertheless, the magnitude of the vertical component of the surrounding liquid velocity is even higher than the one observed in the region immediately surrounding the bubble. Therefore, the wake of the preceding bubble might have an important suction effect on the formation of the following bubble and cannot, *a priori*, be neglected as made in many theoretical models [Tan & Harris (1986); Liow & Gray (1988)].

In the necking phase, just prior to detachment from the needle (t=457.5 ms), the wake effect becomes negligible and bubble growth is governed simply by its own dynamics (surface tension, buoyancy and liquid inertia forces). The bubble lift-off results simultaneously from the upward translation of the bubble and the liquid entrainment at its base, yielding two important recirculation structures on the lateral sides of the bubble.

In the intermediate stages, shown in Fig. 9 (from t=35 ms to t=350 ms), the bubble expands and translates vertically at the same time. Nevertheless, the magnitude of the vertical upward component is always much larger than the horizontal component.

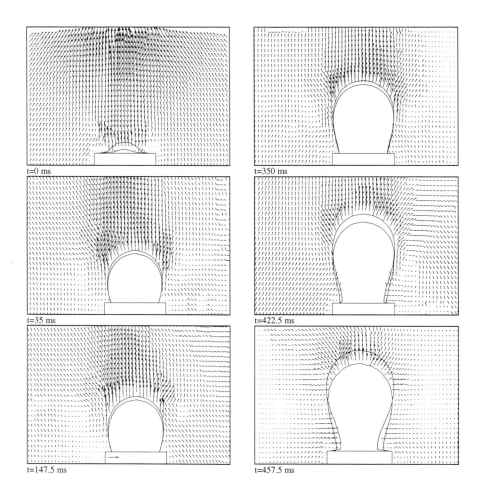

Fig. 9. Liquid velocity field (PIV) and bubble contours (shadow detection method).The two contours shown in each plot are separated in time by 20 ms. Needle external diameter is 4 mm.

Although the bubble shape during formation is symmetric around its axis, the surrounding flow is asymmetric due to the wakes created by the preceding bubbles. Once the bubbles detach from the needle they first rise to the free surface following a straight vertical movement, during a few bubble equivalent diameters, but then they change to a helicoidal trajectory. Therefore, the bubble wake has a three dimensional nature.

The comparison between the liquid velocity profile, measured by means of PIV technique, and the theoretical solution for potential flow surrounding a translating sphere (Batchelor, 1979) can be observed in Fig. 10.

The experimental data presented corresponds to the final phase of bubble growth, as shown in Fig. 9 (t=457.5 ms). The velocity profile is extracted from the region above the top of the bubble, where potential flow assumption is valid, at an angle of 90^0 from the top surface of the bubble. In this position, the radial velocity corresponds practically to the streamwise velocity component.

In the necking phase of the bubble, just prior to detachment, the movement of the top of the bubble can be approximated by the upward translation of a sphere, because bubble expansion is negligible. According to the potential flow theory, provided the speed of the water is small enough for it to be regarded as incompressible, the radial velocity V_r in the water is:

$$V_r = \left(\frac{R}{r}\right)^3 V_T \qquad (3)$$

where R is the radius of the sphere, r is the radial distance from the center of the sphere and V_T is the translation velocity. The input conditions for the determination of the potential flow solution are the equivalent bubble spherical radius R and the translation velocity. These parameters are evaluated from the bubble contour evolution, experimentally obtained with the shadow detection method. The top of the bubble is approximated by a sphere of radius R equal to 1.8 mm and the mean translation velocity is 42.3 mm/s.

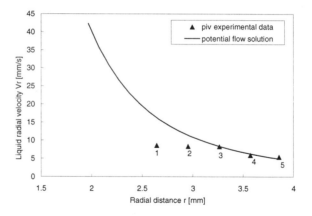

Fig. 10. Comparison of the liquid velocity measured by PIV and predicted from a potential flow solution

It is assumed that the translation velocity is constant during the time interval (20 ms) between the acquisition of the two consecutive bubble contours. To obtain the PIV measurements, the laser is triggered at two different instants separated by 16 ms. The velocity determined is the velocity of the fluid averaged over this period.

However, the theoretical solution is calculated for an instant that is the median between the two laser pulses.

The interpretation of the PIV measurements in the vicinity of the bubble moving interface is schematized in Fig. 11. The numbering of the velocity vectors positions in Fig. 11 corresponds to the PIV data shown in Fig. 10, where the first position is the closest outside of the second contour of the bubble. The velocity vectors number 1 and 2, are determined over final windows that partially include the gas phase. As mentioned previously, external reflections of the particles can be seen inside the bubble contour having a non-determined and erroneous contribution for the evaluated displacement. These vectors cannot be considered as a valid measurement and show a clear underestimation when compared with the potential flow solution. Once the search window is entirely belonging to the liquid phase (vectors 3, 4 and 5), the PIV measurement is valid and is in agreement with the potential flow solution as it can be observed in Fig. 10.

The minimum distance from the air/liquid interface at which PIV measurements can be safely interpreted is half of the final window size. The absolute distance can be reduced using smaller windows, but on the other side the number of particles contributing to the correlation function is also diminished, decreasing the confidence level of the measurement.

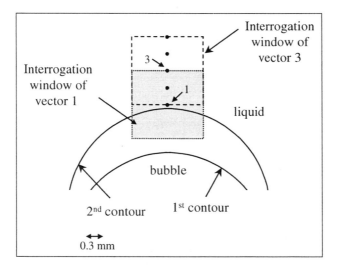

Fig. 11. Schematic representation of the bubble moving interface and PIV measurement locations. The rectangles represent the interrogations windows, and the black dots the measurement locations.

PIV is a valuable technique to analyze the overall features of the liquid flow around a forming bubble, but these results suggest that one should be careful in interpreting the velocity field in the vicinity of the air/liquid interface. It is therefore suitable to analyze the importance of the wake effect in the formation of the subsequent bubble. However, the shadow detection method is preferable for the bubble contour and interface velocity determination. These results emphasize the importance of accurately determining the bubble contour, and therefore the need of a second synchronized CCD-camera.

5 Conclusions

In the present work, a two-phase measurement technique is used to study a bubble formation process. This technique is based on the simultaneous sizing of the bubble and PIV measurements of the surrounding liquid using fluorescent particles. Individual bubbles are formed when air is injected into stagnant water through a single needle.

It is shown that the bubble contour co-ordinates cannot be determined from the PIV image because of the important optical effects that occur at the bubble interface when crossed by a laser light sheet. Therefore, the bubble shadow method is here used as supplementary technique, yielding an accurate contour of the bubble.

The bubble sizing algorithm performs the binarization of the difference image, instead of the original bubble contour image. This method has the advantage of eliminating fine noise effects and of operating with an almost insensitive threshold level for the binarization. The uncertainty in the bubble size and growth velocity determination, using the shadow detection method, is less than 3%.

The evolution of the liquid velocity field around a growing and detaching bubble is shown. The wake of the preceding bubbles might have an important suction effect on the early stages of the bubble formation cycle. Therefore, it cannot *a priori* be neglected as made in many available theoretical models.

PIV is a valuable technique to analyze the overall features of the liquid flow around growing bubbles. However, care is to be taken in interpreting the velocity field in the immediate vicinity of gas/liquid moving interfaces.

References

BATCHELOR, G. K. 1979 *An introduction to fluid dynamics.* Cambridge University Press.

BOW, S. T. 1992 *Pattern recognition and image processing.* Marcel Dekker Inc.

GUI, L. & MERZKIRCH, W. 1996 Phase-separation of PIV measurements in two-phase flow by applying a digital mask technique. *ERCOFTAC* **30**, 45-48.

HASSAN, Y. A. & CANAAN, R. E. 1991 Full-field bubbly flow velocity measurements using a multiframe particle tracking technique. *Exps. Fluids* **12**, 49-60.

HASSAN, Y. A., PHILIP, O. G. & SCHMIDL, W. D. 1993 Bubble collapse velocity measurements using a particle image velocimetry technique. *ASME* **172**, 85-92.

HILGERS, S., MERZKIRCH. W., & WAGNER, T. 1995 PIV measurements in multi-phase flows using CCD- and photo-camera. *ASME* **209**, 151-154.

LIOW, J. L. & GRAY, N. B. 1988 A model of bubble growth in wetting and non-wetting liquids. *Chem. Eng. Sci.* **43**, 3129-3139.

LIU, Z. C. & ADRIAN, R. J. 1993 Simultaneous imaging of the velocity fields of two phases. *Particulate two phase flow.* M.C. Roco ed., Butterworth-Heinemann, Stoneham, M.A.

MEYNART, R. & LOURENÇO, L. M. 1984 Laser speckle velocimetry in fluid dynamics applications. *Digital Image Processing in Fluid Dynamics Lecture Series*. von Karman Institute for Fluid Dynamics, 175-195.

MURAI, Y. & MATSUMOTO, Y. 1995 Three dimensional structure of a bubble plume - measurement of the three dimensional velocity, Flow visualization and image processing of multiphase systems. *ASME* **209**, 187-194.

OAKLEY, T. R., LOTH, E. & ADRIAN, R. J. 1995 Cinematic two-phase PIV for bubbly flows, Flow visualization and image processing of multiphase systems. *ASME* **209**, 123-128.

PHILIP, O. G., SCHMIDL, W. D. & HASSAN, Y. A. 1994 Development of a high speed particle image velocimetry technique using fluorescent tracers to study steam bubble collapse. *Nuclear Engineering and Design* **149**, 375-385.

SCARANO, F. & RIETHMULLER, M. L. 1997 Iterative multigrid approach in PIV image processing with discrete window offset. Accepted for publication in *Exps. Fluids*.

SRIDHAR, G., RAN, B. & KATZ, J. 1991 Implementation of particle image velocimetry to multiphase flow, Cavitation and multiphase flow forum. *ASME* **109**, 205-210.

TAN, R. B. & HARRIS, I. J. 1986 A model for non-spherical bubble growth at a single orifice. *Chem. Eng. Sci.* **41**, 3175-3182.

Acknowledgments

The partial support of "FUNDAÇÃO PARA A CIÊNCIA E TECNOLOGIA - PORTUGAL", program PRAXIS XXI, who provides a Ph.D. fellowship (BD/5921/95) is gratefully acknowledged.

I.6 Measuring Water Temperatures by Means of Linear Raman Spectroscopy
J. Karl, M. Ottmann, and D. Hein

Lehrstuhl für Thermische Kraftanlagen, Technische Universität München,
Boltzmannstr. 15, D-85747 Garching, Germany

Abstract. In many cases it is necessary to measure local water temperatures with high local resolution. The usage of common thermocouples limits the local resolution, affects the temperature field and may cause incorrect readings.

The paper presents a new technique to measure temperatures of liquid water in situ, using an optical measurement method - linear Raman spectroscopy. The shape of liquid water Raman spectra changes with temperature and this allows to measure water temperatures one dimensionally - that means along a laser beam - with good accuracy.

Two dimensional measurements miss one of the major advantages of one dimensional Raman measurements. The spectral information about the fluorescence background will be lost and this reduces accuracy. But it might be a valuable method to locate and evaluate local temperature differences.

Keywords. water temperatures, Raman spectroscopy, hydrogen bonds, condensation heat transfer, interference filter

1 Introduction

Raman spectroscopy is a well established method to measure concentrations in gases or liquids, to investigate the molecular structure of pure substances and to measure temperatures in flames or high temperature gas flows.

There are two different effects which can be used to measure temperatures by means of Raman spectroscopy.

The first effect originates from the ratio between gas molecules which are in a vibrational exited state and gas molecules in the vibrational ground state. This ratio depends on the gas temperature according to Boltzmann's law. The effect requires gas temperatures above 1000 K, because there won't be a significant number of molecules in an exited state otherwise. The most common application of Raman spectroscopy is to measure temperatures in flames, therefore. In this case flame temperatures can be calculated from the ratio between stokes and antistokes peak of a certain component like nitrogen for example or from the shape of the vibrational line of nitrogen for example (Schrader, 1995)

A completely different effect allows to measure the temperature of liquid water by means of Raman spectroscopy.

The shape of Raman spectra of liquid water is quite different to the shape of Raman spectra of water vapor, because hydrogen bonds affect the O-H-stretching vibration. The number of hydrogen bonds depends on the temperature and the shape of liquid water Raman spectra changes with temperature, too, therefore.

Walrafen et al. (1986) described the influence of the hydrogen bonds to the vibrational line near 3652 cm^{-1} already in 1967. He notified a temperature dependence of the shape of the Raman peak in the region between 3000 and 3800 cm^{-1}. Walrafen and Fujita and Ikawa (1989) used this effect to determine the molecular structure of liquid water and to determine the bonding energy of hydrogen bonds by means of Raman spectroscopy and infrared spectroscopy.

Leonard et al. (1979) started a first attempt to use this effect for water temperature measurements in the early seventies. He developed an airborne LIDAR system to measure ocean water temperatures over large ocean areas. He measured water temperatures 100 m below the water surface. Within a depth of 3 m he got an accuracy of ± 1,2 K. Schwaiger (1992) used the effect to measure temperatures of water droplets.

Leonard used Raman spectroscopy to determine water temperatures from discrete points without investigation of local temperature distributions.

In our previous work we applied Raman spectroscopy to measure the water temperature distribution along a laser beam. We measured water temperatures simultaneously with concentration profiles in a steam-nitrogen boundary layer during condensation in the presence of a non condensable gas (nitrogen) to calculate condensation heat transfer coefficients (Karl, 1997). With another experimental set-up we focused the laser beam vertically through a water layer and measured the water temperature distribution along the laser beam.

We obtained one dimensional temperature profiles, therefore, which allowed to calculate local heat transfer coefficients during direct contact condensation (Karl and Hein, 1997)

Our recent work was aimed at possibilities to measure water temperatures in two dimensions to get in situ detailed two dimensional images of a temperature distribution in liquid water flows.

The best way to evaluate Raman signals is to register the entire spectral information using a spectrograph.

The main advantage of Raman spectroscopy for various applications is, that the required Raman signal can be separated from any Mie-, Rayleigh- or Fluorescence signal. But separation of those signal components requires that the spectral information is available completely.

Using a spectrograph together with a CCD camera allows to get one dimensional temperature distribution for example along the laser beam. As shown in Fig. 1 the images provided by the a CCD chip can be used to get spectral information simultaneously to local information. Each pixel row provides a spectrum corresponding to a certain location of the laser beam, then.

Of course it is not possible to get two dimensional information for flow field measurements with such a configuration, simultaneously, because one dimension of the two dimensional CCD detector is necessary to provide the spectral information.

Measuring water temperatures in two dimensions causes the loss of the spectral information and reduces the accuracy inevitably. But in case not absolute temperatures but temperature differences are looked for, two dimensional measures can make valuable results.

2 Raman spectra of liquid water

Raman scattering is an inelastic light scattering process. A molecule may scatter light of any wavelenght either elastically - that means without wavelength shift - or inelastically. In that case the scattered light is shifted to a different wavelength. The first case is called Raleigh scattering and the second case is called Raman scattering. Inelastic scattering of incident light means, that a certain amount of energy remains in the molecule. The energy loss of the scattered light quantum corresponds to the energy of a certain molecule vibration caused by the scattering process.

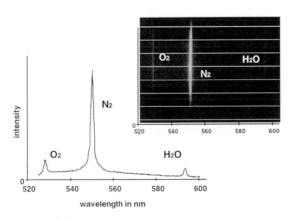

Fig. 1. Raman spectrum of air

Those molecule vibrations, for example stretching or rotational vibrations, are specific for the molecule. This allows to use Raman spectroscopy in quantitative chemical analysis or to investigate the molecule structure.

The Raman shift of a certain molecule depends on the energy of a molecule vibration and on the structure, the atomic distance and the atomic weights of a molecule, therefore.

A typical spectrum of air is shown in Fig. 1. We used an argon-ion laser operating in single line mode at 488 nm to obtain this spectrum. This laser wavelength caused an oxygen peak at 528 nm, a nitrogen peak near 550 nm and a single narrow water-vapor peak near 594 nm. This water peak is caused by the v_s O-H-stretching vibration of the water molecule:

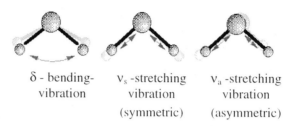

δ - bending-
vibration

v_s -stretching
vibration
(symmetric)

v_a -stretching
vibration
(asymmetric)

Fig. 2. Vibrational modes of water molecules

It is a single narrow peak like the peaks caused by two atomic nitrogen or oxygen molecules.

As shown in Fig. 3 the Raman spectrum of liquid water is quite different. The same O-H-bond is now affected by hydrogen bonds, which increase the atomic distance between the oxygen and the involved hydrogen atom.

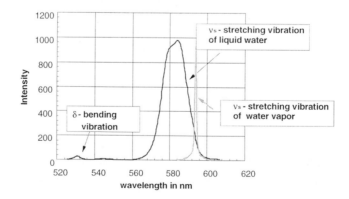

Fig. 3. Raman spectra of Liquid and vaporous water

This effect decreases the energy of the O-H-stretching vibration and reduces the Raman shift. The liquid water peak is much wider then the vapor peak, because the angle between the O-H-bond and the hydrogen bond may differ. The influence of the hydrogen bonds to the O-H-stretching vibration depends on the angle between the hydrogen bond and the O-H-bond.

The shape of the liquid water peak changes with the temperature as shown in Fig. 4

Walrafen explained this effect with a temperature dependent equilibrium between hydrogen-bonded and non hydrogen-bonded O-H stretching oscillators in liquid water. The number of established hydrogen bonds decreases with higher temperatures.

This effect can be used to measure water temperatures with an accuracy of about ± 2 K using an algorithm described in (Karl and Weiss, 1997)

Fig. 4 shows the shape of measured Raman spectra of liquid water at temperatures between 30 and 100 °C. After base line extraction the normalized water peaks cross at the same point (*isosbestic point*) as documented by Walrafen.

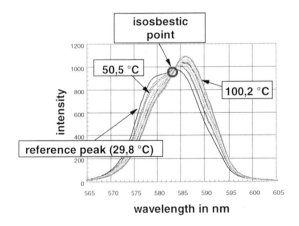

Fig. 4. Temperature dependence of the v_s- Stretching region of liquid water

The real molecular structure of liquid water is quite complicated, but in simplified terms it can be said that the water peak consists of two peaks with a maximum at 577,6 nm and 589,4 which represent the hydrogen bonded O-H-bonds and the non hydrogen bonded O-H bonds.

The shape of those difference spectra proved to be useful to get the intensity ratio between the peak of the hydrogen bonded and the non hydrogen bonded molecules.

Assuming a temperature dependent equilibrium between hydrogen-bonded and non hydrogen-bonded v_1-stretching oscillators, this equilibrium may also be interpreted as an equilibrium between hydrogen-bonded (polymer) and non hydrogen-bonded (monomer) water molecules

monomer   polymer

The concentration ratio of the established bonds represents the equilibrium constant of that equilibrium. This constant can be expressed by the enthalpy of the hydrogen bonds ΔH^0 and the entropy (Atkins, 1986)

$$\ln\left(\frac{[P]}{[M]}\right) = \frac{\Delta H^0}{\Re \cdot T} + \Re \cdot \Delta S^0 \qquad (1)$$

Assuming that the hydrogen-bonded and the non hydrogen-bonded water O-H-bonds cause peaks in the Raman spectra with a differential cross section S_M and S_P, Eq. (1) may be written as

$$\ln\left(\frac{S_M \cdot I_P}{S_P \cdot I_M}\right) = \frac{\Delta H^0}{\Re \cdot T} + \Re \cdot \Delta S^0 \qquad (2)$$

or

$$\ln\left(\frac{I_P}{I_M}\right) = c_1 \cdot \frac{1}{T} + c_2 \qquad (3)$$

A linear least squares fit to measured temperature and intensity data provided the coefficients c_1 and c_2. We used Vant'Hoffs principle to compute the enthalpy of the hydrogen bonds ΔH^0 from coefficient c_1 and obtained a value of $10.47 \cdot 10^3$ KJ/kmol. This value is in good agreement to the enthalpies published elsewhere.

If the intensity ratio $\frac{I_P}{I_M}$ is known from measured data the liquid water temperature can be calculated from

$$T = \frac{c_1}{\ln\left(\frac{I_P}{I_M}\right) - c_2} = \frac{1259.66\,K}{\ln\left(\frac{I_P}{I_M}\right) - 3.92} \qquad (4)$$

3 Temperature evaluation procedures

There are two different ways to obtain that intensity ratio. The first method requests a monochromator or a spectrograph as mentioned above and uses the whole spectral information for temperature evaluation. The spectrograph or monochromator provides a complete Raman spectrum, which allows to calculate the intensity ratio $\dfrac{I_P}{I_M}$ with best reliability.

The main advantage of this method is, that any Rayleigh,- Mie or fluorescence signal can be separated and discharged easily.

A base line extraction can be done automatically using a line fitting procedure which separates the signals of the spectrum. The image processing procedure presented by Karl and Weiss, 1997 provided water temperatures with an accuracy of less than ± 2 K within a temperature range between 20 and 180 °C as shown in Fig. 5.

Raman spectroscopy has proofed to be an adequate method to get one dimensional temperature profiles, therefore. But there are various applications where two dimensional measurements might be more useful. The second method to obtain water temperatures uses narrow band width interference filters. The idea is to use two narrow band bandpass filters which transmit mainly the Raman signals of hydrogen bonded and non hydrogen bonded O-H bonds.

The laser beam can be expanded to a light sheet. A CCD camera samples the scattered light with or without a spectrograph. Two images of the light sheet taken with different intereference filters are necessary to evaluate the water temperature distribution.

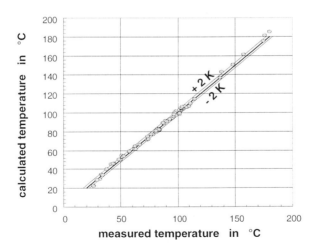

Fig. 5. Water temperatures calculated from Raman spectra (using a spectrograph)

Using two interference filters with a central wavelength at the wavelength of the hydrogen-bonded and the non hydrogen bond Raman wavelength makes it possible to get the intensity ratio $\dfrac{I_P}{I_M}$ from two separate measurements with that filters. Using the blue line of an argon ion laser at 488 nm requires two bandpass filters near 577 nm and 589 nm. We used two interference filters with a central wavelength of 578 an 590 nm and a *full width at half maximum* (FWHM) of ± 10 nm. Fig. 6 shows the liquid water spectra transmitted by those filters:

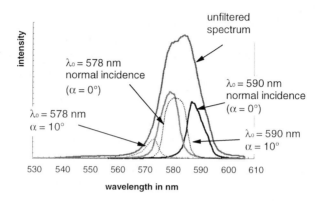

Fig. 6. Spectrum of liquid water transmitted by narrow band intereternce filters

Fig. 7 shows the spectra of water at different temperatures obtained with both filters:

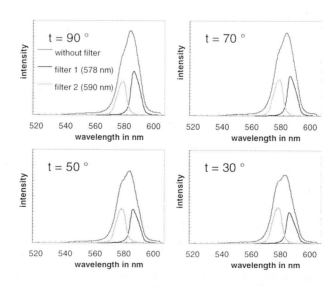

Fig. 7. Influence of the water temperature on the spectra transmitted by the bandpass filters

It is obvious that the ratio of the intensities transmitted through the filters depends on the water temperature. Integration of the spectra gives the total intensity transmitted through the filters. The spectra shown in Fig. 7 represent the light which will meet the pixels of CCD chip corresponding to certain location of the water flow with two dimensional measurements.

This integration of the spectra leads to a characteristic line which can be used to calculate water temperatures from two dimensional intensity data according to Eq. (4). The same characteristic can be achieved without experiments only using the transmission curves of the filters and the knowledge of the shape of the water peak with respect to the temperature. The accuracy is much lower as the accuracy achievable with a spectrograph:

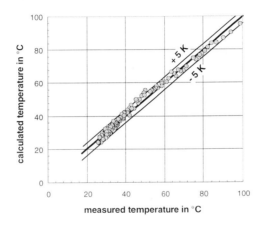

Fig. 8. Water temperatures calculated from Raman signals ('two color method', without spectrograph)

This 'two-color-method' has to deal with two major problems. The first is shown in Fig. 6 and Fig. 9. Narrow band interference bandpass filters are usually Fabry-Perot-filters, which operate with the same principle as the Fabry-Perot interferometer. A Fabry-Perot filter consists of coated surfaces with a well defined distance which corresponds to a certain wavelength. The transmission of light with a certain wavelength depends on this distance. The central wavelength of the filter is given for normal incidence. The real central wavelength depends on the angle of incidence according to the equation

$$\lambda_{0,\alpha} = \lambda_{0,\alpha=0°} \cdot \left[1 - const. \cdot \sin^2(\alpha)\right] \qquad (5)$$

The central wavelength is shifted to lower wavelengths and alters the transmitted spectrum as shown in Fig. 9 for an angle of incidence of 10°. It is obvious that this effect will cause severe errors and it is only valid to calculate water temperature values from the intensity ratio $\frac{I_P}{I_M}$ if the angle of incidence is very small. But especially for laboratory scale applications, if the distance between laser beam and detector is small this angle of incidence might be very large. Optics with small focal lengths cause the angle of incidence to be 10 degree or more as shown in Fig. 9.

The second problem of the two filter method is caused by fluorescence. High fluorescence rates increase the intensities of the 578 nm and 590 nm peaks and distort the intensity ratio $\frac{I_P}{I_M}$, therefore.

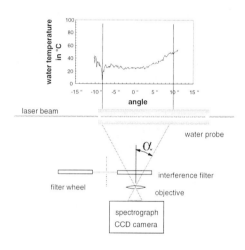

Fig. 9. Influence of the optical set-up to the measured temperature

With pure water most of the fluorescence signal originates from the water surface. Reflection and scattering of laser light on solid surfaces water causes additional fluorescence which enhances the total fluorescence rate.

For any technical applications the third effect will be the most important. Any solvents or particles cause additional fluorescence which might be even more intense then the Rayleigh signal. Dirtiness of the laser windows and mirrors will enhance fluorescence, too.

The best way to eliminate the error caused by fluorescence is to measure it using a spectrograph. For two dimensional measurements there are two methods to get the fluorescence level. The first possibility is to use a third bandpass filter at a suitable wavelength. If the shape of the fluorescence spectrum is given for a certain optical setup, the intensity measured at a single wavelength will be enough to estimate the fluorescence rate which distorts the Raman signals obtained with the 578 and 590 nm filters.

But of course this method increases the number of images which are necessary to get the temperature. If the fluorescence signal does not vary, for example if the particles content or the dirtiness of the water flow keeps constant, it is not necessary to get the fluorescence intensity for each temperature measurement.

A second way which is often used to reduce errors caused by fluorescence can not be used for the presented measurements. Especially for Raman spectroscopy in gases, where high fluorescence or Mie signals superimpose the Raman signal, it might be useful to rotate the polarisation of the laser beam using half wave plates. Mie scattering and fluorescence depend on the polarisation of the laser light only weakly. The difference obtained with two images taken with vertical and horizontal polarisation can be used to eliminate the fluorescence signal, therefore.

But we observed different depolarisation rates for the hydrogen bonded and non hydrogen bonded Raman signals. Rotation of the polarisation plane affected the intensity of both Raman peaks differently. Elimination of fluorescence by rotation of the laser beam polarisation will change the intensity ratio $\dfrac{I_P}{I_M}$ and will cause additional errors, therefore.

The third method to eliminate fluorescence uses the different time scales of Raman and fluorescence processes. The Raman signal arises spontaneous without retardation. The fluorescence process will last some nanoseconds. Short laser pulses and a intensified CCD camera can be triggered with very short time gates to enhance the ratio between Raman and fluorescence signal.

4 Experimental applications

In earlier experiments we used Raman spectroscopy to measure concentration profiles in condensation boundary layers during condensation in the presence of non condensable gases. We investigated the condensation of nitrogen steam mixtures at system pressure levels from 0,2 to 2,0 MPa at a thin water layer. The experiments were aimed at measuring local heat transfer coefficients by measuring local concentration profiles. We designed the pressure vessel shown in Fig. 10 to simulate the boundary conditions of one dimensional film theory.

The laser beam was focused vertically onto the water layer and an objective focused the image of the laser beam to the entrance slit of a 125mm - spectrograph. This spectrograph separated shifted Raman signals from unshifted Rayleigh and Mie signals. A CCD chip of an intensified CCD Camera detected the weak Raman signals and the image processing system provided images of the mass transfer boundary layer as shown in Fig. 10.

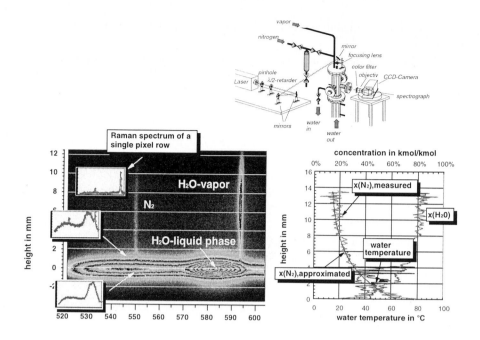

Fig. 10. Raman image of a condensation boundary layer and calculated concentration profiles

Fig. 10 shows two images of the same laser beam, shifted to different wavelengths by the spectrograph. The line located at a wavelength of 550,6 nm is caused by light scattered at nitrogen molecules and the line located at 594 nm is caused by light scattered at water vapor molecules.

Each pixel row corresponds to a certain position at the laser beam and yields an intensity profile or a Raman spectrum, therefore, to calculate local concentrations from measured intensities.

The steam water interface caused an intense fluorescence signal and a very strong Raman signal of the liquid phase which provided the temperature of the water layer.

The concentration profiles calculated from those images are in good agreement to film theory and the presented method has proofed to be suitable to measure heat and mass transfer rates.

With another experiment we tried to get not only a single temperature value but a complete temperature profile in a water layer.

We focused the laser beam of a Coherent SabreFred argon ion laser with an single line output power of about 10 W_{cw} onto a water layer flowing in a rectangular duct inside of an autoclave. The same software which processed the image shown in Fig. 10 calculated temperature profile as shown in Fig.11 from the images provided by the image processing system.

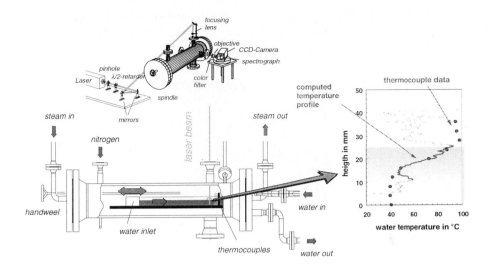

Fig. 11. Experimental set-up to measure temperature profiles in stratified flows

The experiment shown in Fig. 12 was designed to get two dimensional water temperature measurements. We performed the experiments with a small model of a tube bundle heat exchanger and without spectrograph. The CCD camera sampled the scattered light of a light sheet which was focused along the tube plate.

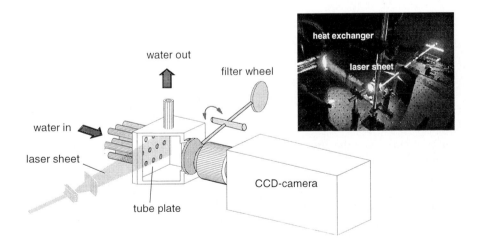

Fig. 12. Tube bundle heat exchanger model

Single tubes were heated and caused different temperatures at those positions where the exit water passed the light sheet. An image of the tube plate without laser light sheet is shown in Fig. 13:

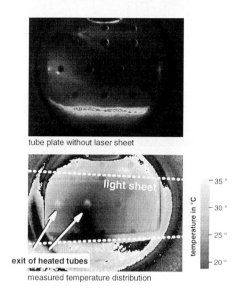

Fig. 13. Image of the tube plate of the heat exchanger model (without laser sheet) and temperature image

The exit temperatures were measured together with the camera readings by thermocouples. The exit temperature differences of the tubes in the middle of the tube plate varied between 1 and 30 K.

We sampled the scattered light with each filter for a certain time and used the characteristic to compute a temperature image from the images taken with different filters. We had to measure for a certain time to reduce the statistical errors caused by particles or variation of the laser power.

It is obvious, that the absolute accuracy of this method is not acceptable. Above all the error caused by the angle of incidence has proved to be very high. This effect made the temperature seem to rise from the middle of the image to the edges. But it is significant that temperature differences can be recognized very precisely. This can be skillful to locate or evaluate local temperature differences.

Fig. 14 shows the influence of measuring time and the variance of the images with respect to the temperature difference. It is possible to measure absolute temperatures only for normal incidence with an accuracy of about ± 5 K but it is possible to detect local temperature differences of 1 K even with measuring times less then 2 minutes.

5 Conclusion

Measuring water temperatures by means of Raman spectroscopy requires an expensive experimental setup. If flame temperatures shall be measured Raman spectroscopy is a good choice, because there are only few other possibilities to measure temperatures of about 2000 K with sufficient accuracy.

Water temperatures can be measured very simple and with few costs by means of thermocouples.

For that reason water temperature measurements by means of Raman spectroscopy will by useful only for special applications, if very detailed local information is needed - in example for heat transfer applications - or if a thermocouple will disturb the measurement, in example with micro scale heat transfer applications.

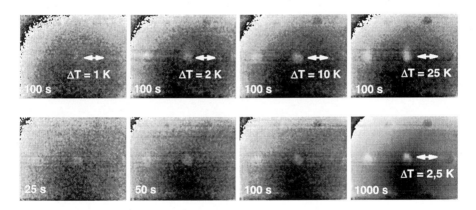

Fig. 14. Temperature images of the water flow in front of the tube plate of the heat exchanger model with different temperature differences and measuring times

If only low accuracy is required two dimensional measurements may by quite useful to supervise any chemical or thermal processes with a 'two color method' set-up. Because the intensity of the Raman signal is very high it is not necessary to use high sensible CCD-cameras or high laser power.

Another possible application of the water temperature effect might be to do remote temperature sensing with a LIDAR set-up for example to get ocean water temperatures.

Linear Raman spectroscopy has proofed to be an optical measurement method which provides the possibility to measure temperatures of liquid water with highest local resolution even in two dimensions without affecting the fluid flow, therefore. The local resolution is only limited by the optical setup and the resolution of the CCD-chip.

I.7. A New Approach to Eliminate the Measurement Volume Effect in PDA Measurements

H.-H. Qui and C.T. Hsu

Department of Mechanical Engineering
Hong Kong University of Science and Technology
Kowloon, Hong Kong

Abstract. A new model of phase-size correlation was introduced to improve the accuracy of sizing large particles in two-phase flows. This model takes both the refractive and reflective mechanisms into consideration and, hence, relaxes the assumption of single scattering mechanism in conventional phase-Doppler anemometry. As a result, the Gaussain beam defect and slit effect is eliminated. This method will need a three-detector PDA system to determine the phases of Doppler signals scattered from a spherical particle in the measurement volume. The results of simulations based on this new method using the Generalized Lorenz Mie Theory (GLMT) were compared with those based on the conventional method. The results of simulation show that the newly developed method can greatly reduce the Gaussian beam defect and the slit effect in a PDA system.

Keywords. PDA, GLMT, scattering

1. Introduction

Laser Phase-Doppler Anemometry (PDA) is nowadays widely used for velocity measurement and particle sizing in experimental studies of dispersed two-phase flows. The new extensions in dynamic range, signal processing and measuring the particle refractive index have made PDA one of the most versatile and accurate techniques available today. One of most important applications of PDA is to determine the local mass flux and concentration in multiphase flows, especially in sprays. Such measurements are essential when studying droplet evaporation, spray deposition and cooling, etc. To this end, the particle size, velocity and size dependent measurement volume (or cross-sectional area) must be determined accurately. Among the above parameters, the accurate measurement of droplet size is the utmost important since the particle volume depends on the third power of the particle diameter. Because PDA is based on the single particle scattering theory, it is often required to decrease the size of measurement volume in order to achieve highly accurate results in dense particle conditions. Unfortunately, a simple reduction in the measurement volume can cause phenomena in conflict with the assumption of a uniform illumination. Consequently, the result of sizing large particles suffers from error due to the nonlinearity in the phase-diameter relationship from the nonuniformity in beam intensity (Saffman 1986, Bachalo and Sankar 1988, Sankar et al. 1992, Grehan et al. 1991, Qiu and Sommerfeld 1992, Qiu and Hsu 1995). This nonlinear effect is usually called as Measurement Volume Effect (MVE) or trajectory ambiguity.

Recent investigation using an extended Phase-Doppler anemometry has shown that the MVE is also large in refractive index measurements (Durst et *al*. 1994). Numerical tools were developed to simulate the Gaussian beam defect by using Geometrical Optics Theory (GOT) and Generalized Lorenz Mie Theory (GLMT) (Grehan et *al*. 1992). For the effect of the slit image in measurement volume, previous studies focused on the vignetting effect from receiving optical aperture and slit (or pinhole) to the effective measurement volume size. The experimental studies by Durst et al. (1994) showed that the nonuniform illumination effect due to the image boundary of the spatial slit filter in receiving optics is even more critical than the Gaussian beam defect. The Fourier Optics Method recently developed by Qiu and Hsu (1996) has been successfully used to simulate the slit effect. Apparently, the MVE, which consists of the Gaussian beam defect and the slit effect, greatly hinder the accuracy in particle sizing. Various solutions to minimize the trajectory ambiguity have been proposed for classical geometry (Qiu and Sommerfeld 1992), or more original designs of planar geometry (Aizu et al 1993) and spatial frequency geometry (Qiu and Hsu 1995). A promising one is the dual mode geometry which integrates a standard PDA (SPDA) with a Planar PDA (PPDA) (Tropea et al. 1994).

One of the important features of PDA in a planar configuration is the temporal separation of the contributions of the scattering reflection and refraction mechanisms to the light received by the photodetectors. By using the SPDA/PPDA phase ratio and the amplitude/size ratio together with the burst-centering technique, the scattering mechanism can be validated. This offers the possible technique to suppress the measurement errors due to the mixing of the two scattering modes, and therefore, makes it possible to perform refractive-index measurements by the comparison of the reflective and the refractive phases. However, all of above methods for the elimination of the measurement volume effect are based on the signal validation scheme. If the phase ratio or amplitude/size ratio is not within the expected tolerance, the measured phase will be invalid. Because of the strong dependence on the validation scheme, the effective measurement volume size or cross-sectional area is reduced in this method for those particles moving in negative Y position (see Fig. 1). This also results in the measurement volume asymmetrically to its intensity distribution. To correct this variation the measurement volume needs to be calibrated. Certainly, this kind of correction is still feasible in a simple directional flow, such as sprays.

However, in complex flow conditions, such as in two-phase swirling flows with arbitrary particle trajectories, the projection of the effective cross-sectional area in each particle size and moving direction becomes very complex. Consequently, the calibration method will be very complicated. Furthermore, as PPDA has relative larger fluctuation in the size-phase relation than SPDA for small particles, the validation method only based on the SPDA/PPDA phase ratio is not sufficient to completely eliminate the measurement volume effect especially the slit effect which is also sensitive to small particles. The main reason of using the phase ratio and amplitude/size ratio for the strong validation is to identify the scattering mechanism. Because only one scattering mechanism was assumed in the conventional PDA scattering model, any other scattering mechanisms are considered as the error sources for size measurements. If a complete scattering model can be developed by taking both the refraction and reflection into consideration, no validation between the scattering mechanisms is necessary.

In this study a scattering model developed recently including both refraction and reflection by Qiu and Hsu (1995) was used and modified for calculating the particle size-phase relation. One of the advantages of this method is the requirement of only a conventional three-detector receiving optics. In this model the signal phases from the three detectors are used to calculate the particle size. Because the fringe patterns from the refraction and reflection are different in space and moving direction, it has been found that there is an optimized optical orientation where the particle size can be solved analytically. To demonstrate the capability of the new developed approach, the model was simulated numerically by using both Geometrical Optics Theory (GOT) and Generalized Lorenz Mie Theory (GLMT). Different optical parameters such as the measurement volume size, the focus lengths of the sending and receiving lenses, the size and shape of the receiving aperture, the particle size and its trajectory, and the phase conversion factors are used to analyze the performance of the new developed method.

2. Analytical Description

The schematics of a conventional phase-Doppler system is shown in Fig. 1, where the receiving optics assembled into a three-detector system is used. The phase difference between the three receiving detectors can be determined from the three Doppler signals. If we assume the scattering light as being dominated by refraction, from the geometrical consideration the relation between the phase difference ϕ of the outer detectors and the particle diameter D are related linearly as following:

$$\phi = C_1 \cdot D \tag{1}$$

The phase conversion factor C_1 can be written as (Bauckhage, 1988):

$$C_1 = \frac{4\pi}{\lambda}\left\{\sqrt{1+m^2 - \sqrt{2}m\sqrt{1+\sin\tfrac{\theta}{2}\sin\psi + \cos\tfrac{\theta}{2}\cos\psi\cos\varphi}}\right.$$
$$\left. - \sqrt{1+m^2 - \sqrt{2}m\sqrt{1-\sin\tfrac{\theta}{2}\sin\psi + \cos\tfrac{\theta}{2}\cos\psi\cos\varphi}}\right\} \tag{2}$$

where m is the particle refractive index and the angles, θ, φ and ψ, are defined in Fig. 1.

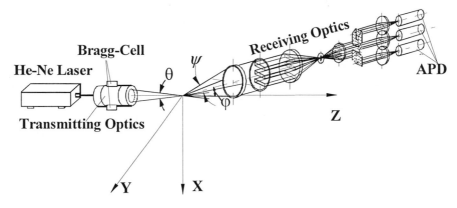

Fig. 1: Optical layout of a phase Doppler anemometry

Equation (1) is true only when the scattering is dominated by the refractive light. When the particle size is not small in comparison with the laser beam diameter in the measurement volume, the light dominance and then the phase-size relation depend on the particle location. This effect is called *Gaussian beam effect* or *trajectory ambiguity,* as documented by Gréhan et al. (1991), Sankar et al. (1992), Qiu and Sommerfeld (1992), Qiu and Hsu (1995).

The Gaussian beam effect is due to the Gaussian distribution of laser beam intensity in the measurement volume. When the particles pass through the focused beams on different trajectories, it can cause a change in balance between the reflected and refracted rays, i.e., the dominant light can occur sequentially depending on the two scattering mechanisms. In either case, the trajectory dependence can lead to significant errors. For example, when the scattering light is dominated by reflection, the phase conversion factor is then written as:

$$C_0 = \frac{4\pi}{\lambda\sqrt{2}}\left\{\sqrt{1+\sin\tfrac{\theta}{2}\sin\psi - \cos\tfrac{\theta}{2}\cos\psi\cos\varphi}\right.$$
$$\left. - \sqrt{1-\sin\tfrac{\theta}{2}\sin\psi - \cos\tfrac{\theta}{2}\cos\psi\cos\varphi}\right\} \tag{3}$$

Because the reflective fringe pattern moves in the opposite direction to that of the refractive pattern, the phase and size relation for the receiving optics is described as:

$$2\pi - \phi = |C_0| \cdot D \tag{4}$$

Therefore, the detection of the wrong scattering component will occur, leading to the misrepresentation of mid-range size particles as being very large. This has a significant effect on size measurements.

3. Combined Scattering Model

According to the model proposed by Qiu and Hsu (1995), the relationship between the measured signal phase and particle diameter for a conventional PDA system taking both refractive and reflective rays into consideration can be described by the following form

$$\phi = 2\arctan\left(\frac{I_1 \sin\left(\frac{C_1 D}{2}\right) - I_0 \sin\left(\frac{C_0 D}{2}\right)}{I_1 \cos\left(\frac{C_1 D}{2}\right) + I_0 \cos\left(\frac{C_0 D}{2}\right)}\right) \tag{5}$$

where the subscript "1" and "0" referred to the refraction and reflection, respectively. If the intensities of refractive and reflective rays can be measured experimentally or determined analytically, the phase-size relation can be uniquely quantified by Eq. (5). Unfortunately, a method to separately measure the intensities of each ray has not been available yet and, therefore, it is impossible to directly calculate the phase-size relation by using equation (5) for a two-detector system. For a three-detector system, however, the signal phases for each detector with different elevation angles can be described as

$$\begin{cases} \phi_1 = 2\arctan\left(\frac{I1_1 \sin\left(\frac{C1_1 D}{2}\right) - I1_0 \sin\left(\frac{C1_0 D}{2}\right)}{I1_1 \cos\left(\frac{C1_1 D}{2}\right) + I1_0 \cos\left(\frac{C1_0 D}{2}\right)}\right) \\ \phi_2 = 2\arctan\left(\frac{I2_1 \sin\left(\frac{C2_1 D}{2}\right) - I2_0 \sin\left(\frac{C2_0 D}{2}\right)}{I2_1 \cos\left(\frac{C1_1 D}{2}\right) + I2_0 \cos\left(\frac{C1_0 D}{2}\right)}\right) \end{cases} \tag{6}$$

In the system equation (6), because there are total five unknowns ($I1_1$, $I1_0$, $I2_1$, $I2_0$, and D), it still can not be solved. In a standard PDA system, the elevation angles of each detectors are quite small, therefore, the magnitude of scattering intensities are almost the same for each detectors. In this case, $I1_1 \approx I2_1 = I_1$ and $I1_0 \approx I2_0 = I_0$ can be assumed which have also been proved by GLMT simulations. Hence, only three unknowns have to be determined now. A further simplification is based on the method suggested by Qiu and Hsu (1995) where the magnitude of the phase conversion factors for refractive and reflective rays are used as the key factor for the optimization process. By taking the same magnitude for the phase conversion factors for refractive and reflective rays, i.e., $C1_1 = C1_0 = C_1$ and $C2_1 = C2_0 = C_2$, Eq.(6) reduce to

$$
\begin{cases}
\phi_1 = 2\arctan\left(\dfrac{I_1 - I_0}{I_1 + I_0}\tan\left(\dfrac{C_1 D}{2}\right)\right) \\[3mm]
\phi_2 = 2\arctan\left(\dfrac{I_1 - I_0}{I_1 + I_0}\tan\left(\dfrac{C_2 D}{2}\right)\right)
\end{cases}
\tag{7}
$$

Hence, the relationship between the measured phases, ϕ_1, ϕ_2, and the particle size, D, is described as

$$
\frac{\tan\left(\dfrac{\phi_1}{2}\right)}{\tan\left(\dfrac{\phi_2}{2}\right)} = \frac{\tan\left(\dfrac{C_1 D}{2}\right)}{\tan\left(\dfrac{C_2 D}{2}\right)}
\tag{8}
$$

which can be used to solve for D, using the two measured phases and optical geometry. A further optimization method has been found if $C_1 = 2C_2$. Finally, the relationship for the determination of the particle size using the measured signal phases under the optimized condition for the combined model is reduced further to

$$
D = \frac{4}{C_1}\arctan\left(\sqrt{1 - \frac{2\tan\left(\dfrac{\phi_2}{2}\right)}{\tan\left(\dfrac{\phi_1}{2}\right)}}\right)
\tag{9}
$$

4. Results of Simulation

To evaluate the performance of the newly developed model by using Eq. (8), the GLMT was adopted to simulate the different particle sizes in comparison with the conventional single scattering mechanism assumption. The simulation was carried out for the system geometry as shown in Fig. 1 with the parameters given in Table 1.

The results of simulation are shown in Figs. 2-4. As shown in Fig. 2, a 30 μm water droplet passing through the measurement volume parallel to the Y-axis at z = 0 µm will yield large phase transition (> 100% in this special case) when the scattering mechanism changed from the reflection to refraction. If the phase is determined by the conventional method, i.e. directly calculated by GLMT, large measurement error will occur. Only very small fluctuation (< 2%) for the phase determination by using the newly developed method and hence the error due to measurement volume effect is eliminated.

Table 1 Optical parameters

Wavelength (λ)	0.5145	µm
Beam Waist Diameter	100	µm
Transmitting Angle	2.2916	degree
Receiving Elevation Angle 1	3.69	degree
Receiving Elevation Angle 2	1.845	degree
Off-Axis Angle	46	degree
Refractive Index	1.33	

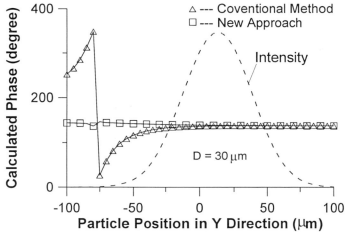

Fig. 2 Comparison of the phase measurement results between the conventional method and new approach (z = 0µm, r_0 = 50 μm)

Fig. 3 and 4 show the similar results for droplets of 50 and 70 µm respectively. Again the scattering changing from the reflection to refraction yields large phase change using the conventional method, and the calculated phase using the new approach is almost constant which is proportional to the particle diameter. The above result clearly demonstrated that the newly developed method can eliminate the measurement volume effect completely even for a conventional three-detector PDA system which is very commonly used in two-phase flow measurements. Because no strong validation scheme is necessary other than the signal-to-noise ratio (SNR), the measurement volume can be accurately determined by the new method and hence also the high accuracy in the particle mass flux and concentration measurements. By applying this method in the dual mode PDA system, this method can further improve the accuracy for the refractive index measurements.

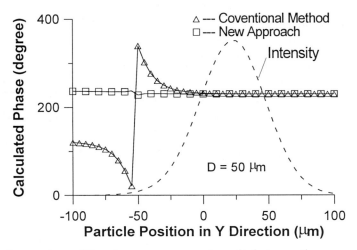

Fig. 3 Comparison of the phase measurement results between the conventional method and new approach (Z = 0μm, r_0 = 50 μm)

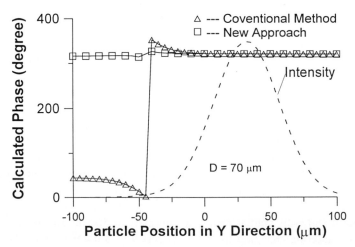

Fig. 4 Comparison of the phase measurement results between the conventional method and new approach (Z = 0μm, r_0 = 50 μm)

5. Conclusion

A new phase-size correlation model is proposed and investigated towards an accurate sizing of large particles. This new system has the same transmitting optics as a conventional three-detector PDA system, but with optimized orientation of receiving units. The key point to this newly developed method is to set $C_1 = C_0$ and $C_1 = 2C_2$. The performance of this new method was simulated by using the GLMT. The results of the simulation show that the new method can effectively eliminate the Gaussian beam effect and slit effect for sizing large particles. The newly developed model as described in Eq. (8) for the phase-size correlation using a geometrical optics method is specially suitable for sizing large particles. Therefore, the application of this new method to a conventional PDA system is expected to have almost with no Gaussian beam defect and slit effect. The next step to this work is to perform validation tests experimentally by using this method.

6. Acknowledgment

This work was supported by the Hong Kong Government and Hong Kong University of Science & Technology under the RGC/ERG Grant No. HKUST 812/96E, HKUST708/95E and G-HK96/97.EG11.

References

Aizu Y, Durst F, Gréhan G. Onofri F, Xu T-H, 1993: "PDA systems without Gaussian beam defects.", *Proc. 3rd Int. Conf. Optical Particle Sizing*, Yokohama, pp461-470.

Bachalo, W. D. & Sankar, S. V., 1988: "Analysis of the light scattering interferometry for spheres larger than the light wavelength, *Proc. 4th Int. Symposium on Applications of Laser Anemometry to Fluid Mechanics*, Lisbon paper 1.8.

Durst F., Tropea C. and Xu T.-H., 1994: "The Slit Effect in Phase Doppler Anemometry", *2nd. Int. Conf. on Fluid Dynamic Measurement and Its Applications*, Beijing, 38-43.

Grehan G., Gouesbet G., Naqwi A., and Durst F., 1991, "Evaluation of Phase Doppler System using Generalized Lorenz-Mie Theory," *Proc. of the Int. Conf. on Multiphase Flows '91-Tsukuba*, 291-294.

Qiu H. -H. and Hsu C. T., 1995: "Optimization of EPDA Parameters for Accurate Material Recognition in Multiphase Flow", *2nd Int. Conf. on Multiphase Flow '95-Kyoto*, IN1-IN9.

Qiu H. -H. and Hsu C. T., 1995: "A New Spatial Frequency Method For Sizing Large Particles in Laser Anemometry", *ASME Fluids Engineering Summer Meeting and 6th Int. Conf. on Laser Anemometry*, South Carolina, vol. FED 229, 81-88.

Qiu H. -H. and Hsu C. T., 1996: "A Fourier Optics Method for the Simulation of Measurement-Volume-Effect By the Slit Constraint", Proc. of *8th International Symposium on Applications of Laser Techniques to Fluid Mechanics*, Lisbon, pp 12.6.1-12.6.8.

Qiu H.-H., and Sommerfeld M., 1992: "The impact of signal processing on the accuracy of phase-Doppler measurements," *Proc. 6th Workshop on Two Phase Flow Predictions*, Erlangen, 167-185.

Saffman,M., 1986: The use of polarized light for optical particle sizing, *Proc. 3rd Int. Symp. on Applications of Laser Anemometry to Fluid Mechanics*, Lisbon.

Sankar, S. V., Inenaga, A.S., and Bachalo, W. D., 1992, "Trajectory Dependent Scattering in Phase Doppler Interferometer: Minimizing and Eliminating Sizing Errors", *Proc. 6th Int. Symp. on Applications of Laser Anemometry to Fluid Mechanics*, Lisbon.

Tropea C., Xu T.-H., Onofri F., Gréhan G. and Haugen P., 1994: " Dual Mode Phase Doppler Anemometry", *Proc. 7th Int. Symp. on Applications of Laser Techniques to Fluid Mechanics*, Lisbon, pp18.31-18.3.7.

I.8. Concentration Measurement by Molecular Absorption Using Narrow Band Tunable Infrared Laser

T. Kawaguchi, K. Hishida, and M. Maeda

Department of System Design Engineering, Keio University
3-14-1 Hiyoshi, Kohoku-ku, Yokohama 223-8522, Japan

Abstract. This paper describes a technique to measure the spatial concentration distribution of species using the absorption characteristics of a molecule in the near infrared light. In the present work, a novel measurement technique employing the spectroscopic detection of molecules was examined in conjunction with two-dimensional Computed Tomography (CT). Since the spectroscopic measurement allows finding the molecule itself, the mass transfer and diffusion can directly be observed without any tracer particles in flow. The detection by absorption realizes to distinguish the interface between different species even in the same liquid-liquid and gas-gas phases when the wavelength of infrared light source was tuned onto the convenient absorption band of each target species. The experimental results of the concentration measurement show that our technique can easily distinguish the invisible interface of species. Furthermore, the results indicate that the multiple sets of projecting optics enable us to perform time-resolved measurement for turbulent flow fields.

Keywords. concentration measurements, molecular absorption, distributed feedback, semiconductor tunable laser, infrared spectroscopy, computed tomography

1 Introduction

Classical shadow-graphy and Schlieren photography visualize the concentration or density of the compressible or high temperature flow field. Holographic measurements of velocity [Royer, 1997], mass transfer [Grosse *et al.*, 1977; Ito *et al.*, 1991] and temperature field [Ito *et al.*, 1987] have been developed in recent years. These techniques visualize the internal structure of flow field; however, quantitative and time-resolved measurements are required.

An analysis of a turbulent structure by measuring temperature and velocity of a jet into a water vessel have been successfully conducted [Sakakibara *et al.*, 1993; Mastorakos *et al.*, 1996] using Laser Induced Fluorescence (LIF) technique, i.e. noting the temperature dependence of fluorescence of

Rhodamine-B in water solution when excited by green light in combination with a PIV technique. Concentration and velocity measurement in an axisymmetric turbulent jet was conducted by using the three-dimensional LIF tomography technique [Merkel et al., 1996]. Although these combination of PIV and LIF techniques enable to investigate the local momentum, heat and mass transfer with the characterized tracer particles, it is difficult to separate the effects of temperature and of concentration to a fluorescent intensity.

On the other hand, absorption-based sensors detect a molecule itself without tracer particles in flow, and have high sensitivity and selectivity when a spectrally narrow light source is used to probe a spectrally narrow feature. Tuning the wavelength of the light source across the absorption bands of molecule distinguishes the isolated feature from background absorption, scattering or excitation effects due to obscuration of the optical path or changes in the total source power coupled onto the receiver. Thus, most applications relevant to gas or fluid dynamics are based on an absorption with a well-resolved transition.

Recent advances in near infrared and visible semiconductor laser sources for telecommunication and high-speed computer networks have simultaneously inspired a new generation of laser measurements based on frequency modulated spectroscopy [Carlisle et al., 1989]. Such measurements for environmental species has been applied in the various forms to near infrared and visible diode lasers for detection of H_2O[Cooper et al., 1986; Silver et al., 1994], CH_4[Lucchesini et al., 1993; Uehara et al., 1997], C_2H_2[e.g. Grosskloss et al., 1994] and NH_3[e.g. Feher et al., 1993] and O_2[Bruce et al., 1990; Phillipe et al., 1993]. In particular for an absorption spectrum between 6300 and 7900cm^{-1} of H_2O, detailed research on the line intensities have been investigated and many absorption bands have been reported by Mandin et al., 1986.

A semiconductor diode laser has many advantages over the usual solid or gas laser devices, although both of them have common structure of a short section of lasing material and a resonator with a pair of mirrors. Such characteristics of a diode laser as operation at room temperature, rapid tuning of wavelength, high spectral purity and long term stability are so attractive that huge dye laser systems have been replaced by portable diode lasers. Especially a Distributed Feedback (DFB) tunable diode laser, which is equipped with a special diffraction grating, has very important properties for spectroscopic sensing. For example, ultra-narrow emission spectrum, singlemode stability, wavelength tuning without mode hopping and so on. Details of a DFB diode laser were described in section 2.

In general, it is necessary to determine the distribution of some physical property (e.g., concentration, absorption coefficient, brightness) of an object under investigation. A Computed Tomography (CT) technique [Hounsfield, 1973] is one of the useful technique to make an internal structure clear from the multiple "projection"s which are the set of line integrals at a particular angle of view. In order to realize time-resolved CT measurement, the number of projections per image must be limited so that we can acquire multiple projected images instantaneously. This kind of inverse problem of tomography for scalar properties has been discussed in a variety of ways [e.g. Nicolas, 1997], we adopted a powerful method based on an Algebraic Reconstruction Technique (ART) to solve ill-conditioned problems.

The objective of the present contribution is to establish the technique to measure the two-dimensional concentration field applying the light absorption and two-dimensional CT technique. Both the spectroscopic sensing and the data processing of ART reconstruction are described in section 2, which is followed by the experimental result in section 3.

2 Principle

2.1 Spectroscopic detection of molecules

The scope of this work focuses primarily on liquid phase flow, however it is worthwhile to briefly note some highlights of work in the gas phase [Mark, 1998]. Recently, by using a variety of lasers including a semiconductor diode laser in absorption-based measurements, we are able to identify the fundamental parameters of various molecules in UV, visible, and infrared region. Water vapors, as droplets or liquid exist in trace quantities in atmospheric air and are the major species of most industrial flows.

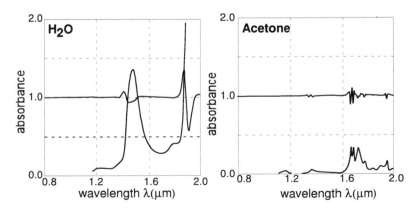

Figure 1: Comparison of spectral absorption coefficients of a liquid water (H_2O) and acetone (CH_3CHOCH_3) per cm; 0.8 - 2.0μm at T = 293[K], P = 0.1[MPa]

Figure 1 shows the absorption spectrums of H_2O and acetone between $0.8\mu m$ and $2.0\mu m$. These profiles show that the spectral absorption coefficient of H_2O is significantly greater than that of acetone at $1.4\mu m$, and indicates that the light source at this frequency can distinguish water from acetone. A molecule of water has three types of fundamental vibrations as shown in Figure 2. There are many absorption bands corresponding to the overtone and combination vibration in this region. In fact, the $\nu_1 + \nu_3, 2\nu_1$, and $2\nu_3$ absorption bands correspond to the $1.3\mu m$ - $1.4\mu m$ include well-resolved lines at the technically important wavelength.

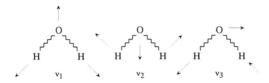

Figure 2: Fundamental vibration of H-O-H bond. $\nu_1 = 3657\text{cm}^{-1}$, $\nu_2 = 1595\text{cm}^{-1}$, $\nu_3 = 3756\text{cm}^{-1}$,

In order to detect molecules by absorption, a light source must have a very narrow emission spectrum, as well as a wavelength that is tunable onto the absorption line accurately. The wavelength of a tunable diode laser can be controlled in two ways.

One way is by adjusting the temperature of the diode element, so that the effective optical index of the waveguide is changed; thus adjusting the resonant condition of the laser cavity. Using this approach, each laser diode can be tuned from 3 to 5 nm between 275K and 325K. For Fabri-Perot lasers, this tuning range is not continuous and exhibits hysteresis and regions of multimode instability. In the continuous tuning range, the mode hops may be as small as 100 pico-meters. Because of the thermal mass of the diode and heater/cooler element, this type of tuning is usually restricted to a few Hz of maximum bandwidth.

The second mode of wavelength tuning involves the injection of current density into the gain section. At any given temperature, the injection current can be varied to modulate the optical index. Of course, the output power is also modulated, but the laser wavelength may be tuned quickly. In fact, the modulation bandwidths of the injection current on the order of 10GHz are achieved in telecommunications lasers.

Figure 3 shows a diagram of a laser diode power supply with Automatic Temperature Control(ATC) and Automatic Power Control (APC) which control the temperature of the laser cavity and the injection current to stabilize the output power and the wavelength simultaneously.

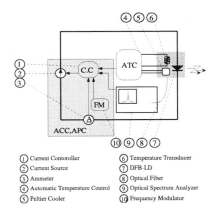

Figure 3: Schematic diagram of the principal components of the laser diode power supply.

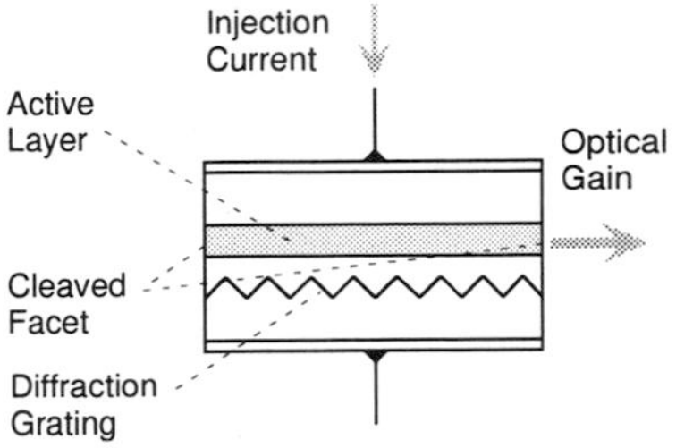

Figure 4: Structure of Distributed Feedback(DFB) laser diode.

In the present study, we adopted an InGaAsP/InP DFB [Nagai *et al.*, 1986; Ventrudo *et al.*, 1990; Morris *et al.*, 1995] single-mode tunable diode laser, that has important characteristics: 1) very narrow emission spectrum($\sim$ 30MHz typically) and 2) high Side Mode Suppression Ratio(SMSR) without mode hopping. Figure 4 shows the fundamental structure of a DFB laser diode. The detailed specifications are shown in Table 1. The special feature of a DFB laser diode is the diffraction grating under the cavity that enhances the single-mode stability and suppresses mode hopping. It has been developed at wavelength as short as 760nm and are commonly available as InGaAsP/InP devices out to 2μm. They can also be temperature tuned over 3-5nm about the normal operating wavelength although the injection current tuning rate is typically below $10^{-2}[\text{nm} \cdot \text{mA}^{-1}]$, and $10^{-1}[\text{nm} \cdot \text{K}^{-1}]$.

Table 1: Optical and electrical characteristics of Anritsu DFB laser diode module at $T_c = 298[K]$

Item	Symbol	Value(Typ.)	Unit
Optical Output Power	P_0	4	mW
Spectral Half Width	$\delta\lambda$	30	MHz
Side Mode Suppression Ratio	SMSR	35	dB
Parallel Beam Divergence	$\theta_{//}$	35	deg
Perpendicular Beam Divergence	$\theta_\perp$	40	deg.
Rise time	t_r	0.5	ns
Fall Time	t_f	0.7	ns

2.2 Computed Tomography(CT)

2.2.1 Lambert-Beer's law

Absorbance is defined as the ratio of intensity between initial and transmitting beams through an absorbing medium. As is well known, the Lambert-Beer's equation describes the relationship between the absorbance and the concentration. If there is a spatial distribution of concentration, the absorbance would correspond to the integral value of absorbance along the probe beam expressed as

$$A(\nu) = -\log \frac{I(\nu)}{I_0(\nu)} = c_m d\varepsilon(\nu) \tag{1}$$

where A (ν) is the absorbance at frequency ν, $\varepsilon(\nu)$ is the spectral molar absorption coefficient. I_0 is the incident intensity of the probe beam and $I(\nu)$ is the intensity transmitting through the length d of the absorbing medium.

2.2.2 CT reconstruction based on Algebraic Reconstruction Technique(ART)

The projection of two-dimensional tomography technique is based on Radon transforms as follows.

$$p(r,\theta) = \int \rho(r\cos\theta - s\sin\theta, r\sin\theta + s\cos\theta)\ ds \tag{2}$$

where p is a projection and ρ is a scalar value of a target field which corresponds to a concentration in this case.

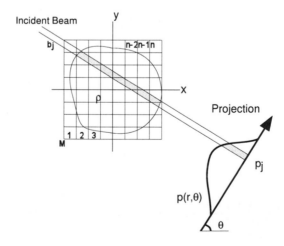

Figure 5: Geometry of tomographic projection system.

Figure 5 shows the simplified geometry of the projection system. M is a two-dimensional reconstruction area, which is divided into n small square elements $M_i, i = 1, ..., n$. Each of square element has a value of the unknown density function $\rho(\vec{r}), \vec{r} = r(x, y)$. Also, $p_j, j = 1, ..., m$, represent all the projection elements. Let $\vec{r}$ represent a point in M, such that

$$\int_{b_j} \rho(\vec{r}) \, d\vec{r} \approx p_j \qquad j = 1, ..., m \qquad (3)$$

and p_j represents the jth projection element. The region $M_i \cap b_j$ is defined as the intersection of beam passage b_i with the each element M_i. We can thus rewrite Equation 3 as,

$$p_j \approx \sum_{i=1}^{n} \int_{M_i \cap b_j} \rho(\vec{r}) \, d\vec{r} \qquad j = 1, ..., m \qquad (4)$$

If we assume that ρ_i is the average value of $\rho(\vec{r})$ over M_i

$$\rho_i \approx \frac{\int_{M_i} \rho(\vec{r}) \, d\vec{r}}{\int_{M_i} d\vec{r}} \qquad i = 1, ..., n \qquad (5)$$

Equation 4 can then be rewritten as,

$$p_j \approx \sum_{i=1}^{n} w_{ij} \, \rho_i \qquad j = 1, ..., m \qquad (6)$$

where w_{ij} is the shaded part of the small squares representing $M_i \cap b_j$

$$w_{ij} = \frac{\int_{M_j \cap b_i} d\vec{r}}{\int_{M_i} d\vec{r}} \qquad i = 1, ..., n \qquad (7)$$

In order to obtain a satisfactory result, we took some constrained conditions within the iterative calculation [e.g. Herman *et al.*, 1973]. We considered the concentration being reconstructed should be non-negative and is known to have a value less than F, which is decided for an experimental condition. In this case, F corresponds to an initial concentration.

$$\rho_i^{q+1} = \min\left[F, \, \max\left[0, \, \rho_i^q + \Delta w_{ij}(p_j - p_j^q)/N_j\right]\right] \qquad (8)$$

resulting in $0 \leq \rho^q \leq F(= C_{\text{initial}})$, where Δ is a damping factor, equal to 0.3 in our calculation. It is reported that the dumping factor improve the reconstruction when $\Delta \leq 0.5$ [e.g. Sweeney *et al.*, 1973]. There are additional features in this system.

- The equation are highly underdetermined, i.e., $m \ll n$.

- The rank of the matrix w_{ij} is unknown.

- The matrix w_{ij}, projection p_j and unknown function ρ_i is usually non-negative.

3 Experiment

In the experiments, H_2O was used as the absorbing species and acetone as the environmental medium. As mentioned above, one advantage of a spectroscopic sensing to utilize a DFB diode laser is its sensitivity to rare species.

3.1 Absorption Characteristics

Initially we confirmed the relationship between the absorbance $(= A(\nu))$ and the concentration of H_2O $(= C_{H_2O}[mol/l])$. There were rare H_2O molecules in the acetone environment; concentration of H_2O was up to 1.5[mol/l]. The intensity of the laser beam through a sufficiently mixed solution was detected by the InGaAs photodiode and the intensity of the incident light was measured in advance. From the result as shown in Figure 6, we confirmed that this combination of environmental and absorbing mediums satisfied the Lambert-Beer's relation.

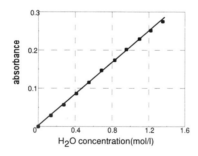

Figure 6: Relation between absorbance of water at $1.4\mu m$ and its concentration in acetone environment.

3.2 Concentration Distribution Measurements

In this section, we describe the experimental setup of CT measurement and discuss the result of experiments. At first, we tried to measure the internal concentration distribution in the steady round jet which contained H_2O molecules as the tracing species by using a pair of rotating optics consists of transmitting and receiving equipment. It is required to obtain one-dimensional projections to measure the two-dimensional sectional image. Figure 7 shows the experimental apparatus to obtain the projections. It consists of a DFB-LD light source, power supply for the laser diode, an InGaAs infrared line camera located on the opposite side of the light source, signal processor and host computer.

For a receiver an infrared camera with a Peltier cooler was adopted to suppress the dark current and enhance the signal to noise ratio (SNR). This camera has high sensitivity in the near infrared region. Figure 8 shows the spectral property and Table 2 details other specification of it. The projected images onto the sensor were firstly converted to analog video signal. These were logarithmically amplified by a signal processor in order to convert the intensity information to absorbancies and to extend the dynamic range, and then transferred into a personal computer through a 60 MSPS, 12bit high speed Analog to Digital converter. The original images were reconstructed from the projected images on ART algorithm by central processing units on the computer.

Figure 9 illustrates the example of iterative results under reconstruction based on Equation 8. A symbol 'q' is the number of iteration. Each of results ρ^q in progress indicates that the convergent result can be obtained within 30 times of iteration typically. It was considered that the constraint conditions in Equation 8 surpressed divergence successfully. Equally, it was important that we took no axisymmetric assumption; the measured distributions were reconstructed from only 3 projections on Equation 8.

Figure 10 illustrates a simplified setup for CT measurements of the round jet. We performed the experiment by changing the initial concentration of H_2O ($= C_{H_2O}$), which was considered to be uniform in nozzle, from 0.01 to 0.09[mol/l]. The other conditions were fixed; nozzle diameter $D = 1.0$[mm], average velocity $U = 3.2 \times 10^{-3}$[m/s].

Figure 11 shows the experimental result of the CT measurement. The profiles show the concentration in the central section of the round jet at x/D=1, where D is defined to be the diameter of the nozzle. Measured profiles of concentration corresponded clearly to the initial condition. The result indicated that the shape of distributions were almost uniform, for the diffusion of H_2O was quite slow in an acetone environment in contrast to the principal velocity of the jet (~ 10[mm/s]). The average error of the measured value was 8% in typical, did not exceed 15%.

4   Conclusion

In the present work, we successfully measured the two-dimensional spatial distribution of the concentration inside a round jet with a pair of near infrared Distributed Feedback (DFB) tunable diode laser and an infrared line camera.

(1) The spectroscopic detection technique in the near infrared region has high sensitivity, as the concentration is estimated quantitatively from measured absorbance. The present measuring system can be utilized for various species

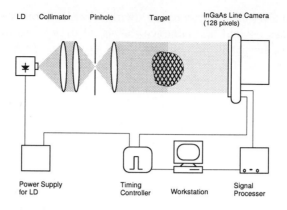

Figure 7: Experimental CT apparatus

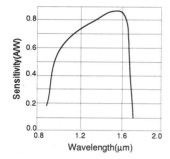

Figure 8: Spectral property of a line camera at T=263[K].

Table 2: Optical and electrical properties of line camera.

Item	Sym.	Val. W/m^2
Pixel Width	W	50 μm
Pixel Height	H	200 μm
Channels	-	128
Pixel Rate	f	$\leq$1 MHz
Sensitive Wavelength	λ	0.9 - 1.7 μm
Dark Current	I_D	3 pA/pix.
Cooling Temperature	T_S	263 K

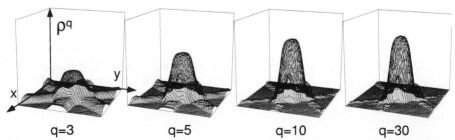

q=3   q=5   q=10   q=30

Figure 9: Iterative results under reconstruction. 'q' is the number of iteration.

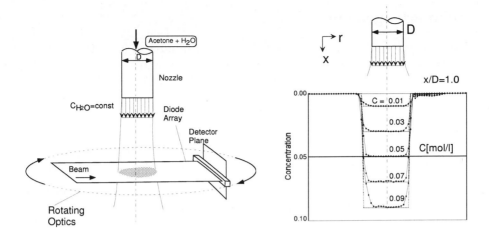

Figure 10: Experimental setup for CT measurements of the round jet including H_2O molecules.

Figure 11: Profile of H_2O concentration at the central section of the jet.

of interest by choosing the absorption bands corresponding to overtone and combinational vibration.

(2) The sectional images of the concentration field are reconstructed from three projected images (projections) in combination with a Computed Tomography technique based Algebraic Reconstruction Technique. We obtained reasonable results from three projections through an iterative calculation of reconstruction with some constrained conditions in spite of a quite ill-conditioned inverse problem.

(3) The proposal CT method can be easily extended to three dimensions. As a multiple projected matrix, i.e. two-dimensional projections were obtained, we could reconstruct the volumetric distribution of the concentration field with a three dimensional CT technique.

(4) In this experiment, we applied the techniques to the steady axisymmetric jet and resulted that the invisible interface of species could be distinguished clearly. There is every possibility of measuring such the mass transfer phenomena as turbulent mixture and diffusion of species when the multiple projected images are obtained simultaneously by using pairs of fixed projecting optics instead of a rotating optics.

Acknowledgement

The authors would like to acknowledge the useful advices from Professor K. Uehara at Keio University, department of physics, and also to thank AN-RITSU Co. Ltd. for the supply of Distributed Feedback laser diodes.

References

BRUCE,D.M. AND CASSIDY,D.T. 1990 Detection of oxygen using short external cavity GaAs semiconductor diode laser. Appl. Opt., 29, 1327-1332.

CARLISLE,C.B. AND COOPER,D.E. 1989 Tunable-diode-laser frequency-modulation spectroscopy using balanced homodyne detection. Opt. Lett., 14, 1306-1308.

COOPER,D.E. AND WATJEN,J.P. 1986 Two-tone optical heterodyne spectroscopy with a tunable lead-salt diode laser. Opt. Lett., 11, 606-608.

FEHER,M., MARTIN,P.A., ROHRBACHER,A., SOLVA,A.M. AND MAIER,J.P. 1993 Inexpensive near-infrared diode-laser-based detection system for ammonia. Appl. Opt., 32, 2028-2030

GROSSE,W.G. AND UHLENBUSCH,J. 1977 Measurement of local mass-transfer coefficients by holographic interferometry. Int. J. Heat Mass Transfer. Vol.21, 677-682.

GROSSKLOSS,R., KERSTEN,P. AND DEMTRODER,W. 1994 Sensitive amplitude- and phase-modulated absorption spectroscopy with a continuously tunable diode laser. Appl. Phys., B58, 137-142.

HERMAN,G.T., LENT,A. AND ROWLAND,S. 1973 ART:Mathmatics and applications. A report on the mathematical foundations and on the applicability to real data of the Argebraic Reconstruction Technique. J. Theor. Biol., 42(1), 1-32.

HOUNSFIELD,G.N. 1973 Computerized transverse axial scanning (tomography). Brit. J. Radiol., 46, 552, 1016-1022.

ITO,A., AND KASHIWAGI,T. 1987 Measurement technique for determining the temperature distribution in a transparent solid using holographic interferometry. Appl. Opt., Vol.26, No.5.

ITO,A., MASUDA,D. AND SAITO,K. 1991 A study of flame spread over alcohols using holographic interferometry. Combustion and Flame, 83, 375-389.

LUCCHESINI,A., LONGO,I., GABBANINI,C., GOZZINI,S. AND MOI,L. 1993 Diode laser spectroscopy of methane overtone transitions. Appl. Opt., 32m 5211-5216

MANDIN,J.Y., CHEVILLARD,J.P., CAMY-PEYRET,C. AND FLAUD,J.M. 1986 Line Intensities in the $\nu_1+2\nu_2$, $2\nu_2+\nu_3$, $2\nu_1$, $\nu_1+\nu_3$, and $2\nu_1+\nu_2+\nu_3-\nu_2$ Bands of $H_2^{16}O$, between 6300 and 7900 cm^{-1}. J.Mol.Spectrosc., 118, 96-102.

MARK,G.A. 1998 Diode laser absorption sensors for gas-dynamic and combustion flows. Meas. Sci. Technol. Vol.9 No.4, 545-562.

MARKEL,G.J., DRACOS,T., RYS,F.S. AND RYS,P. 1996 Some statistical result from concentration and velocity field measurements using Laser Induced Fluorescence tomography. 8th Int. Symp. on Applications of Laser Techniques to Fluid Mechanics, Lisbon, Portugal, 4.6

MASTORAKOS,E., SHIBASAKI,M. AND HISHIDA,K. 1996 Mixing enhancement in axisymmetric turbulent isothermal and buoyant jets. Exp. in Fld., 20, 279-290.

MORRIS,N.A., CONNOLLY,J.C., MARTINELLI,R.U., ABELES,J.H. AND COOK,A.L. 1995 Single-mode distributed-Feedback 761-nm GaAs-AlGaAs quantum-well laser absorption. IEEE Photon., Technol. Lett.7, 455-457.

NAGAI,H., MATSUOKA,T., NOGUCHI,Y., SUZUKI,Y. AND YOSHIKUNI,Y. 1986 InGaAsP/InP distributed feedback buried heterostructure lasers with both facets cleaved structure. IEEE J.Quantum Electron., QE-22 No.3, 450-457.

NICOLAS,R., GEORG,P. AND DIETER,M. 1997 Reconstruction of limited view problems using adapted coordinates. ASME., FEDSM 97-3101.

PHILLIPE,L.C. AND HANSON,R.K. 1993 Laser diode wavelength-modulation spectroscopy for simultaneous measurement of temperature, pressure, and velocity in shock-heated oxygen flows. Appl. Opt., 32, 6090-6103.

ROYER,H. 1997 Holography and particle image velocimetry. Meas. Sci. Technol. Vol.8 No.12.

SAKAKIBARA,J., HISHIDA,K. AND MAEDA,M. 1993 Measurement of thermally strayfied pipe flow using image-processing techniques. Exp. in Fluid, 16, 82-96.

SILVER,J.A. AND HOVDE,D.C. 1994 Near-Infrared diode laser airborne hygrometer. Rev. Sci. Instrum., 65, 1691-1694.

SWEENEY,D.W. AND VEST,C.M. 1973 Reconstruction of three-dimensional refractive index fields from multidirectional interferometric data. Applied Optics, Vol.12 No.11, 2649-2664.

UEHARA,K., TAI,H. AND KIMURA,K. 1997 Real-time monitoring of environmental methane and other gases with semiconductor lasers : a review. Sensors and Actuators B38-39, 136-140.

CHAPTER II.   TURBULENT SHEAR FLOWS

II. 1 The Kármán Vortex Street – LDV and PIV Measurements Compared with CFD

O. Pust and C. Lund

Department of Fluid Mechanics, Faculty of Mechanical Engineering
University of the Federal Armed Forces Hamburg, D–22039 Hamburg, Germany

1 Introduction

The laminar flow of an incompressible newtonian fluid around a circular cylinder is well known and described in literature as *Kármán vortex street*. For its extensive investigations in the past this flow is frequently used as a test case for numerical simulation (CFD). In spite of that, direct comparisons of the time dependent development of the velocity components between experiment and simulation still seem to be missing. Until now to our knowledge just comparisons for the mean values have been reported. In this work results for the time dependent flow obtained with Laser Doppler Velocimetry (LDV) and Particle Image Velocimetry (PIV) are reported and compared to results of a numerical simulation. The qualitative and quantitative correspondences and differences are presented and discussed.

2 Experimental set-up

Figure 1 shows the side view and the partial top view of the set-up. The dotted area comprising the cylinder and its wake—where vortex shedding is expected—represents the region of interest both for experiment and numerical simulation. Details are shown in figure 2 with (ξ, ψ) defining a dimensionless coordinate system in which the cylinder diameter is used as a reference length. The region of interest is 352 mm in length, 65.6 mm in height and 328 mm in depth. The cylinder has a diameter of 16 mm and extents over the whole channel depth with 328 mm in length. The measurements are conducted at half channel depth where the flow is taken to be two-dimensional[1]. At the beginning of the channel a honeycomb flow guide is installed in order to reduce inflow effects that are caused by the 90° change in direction of the flow from the reservoir to the channel.

The fluid (water) is pumped into the reservoir by a centrifugal pump. The flow rate is adjusted to its desired value by an electronically controllable valve and measured simultaneously. Behind the valve the water is stored in an open tank from which it is pumped back into the reservoir. So we have a circular flow that allows us to seed the fluid only once before the measurements start. At all times we pump more water into the reservoir than is needed for the desired maximum flow rate. The unused water flows back into the tank via a wasteway. So the water level in the

---

[1] The geometrical ratios of this two-dimensional problem are taken from [3].

130

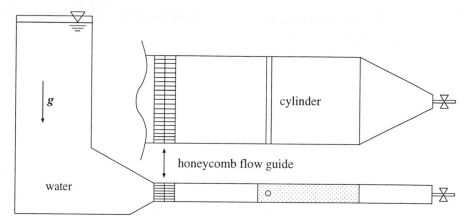

Fig. 1. Sketch of the set-up (side view and partial top view)

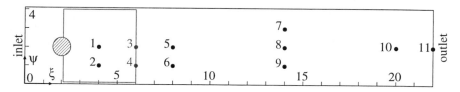

Fig. 2. Geometry of the channel and location of the measuring points

reservoir is kept constant, thus ensuring an also constant pressure gradient along the channel.

For the LDV measurements a two component backward scattering system (DAN-TEC) with an Argon-Ion-Laser is used. The measuring volume (0.2 mm × 0.2 mm × 2 mm) is traversed within a plane in the middle of the channel that is orientated normally to the cylinder axis. The location of the measuring points can be seen in figure 2. The tracers used are titanium-oxid particles with a mean diameter of 5 μm.

For the PIV measurements a system (Optical Flow Systems Ltd.) is used that mainly consists of a double oscillator Nd:YAG laser (30 mJ per cavity), a cross-correlation CCD-camera with a 1 K by 1 K sensor and a PC equipped with 512 MB RAM in order to record sufficiently long time series of the flow. The area recorded by the CCD-camera is described by a square with its lower left corner at ($\xi = 2.1, \psi = 0.1$) and its upper right corner at ($\xi = 6.0, \psi = 4.0$). For better light scattering efficiency the tracers used are silver-coated hollow glass-spheres with a mean diameter of 10 μm.

We define a Reynolds number that characterizes the stationary flow in the middle plane as

$$\text{Re} = \frac{u_m D \rho}{\mu} \tag{1}$$

with the average horizontal velocity u_m, the newtonian dynamic viscosity μ, the fluid density ρ and the cylinder diameter D. The value of u_m is determined by integrating the inflow velocity profile ($\xi = -2$) (see figure 3(a)) over the channel height thus giving a Reynolds number of 100.

Further the Strouhal number Sr and the dimensionless time $\tilde{t}$

$$\text{Sr} = \frac{Df}{u_m} \quad , \quad \tilde{t} = \frac{t\mu}{\rho D^2} \tag{2}$$

with the separation frequency f of the vortices are used. Thus the dimensionless time is calculated to $\tilde{t} = t/256$ s. All velocities are made dimensionless with u_m.

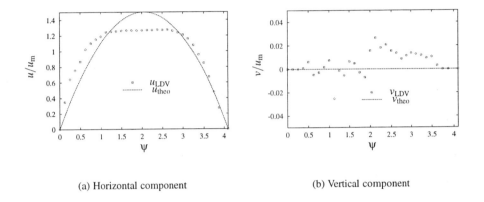

(a) Horizontal component　　　　　　(b) Vertical component

Fig. 3. Inflow velocity profile at $\xi = -2$

In the stationary case the inflow should ideally be fully developed, time independent and of parabolic shape across the channel height. As one can see in figure 3 the inflow is not of parabolic shape and shows a small vertical component ($\approx 2\%$ of u_m). The reason for the non-parabolic shape obviously is that the length of the inlet is too short to allow the flow to fully develop. The vertical component that results from the 90° change in direction of the flow from the reservoir to the channel is limited to an acceptable value by the already mentioned flow guide at the beginning of the inlet. Nevertheless this inflow profile is used for the numerical simulation so that direct comparison is possible.

3　Numerical Simulation

The primary aim of the simulation is to compute the velocity in the mid-plane area shown in figure 2, where the flow is assumed to be two-dimensional. Using a primitive-variable formulation with cartesian velocity $\mathbf{v}(u,v)$ and pressure p, the simulation is based on the equation of motion for an incompressible isothermal fluid

$$\rho \left(\frac{\partial \mathbf{v}}{\partial t} + \mathbf{L} \cdot \mathbf{v} \right) = -\text{grad } p + \text{div } \mathbf{T} + \mathbf{f} \tag{3}$$

and the continuity equation

$$\text{div } \mathbf{v} = 0. \tag{4}$$

Here t represents the time, ρ the fluid density, $\mathbf{L}$ the velocity gradient ($\mathbf{L} = \text{grad } \mathbf{v}$) and $\mathbf{f}$ the body force per unit volume (which in this case is set to zero). For a newtonian fluid the extra-stress tensor $\mathbf{T}$ is connected with the strain rate tensor $\mathbf{D} = \frac{1}{2}(\mathbf{L} + \mathbf{L}^{\mathbf{T}})$ by $\mathbf{T} = 2\mu\mathbf{D}$, where μ is the fluid viscosity.

The simulation uses a standard Galerkin finite element method based on the isoparametric triangular Taylor-Hood element with continuous piecewise quadratic shape functions for the velocity and continuous piecewise linear ones for the pressure. The unstructured mesh used to discretize both the entrance and the mid-plane area of figure 2 is shown in figure 4. It describes the area $-2 \leq \xi \leq 22$ and $0 \leq \Psi \leq 4.1$. A close-up of the cylinder section can be seen in figure 10.

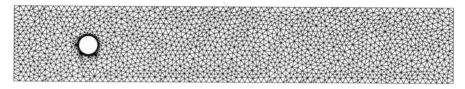

Fig. 4. Finite element mesh with ca. 17000 unknowns (u,v,p)

The above equations are supplemented with initial and boundary conditions all selected in such a way, that the resulting simulation will follow the experimental set-up as close as possible. Therefore the initial condition is a fluid at rest with zero velocities and zero pressure gradient. A no-slip boundary condition is used for all walls. It is assumed the time-varying inlet velocity can be approximated by

$$\mathbf{v}_{in}(y,t) = \mathbf{v}_s(y) \, f_p(t) \tag{5}$$

where index s refers to the stationary velocity of figure 3 as measured with the LDV equipment. The process function $f_p(t)$ describes the change in inlet velocity from 0 to $\mathbf{v}_s$ and is in this case assumed to be linear. It thus roughly approximates the quasi-linear curve in figure 9 as seen with LDV at $\xi = -2, \Psi = 2$. Different functions $f_p(t)$ will result in a different fluid structure only during the start-up, but not after the vortex shedding has started and the inlet velocity remains unchanged. The outflow condition is somewhat more complicated since no velocity data is available and the flow itself is, due to the vortex shedding, suspected to be very complex. A useful condition is to enforce a vanishing vertical velocity in combination with a vanishing horizontal tension vector, viz. $\mathbf{v} \times \mathbf{n} = 0$ and $\mathbf{t} \cdot \mathbf{n} = 0$ where $\mathbf{n}$ is the surface normal vector and $\mathbf{t} = -p\,\mathbf{n} + \mathbf{T} \cdot \mathbf{n}$ the tension vector. As shown in figure 6 this condition has no visible influence in upstream direction.

The spatial discretization leads to a semi-discrete coupled system of nonlinear differential ordinary equations. Applying an implicit time discretization (e.g.

Crank-Nicholson, second order in time) this results in an algebraic nonlinear indefinite system

$$\begin{pmatrix} N(v) & C \\ C^T & 0 \end{pmatrix} \begin{pmatrix} v \\ p \end{pmatrix} = \begin{pmatrix} n \\ c \end{pmatrix} \quad (6)$$

which then has to be solved at all timelevels. Formally solving the first equation for v and inserting the result in the second equation gives two definite nonlinear problems,

$$S p = C^T N^{-1} C p = C^T N^{-1} n - c \quad (7)$$

$$N v = n - C p \;, \quad (8)$$

with the matrix S however given implicitly only. The solution algorithm solves the first equation using a variant of the well known Richardson-iteration, involving a suitable preconditioner for S. After the solution for p is known, the second equation can be solved for v. In practice, both equations are solved in a successive manner within the iteration process. A detailed description of the algorithm is given in [2].

The timestep size Δt is automatically selected using a local error estimation based on two similar integrations of different timesteps (one step with Δt, two steps with $\Delta t/2$). At the begin a strong increase in the timestep length is seen in figure 5. It is interesting to note that later, with the vortex street fully developed, still a periodic change of then small steps is used.

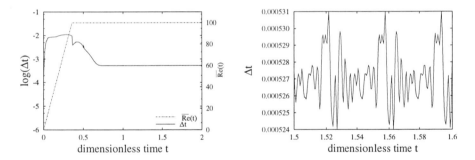

Fig. 5. Adaptive selection of the simulation timestep

The amount of computational time required for $0 \le t \le 2$ is ca. 28 h on a Sun Ultra 1 workstation using a workspace of ca. 12 MB. As an example for the flow computed by the simulation, figure 6 shows several streaklines during vortex shedding. The greyscales refer to different positions where tracer particles of zero mass have been added into the flow. Figure 7 gives a closer view of the quasi-stationary flow around the cylinder.

Besides the calculation of the velocity and pressure field, the simulation is also able to predict not easily measurable data, like drag and lift forces on the (two-dimensional) cylinder as a function of time. Figure 8 shows the corresponding dimensionless coefficients, which are defined as usual, viz. $C_D = F_D/(\rho/2u_m^2 D)$.

Fig. 6. Streaklines during vortex shedding

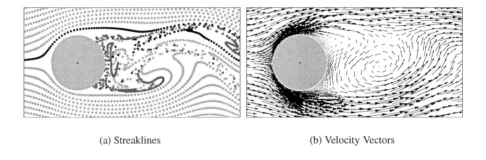

(a) Streaklines   (b) Velocity Vectors

Fig. 7. Quasi-stationary streaklines and velocity vectors

Using such integral data the program code had previously been verified by computing the so called benchmark-flow described in [3] within the geometry of figure 2. As shown in [2], a very good agreement has been found. In fact, the investigation described here was inspired by that problem, since integral data alone, even if sensitive like the lift coefficient, is not sufficient for a complete validation of a simulation. In view of this there is a need for time-varying field information $(\mathbf{v}, p)$ for other than pure academic problems. We hope the data presented here will help to somewhat narrow this gap.

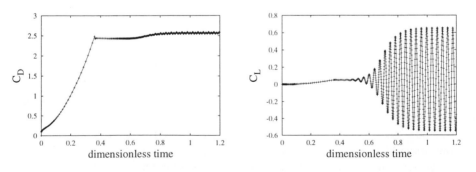

Fig. 8. Drag and lift coefficient C_D and C_L

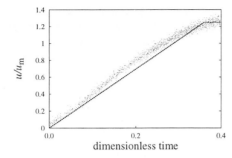

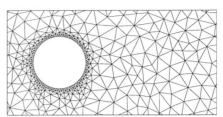

Fig. 9. Start-up of the inflow at $\xi = -2, \psi = 2$ **Fig. 10.** Finite element mesh near cylinder

4 Results

4.1 Instationary Flow

We start the flow by opening the valve shown in figure 1. The mean velocity of the flow entering the investigation area increases nearly linearly from zero to its maximum value during a time intervall of $\Delta \tilde{t} = 0.35$. It is measured at ($\xi = -2, \psi = 2$) with LDV (see figure 9, points: LDV-data, line: approximation for simulation). Afterwards the mass flow is kept constant and then we measure the inflow velocity profile at ($\xi = -2, \psi$) with LDV (see figure 3). As described previously, both data sets are used for the simulation.

In figure 11 we show the velocity at point 4 as measured by LDV. The velocity increases during start-up and shows oscillations immediateley after the mass flow is constant. Comparing this curve to the corresponding simulation data, we find very good agreement for both the amplitude and the frequency. The earlier start of the oscillations in the experiment may be caused by small disturbances in the set-up.

Results of PIV-measurements during start-up are shown in figure 12. As examples for instationary and non-periodic flow fields streamlines and vorticity (left) and velocity vectors (right) are shown for two different moments when vortex shedding has not yet started. One can observe an increasing area of slack water where the velocities are close to zero. The vorticity intrudes into the flow from the cylinder and the channel walls and increases in magnitude as it is indicated by the darker gray of the contour plot.

4.2 Quasistationary Flow

When constant mass flow is reached, we measure the velocity with LDV at points 1–11 and take images for PIV within the rectangle shown in figure 2. With LDV as a point measurement technique we are not able to capture the velocities at different points at the same instant as it can be done with the simulation or with PIV. Therefore we choose an arbitrary time interval of $\Delta \tilde{t} = 0.2$ and compare LDV and PIV data to the simulation data, keeping in mind that the phase of the velocity oscillations will be different. Further on, for LDV the shape (the temporal resolution) of the curves strongly depends on the number of particles passing the measuring volume, whereas for PIV and the simulation the shape of the curves strongly depends on the image capturing freqency respectively the simulation timestep. In these experiments PIV image pairs are taken with a frequency of 1.3 Hz, which is sufficient to capture the main flow structures caused by the shedding of the vortices (according to the NYQUIST criterion). The frequency can be increased up to 15 Hz so that even faster oscillations can be observed with PIV.

In the following results of PIV, LDV and simulation for two measuring points (3 and 4) are reviewed. In general, experiments and numerical simulation agree well with respect to both qualitative and quantitative behaviour. The shape of the oscillations are given likewise with all methods (see figures 13 – 18). Even striking fine structures of the vertical velocity are found in both experiments and simulation (see figures 13(b), 14(b) and 15(b)). The marks in the figures represent measured particles (LDV), results of cross-correlating successive image pairs (PIV) and timesteps (simulation). The different amplitudes of the horizontal velocity may be caused by an uncertainty with which the actual location of the measuring points can be accessed in the experiment. The Strouhal number derived from the measurements is between 0.271 (PIV) and 0.281 (LDV) and computed to be 0.283 in the simulation.

Further on, PIV is used to gain experimental data of the time dependent velocities within a plane orientated normally to the cylinder axis. It is not possible to get similar velocity maps of flows like the instationary vortex street with LDV because as a point measurement technique LDV can only be used to record velocities at a single point over time. Figures 19 and 20 show the contour plot of the absolute velocity and streamlines of the flow. To reduce the amount of presented data only every second PIV image pair ($\Delta \tilde{t} = 6 \cdot 10^{-3}$) and the corresponding result of the simulation are shown. The unusual means of streamlines for instationary flows is chosen in order to visualize the positions of the vortices. The 20 grayscales for the dimensionless absolute velocity represent a value of 0 with white and a value of 2.0 with dark gray. One can find very good agreement in the position of the vortices and the form of the streamlines.

Figures 21 and 22 show a detailed view around measuring point 1 for the same points in time as in figures 19 and 20. One can observe a vortex core moving from the upper left corner down to the lower right corner. Again very good agreement between experiment and simulation is shown.

5 Conclusions

The PIV and LDV measurements make experimental information on the time dependent development of the velocities in the *Kármán vortex street* available for a whole cross-section and at discrete points for comparison with numerical simulations. In spite of the actual three dimensionality and the wall influences[2] excellent agreement between experiments and two dimensional simulation can be found.

The results of the PIV measurements prove that it is possible to extend the application of PIV away from statistical evaluation of turbulent flows towards to the recording of instationary flows that cannot be triggered to external events—like it is possible in turbo machinery—and therefore cannot be recorded by LDV.

References

[1] EISENLOHR, HOLGER und HELMUT ECKELMANN. *Vortex splitting and its consequences in the vortex street wake of cylinders at low Reynolds number.* Physics of Fluids A, **1** (2): (1990) 189–192.

[2] LUND, CHRISTOPH. *Ein Verfahren zur numerischen Simulation instationärer Strömungen mit nichtlinear-viskosen Fließeigenschaften.* VDI-Verlag GmbH, Düsseldorf, 1998.

[3] TUREK, S. und M. SCHÄFER. *Benchmark computations of laminar flow around cylinder.* In E. H. Hirschel (Hg.), *Flow Simulation with High-Performance Computers II*, Bd. 52 von *Notes on Numerical Fluid Mechanics.* Vieweg, 1996: S. 547–566. (support of F. Durst, E. Krause, R. Rannacher).

[4] WILLIAMSON, C. H. K. *Oblique and parallel modes of vortex shedding in the wake of a circular cylinder at low Reynolds numbers.* Journal of Fluid Mechanics, **206**: (1989) 579–627.

---

[2] See [4] and [1] for the sensitiveness of the *Kármán vortex street* against three dimensional effects.

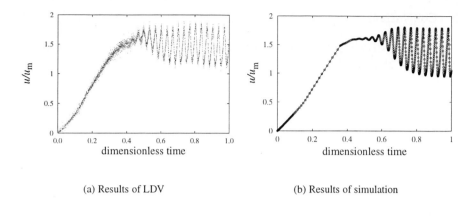

(a) Results of LDV

(b) Results of simulation

Fig. 11. Horizontal velocity at point 4 during start-up

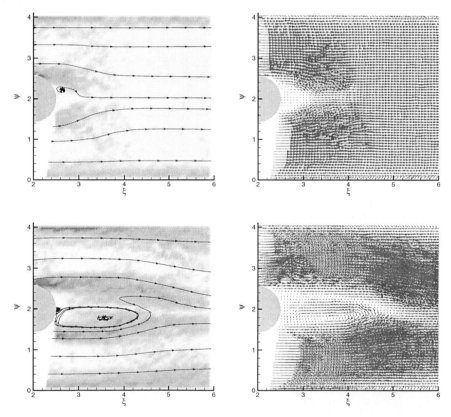

Fig. 12. PIV results during start-up (top: $\tilde{t} = 0.144$, right: $\tilde{t} = 0.324$); streamlines and vorticity (left), velocity vectors (right)

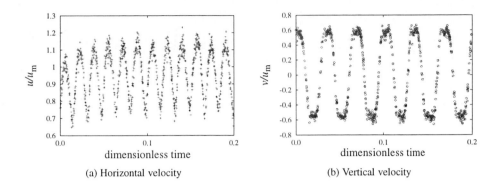

(a) Horizontal velocity

(b) Vertical velocity

Fig. 13. Results of LDV at point 3

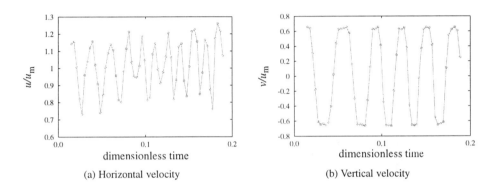

(a) Horizontal velocity

(b) Vertical velocity

Fig. 14. Results of PIV at point 3

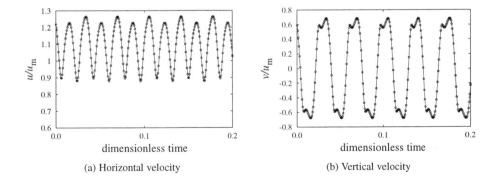

(a) Horizontal velocity

(b) Vertical velocity

Fig. 15. Results of simulation at point 3

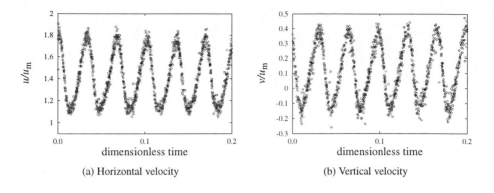

(a) Horizontal velocity

(b) Vertical velocity

Fig. 16. Results of LDV at point 4

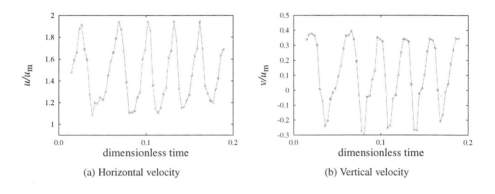

(a) Horizontal velocity

(b) Vertical velocity

Fig. 17. Results of PIV at point 4

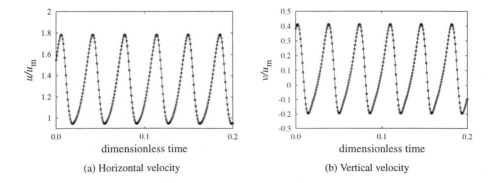

(a) Horizontal velocity

(b) Vertical velocity

Fig. 18. Results of simulation at point 4

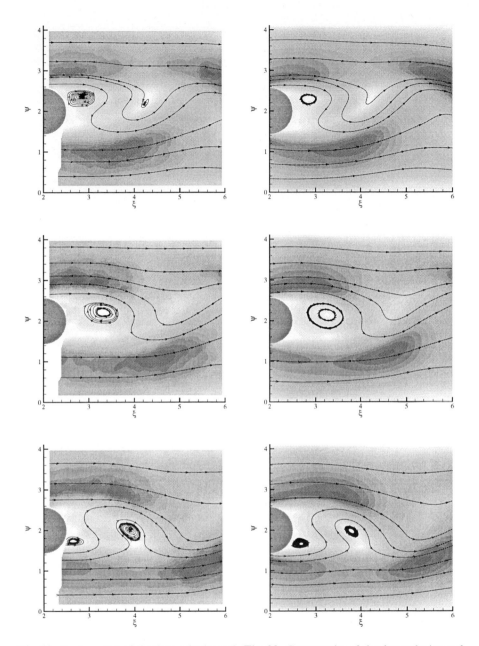

Fig. 19. Contour plot of absolute velocity and streamlines—PIV

Fig. 20. Contour plot of absolute velocity and streamlines—simulation

142

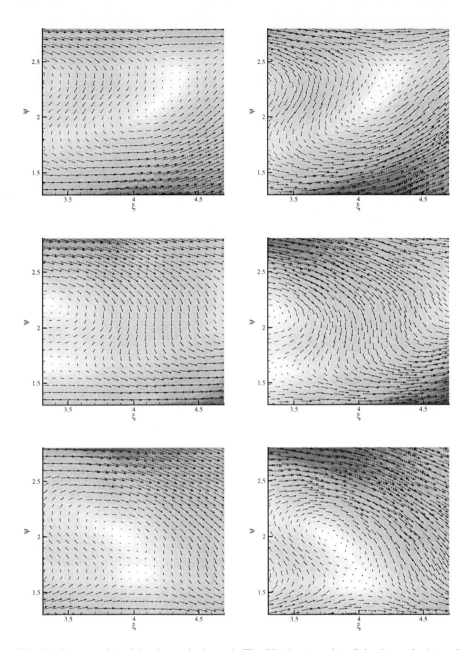

Fig. 21. Contour plot of absolute velocity and velocity vectors—PIV

Fig. 22. Contour plot of absolute velocity and velocity vectors—simulation

II.2. A Flow Structure at Reattachment Region of a Two-Dimensional Backward-Facing Step by Using Advanced Multi-Point LDV

N. Furuichi[1], T. Hachiga[2], K. Hishida[3], and M. Kumada[1]

[*1] Department of Mechanical System Engineering, Gifu University, Yanagido 1-1, Gifu 501-1193, Japan

[*2] Department of Automotive Engineering Takayama College, Shimobayashi 1155, Takayama, Gifu 506-8577, Japan

[*3] Department of System Design Engineering , Keio University, Hiyoshi 3-14-1, Kohoku-ku, Yokohama 223-8522, Japan

Abstract. A temporal data of instantaneous velocity profile of a two-dimensional backward-facing step flow was measured by multi-point LDV, and was discussed with a special focus on the details of the motion of the separated shear layer and the phenomena of reattachment. As a result, The phenomenon of reattachment was seen to follow two patterns, one when the separated shear layer reattaches, the other when the separated shear vortex reattaches. The dividing stream line oscillates because of vortex concentration in the separated shear layer and has a quasi-periodical behavior as concerned to reattachment.

Keywords. Multi-point LDV, Two-dimensional backward-facing step, Separated shear layer, Reattachment

1. Introduction

Experiments have been performed to study the flow structure around the reattachment region behind a two-dimensional backward-facing step with a turbulent shear layer (Eaton & Johnston, 1981 and Armly et al., 1983). This flow field exhibits a three-dimensional, unsteady structure involving oscillation of the reattachment point and the large scale vortex (Eaton et al. 1980). Very few other studies on this configuration have been performed with the aim of investigating the unsteady structure of the flow field. Hijikata et al. (1991) visualized the pressure field in a backward-facing step using a holographic method and velocity-pressure cross-correlation. They clarified that the motion of the high cross-correlation region corresponds to the motion of the pressure fluctuation lumps obtained in the holographic visualization. By direct numerical simulation, Le et al. (1997) have shown that the fluctuation in the reattachment location was caused by a large-scale roll-up of the shear layer extending to the reattachment region, which is composed of many small counter-rotating vortices. In terms of spatial measurement, Kasagi & Matsunaga (1995) measured velocity vector using a 3-D PTV, then calculated detailed turbulence statistics and energy budget.

As mentioned above, many studies have shown that the three-dimensional flow structure caused by the transformation of the shear layer vortex governs the fluid dynamics near the reattachment region. However, none of these studies provided quantitative experimental data on the three-dimensional flow structure. This lack of quantitative data was a result of the absence of a method of measurement for spatial and temporal velocity distributions over an unsteady, three-dimensional flow field. Measuring simultaneous and spatial velocity profiles a cross the dividing stream line in the separated shear layer is especially important for clarifying the process of vortex concentration, shedding of large-scale vortices and vortex deformation near the reattachment point. Up to now, there exist several methods for measuring multi-point velocity, for example, ultrasonic velocity profile method (Takeda, 1991), the particle image velocimeter (Sakakibara et al., 1993), and others. But these methods have limitation in terms of sampling interval and spatial measurement.

We developed the advanced multi-point LDV using a 1-bit FFT approach that had been described in a previous paper (Hachiga et al. 1998), and were able to improve the measurement of reverse flow in the present study. We accumulated the instantaneous velocity profile by setting up this LDV system spatially in the flow field and considering the flow structure of the 2-D backward-facing step. We paid particular attention to clarifying the fluid flow motion of the separated shear layer and the phenomenon of reattachment.

2. Experimental Apparatus

2.1 Backward-Facing Step

The flow field and coordinate system are shown in Fig.1. The closed-loop water channel used in this experiment has a working section of 240×60mm in cross-sectional area and 2300mm in length. The backward-facing step (with a step height h of 20mm) was formed below the floor plate at a point 250mm downstream from the construction nozzle exit. The expansion ratio is ER=1.5, and the aspect ratio is AR=12. The main flow velocity is fixed at Uc=0.25m/s in all experiments (Reynolds number of Re_h=5000). The turbulence intensity of the main flow is Tu=0.6%. The distribution of u-component mean velocity upstream of the step is in agreement with Blaussius theory. The boundary thickness upstream of the step is about 4.6mm.

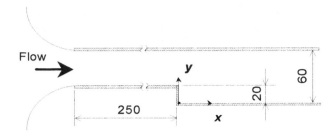

Figure 1. Experimental apparatus of two-dimensional backward-facing step

2.2 Multi-Point LDV System

A velocity profile was measured by an advanced multi-point LDV system which was reported in a previous paper (Hachiga et al., 1998). For measuring the backward-facing step flow, this LDV has been improved on two points mentioned below. One is that this LDV can measure reverse flow by double Brugg cell system and another one is that can be measure simultaneously 24-point data.

The optical system of this LDV is shown in Fig.2. This LDV is made very compact by using semiconductor laser with a maximum power of 40mW and a wavelength of 685nm, and optical fiber unit with 96 plastic fibers with a diameter of ϕ 0.25mm. Two Brugg cells are used to introduced frequency shifts of 80MHz and 79.9MHz respectively. Therefore Doppler burst signal is shifted 0.1MHz. Two Brugg cells role like a beam splitter by using two shifted beam. After two shifted beam are detached by two mirrors, the beam are deformed to a fan-like shape by rod lens. Non-shifted beam is cut before the rod lens. The process after two shifted laser beam of fan-like shape are made into rectangular-like light sheet by cylindrical lens is similar to the process in the previous paper.

The hardware system block diagram of an advanced multi-point LDV used in the present study is shown schematically in Fig.4. The hardware system was modified that 24-point scattered light signal could be measure simultaneous as mentioned above. A scattered light signal is converted into an individual Doppler burst signal by Si-APD. This signal is digitized to one bit using a comparator. The sampling frequency of a raw one-bit data is 250KHz, and the sampling interval of the velocity data is 1.024ms with frequency analysis by FFT of 256 points. Therefore, data sampling rate is about 976/s. The data processing is similar to that performed in the previous paper (Hachiga et al., 1998). In present study, a lack of data is interpolated by adjacent measuring point in spatial and temporal. The rate of capture for temporal velocity data in all experimental data is over 70% and if there was a lack of data which is 3mm continuous in spatial or 12msec continuous in temporal, that data was not used because it could not detect micro scale eddy. The spatial resolution of each point set up in the x-z plate is 0.22×1.40mm which

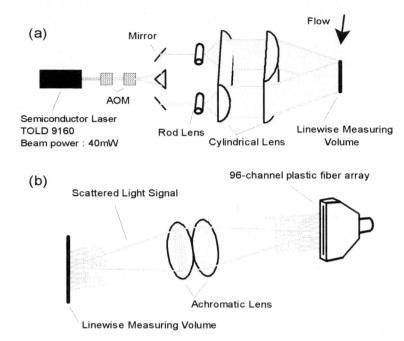

Figure 2. Arrangement of (a) transmitting optics (b) receiving optics of a Multi-point LDV

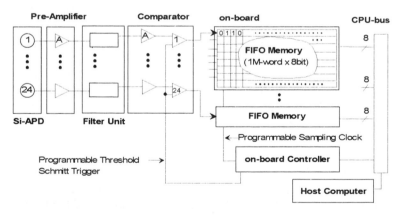

Figure 3. Hardware system block diagram

decided by the cross angle of the laser light sheet, and in the y plate it is ϕ 0.24mm, which corresponds to the core diameter of the receiving optical fiber. In this experiment, the optical fiber for measurement is selected 24-point out of 96-point, and the distance between adjacent measuring points is 1mm. Besides about seeding, polystyrene particles of 5 μ m in mean diameter were used as the tracer particles for measuring of velocity.

3. Results and Discussion

3.1 Instantaneous Velocity Profile

Typical time series velocity profiles at x/h=3 and x/h=6 are shown in Fig.4. Each profile interval is 0.5 in dimensionless time tUc/h (about 40msec) and the dotted lines mean 0 point of each profile. The distribution of u-component mean velocity and turbulent intensity behind the step which calculated by the velocity profile is shown in Fig.5. In this figure the dotted line represents the time-averaged dividing stream line. Each distribution is in good agreement with the other previous studies (Kasagi&Matsunaga,1995). The time-averaged reattachment length, determined by the fraction of forward flow at y/h=0.05, is x_R=6.0 in this experimental apparatus.

Contour maps of instantaneous u-component velocity are shown in Fig.6, which was measured at streamwise locations of x/h=1-10. The measuring point of the bottom is where y/h=0.05 and the top is where y/h=1.15, therefore the measuring width is 22mm. Time progresses from left to right.

At x/h=1, an oscillatory motion of the separated shear layer, caused by shear vortex concentration, can be observed. Under the step height, the flow is almost stagnated, the flow direction is negative at y/h=0.4-0.8, and the flow direction is only slightly positive at y/h<0.3 near the step wall. In the vicinity of y/h>0.8 in the separated shear layer, the flow direction is seen to change periodically due to the

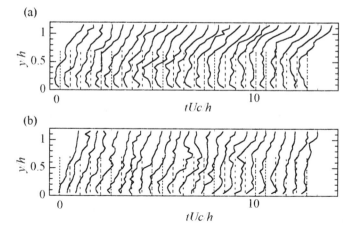

Figure 4.   Typical time series velocity profiles (a)x/h=3, (b) x/h=6

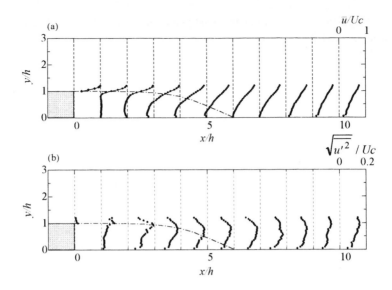

Figure 5. (a) Mean velocity profile (b) Turbulent intensity

shear vortex concentration which can seen as the stripe-shaped structure which forms the dark red section of the figure. At $x/h=2$, the velocity distribution of the separated shear layer is almost identical to that at $x/h=1$. However, the structure in the separation bubble at $x/h=2$ is different from that at $x/h=1$ on two points mentioned below, the stripe-shaped structure grows as the separated shear vortex increases, and the flow direction near the step wall is almost negative.

The governing frequency of velocity fluctuation in the separated shear layer ata the upstream of $x/h=2$ is about 4Hz which caused by the concentration of the shear vortex. But the governing frequency can not be observed clearly at the downstream of $x/h=3$, and it disappears at the downstream of the reattachment region. This means that it is difficult to extract the periodicity of the velocity fluctuation as measured at only one point in the measuring area, because the velocity fluctuation in spatial has a largely oscillating at the downstream of $x/h=3$, as seen in the figure.

The separated shear layer with vortex concentration is largely oscillating with the increasing reverse flow in the separation bubble. The forward flow of the stripe-shaped structure of the separated shear layer is inclined to streamwise and reaches the step wall more frequently. Time of forward flow in succession is longer at the step wall than upstream. This fluid motion indicates the dynamic phenomena of reattachment. A structure of the stripe-shaped structure is not detected at the downstream of $x/h=7$. On the other hand, large-scale velocity fluctuation of forward flow is detected over the measuring cross-section.

It can be seen a phenomenon of reattachment by paying attention to a behavior of reverse flow near the step wall around the time-averaged reattachment region. The area ① in Fig.6 (e) shows a flow motion that the separated shear layer

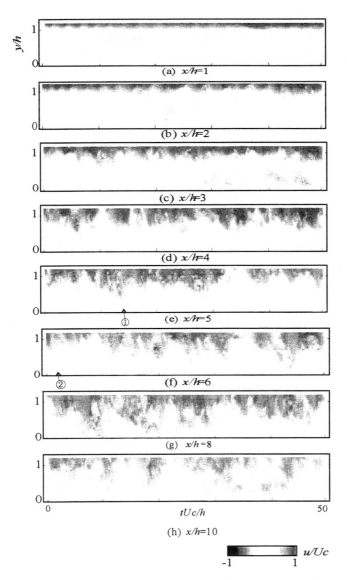

Figure 6. Contour maps of instantaneous velocity at each streamwise location

reattaches to the step wall. It is found clearly that the separated shear flow which approach to the step wall reattaches maintaining velocity gradient with decreasing a reverse flow near the step wall. The area ② in Fig.6 (f) shows a flow motion that the separates shear vortex reattaches to the step wall. It should be noted that reverse flows are detected at two region where near the wall and at $y/h=0.5$ at the same time, as shown in the velocity profile. This phenomenon is suspected to result which the reattachment of the separated shear vortex. The reverse flow region of measuring line is in the separation bubble. Furthermore the origin of the up side reverse flow is the separated shear vortex, and down side reverse flow is the reattachment point downstream from the measuring point. It should be noted that a forward flow can be observed sometimes vicinity of the step wall in spite of measuring point existing in the separation bubble.

As mentioned above, the phenomenon of reattachment have two patterns. One is the case that separated shear layer reattaches and the other one is the case that separated shear vortex reattaches.

The phenomena of reattachment can be observed only by spatial measurement. Especially the definition of time-averaged reattachment point which decide forward flow rate does not indicate the time-averaged instantaneous reattachment which decide forward flow rate does not indicate the time-averaged instantaneous reattachment correctly, because forward flow can be observed vicinity the step wall when the measuring region is separation bubble.

In this experiment, continuous time over of measurements of velocity profile is 4.2msec. If there is large scale fluctuation in the flow dynamics beyond this measuring time, it is impossible for the frequency of this fluctuation to be detected. Eaton&Johnston(1980) investigated low frequency fluctuation around 0.066 in dimensionless frequency. In our experiment, low frequency peak is not detect clearly but separated shear layer reattaches two or three times at reattachment region in Fig. 6, so that it is suspected that it can be measured quasi-periodical fluctuation of low frequency in one experiment.

3.2   Spatial correlation

To clarify the flow structure of the separated shear layer, a two-point correlation of velocity fluctuation in space is given by the following formula,

$$R_{ui}(x,y_1,y_2,\tau) = \frac{\overline{u_1'(t,x,y_1)u_2'(t+\tau,x,y_2)}}{\sqrt{\overline{u_1'(t,x,y_1)}}\sqrt{\overline{u_2'(t+\tau,x,y_2)}}}$$

where τ is a lag time, and y_1 and y_2 are the vertical locations of a fixed-point and a moving-point, respectively. A two point correlation map that fixed-point is located at y/h=1 corresponding to the step height are shown in Fig.8. Solid and dotted line contours represent positive and negative signs, respectively. The interval between contours is set at 0.1. A correlation time is about tUc/h=10 corresponding to twice time length of the shedding frequency of the separated shear vortex to detect typical structure.

At x/h=2, the positive high correlation area around the fixed-point is relatively small, and a negative correlation area is detected around y/h=0.8. This means that the velocity fluctuation in the separated shear layer and in the separation bubble are correlated with 180° out-of-phase. A fluctuation of velocity is relatively frequent so that the negative correlation region appears $\tau Uc/h = 3$ at y/h=1 which means auto-correlation. However, it is difficult to detect this periodical motion

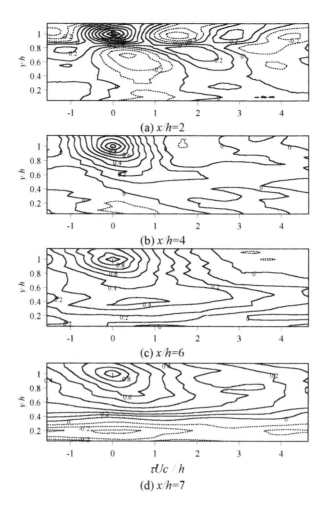

(a) $x/h=2$

(b) $x/h=4$

(c) $x/h=6$

$\tau Uc / h$

(d) $x/h=7$

Figure 7. Spatial correlation between fixed-point $y/h=1$ and moving point over a measuring cross-section

downstream of $x/h=3$, as mentioned above.

The positive high correlation area of $R_{uu}>0.4$ around a fixed-point remains roughly at the same size until $x/h=5$. At the downstream of $x/h=6$, this area gets gradually larger. As a whole, much as positive correlation area approach to the step wall , it is suspected that the flow structure does not change at the upward of separated shear layer as the flow proceeds downstream to the reattachment region, and the flow structure changes downstream around reattachment region.

A negative correlation region is detected near the step wall at $x/h=7$. This correlation means that there is a large scale structure corresponding to the step height downstream of the reattachment region.

3.4 Dividing Stream Line

Temporal variation of dividing stream line are shown in Fig.9 which correspond to Fig.6(a)-(f), with the boundary point between forward flow and reverse flow of instantaneous velocity that is furthest from the step wall (hereafter, this point will be referred to as the 0 velocity point). The distance between the 0 velocity point and the dividing stream line, which represents, so to speak, a thickness of the separated shear layer, so that the motion of the separated shear layer with the vortex concentration is clear. The dividing stream line is defined as the point at which the integrated value of the instantaneous velocity from the step wall upward changes from minus to plus at first as follow dividing stream line definition. Note that there are several velocity profiles for which it is difficult to calculate the dividing stream line. For example, at $x/h=1$, there exists a "dividing stream line" in the separation bubble because the flow direction is forward near the step wall. Although this profile includes these problematic points, the motion of the separated shear layer is still represented well.

At $x/h=2-3$, the dividing stream line oscillates around $y/h=1$ in the separated shear layer. The interval between the dividing stream line and the 0 velocity point becomes narrow when the 0 velocity point and the dividing stream line go down immediately. This means that the reverse flow in the separation bubble becomes slow when a forward flow simultaneously enters the separation bubble from the separated shear layer. This phenomenon of the dividing stream line immediately going down can also be observed at $y/h=4$. The thickness of separated shear layer grows gradually larger as the flow proceeds downstream to $x/h=3$, and the time averaged dividing stream line approaches the step wall gradually.

At $x/h=4-5$, there is a tendency that the dividing stream line, that is oscillating around $y/h=0.5-0.8$ for some time to approaches the step wall immediately. When the dividing stream line can not be detected for a time, this means that the dividing stream line reattaches to the step wall behind the measuring point. The velocity at which the dividing stream line descend to the step wall is about 100mm/s. After this reattachment, the dividing stream line ascend immediately to $03-0.8h$ from the step wall. These flow dynamics mean that the instantaneous reattachment point moves to the downstream of the measuring point.

At $x/h=6$, the oscillations at $y/h=0.5-0.8$ of dividing stream line that appeared show at $x/h=4,5$ are not detected. Much as an approach to the step wall of dividing stream line is detected as approach to the separated shear layer, there is a tendency for the dividing stream line to immediately ascend to $y/h=0.5-0.8$ from the step wall and to reattach after 5 times in dimensionless time (about 400msec).

A thickness of the separated shear layer grow till $x/h=3$ and decrease as flow proceeds downstream. As a whole, it remains unchanged at each measuring section in spite of the oscillation of the separated shear layer. The standard deviation of a fluctuation of the thickness of the separated shear layer between $x/h=1$ and $x/h=6$ is $\sigma/h \fallingdotseq 0.1$.

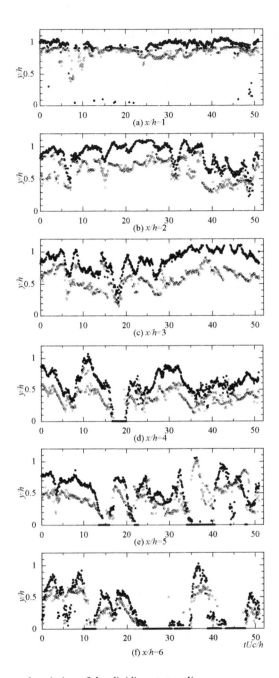

Figure 8. Temporal variation of the dividing stream line
●: dividing stream line, ○: boundary point between forward flow and reverse flow

The probability density of the dividing stream line for each streamwise location are shown in Fig.10. As mentioned above, it must be noted that the dividing stream lines are detected in the separation bubble at x/h=1-2 as a result of the velocity profile in which it is difficult to define the dividing stream line. The oscillation width of the dividing stream line is limited near the separated shear layer until the point where x/h=3. This width becomes larger as the flow proceeds downstream from the step wall to upper side of the separated shear layer. The highest peak is on the time-averaged dividing stream line at x/h=1-2, but it is a slight down side at x/h=3. The most notable point is that two peaks appear at x/h=5, where the time-averaged dividing stream line is on a saddle of them. The vertical location of the peaks are almost same that the peak of x/h=3,4 is y/h=0.8 and y/h=0.75 respectively, and upside peak of x/h=5 is y/h=0.7. These peak indicate that the probability density of pass of the separated shear vortex is large in their region. The downside peak at x/h=5 slightly approach to the step wall as flow to reattachment region. This histogram profile means that the dividing stream line that oscillates around the time-averaged dividing stream line approaches to the step wall immediately which correspond to reattachment of separated shear layer, as mentioned above. Therefore it is suspected that there is quasi-periodical behavior between the case of the separated shear layer reattachment and the separated shear vortex reattachment.

4.   Conclusion

A temporal data of instantaneous velocity profile between the region of separation bubble and the region separated shear layer behind a two-dimensional backward-facing step was measured by multi-point LDV, and was discussed with a

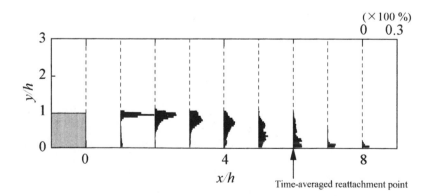

Figure 9.   Probability density of the dividing stream line

special focus on the details of the motion of the phenomena of reattachment and the behavior of the separated shear layer.

As a result, the phenomenon of reattachment have two patterns which characterize flow motion vicinity the step wall. One is the case that separated shear layer reattaches and another one is the case that separated shear vortex reattaches. The phenomena of reattachment can be observed only by spatial measurement. Especially the definition of time-averaged reattachment point which decide forward flow rate does not indicate the time-averaged instantaneous reattachment correctly, because forward flow can be observed vicinity of the step wall when the measuring region is separation bubble.

The scale of separated shear layer vortex increase slightly till upstream of the time-averaged reattachment point and rapidly increase at the reattachment region. At downstream $1h$ of the reattachment region, there is a large scale structure corresponding to step height. Therefore, it is suspected that the reverse flow in that region cause by the large scale structure, not cause by reattachment.

The dividing stream line that was calculated in instantaneous velocity profile oscillates as vortex concentration in the separated shear layer. But the thickness of the separated shear layer does not show large oscillation in spite of the dividing stream line oscillation. The most notable features of the dividing stream line oscillation are that the dividing stream line that oscillated around the time-averaged dividing stream line approached to the step wall immediately as concerned to separated shear layer reattaches, and that standard deviation of the dividing stream line is almost same between $x/h=4$ and $x/h=6$. It is suspected that there is quasi-periodical behavior between the case of the separated shear layer reattachment and the separated shear vortex reattachment.

5. Reference

Armly, B. F., Durst, F., Pereira, J. C. F., and Schonung, B., "Experimental and theoretical investigation of backward-facing step flow", J. Fluid Mech. 127(1983), pp.473-496

Eaton, J., and Johnston, J. P., "Turbulent flow retttachment : An experimental study of the flow and structure behind a backward-facing step", Rep. MD-39, Thermoscience Division, Dept. of Mech. Eng., Stanford University (1980)

Eaton, J., and Johnston, J. P., "A review of research on subsonic turbulent flow reattachment", AIAA J., 19(1981), pp.1092-1100

Hachiga, T., Furuichi, N., Mimatsu, J., Hishida, K., and Kumada, M., " Development of multi-point LDV by using semiconductor laser with FFT-based multi-channel signal processing", Experimental in Fluids 24(1998), pp.70-76

Hijikata, K., Mimatsu, J., and Inoue, J., "A study of wall pressure structure in a backward step flow by a holographic / velocity-pressure cross-correlation visualization", Experimental and Numerical Flow Visualization, 128(1991), pp.61-68

Le, H., Moin, P., and Kim, J., "Direct numerical simulation of turbulent flow over a backward-facing step", J. Fluid Mech., 330(1997), pp.349-374

Kasagi, N., and Matsunaga, A., "Tree-dimensional particle-tracking velocimetry measurement of turbulence statistics and energy budget in a backward-facing step flow", Int. J. Heat and Fluid Flow, 16(1995), pp.477-485

Sakakibara, J., Hishida, K., and Maeda, M., "Measurement of thermally stratified pipe flow using image-processing techniques", Experimental in Fluids 16(1993), pp.82-96

Takeda, Y., "Development of an ultrasound velocity profile monitor", Nuclear Engineering and Design, 126(1991), pp.277-284

II.3. PIV Measurements of Turbulence Statistics in the Three-Dimensional Flow Over a Surface-Mounted Obstacle

J.M.M. Sousa[1], C. Freek[2], and J.C.F. Pereira[1]

[1] Instituto Superior Técnico / Technical University of Lisbon, Mechanical Engineering Department, Av. Rovisco Pais, P-1049-001 Lisboa, Portugal

[2] Volkswagen AG, Research and Development, EZMM-1785, D-38436 Wolfsburg, Germany

Abstract. The primary objective of this work is to show that Particle Image Velocimetry (PIV) can be used to provide reliable turbulence measurements in flows where homogeneous directions do not exist. As in any other measuring system based on sampling, the key to the success of such procedure lies in the acquisition of a statistically large number of samples. The use of compression of digital PIV images strongly facilitates this task. In the present work, the flow over a surface-mounted cube is studied employing digital PIV. Turbulence data is computed by ensemble averaging, based on 2000 PIV images acquired at a frequency of 25-Hertz. Instantaneous views of the flow field provide additional information regarding large-scale dynamics of turbulent structures.

Keywords. Shear flows, surface-mounted obstacles, particle image velocimetry

1 Introduction

Experimental studies of the flow over sharp-edged obstacles mounted in channels have been mainly focused on two-dimensional geometries. These are obviously easier to handle and, consequently, there already exist a large number of reported investigations on two-dimensional surface-mounted obstacles. In contradistinction, quantitative data for turbulent flows over three-dimensional obstacles are still scant. Nevertheless, a very significant amount of information on the topology and turbulent characteristics of the flow around cuboids with various shapes has been provided by Castro and Robins (1977) and Hunt *et al* (1978), for a boundary-layer-type of approach flow, and by Martinuzzi and Tropea (1993), for plane channel flow. Despite their geometrical simplicity, these flows exhibit remarkable topological complexity, making them very attractive to basic fluid mechanics researchers. On the other hand, the dynamics of large-scale turbulent structures, typically present in the aforementioned type of flows, is a matter of potential practical application in various engineering processes.

   As a rule in the assessment of turbulent flow characteristics, a statistical analysis is required when employing any measuring device based on a sampling procedure. Regardless of the fact that Particle Image Velocimetry (PIV) has been traditionally used in the characterization of instantaneous flow patterns, this technique does not constitute an exception to that rule when the aim is to quantitatively describe a turbulent flow. It must be noted that very few attempts have been made in this direction in virtue of the large quantity of data involved in such a statistical analysis of PIV images. However, Freek *et al* (1997) have recently shown that Digital Image Compression can be successfully applied as a sensible solution to abate the inherently high data stream and storage requirements. As a result, full statistical characterization of turbulent flows may be easily accomplished, retaining all the advantages of PIV techniques.

The recent awakening of numerical researchers towards the application of Large Eddy Simulation (LES) to complex geometries has also set an urgent need for detailed experimental data on test cases that may be successfully tackled by this technique. This fact is substantiated by the organization of a number of workshops, such as the "Workshop on Large Eddy Simulation of Flows past Bluff Bodies" by Rodi *et al* (1997) and the "Sixth ERCOFTAC/IAHR/COST Workshop on Refined Flow Modelling" by Hanjalic and Obi (1997). It follows that the requirements on experimental data to validate LES (spatial resolution, temporal resolution, …) can, in general, be easily met by PIV. As a consequence, a fruitful symbiosis will certainly be established between these two techniques in a near future.

In the present work, turbulence statistics are calculated from PIV measurements in the three-dimensional flow over a cube mounted on the floor of an open-surface water channel. Instantaneous views of the flow field are presented as well. Section 2 provides information regarding the experimental setup. The results are shown in Section 3. Finally, Section 4 summarizes the main findings of this investigation.

2 Experimental Arrangement

The unperturbed flow at the test section was a turbulent boundary-layer characterized by a Reynolds number based on momentum thickness $Re_\theta = 770$. The mass flow rate recirculated inside the channel produced a surface velocity $U_o = 0.07$ m/s. A detailed description of the open-surface (water) channel facility employed in these experiments was given by Freek *et al* (1997). The surface-mounted obstacle was a 4-cm cube made of Perspex. Thus, the presently investigated obstacle flow was characterized by a Reynolds number $Re_L = 3210$.

The digital PIV system employed to map the flow field is based on CCD camera recording and Ar-ion laser illumination. MJPEG compression of PIV images is also implemented with the aim of easing the handling of large quantities of data. A thorough description of the present system as well as a meticulous discussion on the advantages and pitfalls of using MJPEG compression was presented by Freek *et al* (1998).

A sequence containing 2000 PIV images was collected at the channel centerplane in the test section area, as shown in Figure 2.1. Digital image compression allowed on-line recording of the data at a sampling frequency of 25-Hertz (only 80 seconds to acquire all the data), maintaining storage needs low. Flow maps were obtained by off-line processing of PIV images interrogated in sub-images of size $IA = 32 \times 32$ pixels. The spacing between interrogation windows was prescribed as 16 pixels (i.e., a 50% IA overlap), producing 47×35 vectors per vector field. An overview of relevant experimental parameters is shown in Table 2.1.

Table 2.1. Summary of relevant experimental parameters

Channel:			
type	open-surface		
length		2.5	m
width		0.20	m
Flow:			
fluid	water		
temperature		28	$^{\circ}$C
depth		0.11	m
surface velocity, $U_{\circ}$		0.07	m/s
obstacle height, L		0.04	m
Reynolds number, Re_{θ}		770	
Reynolds number, Re_{L}		3210	
Seeding:			
type	polyamid		
nominal diameter		60	μm
Illumination:			
type	light sheet		
source	cw Ar^{+} laser + Bragg cell		
maximum power		5	W
thickness		1	mm
pulse separation		10	ms
number of exposures		2	
Recording:			
type	electronic (CCD)		
resolution		576 $\times$ 768	px
lens focal length		26	mm
numerical aperture		2.0	
image magnification		0.21	
image compression		5.5	
Interrogation:			
resolution, IA		32 $\times$ 32	px
spacing		16	px
multi-grid levels		2	
Data set:			
vectors/image		1645	
images/data set		2000	
size		160	Mb

Turbulence statistics were calculated by using single-point, ensemble averaging. Following Raffel *et al* (1998), the total measurement error in displacement PIV vectors can be written as the sum of a bias error ε_{bias} and a measurement uncertainty ε_{rms} (random error). Unbiased values of the correlation function were obtained by dividing out the corresponding weighting factors prior to estimation of fractional displacement, as suggested by Westerweel (1993). A Gaussian estimator of fractional displacement was used to achieve sub-pixel accuracy, leading to negligible values of *rms* tracking error associated with the estimator. The use of high seeding densities has also allowed minimizing the error due to the presence of displacement gradients in the images. Additionally, spurious measurements were removed from all vector fields employing a median-based procedure similar to that indicated by Westerweel (1994). Increased values of dynamic spatial range and data yield have been obtained by the implementation of window offset coupled with a multi-grid/pass procedure (see, e.g., Raffel *et al*, 1998). This has allowed further reductions in measurement noise. Nevertheless, the statistical evaluation of turbulent flow quantities introduces additional uncertainties. Based on the analysis proposed by Yanta and Smith (1973), the statistical uncertainties in mean and variance values for a 95% confidence level can be quantified as 0.4% and 3%, respectively.

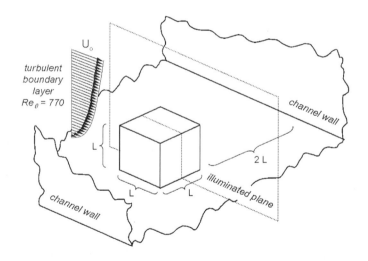

Fig. 2.1. Geometry of test section and measured plane

3 Results and Discussion

First, the mean flow topology obtained by the application of the digital PIV system is analyzed. All flow quantities have been non-dimensionalized using U_o and L, respectively as velocity and length scales. Figure 3.1 shows the mean flow streamlines in the measured plane, computed from *U*- and *V*-velocity components. The overall mean streamline pattern agrees satisfactorily with the sketches reported by Hunt *et al* (1978). Differences in complexity, namely with respect to the horseshoe vortex system, can be attributed to distinct characteristics of the approaching boundary-layer flow. The present experiments have been carried out at a much lower Reynolds number and the obstacle was totally immersed in the boundary-layer, which markedly differs from the flow conditions indicated by the aforementioned authors. Surprisingly, better agreement was found between the presently studied flow field and that corresponding to a surface-mounted cube placed in a turbulent channel flow. The latter geometry was investigated in detail by Larousse *et al* (1991) and by Martinuzzi and Tropea (1993), utilizing laser-Doppler anemometry (LDA) and flow visualization. Yet, they could not find any evidence supporting the existence of the hierarchy of horseshoe vortices reported by Hunt *et al* (1978) either, which is in accordance with the flow features depicted in Figure 3.1. Even the secondary corner vortex (upstream the obstacle) is not clearly observed in the figure, due to the small dimensions displayed by this structure at low Reynolds number.

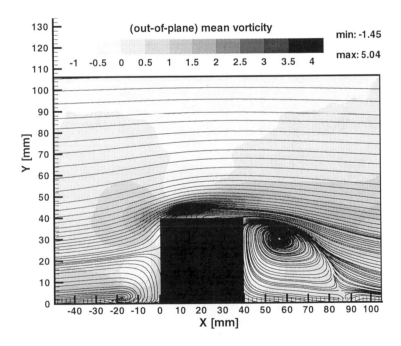

Fig. 3.1. Mean streamline pattern and vorticity contours

A direct comparison between the detailed LDA mean flow maps provided by Larousse *et al* (1991) and the mean streamline pattern represented in Figure 3.1 enhances the main difference between these two flows. It essentially results from the strong flow acceleration promoted by the sudden contraction (2:1) formed by the obstacle and the top channel wall in the former case. As a consequence, the size of separated flow regions over and behind the cube is significantly smaller for a boundary-layer-type of approach flow. Additionally, the corresponding vortex core centers are located closer to bottom walls in this case also.

Contours of the out-of-plane component of vorticity have been superimposed to mean flow streamlines in Figure 3.1. It must be noted that the calculation of vorticity involves differentiation of basic flow quantities. However, vorticity estimates were obtained here by employing Stokes' theorem of circulation (see, e.g., Raffel *et al*, 1998), which seemed to perform better than conventional schemes. Important values of positive vorticity were found in the separated shear-layers formed over and behind the obstacle. It can be seen that the area of the horseshoe vortex is not characterized by large values of mean vorticity, which indicates that this mean flow structure is not very intense at the present Reynolds number.

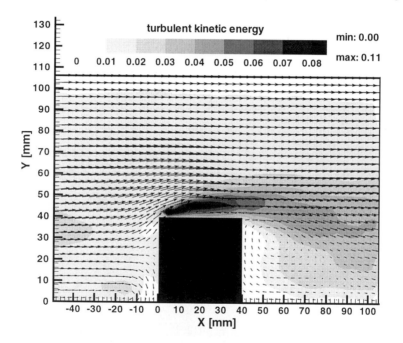

Fig. 3.2. Contours of turbulent kinetic energy

The maxima in turbulent kinetic energy were also found in the separated shear-layers, as shown in Figure 3.2. In order to keep track of the relative position of mean flow structures, contours maps are presented here simultaneously with mean velocity vectors (this procedure will be used throughout the remainder of the paper). It must also be mentioned that, as only in-plane quantities were measured by the PIV system, the values of turbulent kinetic energy in the flow were estimated by the use of the following expression:

$$k = \frac{3}{4}\left(\overline{u'^2} + \overline{v'^2} \right).$$

(1)

The contribution of normal-stresses $\overline{u'^2}$ and $\overline{v'^2}$ to k is depicted in Figures 3.3a and 3.3b, respectively. It can be seen that, while the longitudinal component of normal-stresses reaches peak magnitudes in the first stages of development of the separated shear-layer over the cube, the vertical component is characterized by moderate values in this area. The intense mean shear established as the fluid flows over the obstacle is the main responsible by the large values associated with the longitudinal component. In contradistinction, the vertical component reaches its peak values (always smaller than those for the longitudinal component) farther downstream, essentially as a result of large-scale unsteadiness. This observation will be further substantiated later in the paper. Thus, two major sources of large-scale structures can be identified in the flow field, corresponding to the recirculation regions over and behind the cube. Again, little energy seems to be contained in the flow structures defining the horseshoe vortex upstream the obstacle. The contours of (in-plane) shear-stress corroborate the above assertions (see Figure 3.3c). Two large local maxima can be found slightly upstream the peaks in the vertical component of normal-stresses, indicating intense turbulence production in these areas. Consequently, it seemed reasonable to conclude that

a)

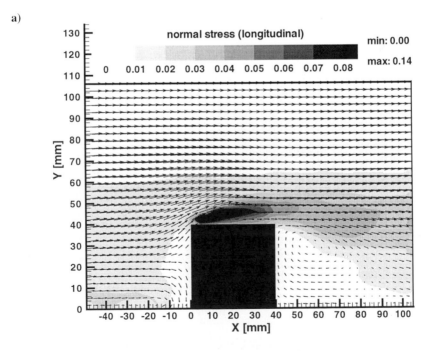

(see next page for caption)

164

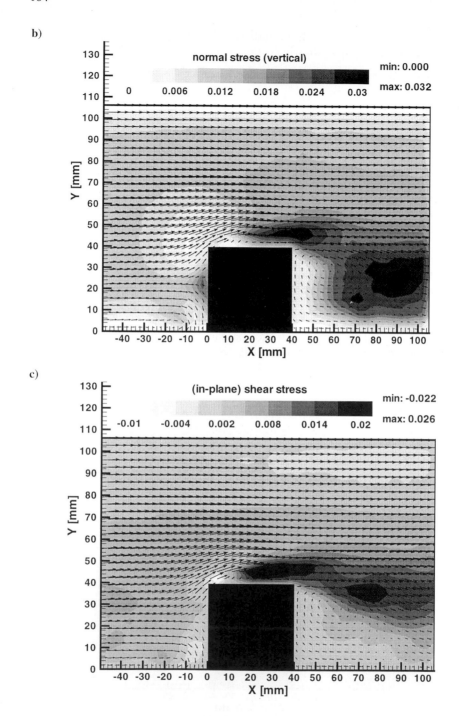

Fig. 3.3. Contours of Reynolds-stresses: a) & b) normal-stress components; c) shear-stress

very organized motions can be found there, leading to highly correlated u- and v-velocity fluctuations. On the other hand, it is also interesting to note that a sharp peak of negative shear-stress occurs in the vicinity of the upper left corner of the cube.

Aiming to analyze some of the flow features with better detail, transverse profiles of mean and turbulent quantities, respectively at $X = 20$ mm and $X = 55$ mm (i.e., crossing the recirculation regions on top and behind the cube, respectively), have been represented in Figures 3.4a, 3.4b, 3.5a and 3.5b. These profiles also facilitate a discussion of the results with reference to measurements reported in the literature.

As noted before, Figure 3.4a shows that the flow acceleration over the cube is much more modest in the present flow configuration than in channel flow, which naturally influences the strength of the recirculation eddy traversed by this profile. At this location ($X = L/2$), Martinuzzi and Tropea (1993) measured longitudinal velocities that were about 30% in excess of the reference velocity. Additionally, the magnitude of reversed flow was reported to reach 50% of the reference velocity. Present values do not exceed 10% of the surface velocity in either case. Martinuzzi and Tropea (1993) did not compute the vorticity field. However, it is quite obvious that the aforementioned differences in longitudinal velocity component would be reflected by vorticity maxima as well (note that vorticity values have been scaled down to one third in Figure 3.4a, i.e., $W/3$, for the sake of figure clearness only). As immediate consequence of these smaller mean gradients, Reynolds-stresses levels in the separated shear-layer over the obstacle (see Figure 3.4b) are also characterized by lower values than in channel flow. Nevertheless, it is important to emphasize that all present peak values closely preserve the various ratios between Reynolds-stress components indicated by the above-referred LDA measurements (a scaling factor of approximately 3.5 was found). The foregoing result strongly supports the credibility of these PIV measurements of turbulence statistics.

A direct comparison for profiles located inside the recirculation region behind the cube is much more difficult, due to the difference in the values of reattachment length. The profiles shown in Figures 3.5a and 3.5b are located slightly upstream $X = 3L/2$ (where LDA data are available for channel flow), in an attempt to account for a smaller reattachment length. It is not surprising that, once again, the measured velocity excess in the longitudinal component is clearly lower than the characteristic values for channel flow reported by Martinuzzi and Tropea (1993). The maximum intensity of the reversed flow is, of course, larger in this area, but still limited to approximately 50% of the corresponding values obtained in channel flow. However, the main point of interest in Figure 3.5a is the "kinked" profile of mean vorticity, which indicates a multiple layer structure. An explanation for such shape can only be found by recognizing this feature as the reminiscence of the existence of two individual (separated) shear-layers. In fact, besides portraying the trend of convergence between the peak values of measured normal-stresses (again in total agreement with the findings for channel flow), Figure 3.5b also shows that the multi-layer structure is exhibited by the vertical component of those stresses as well. This feature can also be detected in some of the LES calculations performed

166

by M. Breuer in the framework of the "6th ERCOFTAC/IAHR/COST Workshop on Refined Flow Modelling" (Hanjalic and Obi, 1997).

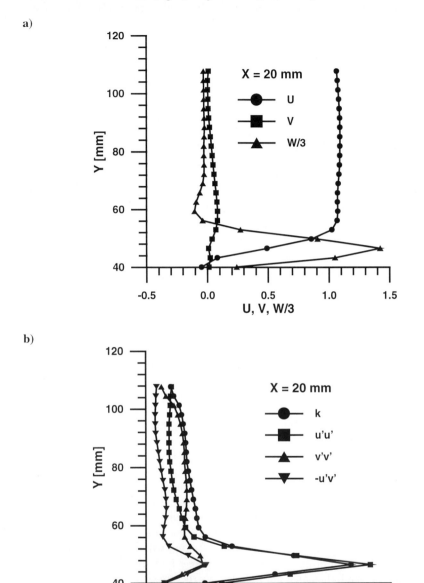

Fig. 3.4. Transverse profiles at $X = 20$ mm: a) mean velocities and vorticity; b) turbulent kinetic energy and Reynolds-stresses

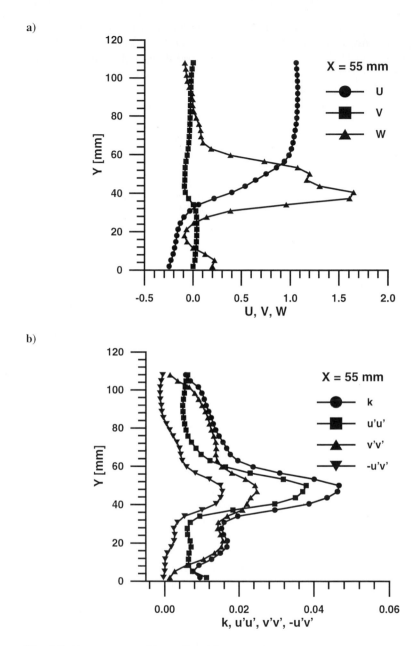

a)

b)

Fig. 3.5. Transverse profiles at $X = 55$ mm: a) mean velocities and vorticity; b) turbulent kinetic energy and Reynolds-stresses

As earlier asserted in this paper, organized large-scale motions associated with vortex-shedding from recirculation regions are in the origin of the process. Figure

3.6 illustrates the formation of notably coherent turbulent structures in a series of snapshots portraying instantaneous velocity fields (75% *IA* overlap was used). During the depicted time instants, one eddy drags over the top of the obstacle. The region in the back of the cube sheds another eddy, which is convected downstream closely following the path defined by the curved shear-layer. The vertical component of normal-stresses naturally receives an important contribution from these two "springs" of organized motions that, meeting in the near-wake of the obstacle, ultimately give rise to the establishment of a multi-layer structure in this area of the flow.

One may still note that the flow in the region of the horseshoe vortex is not quiescent either. However, the excursions traced out by the associated large-scale structures are rather limited in the present case, which explains the moderate values of Reynolds-stresses measured in this portion of the flow field. From this observation it can be inferred that, in contrast with the findings of Larousse *et al* (1991) for a much higher Reynolds number, the horseshoe vortex does not constitute here the dominant source of large-scale unsteadiness. In fact, the spectral analysis of a time series acquired in this area produces a broad peak, without bringing out any dominant frequency (Strouhal number in the graph). Figure 3.7a shows the above result and Figure 3.7b portrays the corresponding time-trace of the (fluctuating) longitudinal velocity component u'. Similar measurements taken in the separated shear-layer over the cube exhibited a bi-modal character in probability-density functions (PDF) of vertical velocity, heralding the appearance of a dominant frequency (or range of frequencies, as broadband behavior is also a characteristic of natural instability of free shear-layers; see Husain and Hussain, 1983).

In addition, a few more observations can be made regarding Figure 3.7. Firstly, one may note that a significant portion of the power spectrum in Figure 3.7a closely reproduces an evolution consistent with the so-called inertial sub-range of turbulence. Although good results have been obtained in the present flow, which was characterized by low fluid velocities, the sampling frequency imposed by the video system is a strong limitation for a more general application of this analysis. It becomes particularly stringent if we keep in mind that Adrian and Yao (1987) suggested that, as a guideline for LDA measurements, the mean data rate should be about twenty times the largest frequency at which undistorted measurements are desired. Secondly, the measured time-trace in Figure 3.7b firmly suggests that the PDF of the (fluctuating) longitudinal component of velocity at the indicated location must be strongly skewed. Aiming to investigate this issue in further detail, higher-order moments of the probability distributions were computed. The results have been expressed in non-dimensional form, in terms of skewness and kurtosis factors.

Figures 3.8a and 3.8b show contour maps of skewness factor, respectively for longitudinal and vertical velocity components. With respect to the former component, it can be seen that the bulk of the flow field is characterized by values close to zero, which corresponds to a normal random process. However, clearly negative values can be found in the separated shear-layer evolving over the obstacle, while even stronger positive values occur in a very limited area in the

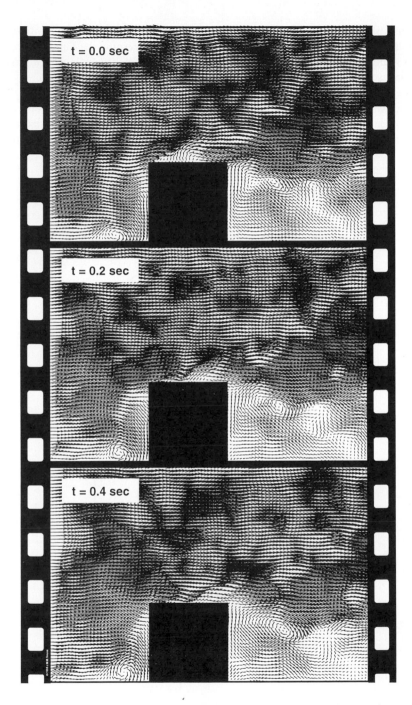

Fig. 3.6. Sequence of snapshots showing the formation of organized large-scale motions (75% *IA* overlap)

vicinity of the lower corner upstream the cube. This last observation confirms the hypothesis formulated in the previous paragraph. Concerning the latter component of velocity (see Figure 3.8b), it can be seen that the maximum levels of positive skewness are always more modest than those observed in Figure 3.8a.

The contour maps of kurtosis factor are shown in Figures 3.9a and 3.9b, respectively for longitudinal and vertical velocity components. Contour levels have been chosen with the aim of evincing the so-called "excess". For this reason, the scale of contour levels always starts at the value reflecting a normal random

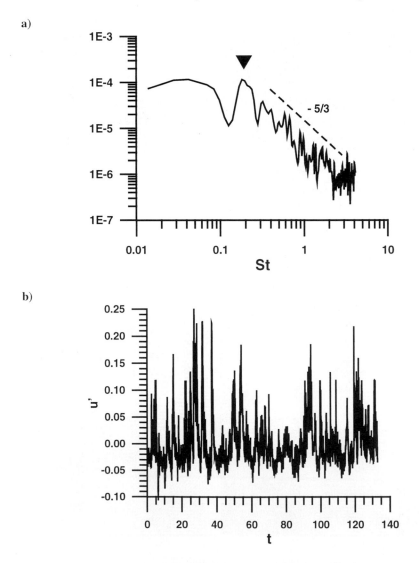

Fig. 3.7. Analysis of u'-velocity in the vicinity of the lower corner upstream the obstacle: a) power spectrum (power scale is arbitrary); b) time series

a)

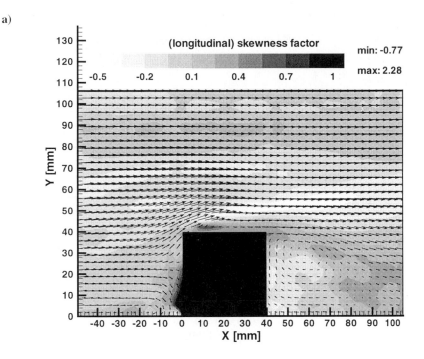

b)

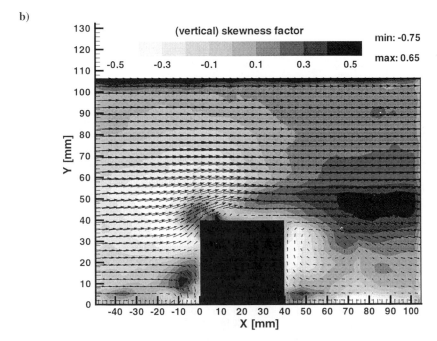

Fig. 3.8. Contours of skewness factor: a) longitudinal velocity; b) vertical velocity

a)

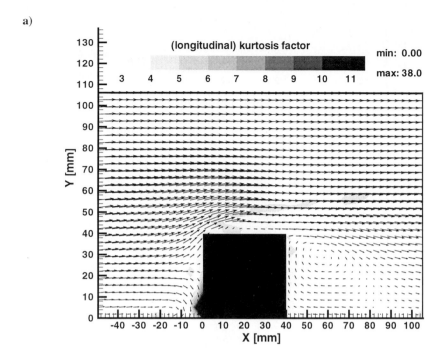

b)

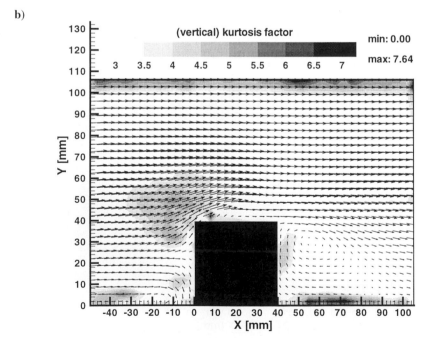

Fig. 3.9. Contours of kurtosis factor: a) longitudinal velocity; b) vertical velocity

process (i.e., at the value of three). Consistently with the observations made for the skewness factor of longitudinal velocity distributions, very large values of "excess" can also be found in the lower corner upstream the cube. On the other hand, with respect to vertical velocity component, the regions of maximum "excess" are located inside the back recirculation zone adjacent to the reattachment at the wall (negative skewness; see Figure 3.8b) and encircling the front edge of the obstacle (essentially positive skewness). Naturally, for both velocity components, large values of kurtosis are always observed where extreme values (negative or positive) of skewness occur.

From the viewpoint of a statistical description of turbulent flows, the main advantage of PIV over LDA probably resides in the fact that the former technique may readily provide estimates of spatial correlations. In the present study, some estimates of Taylor microscales have been obtained. Hence, an assessment of the size of the eddies mainly responsible for dissipation in the flow under investigation could be carried out. The definition of the various dissipation scales is given, e.g., by Hinze (1975). In this paper, only λ_{fu} and λ_{gu} were evaluated, using the following expressions:

$$\lambda_{fu}^2 = \frac{2\overline{u'^2}}{\overline{\left(\dfrac{\partial u'}{\partial x}\right)^2}} \quad ; \quad \lambda_{gu}^2 = \frac{2\overline{u'^2}}{\overline{\left(\dfrac{\partial u'}{\partial y}\right)^2}} \; . \tag{2}$$

The results obtained for these two quantities are shown in Figures 3.10a and 3.10b. At this stage, one should recall that the reference scale employed here in the non-dimensionalization of lengths was $L = 0.04$ m. A general conclusion drawn from the examination of the contour maps is that the spatial resolution set in the interrogation of the video images ($\Delta x = \Delta y \approx 0.08$) seems adequate. However, this fact was confirmed "a posteriori" only. Additional findings can be deduced through a closer inspection of the aforementioned figures. Namely, it can be seen that, as expected, a good correlation can be found between the locations where λ_{fu} reaches minimum magnitudes and the areas where peak values of negative shear-stress were measured. The minima of λ_{gu} occur in the vicinity of the edges of the cube and in the area where the separated shear-layer develops over the obstacle. It is also interesting to note that the strong decrease in the measure of the latter dissipation scale over the obstacle, mainly as a result of longitudinal stretching, is accompanied by the formation of a very localized area of large λ_{gu} in the precise location where the minimum in shear-stress was found (see Figure 3.3c).

Finally, Figure 3.11 illustrates the spatial distribution of the ratio $\lambda_{fu} / \lambda_{gu}$. The contour map shows that, except for the regions where shear-layers are present, the values are not far from the known relation for isotropic turbulence, i.e.,

a)

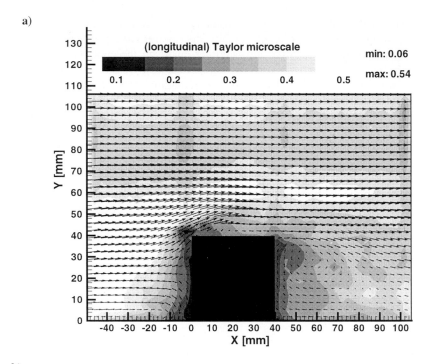

b)

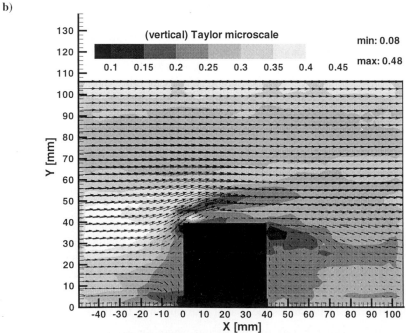

Fig. 3.10. Contours of magnitude of Taylor microscales: a) λ_{fu} ; b) λ_{gu}

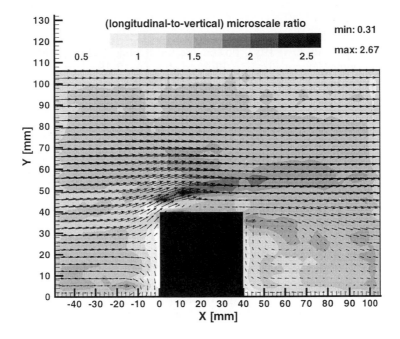

Fig. 3.11. Contours of the ratio $\lambda_{fu} / \lambda_{gu}$

$\lambda_f = \sqrt{2}\lambda_g$. In fact, it can be seen that this ratio becomes notably different from the aforementioned value only in areas dominated by attached shear-layers (smaller) and separated shear-layers (larger). The maximum magnitude can be found in the separated shear-layer developing over the cube.

4 Conclusions

The turbulent flow over a cubical obstacle mounted in the bottom of an open-surface, low-Reynolds number, water channel was thoroughly investigated. A digital PIV system employing MJPEG image compression was the tool that allowed whole-field quantification of turbulence characteristics in the centerplane of the channel, namely, mean velocities and vorticity, Reynolds-stresses, skewness factors, kurtosis factors and Taylor microscales. The study of the time-dependent characteristics of the flow was carried out using spectral analysis of time series and visualization of large-scale dynamics of turbulent structures.

The feasibility of using digital PIV in the statistical investigation of turbulent flows without homogeneous directions was demonstrated in the paper. Furthermore, a detailed characterization of turbulence statistics allowed bringing out additional insight about turbulent flows over surface-mounted, sharp-edged obstacles.

References

Adrian, R.J. & Yao, C.S. 1987, Power Spectra of Fluid Velocities Measured by Laser Doppler Velocimetry, Exp. Fluids, vol. 5, pp. 17-28.

Castro, I.P. & Robins, A.G. 1977, The Flow Around a Surface-Mounted Cube in Uniform and Turbulent Streams, J. Fluid Mech., vol. 79, pp. 307-335.

Freek, C., Sousa, J.M.M., Hentschel, W. & Merzkirch, W. 1997, Digital Image Compression PIV: a Tool for IC-Engine Research, Proc. 7th Int. Conf. on Laser Anemometry Advances and Applications, Karlsruhe, pp. 455-464.

Freek, C., Sousa, J.M.M., Hentschel, W. & Merzkirch, W. 1998, On the Accuracy of a MJPEG-based Digital Image Compression PIV-system, to be published in Exp. Fluids.

Hanjalic, K. & Obi, S. 1997, Flow Around a Surface-Mounted Cubical Obstacle, Proc. 6th ERCOFTAC/IAHR/COST Workshop on Refined Flow Modelling, Delft, vol. 4.

Hinze, J.O. 1975, Turbulence, McGraw-Hill.

Hunt, J.C.R., Abell, C.J., Peterka, J.A. & Woo, H. 1978, Kinematical Studies of the Flows Around Free or Surface-Mounted Obstacles; Applying Topology to Flow Visualization, J. Fluid Mech., vol. 86, pp. 179-200.

Husain, Z.D. & Hussain, A.K.M.F. 1983, Natural Instability of Free Shear Layers, AIAA J., vol. 21, pp. 1512-1517.

Larousse, A., Martinuzzi, R. & Tropea, C. 1991, Flow Around Surface-Mounted, Three-Dimensional Obstacles, Proc. 8th Symposium on Turbulent Shear Flows, Munich, vol. 1, pp. 14.4.1-14.4.6.

Martinuzzi, R. & Tropea, C. 1993, The Flow Around Surface-Mounted, Prismatic Obstacles Placed in a Fully Developed Channel Flow, J. Fluids Engng., vol. 115, pp. 85-92.

Raffel, M., Willert, C. & Kompenhans, J. 1998, Particle Image Velocimetry: a Practical Guide, Springer-Verlag.

Rodi, W., Ferziger, J.H., Breuer, M. & Pourquié, M. 1997, Status of Large Eddy Simulation: Results of a Workshop, J. Fluids Engng., vol. 119, pp. 248-262.

Westerweel, J. 1993, Digital Particle Image Velocimetry: Theory and Application, Ph.D. thesis, Delft University of Technology, the Netherlands.

Westerweel, J. 1994, Efficient Detection of Spurious Vectors in Particle Image Velocimetry Data, Exp. Fluids, vol. 16, pp. 236-247.

Yanta, W.J. & Smith, R.A. 1973, Measurements of Turbulent-Transport Properties with a Laser Doppler Velocimeter, 11th Aerospace Science Meeting, Washington, AIAA paper 73-169.

II.4. Turbulence Analysis of the BFS with Iterative Multigrid PIV Image Processing Multigrid PIV Image Processing

F. Scarano[1,2] and M.L. Riethmuller[1]

[1] Department of Environmental and Applied Fluid Dynamics, von Kármán Institute for Fluid Dynamics, Chaussèe de Waterloo 72, 1640-B Rhode-St-Genèse, Belgium
[2] DETEC - Department of Applied Energetics and Thermo-Fluid Dynamics to Environmental Control, University of Naples, Facolta' di Ingegneria, P.le V. Tecchio, 80 I-80125 Napoli, Italy

Abstract. An advanced PIV processing technique is applied to investigate the turbulent flow over a Backward Facing Step (BFS). The results are compared with Direct Numerical Simulation (DNS) and LASER Doppler Velocimetry (LDV) data available on the subject.

The work proposes an appropriate algorithm to evaluate the PIV recordings. This procedure performs a local relative displacement of the areas to be correlated and optimises it with an iterative procedure. The local particle displacement, due to the in-plane motion, is compensated by means of such a discrete offset.

Moreover, during the iterative procedure the size of the interrogating areas is gradually reduced yielding a finer resolution in space if compared with one step interrogation methods.

A brief section explains the expected improvements in terms of dynamic range and resolution. The accuracy is assessed analyzing reference images of known displacement.

Keywords. PIV; Turbulence; BFS

1 Introduction

To perform PIV measurements in turbulent flows, it is necessary to reach a resolution in space that will allow the small scales of the turbulence spectrum to be detected. In addition to this, a high accuracy is required to correctly measure the small velocity fluctuations. Finally, in order to determine the statistical flow properties, it is necessary to acquire and analyse a large number of images.

To fulfil the first requirement, the primary step is to reduce as much as possible the size of the measurement domain that in PIV corresponds to the interrogation areas where the cross-correlation is to be calculated.

The problem of loss-of-pairs as shown by Keane and Adrian (1991) strongly limits the possibility of reducing the size of the interrogation areas once the time interval Δt between the subsequent acquisitions is chosen.

The displacement of the second exposure interrogation area as already proposed by Keane and Adrian (1993) and further deeply analysed by Westerweel et al.

(1997) allows retrieving most of the particle image pairs lost for the in-plane motion.

Single-step processing methods are not appropriate to take advantage of the local displacement. Many works proposing advanced interrogation algorithms are iterative. Huang et al. (1993) and Jambunathan et al. (1997) proposed techniques for a complete compensation of the in plane motion, Keane et al. (1995) introduced the Super-Resolution concept with a coupled PIV/PTV algorithm. The present work proposes the application of a *multigrid* analysis to the PIV images.

The multiple interrogation with refinement of the window size allows maximising the spatial resolution for the cross-correlation analysis of images.

The separated flow over a BFS is frequently proposed as a test case for techniques under development as well as for code validation. This is motivated by the simple geometry, the stability of the separation region and the high turbulence levels experienced in the shear layer. The results of the present work are compared with DNS (Le et al. 1997) and LDV (Jovic and Driver 1994) available on this subject with a similar test configuration.

2 PIV ANALYSIS ALGORITHM

WIDIM stands for *Window Displacement Iterative Multigrid*. The ratio between the signal peak and the noise level in the correlation map is a crucial parameter in PIV measurements; it has already been established its dependence of the loss-of-pairs due to the in-plane motion (Keane and Adrian 1993). Nevertheless as referred by many authors, other parameters do contribute to the degradation of the signal like high velocity gradients, turbulent motion and loss-of-pairs due to out-of-plane motion. The present method has been conceived with the purpose of compensating for the loss-of-pairs due to in-plane motion. The correlating windows are provided with a degree of freedom in terms of pure translation one respect to the other.

A schematic of the situation is shown in figure 1: regions *a* and *b* correspond to the same physical area where particles are detected at two successive instants of time. Due to the motion some particles (black) present in *a* are no more included in *b* and some new particles (white) enter this last.

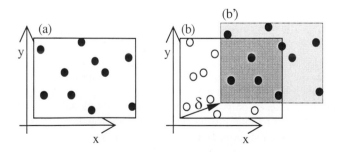

Fig. 1. Principle of the window displacement.

The local displacement of each interrogation area (δ in Fig. 1) is made on the basis of a flow pattern prediction. The predicted displacement is obtained by a previous interrogation of the set of two images. Therefore it is necessary to adopt an iterative procedure (Fig. 2).

Each iteration starts from a predicted displacement field (exception made for the first interrogation which is made with an appropriate window size) and yields a correction for this. Each iteration step halves the size of the interrogation areas leading to the multigrid concept (Fig. 3).

The displacement field is obtained as a sum of the predicted value and a small correction:

$$\delta(x, y) = \delta_p(x, y) + \delta_c(x, y) \tag{1}$$

In the first box the *windowing* consists in splitting each of the two pictures into sets of interrogation areas. In particular, the application of the displacement predictor on the second picture translates each window according to the distribution of $\delta_p(x, y)$. In the second box, the cross-correlation is applied and yields the correction distribution $\delta_c(x, y)$. The result is validated and can either constitute the final output, or the input for a new interrogation based on a finer grid. Hence the iterative form for eq. 1 is

$$\delta^k(x, y) = \delta_p{}^k(x, y) + \delta_c{}^k(x, y) \tag{2}$$

Where the predictor is updated by the relation

$$\delta_p{}^{k+1} = \mathrm{integer}(\delta^k) \tag{3}$$

At the start of the process, no information on the flow pattern is considered available and the first predictor is set uniformly to zero (no relative offset is applied). The window size for this interrogation is set respecting the *one-quarter rule* for the in-plane motion (maximum in-plane displacement < ¼ window linear size).

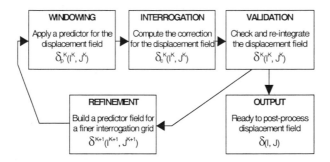

Fig. 2. Image processing flow diagram.

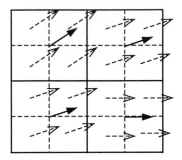

Fig. 3. Coarse interrogation grid result (continuous line vectors and areas) and finer displacement predictor (dotted line vectors and areas).

After the interrogation, the coarse results (solid line arrows and windows in Fig. 3) will be used as a predictor. A finer windowing is made halving the windows in both directions and the predictor is applied to the window offset by means of simple substitution of the previous iteration result.

As a consequence, in the subsequent steps the *one-quarter rule*, related to the in-plane displacement, does not limit anymore the size of the windows and the pictures can be interrogated with a better resolution.

The end of the process is reached when, after various refinements, the maximum possible resolution is achieved (according with the limits of application for the cross-correlation operator on PIV images).

2.1 Resolution and Dynamic Range

The main improvement obtained with this method consists in the de-coupling of two important parameters namely maximum in-plane displacement and interrogation window size. This last is no more constant during the whole processing and a subscript will indicate the refinement step to which it is referred.

The dynamic range of the technique refers to its ability to represent large velocity differences in the flow field. The lower end of the dynamic range is determined by the requirement that the corresponding displacement will be distinguished from the one corresponding to the noise level. Calling l_{min} such a minimum displacement that can be resolved, the consequent minimum measurable velocity is given by:

$$U_{min} = \frac{l_{min}}{\Delta t} \qquad (4)$$

For single-step methods, the limit on the maximum measurable velocity is therefore related to the ratio c_1 between the maximum in-plane displacement l_{max} and the window linear size W_s

$$c_1 = \frac{l_{max}}{W_s} \tag{5}$$

Such a coefficient should not exceed values of 0.2~0.3 if a high confidence level is required in the measurement (Keane and Adrian 1993), thus:

$$U_{max} = \frac{l_{max}}{\Delta t} = \frac{c_1 \cdot W_s}{\Delta t} \tag{6}$$

Combining expressions (4) and (6) one obtains:

$$\frac{U_{max} - U_{min}}{U_{min}} = c_1 \frac{W_s}{l_{min}} - 1 \tag{7}$$

This expression reads as the accuracy-resolution trade off. In fact, the dynamic range can be increased with a larger window size, but lowering the spatial resolution.

Other authors (Adrian 1991, Willert and Gharib 1991) already investigated the dependence of l_{min} from the experimental parameters. It has been found to be associated with the particle image size, with the number of particles within the window and, in digital PIV, with the type of sub-pixel interpolation that is used to determine the location of the correlation peak.

A typical value of $l_{min} = 0.1$ pixel can be assumed in agreement with Willert and Gharib (1993), (three-point gaussian fit, particle image two or three pixels diameter, displacements not exceeding 10 pixels and window size of 32x32 pixels). With the previous considerations eq.7 yields the following result: the choice of a window size of 32x32 pixels will allow distinguishing about 100 different velocity levels.

The method being presented aims at de-coupling space resolution and accuracy leading to a significant increase of the dynamic range of the technique.

Eq. 7 yields for the present case

$$\frac{U_{max} - U_{min}}{U_{min}} = \frac{l_{max}}{l_{min}} - 1 = \frac{c_1 \cdot W_0}{l_{min}} - 1 \tag{8}$$

Where W_o represents the window size at the first interrogation. A refinement ratio can be defined as

$$R(k) = \frac{W_0}{W_k} \qquad k=1,2,...K$$

Eq. 8 is rewritten as

$$\frac{U_{max} - U_{min}}{U_{min}} = \frac{l_{max}}{l_{min}} - 1 = \frac{c_1 \cdot R \cdot W_K}{l_{min}} - 1 \tag{9}$$

It is possible to appreciate that, once fixed the window size, the dynamic range can be amplified in proportion with the refinement ratio R.

Moreover, the maximum displacement can even exceed the window linear dimension (exception made for the first iteration). Of course other factors may limit the maximum displacement: out-of-plane motions and/or high vorticity levels constitute the principal sources for loss-of-pairs. Huang et al. (1993) investigated the limitations in vorticity measurement with PIV and proposed an algorithm that compensates for the relative deformation between the two interrogation windows. Further developments of the current work will be directed to couple these corrections with the window refinement.

Comparisons with the performances of a basic interrogation method (see next paragraph) showed that with the *WIDIM* algorithm the maximum displacement could be increased by a factor four ($R=4$ was achieved).

A further extension of the dynamic range is expected on the lower end too; previous studies conducted by Westerweel et al. (1997) demonstrated how the discrete window offset has a beneficial effect on the accuracy of the determination of the peak location in the cross-correlation analysis.

The above-cited authors found that for highly turbulent flows, the application of a discrete offset to window location yields a noise reduction (NR) of about three. Therefore the superposition of the extension of l_{max} and the refinement of l_{min} can lead to a widening for the dynamic range of about one order of magnitude.

2.2 Accuracy Assessment

The procedure presented hereafter aims at analyzing the effect, on the measurement accuracy, of a compensation for the *in-plane* motion of the tracers.

Starting from PIV images taken under ordinary experimental conditions, synthetic displacement fields were obtained for simulating uniform flows. The reason for this choice was to obtain information related as close as possible to the real experimental conditions. The first image is a real PIV image while the second image was obtained as a 45° shifted version of the first one. The seeding density in the light sheet is of about 15 (particles/mm^2). To allow displacements of non-integer number of pixels, bilinear interpolation was applied to the first image.

A range of displacements from zero to ten pixels was covered with steps of 0.1 pixels.

After the displaced images were generated, the basic and advanced interrogation methods were applied. The results are compared with the expected data.

The correlation peak position is established using a three-point gaussian interpolation for both methods. Areas of 32 by 32 pixels overlapped for 50% of their extension were used as interrogation windows.

The average error relative to the *WIDIM* method exhibits an oscillation with a wavelength of one pixel. This behavior is documented by Westerweel (1993, 1998) and is due to the *peak-locking* effect. A very good agreement on the measurement uncertainty is found with a study conducted by Freek et al. on the same subject (1998). Fig. 4 shows the result of a moving average filter applied to the averaged patterns of the error. For the basic method, a bias error is present. The same is not found when the improved algorithm is applied. Even though such a bias does not exceed one tenth of a pixel, it leads to underestimate systematically the displacements. Lecordier (1997) also reports the presence of this effect comparing the cases of discrete window offset and sub-pixel window transformations. The asymptotic trend shows that even for large imposed displacements this component of the relative error does not vanish. One should therefore expect an increase of the absolute bias error with the displacement. The source for this can be found in the distortion of the peak shape due to the *loss-of-pairs*. The less the number of particles which contribute to build the correlation peak, the lower the corresponding intensity in the peak region.

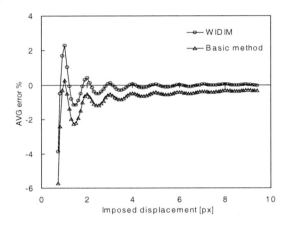

Fig. 4. Measurement error: U_{meas}-U_{exact}/U_{exact}. DPIV results relative to a uniform spatial shift in both directions.

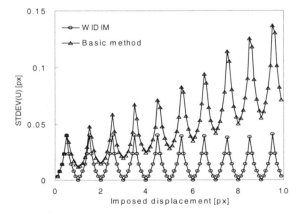

Fig. 5. Measurement uncertainty: standard deviation of the measured displacement distribution.

A straight conclusion is that a higher level of accuracy can be reached enlarging the maximum displacement as much as possible.

The higher limit to the maximum displacement (viz. Δt) comes from other sources of error: in presence of a non-zero acceleration level in the flow field, the error associated to the approximation of the Eulerian velocity in PIV is a linear function of the particle displacement (Boillot and Prasad 1996).

Out-of-plane motions cause loss-of-pairs and degradation of the signal peak (Keane, Adrian and Zhang 1995).

Loss-of-pairs is also due to strong vorticity levels in the flow (Huang, Fielder and Wang 1993).

Comparing the two patterns in Fig. 5 for the basic method, at the higher end of the analyzed range of displacements, the standard deviation can exceed one tenth of a pixel; while the value corresponding to the application of WIDIM keeps below 0.04 pixels.

3   EXPERIMENTAL INVESTIGATION

3.1   Apparatus

Fig. 6 shows a sketch of the experimental section.
The primary elements of the facility are:
- wind tunnel
- LASER cavities
- seeding device
- acquisition system

3.1.1   Wind tunnel and model

The experiments have been conducted in a VKI subsonic wind tunnel 20x20-cm^2 test section. A horizontal flat plate divides the test section into two channels: the backward facing step is installed in the upper channel (Fig. 6).

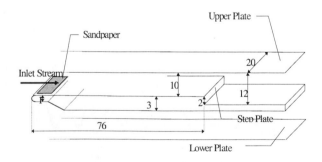

Fig. 6. Geometry of the test section (quotes in cm).

The resulting test section is 20(w)x10(h) cm^2. A centrifugal fan driven by an asynchronous motor supplies the flow. Air is blown through filters into a settling chamber. The contraction ratio between the sections of this chamber and that of the wind tunnel is equal to 6. The nominal turbulence intensity at the inlet is of 0.3%.

At the leading edge of the flat plate, rough sandpaper is placed to trigger the transition of the boundary layer. The step occurs at 76 cm from the leading edge of the plate. The step height is h=2-cm .The expansion ratio is defined as:

$$ER = \frac{L_y}{L_y - h} \tag{10}$$

Where L_y is the channel height downstream the step. For the present case ER=1.2.

The surface of the step and downstream of it is made of polished stainless steel in order to fulfill two needs: 1) to avoid that the laser beam penetrates into the Plexiglas, 2) to reflect the laser sheet back into itself. This also allows increasing the light intensity in the measurement region yielding a final enhanced contrast of the pictures.

3.1.2   Laser and optics

The light source was a double cavity pulsed Nd:Yag LASER. A 5mm diameter light beam passes through a first semi-cylindrical lens that expands the beam in the horizontal plane, then a spherical lens of appropriate focal length makes the light paths parallel to the optical axis, but it also concentrates the beam in the vertical plane. A second semi-cylindrical lens, turned at 90° with respect to the first one, deviates the rays in parallel directions. The laser sheet, horizontal at first, is finally reflected vertically towards the test section by a prism.

3.1.3   Seeding device

A VKI built-in oil smoke generator has been used. A nozzle sprays oil droplets on a hot plate (~150 C), the oil vaporizes, exits the chamber and enters the wind tunnel blower through a flexible pipe. The condensed oil droplets in the test section have a mean diameter of 1 μm.

3.1.4   Acquisition system

To acquire, display and record digital images, use has been made of a commercial PIV system "INSIGHT", developed by TSI. The system is composed of: CCD video sensor (640x480pixels, 256 gray levels) able to acquire couples of images at a rate of 10 or 15 Hz; PC with acquisition board and equipped with a dedicated PIV software; LASER synchronizer.

3.2   Measurement Conditions

A sketch of the region over which the measurements have been performed is shown below.

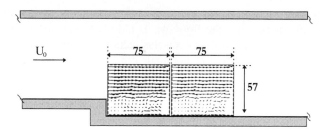

Fig. 7. The two measurement areas (quotes in mm).

In order to reach a high resolution and to adapt the aspect ratio of the sensor (4/3) to the one of the region of interest, the acquisition over the entire measurement domain was split into two sessions pointing the sensor on two adjacent regions as shown in Fig. 7. Composing the data coming out of the two acquisitions all the region of re-circulation has been investigated, including the region of reattachment.

The Reynolds number for this configuration is defined as:

$$R_{e_h} = \frac{U_o \cdot h}{\nu} \tag{11}$$

During the experiments $Re_h=5000$, $U_o=3.8.m/s$ and $\nu=1.52\cdot10^{-5}$ m^2/s.

The separation between the two LASER pulses has been chosen in order to obtain a relatively large particle displacement distribution between the two pictures.

This is a fundamental requirement to achieve a good accuracy. With a pulse separation of $\delta t=300$ μs, the corresponding maximum displacement over the whole field was of about ten pixels.

4 RESULTS

The above-described PIV measurement procedure allowed obtaining data sets with high resolution and accuracy. This made it possible to compare extensively the results with the ones coming from the most recent and sophisticated computational techniques.

In the following paragraphs statistical results from the PIV measurements will be presented and a special attention is devoted to the quantitative comparison of the experimental data with the ones coming from DNS (Le et al.) and from LDV (Jovic and Driver) who studied the same case.

4.1.1 Inlet conditions

The mean inlet velocity and the turbulence intensity profiles have been experimentally determined by means of Hot Wire measurement ($x = -3h$). The boundary layer thickness was found to be $\delta_{99}=1.2h$ with a maximum turbulence intensity level $TI_{bl}=0.11$ inside the BL and a free stream level $TI_{fs}=0.008$. These values find a good correspondence with both DNS and LDV inlet data.

4.1.2   Processing parameters

The processing of the acquired images makes use of a three point gaussian interpolation for the estimation of the peak location with sub-pixel accuracy and the following sequence of window sizes in the grid refinement: (I) 64x32 pixels, (II) 32x16 pixels, (III) 16x8 pixels. A 50% window overlap factor is applied at the last iteration yielding a grid spacing of 8x4 pixels ($0.9x0.48$ mm^2).

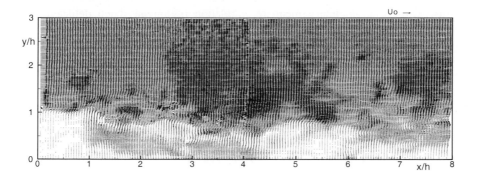

Fig. 8. Instantaneous velocity distribution.

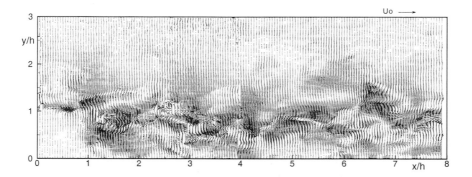

Fig. 9. Instantaneous fluctuating velocity distribution.

4.1.3   Instantaneous velocity field

A typical instantaneous flow pattern is shown in Fig. 8. The free stream velocity U_o normalizes the velocity. Intense span-wise vortices are visualized within the shear layer region. The velocity fluctuation u' associated to these events can reach levels of 0.4 times U_o. Fig. 9 describes the fluctuating velocity field obtained through subtraction of the mean velocity field from the instantaneous one.

4.1.4 Statistical flow quantities

A statistical analysis of the turbulence has been conducted considering 208 available sample fields. The measurement conditions allow considering each velocity field independent of the following and preceding one. Fig. 10 shows the average velocity and vorticity fields. It is possible to visualize the mean shear inversion at the wall (x/h~6). A small set of stream traces aid detecting the mean reattachment region. A weaker inversion is also found in the bottom step corner (x/h<1) where a secondary mean re-circulation region has been detected. The boundary layer re-development downstream the reattachment is visible as a vorticity stripe concentrated at the wall. On the left side of the reattachment a counter-clockwise wall bounded shear develops due to the main circulation bubble.

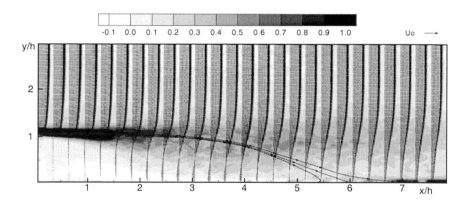

Fig. 10. Mean velocity and span-wise vorticity. Velocity profiles are represented one over six for clarity. Vorticity values are normalized with the field maximum value. Positive values refer to clockwise vorticity.

The comparison between PIV, LDV and DNS data is presented in figs. 11, 12 and 13 where vertical profiles of mean and fluctuating quantities have been extracted from the whole field PIV data available. It is possible to appreciate that the resolution reached with the PIV analysis leads to the detailed description of the shear region downstream the step.

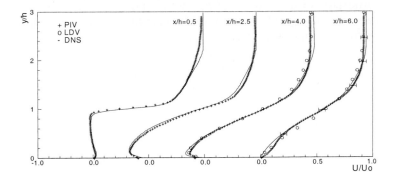

Fig. 11. Mean stream-wise velocity component at different abscissae.

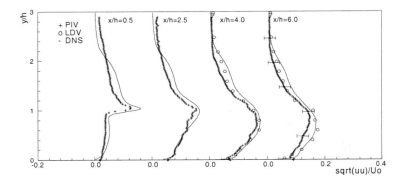

Fig. 12. Stream-wise component turbulence intensity.

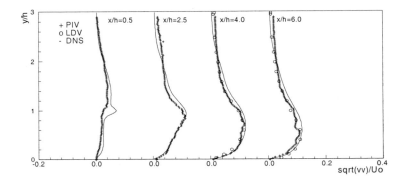

Fig. 13. Vertical component turbulence intensity.

This is particularly evident observing the steep velocity profile at x/h = 0.5. PIV data agree with DNS as well as LDV results within the estimated uncertainty level (3% of the free stream velocity U_o).

The streamwise turbulence intensity (Fig. 11) exhibits a maximum of about 18% at $x/h=5$. All the profiles show the local maxima roughly aligned at one step unit and moving toward the wall downstream. A sharp peak at the smallest x/h location is observed. The shear layer instability leading to spanwise vortex shedding is responsible for this behavior. The vertical turbulence profiles are similar to the previously analyzed, but the peak at $x/h=0.5$ is significantly lower. This is due to the fact that vertical velocity component fluctuations are weakly linked to the shear flapping. The absolute maximum is encountered at $x/h=5$ with 16% intensity.

5   Conclusions

An improved algorithm to interrogate digital PIV recordings has been proposed. The performances of the technique in terms of dynamic range and spatial resolution have been evaluated. The application of successive (cross-correlation) analysis steps with window refinement showed an enlargement of the dynamic range taking as a reference a one-step algorithm.

The measurement of the turbulent flow on the BFS allowed the detailed characterization of the flow field over the whole separation region. The estimated uncertainty of the measurement of the instantaneous velocity is of 0.5% of the free stream value. The measurement error increases for the statistical quantities (3% for the mean quantities and 6% for the Reynolds stress terms) due to the limited number of flow samples.

The spatial resolution of 0.9x0.48-mm^2 grid spacing and the above-mentioned uncertainties assessed the capability of the technique of analyzing successfully this kind of flow. The comparison with numerical and experimental data available on the same test case showed a good quantitative agreement.

The whole image processing procedure did not make use of image pre-processing. As a consequence the additional computational load due to the multiple iterations grows linearly with the number of iterations and a speed of about 300 vectors/s was achieved with an ordinary PC (Pentium 200).

The authors are currently working on further improvements of the technique. Schemes allowing full windows transformations (translation + rotation + stretching) are in assessment phase.

References

ADRIAN R.J. 1991, Particle-imaging techniques for experimental fluid mechanics. *Ann. Rev. Fluid Mech.* **23**, 261-304.

BOILLOT A. & PRASAD A. K. 1996, Optimization procedure for pulse separation in cross-correlation PIV. *Exp. Fluids* **21**, 87-93.

HUANG H. T., FIELDER H. F. & WANG J. J. 1993a Limitation and improvement of PIV, part I. Limitation of conventional techniques due to deformation of particle image patterns. *Exp. Fluids* **15**, 168-174.

HUANG H. T., FIELDER H. F. & WANG J. J. 1993b Limitation and improvement of PIV, part II. Particle image distortion, a novel technique. *Exp. Fluids* **15**, 263-273.

JAMBUNATHAN K., JU X. Y., DOBBINS B. N. & ASHFORT-FROST S 1997 An improved cross correlation technique for particle image velocimetry. *Meas. Sci. Technol.* **6**, 507-514.

JOVIC S. & DRIVER D. M. 1994 Backward facing step measurement at low Reynolds number, Re_h=5000. *NASA technical memorandum* **108807**.

KEANE R. D. & ADRIAN R. J. 1993 Theory of cross-correlation analysis of PIV images. *Flow visualization and image analysis*, 1-25. Kluwer academic Publishers.

KEANE R. D. ADRIAN R. J. & ZHANG Y. 1995 Super-resolution particle imaging velocimetry. *Meas. Sci. Technol.* **6**, 754-768.

LE H., MOIN P. & KIM J. 1997 Direct numerical simulation of turbulent flow over a backward facing step. *JFM* **330**, 349-374.

LECORDIER B. 1997 Etude de l'interaction de la propagation d'une flamme prémélangée avec le champ aérodynamique, par association de la tomographie Laser et de la Vélocimétrie par Images de Particules. *Thése de doctorat*, Université de Rouen.

WESTERWEEL J. 1993 Particle image velocimetry, theory and application. *Ph.D. Thesis*, Delft University.

WESTERWEEL J., DABIRI D. & GHARIB M. 1997 The effect of a discrete window offset on the accuracy of cross-correlation analysis of digital PIV recordings. *Exp. Fluids* **23**, 20-28.

WILLERT C. E. & GHARIB M. 1991 Digital particle image velocimetry. *Exp. Fluids* **10**, 181-193.

II.5. Characterization of a Bluff Body Wake Using LDV and PIV Techniques

Ö. Karatekin, F.Y. Wang, and J.M. Charbonnier

Department of Aerospace and Aeronautics, von Kármán Institute for Fluid Dynamics, Chaussèe de Waterloo 72, 1640-B Rhode-St-Genèse, Belgium

Abstract. Steady and unsteady near-wake characteristics of a sphere-cone bluff body in incompressible flow were investigated experimentally in the 10^4 Reynolds number range and their salient features discussed. Three-dimensional time averaged measurements were made with LDV while the unsteady flow features were examined with PIV. Comparison of mean velocity and turbulence quantities obtained with the two techniques indicated that PIV results, compiled from two orders of magnitude less sampling points, are in good agreement with the mean flow field revealed by LDV but less so for the values related to second moments. This suggests that although the number of PIV images are deemed sufficient to duplicate the mean flow according to the usual variance based statistical criteria, additional samplings are required to ensure a better quantification of higher moment turbulence quantities.

Keywords. PIV, LDV, Bluff bodies, Wake flow, Unsteady flow

1 Introduction

The characterisation of bluff body wakes has been the subject of many investigations (Bearman, 1997; Williamson, 1997; Oertel, 1990). With the advent of Particle Image Velocimetry (PIV), the unsteady nature of the flow field can be captured in a whole field. However, the temporal and spatial resolutions of PIV are often inadequate for a detailed statistical analysis at high velocities. On the other hand, point measurements with Laser Doppler Velocimetry (LDV) provide detailed statistical information as large amount of data can be collected without placing a heavy demand on computer storage resource. However, the flow field revealed by LDV can be misleading in periodic wakes due to the presence of large scale motion which should be considered separately from the background turbulence. In such cases, Grosjean et al. (1997) proposed to combine LDV and PIV for turbulence measurements in unsteady flows, for the two sets of data would complement each other in providing an enhanced understanding of the flow field.

The capture of large scale coherent structures is crucial for the proper modelling and understanding of bluff body wakes as they play the principal role in the mass and energy transport phenomena. Hot-wire or LDV with conditional or phase-averaging sampling techniques had been applied in past. These studies have been recently replaced by PIV technique which provides instantaneous velocity field from which the structures can be directly identified (Lourenco et al., 1997). However, the application of PIV technique to obtain time dependent motion of coherent structures at high velocities is not straightforward due to the limited data acquisition frequency available from a typical system. Double cavity YAG lasers are often employed for PIV measurements and they can achieve sampling frequencies on the order of 10-15 Hz. This acquisition rate is not sufficient for a time dependent analysis of bluff body wakes of reasonable Reynolds numbers in wind tunnels. For example, the wake of a cylinder with a typical Strouhal number St=0.2, a diameter D=0.03 m and a free-stream velocity Uo=20m/s already yields a vortex shedding frequency of 133 Hz which is one order of magnitude higher than the capability of YAG lasers in a Re=4.5x10^4 flow. The utilisation of special lasers such as Eximer with a pulse repetition rate up to several hundred Hz or high speed cameras in combination with a continuous laser are some of the costly options. For the second alternative mentioned, the need for a very powerful and continuous illumination source presents an additional problem.

The present study involves a model of sphere-cone combination representing a planetary entry capsule. Capsule designs are based on a blunt heat shield to resist the high aerothermodynamic loads during the early hypersonic phase of the planetary entry. This design is however prone to severe aerodynamic instabilities at lower altitudes which could lead to the lost of vehicle in the worst situation. The origin of these instabilities is believed to be partly induced by the unsteady wake which results in an amplification of small angular disturbances around the position (Baillion, 1995).

In the schematic of the model presented in Fig. 1, the origin of the co-ordinate system is situated at the nose of the model. For all the measurements, the model was supported by a 3 mm. diameter cantilever rod from the tunnel side wall.

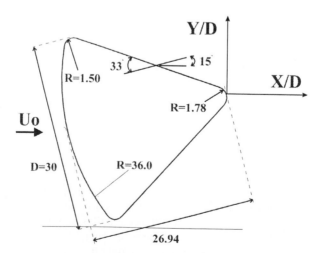

Fig. 1. The model drawing (dimensions in mm)

2 LDV Measurements

2.1 Set-up

The complete three-dimensional wake flows were mapped by using a TSI™ two component backward scattering LDV system. The fiber optical probe of LDV system was mounted on a three axis programmable traversing system. The doppler frequency information is extracted from the scattered light by an IFA 750 Digital Burst correlator and analysed by the commercial data analysis software FIND.

LDV measurements were conducted in the VKI L7 low speed wind tunnel at a Reynolds number of 4.5×10^4 and a blockage ratio of 2.7%. The flow was seeded by vaporising fog oil droplets which produce seeding particles on the order of 1μm in diameter. The typical uncertainty in the measured velocity is less than one percent (Wang et al., 1998).

2.2 Results

Three-dimensional time averaged flow field was obtained by successive measurements of velocity profiles in two orthogonal, namely X-Y and Z-Y planes. The measurements were performed in only one half of the flow field ($Z \leq 0$) due to flow symmetry. Spatial resolutions in the Y and Z directions were 2 mm while the spatial resolution in the X direction was ranged from 3 to 9 mm.

During the measurements, the LDV system was set to collect either 2048 samples per measurement location or as many data points as possible within a 90 seconds window, depending on the condition first reached. In the latter scenario, a minimum of 800 data points was collected at each measured location. The corresponding three-dimensional flow fields are presented in the following sections.

Average flow field in the symmetry Plane: The velocity profiles normalised by the free-stream velocity (Uo) in the symmetry plane are presented in Fig. 2. The measurements in the symmetry plane where the average in-plane velocity is zero, provided certain characteristic features of the flow field and furnished a reference data set for the comparison with PIV measurements. The flow separated from the shoulder of the model's spherical part creates two large counter rotating vortices in the downstream of the model. The wake closure takes place at approximately X/D=1.2 downstream of the origin

General velocity field: Three-dimensional volumetric data are plotted for $0.2 \leq X/D \leq 2.0$, $-1.13 \leq Y/D \leq 0.8$ and $-0.93 \leq Z/D \leq 0.93$. The results are shown in Fig. 3 to 9 where the streamwise axis has been exaggerated for clarity. The streamtrace feature of the commercial software Tecplot™ was used to visualise the velocity data. It is important to note that these streamtraces were constructed from the two-dimensional velocity field in the selected planes and do not represent the streamlines in general.

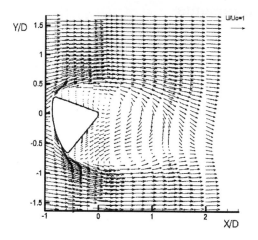

Fig. 2. Time averaged velocity vectors in the symmetry plane (Z/D=0)

The surfaces of constant axial velocity normalised by the free-stream velocity are presented in Fig. 3. The recirculation zone represented by negative velocity contours is shown to occupy a larger spatial extent in the lower side of the model.

Some important features of three-dimensional mean flow field can be observed in Fig. 4 where the streamtraces in four measurement planes are shown (i.e. Y/D=0, X/D=2, Z/D=0.93 and Z/D=0). The streamtraces in the symmetry and Y/D=0 planes illustrate the recirculation area confined behind the body. The plane

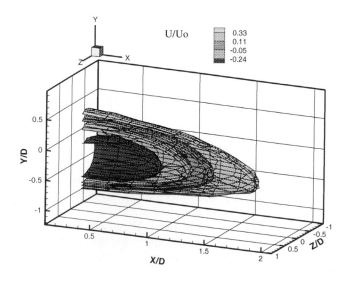

Fig. 3. Axial velocity contours

Z/D=0.93 representing the outer limit of the measurement volume is outside of the recirculation region as suggested by the streamtraces being parallel to streamwise axis. A large scale clock-wise swirling structure is present in the exit plane (i.e. X/D=2), revealing a out-plane flow motion from the Z/D=0.93 plane.

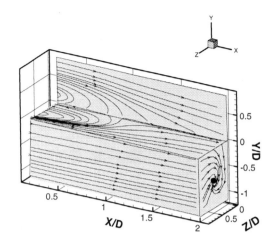

Fig. 4. Three-dimensional streamtrace representation of flow field

Streamtraces in the symmetry plane and two cross planes, (i.e. inlet and outlet planes) are drawn in Fig. 5. The centre of the clock-wise swirl situated at Y/D≈0 in the inlet plane is moved to Y/D≈-0.6 and further away from the symmetry

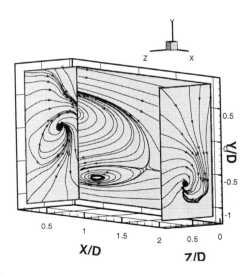

Fig. 5. Streamtraces in the symmetry and cross flow planes

Vorticity: The components of vorticity defined by Eq. (1) were constructed from the measured velocity data by applying the central difference scheme.

$$\vec{\omega} = \vec{\nabla} \times \vec{V} = \vec{\omega}_x + \vec{\omega}_y + \vec{\omega}_z \qquad (1)$$

where:

$$\omega_x = \left(\frac{\partial w}{\partial y} - \frac{\partial v}{\partial z} \right), \quad \omega_y = \left(\frac{\partial u}{\partial z} - \frac{\partial w}{\partial x} \right), \quad \omega_z = \left(\frac{\partial v}{\partial x} - \frac{\partial u}{\partial y} \right)$$

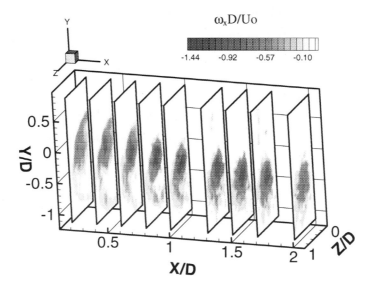

Fig. 6. Non-dimensional streamwise vorticity contours

The iso-streamwise vorticity contours, ω_x, at various downstream locations are plotted in Fig. 6. The location of the maximum streamwise vorticity gradually moves downward, passing through the neighbourhood of wake closure point location where its magnitude is maximum. The existence of streamwise vorticity, its trajectory and distribution were reflective of positive lift force generated by the model at this angle of attack (Wang et al., 1998).

The measurement volume and its mirror image are shown together in Fig. 7 where the contours of the total vorticity magnitude are presented in six successive cross planes. The maximum vorticity is concentrated close to the boundaries of the

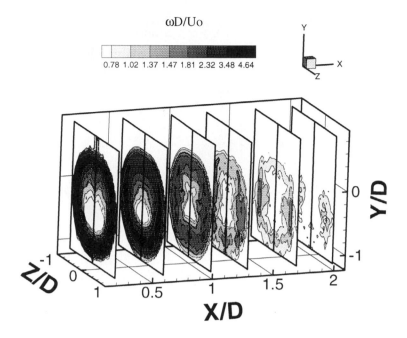

$\omega D/U_0$

0.78 1.02 1.37 1.47 1.81 2.32 3.48 4.64

Fig. 7. Non-dimensional total vorticity contours

recirculating region while the total vorticity magnitude decreases rapidly outside of the recirculation area where, ω_x, remains as the significant component of the vorticity.

Turbulence intensities: Fig. 8 represents the turbulence intensity contours calculated according to Eq. (2) superimposed on the streamtraces. The location of the maximum turbulence intensity follows that of the shear layer, reaching as high as 30% in the vicinity of wake closure point where all the fluids impinged upon each other and abruptly diverted subsequently.

$$I = \frac{\sqrt{1/3\ (\overline{u'^2} + \overline{v'^2} + \overline{w'^2})}}{U_\infty} \qquad (2)$$

The turbulence intensity contours presented in successive cross planes (Fig. 9) confirm the location of the highest turbulence intensity levels in the lower shear layer in the symmetry plane. The maximum intensity occupies an area much larger in the neighbourhood of wake closure point than the other cross planes. The close examination of the data revealed a strong anti-isotropy in the near wake.

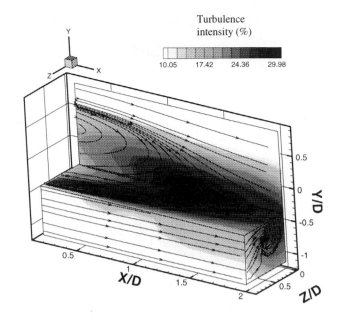

Fig. 8. Three dimensional representation of turbulence intensity

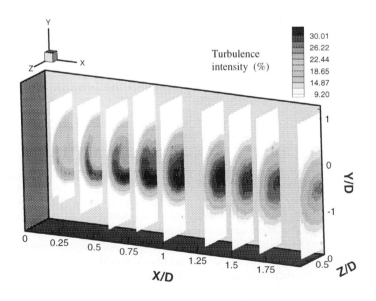

Fig. 9. Turbulence intensity in cross planes

3 PIV Measurements

3.1 Set-up

The measurements were conducted in the VKI-L7+ tunnel at a Reynolds number of 3.2×10^4 with a blockage ratio of 0.7%. A Neodymium YAG laser double pulsing at 10 Hz. was used to illuminate the flow field which was seeded by the same method as for LDV investigation. The images were recorded with a TSI™ cross correlation CCD camera having 640x480 pixels resolution and stored in a personal computer. A TSI™ synchroniser was used to control the timing of the laser and the camera. This measurement set-up allows the acquisition of 6 instantaneous velocity fields at 0.1 seconds apart. For the calculation of statistical quantities, this acquisition procedure will be repeated until the desired number of samples is obtained.

The images were treated with a cross correlation scheme developed at VKI (Scarano and Riethmuller, 1998). The programme optimises the displacement of the second exposure interrogation area to compensate for the loss-of-pairs due to the in-plane motion with an iterative procedure. The processing of the acquired images uses a Gaussian three-point-fit for the sub-pixel interpolation peak determination. Successive cross correlation steps with window refinement are applied and the final interrogation area is 24x24 pixels which yields a spatial resolution of 1.8x1.8 mm (i.e. 0.06Dx0.06D). For the images processed, the amount of non-validated vectors was below 5%. The uncertainty in these measurements was mainly due to the errors in calculating the displacement within a given interrogation area which was 0.1 pixel, yielding an uncertainty of 3.7% in free-stream velocity.

3.2 Instantaneous PIV Images

At high Reynolds numbers the near wake of a bluff body is characterised by two types of quasi-periodic shedding of vorticity. The high frequency mode is a small scale instability associated with the rolling up of shear layers which manifests itself in elliptical vortex rings. The low frequency mode is connected with the alternating motion of the shear layers which causes a "fish tail" flapping of the wake. Downstream of the rear stagnation point, large scale vortical structures are created by mutual interactions between the shear layers. The Strouhal number of this movement was reported to be 0.16, whereas the high mode frequency is one order of magnitude higher, namely $St \approx 2.1$ (Karatekin et al., 1998).

Instantaneous PIV images allow the observation of large and small scale modes concurrently. In Fig. 10, the small regions of high ω_z concentration correspond to high mode instabilities, while the directions of velocity vectors in the exit plane demonstrate the flapping motion of the wake.

It is not possible in the current set-up to track the temporal evolution of flow structures since the maximum data acquisition of the set-up is limited to 10 Hz. whereas the low mode instability at St=0.16 corresponds to a frequency of 80 Hz. To obtain such time resolved data while maintaining reasonably high Reynolds numbers, water tunnel experiments present an interesting alternative to conventional wind tunnels, since for the same Reynolds and Strouhal numbers the characteristic frequency can be significantly reduced.

The dye flow visualisations performed in VKI water tunnel at $Re_D \approx 2000$ provide quantitative time dependent information of vortex rings associated with the high mode instability (Fig. 11). Although the Reynolds number is one order of magnitude smaller than for the wind tunnel tests, the flow field features expected to be comparable due to the separation location fixed at the shoulders between the spherical and conical parts, as well as attributed to the fact that bluff body wakes between $1.2 \times 10^3 < Re_D < 10^5$ are considered to exhibit similar characteristics (Williamson, 1997). This regime of Reynolds number is called shear layer transition in which the free shear layers become unstable due to Kelvin-Helmholtz instabilities.

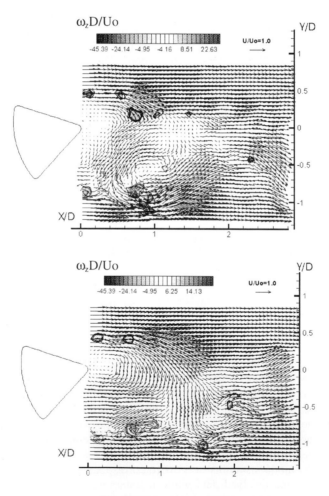

Fig. 10. Instantaneous PIV images

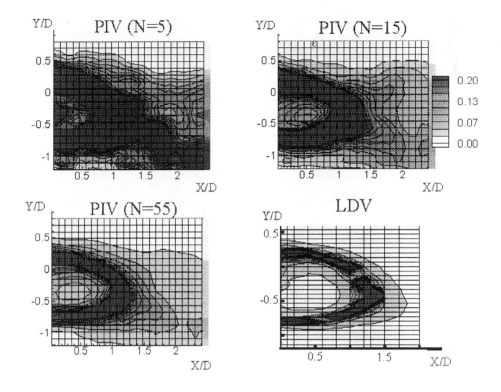

Fig. 12. Spatial distribution of ε as a function of sampling number

The mean and variance were calculated for axial velocity at each measurement node of PIV and LDV data. The associated error, ε, is presented in Fig. 12. The exact number of samples per node in the LDV data is varied in the range of 800<N<2048 according to the sampling criterion chosen by the measurement system (section 3.1). The estimated error associated with LDV measurements is therefore calculated by using the lowest sampling size, i.e., N=800.

As the number of samples increases, the calculated mean approaches the true value. Since the error varies with the square root of 1/N, the calculated error is therefore more significant for small values of N. For N>15, maximum values of ε are found to concentrate in a small area around the recirculation zone and change slowly with increasing N. Although in section 2 the flow behind the model is shown to be highly turbulent, this error analysis suggests that for N>15, it is possible to obtain reasonable averaged flow quantities.

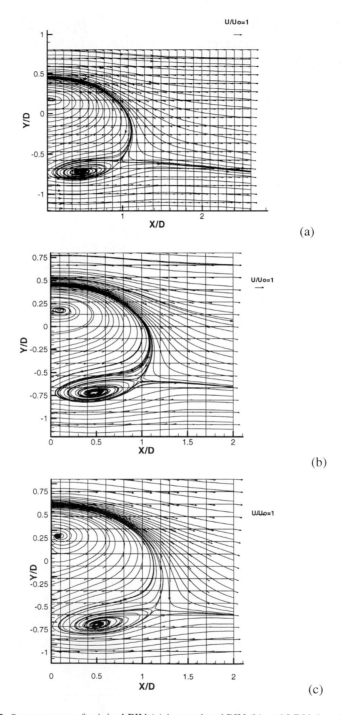

Fig. 13. Streamtraces of original PIV (a) interpolated PIV (b) and LDV data (c)

4   PIV AND LDV FLOW FIELDS COMPARISON

4.1   Interpolation

In order to perform a node-by-node comparison, the LDV and PIV experiments should ideally employ the same measurement mesh. This procedure was however not practical during experiments. Therefore PIV data were interpolated into a grid matching that used by the LDV measurements. Average and root mean square values of u and v velocities were interpolated successively in X/D and Y/D directions by the natural cubic spline method to a final 16x10 grid. The interpolation did not alter the original flow field as presented in Fig. 13 (a), (b).

4.2   Comparison of Critical Point Locations

The comparison of topologically critical locations, i.e. vortex centers and saddle points based on the streamtraces presented in Fig. 13(b) and (c), indicates no significant change for the upper and lower vortex centers. However the wake closure point revealed by PIV data is situated approximately 0.15D upstream. The exact cause of this discrepancy was not examined.

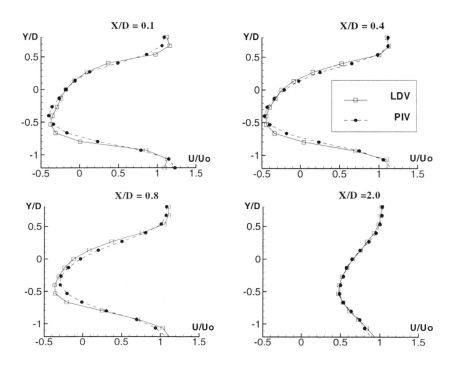

Fig. 14. Comparison of velocity profiles

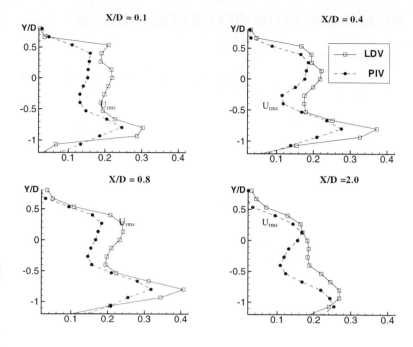

Fig. 15. Comparison of normalized U_rms profiles

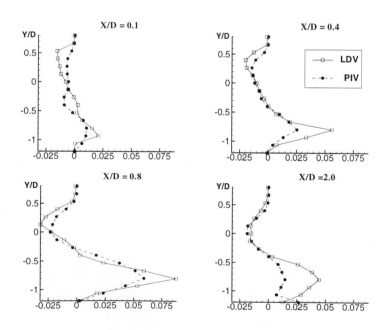

Fig. 16. Comparison of normalized Reynolds shear stress profiles

4.3 Comparison of Profiles

Normalised axial velocity U/Uo as well as velocity fluctuations, U_{rms}, and Reynolds shear stress profiles are calculated according to the Eq. (4).

$$\text{Reynold shear stress} = \frac{\overline{u'v'}}{U_o^2}, \quad U_{rms} = \sqrt{\frac{\overline{u'^2}}{U_o^2}} \tag{4}$$

These profiles are plotted in Fig. 14 to 16 at four locations downstream of the model. The first three profiles are in the recirculation area, with the third being in the vicinity of wake closure location. The last profile is located two body diameters downstream of the origin.

Fig. 14 shows that the agreement between the averaged axial velocity profiles, U/Uo, obtained by LDV and PIV is very good, thereby confirming the conclusion of the error analysis made in section 3.3. As mentioned in the previous section, only a small difference is expected at X/D=0.8 close to the wake closure location.

While the mean flow depends on the first moment of the data, the variance and correlation are of the second moment quantities. Therefore averages from 55 PIV samples are expected to differ more significantly from LDV results for Reynolds stress profiles. Normalised root mean square of axial velocity and Reynolds shear stress profiles are plotted in Fig. 15 and 16 according to Eq. (4).

Although the shape of the profiles is in good agreement, PIV results consistently under-estimate the magnitude of U_{rms} up to 20%. Outside of the recirculation region, the agreement is better for Y/D>0. Reynolds shear stress profiles show better agreement but PIV data again tend to underestimate the shear stress levels in lower shear layer compared to LDV measurements.

5 CONCLUDING REMARKS

LDV and PIV techniques were used complimentarily to map out the three-dimensional unsteady flow field. Time averaged measurements obtained by LDV showed that for the configuration examined, turbulence intensity and vorticity have their maxima located in the lower region of the recirculation area and in the vicinity of wake closure point. Spatial evolution of total vorticity illustrates that the streamwise vorticity is the predominant component beyond the wake closure point.

PIV measurements were performed in the symmetry plane to explore the unsteady nature of the near wake which allowed the observation of a low frequency motion downstream of the wake closure point as well as a high frequency motion created by the shear layer instabilities. Comparison of the non-dimensional mean velocity profiles obtained by two measurement methods in two similar wind tunnels indicates a very good agreement. However comparison of second order statistical quantities such as U_{rms} or Reynolds shear stress are less favorable, indicating that additional PIV samples are required to ensure a better quantification of turbulence quantities.

References

Baillion M., (1995): Blunt Bodies Aerodynamic Derivatives. *AGARD/VKI Special Course on Capsule Aerodynamics*, von Karman Institute for Fluid Dynamics, Belgium.

Bearmann P.W., (1997): Near Wake Flows Behind Two-and Three-Dimensional Bluff Bodies. *J. Wind Eng. & Ind. Aero.* , **69-71**, 33-54.

Grosjean N., Graftieaux L., Michard M., Hübner W., Tropea C., Volkert J., (1997): Combining LDA and PIV for Turbulence Measurements in Unsteady Swirling Flows. *Meas. Sci. Technol.* **8**, 1523-1532.

Karatekin Ö., Wang F.Y., Charbonnier J-M., (1998): Visualisation of an Unsteady Wake. *8th International Symposium on Flow Visualization*, Sorrento, Italy.

Lourenco L., Subramanian S., Ding Z., (1997): Time Series Velocity field Reconstruction From PIV Data, *Meas. Sci., Technol.*, **8**, 1533-1538.

Oertel H., (1990): Wakes Behind Blunt Bodies. *Ann. Rev. Fluid Mech.*, **22**, 539-564.

Olivari D., (1994): Signal Processing. *Measurement Techniques in Fluid Dynamics, von Karman Institute for Fluid Dynamics Annual Lecture Series*, 301-366.

Scarano, F., Riethmuller M.L., 1997, Iterative multigrid approach in PIV image processing with discrete window offset. Accepted for publication in *Exps. Fluids.*

Wang F.Y., Karatekin Ö., Charbonnier J-M., (1998): An Experimental Study of the Flow Field Around an Apollo Capsule at Low Speed. AIAA-98-03193, *36th Aerospace Sciences Meeting and Exhibit*, Reno, USA.

Williamson C.H.K., (1997): Advances in Our Understanding of Vortex Dynamics in Bluff Body Wakes. *J. Wind Eng. & Ind. Aero.* **69-71**, 3-32.

Acknowledgments

The support from ESA's GSTP Programme - Micro-Aerodynamics of Complex External and Internal Configurations is acknowledged with appreciation. The Scientific and Technical Research Council of Turkey for the first author and the NSF-NATO Postdoctoral Fellowship Program in Science and Engineering (Grant Number DGE-9633933) for the second author are likewise gratefully acknowledged

II.6. Reynolds Number Dependence of Near Wall Turbulent Statistics in Channel Flows

M. Fischer, F. Durst, and J. Jovanovic

Lehrstuhl für Strömungsmechanik, University of Erlangen-Nürnberg
Cauerstr.4, 91058 Erlangen, Germany

Abstract.   Theoretical coniderations show that the turbulent fluctuations and the dissipation rate are not universal in inner scaling. The Reynolds number dependence at the wall results from a sink term in the dissipation rate equation. Away from the near-wall region the Reynolds number dependence originates form the streamwise pressure gradient which enters into the equations for the turbulent kinetic energy and turbulent dissipation rate through the gradient production. A quantification of the Reynolds number effects from theoretical considerations is not possible at present. A new method for the statistical interpretation of laser-Doppler signals is derived and applied to investigate the wall-limiting behaviour of turbulent quantities. New experimental results using the LDA measuring technique are presented and compared with Direct Numerical Simulation-data from the literature.

Keywords.   Reynolds number effects, channel flows, near-wall measurements, laser-Doppler anemometry

Nomenclature.

U	Streamwise velocity component
$u, v, w(u_1, u_2, u_3)$	Streamwise, normal and spanwise fluctuating velocity component
$x, y, z(x_1, x_2, x_3)$	Streamwise, normal and spanwise position
p	Pressure
A_w	Value of A at the wall
$\overline{A}$	Time averaging of A
A_{c_v}	Measured value A (A averaged over the whole control volume)
A_c	Value A in the centre of the control volume
A^+	Normalization of A with inner variables (velocities normalized with $u_\tau = \sqrt{\nu(\partial U/\partial y)_w}$, lengths normalized with ν/u_τ)
$\underline{A}$	(A_1, A_2, A_3)

1 INTRODUCTION

Fully developed, turbulent channel flows have been discussed extensively in the literature with regard to the mechanisms of wall bounded turbulence. The mean flow properties and the turbulence quantities of the flow close to the wall have often been assumed to be Reynolds number independent when normalized with inner variables. When measurements of mean velocity, RMS-values and higher order moments did not fullfill this 'Law of the Wall', the deviations were attributed to experimental errors inherently present when hot-wire or laser-Doppler anemometry (LDA) measurements were carried out in the near wall region of a wall bounded flow. These errors increase with decreasing distance from the wall and show a strong Reynolds number dependence. For a long time it was felt that there was no way to answer the question of the universality of turbulent fluctuations close to a wall. This belief came to an end when numerical results from DNS became available. These data clearly reveal a Reynolds number dependence close to the wall for the time mean velocity, the RMS-values of turbulence velocity fluctuations, the Reynolds shear stress, etc. To date, no physical explanation has been given for this finding. This paper presents a theoretical approach to the problem and an experimental clarification.

2 THEORETICAL CONSIDERATIONS

Integration and time-averaging of the momentum equation for an incompressible fluid and application to two-dimensional channel flow yields:

$$1 - \frac{y^+}{Re_\tau} = \frac{d\overline{U}^+}{dy^+} - \overline{u^+v^+}, \quad \text{with} \quad Re_\tau = \frac{u_\tau H}{2\nu} \tag{1}$$

Renewed integration of Eq. (1) results in an expression for the mean velocity that shows a Reynolds number dependence in the form of two substractive terms, one that can be connected to the pressure drop and the other one due to the Reynolds stress:

$$\overline{U}^+ = y^+ - \underbrace{\frac{y^{+2}}{2Re_\tau}}_{\text{pressure drop term}} - \underbrace{\int_0^{y^+} \overline{u^+v^+}dy'^+}_{\text{Reynolds stress term}} \tag{2}$$

Using Taylor series expansions for the fluctuation components, one can show that the Reynolds stress is proportional near the wall to the third power of the distance from the wall:

$$\left.\begin{array}{ll} u^+ \approx a_1^+ y^+ & +a_2^+ y^{+2} \\ v^+ \approx & b_2^+ y^{+2} \\ w^+ \approx c_1^+ y^+ & +c_2^+ y^{+2} \end{array}\right\} \rightarrow \overline{u^+v^+} \approx \overline{a_1^+ b_2^+} y^{+3} \tag{3}$$

At the wall, by definition, there exists no Reynolds number dependence. With increasing y^+, first the pressure drop term is expected to dominate the Reynolds number dependence and later the Reynolds stress term. In the limit of high Reynolds number, the Reynolds number influence vanishes, with the consequence that deviations from the "Law of the Wall" decrease with increasing Reynolds number.

Changes of streamwise turbulent fluctuations are reflected in the equation of the turbulent kinetic energy $k^+ = \frac{1}{2}\overline{u_i^+ u_i^+}$, which consists of the production (P_k), turbulent (T_k) and pressure transport (Π_k), dissipation (ϵ) and viscous diffusion (D_k):

$$0 = -\underbrace{\overline{u^+ v^+}\frac{\partial \overline{U}^+}{\partial y^+}}_{P_k^+} - \underbrace{\frac{\partial \overline{u_i^+ u_i^+ v^+}}{\partial y^+}}_{T_k^+} - \underbrace{\overline{u_i^+ \frac{\partial p^+}{\partial x_i^+}}}_{\Pi_k^+} - \underbrace{\overline{\frac{\partial u_i^+}{\partial x_k^+}\frac{\partial u_i^+}{\partial x_k^+}}}_{\epsilon^+} + \underbrace{\frac{\partial^2 k^+}{\partial y^{+2}}}_{D_k^+}. \tag{4}$$

In the buffer layer the Reynolds number dependence is governed by the production rate term P_k^+ which can be expressed as follows:

$$P_k^+ = \frac{\partial \overline{U}^+}{\partial y^+} - \frac{y^+}{Re_\tau}\frac{\partial \overline{U}^+}{\partial y^+} - \left(\frac{\partial \overline{U}^+}{\partial y^+}\right)^2 \tag{5}$$

The Reynolds number variation of P_k^+ is caused effectively by the pressure drop. However, in the very near wall region the whole physics of Reynolds number effects is contained in behaviour of the dissipation rate term ϵ^+. Inserting Eq. (3) into Eq. (4) results in a direct relationship between k and ϵ:

$$\frac{2(k^+)_w}{y^{+2}} = (\epsilon^+)_w \tag{6}$$

Applying the 2-point-correlation technique (Jovanović et al. 1995) and considering the spatial derivations of the product of velocity fluctuations shows finally that the dissipation equation can be subdivided into an inhomogenous and a homogenous part, which are identical at the wall.

$$\epsilon^+ = \overline{\frac{\partial u_i^+}{\partial x_k^+}\frac{\partial u_i^+}{\partial x_k^+}} = \underbrace{\frac{1}{2}\Delta^+ k^+}_{\text{inhomogenous}} + \underbrace{\epsilon_h^+}_{\text{homogenous}} \tag{7}$$

The behaviour of the turbulent dissipation rate is entirely defined by its homogenous part which can be modelled according to Eq. (8) (see Jovanović et al. 1996) with $Re_\lambda = \frac{\lambda\sqrt{2k}}{\nu}$, λ =Taylor micro scale.

$$0 = \underbrace{-2A\frac{\epsilon_h^+ \overline{u^+ v^+}}{k^+}\frac{\partial \overline{U}^+}{\partial y^+} - 20B\frac{k^+}{R_\lambda}S_{12}^+\frac{\partial \overline{U}^+}{\partial y^+}}_{\text{production term}} - \underbrace{\Psi\frac{\epsilon_h^{+2}}{k^+}}_{\text{sink term}}$$

$$+ \underbrace{\frac{\partial}{\partial y^+}\left[C_\epsilon \frac{k^+}{\epsilon_h^+} \overline{v^{+2}} \frac{\partial \epsilon_h^+}{\partial y^+}\right]}_{\text{transport term}} + \underbrace{\frac{1}{2}\triangle^+ \epsilon_h^+}_{\text{viscous destruction}} \tag{8}$$

Near the wall, besides the viscous destruction, the production term and the sink term show a significant Reynolds number dependence. The viscous destruction plays a rather passive role as it has just to balance the sum of all other terms. The additional production term acts to raise up the level of ϵ with increasing Reynolds number in the buffer layer. At the wall the influence of the additional production vanishes. In the region of viscous sublayer the sink term $(-\psi\epsilon_h^2/k)$ of the dissipation rate equation governs the Reynold number dependence. Analysis of the data extracted from numerical simulations indicates that the anisotropy of turbulence at the wall decreases with increasing Reynolds number. This trend implies decrease of ψ and ϵ at the wall with increasing Reynolds number. As the wall is approached, to the first approximation, turbulence might be considered as two-component and two-dimensional. For such state ψ may be approximated in the form $\psi_{2C-2D} \simeq 0.02R_\lambda$. Analysis of DNS-data demonstrates that the behaviour of $(\psi)_{2D-2C}$ very close to the wall is responsible for raise up of ϵ at the wall with increasing the Reynolds number.

3 Experimental Setup

To study low Reynolds number effects on fully developed plane channel flows, a water flow facility was set up that permitted mean velocities of up to 2.5 m/s to be obtained. Figure 1 shows a schematic of the test rig used for the present investigations. To drive the flow at a constant, pre-chosen flow rate, a total head tank was installed with a water head of approximately 6 m. Experimental investigations on how to set up fully developed low Re-number flows were performed in the two- dimensional channel with dimensions $L \times B \times H = 1 \times 0.18 \times 0.01$ m. The channel test section was proceeded by a rectangular contraction (0.15x0.18 m), providing a defined profile at the channel inlet. Upstream of the contraction, honeycomb and grids were installed to improve the flow uniformity.

The developing boundary layers were tripped by plates inserted at the entrance of the channel with an optimised height of 0.75 mm, giving a blockage ratio of 15%, as determined by Durst et al. (1998). The measurements reported in this paper were performed 71 channel heights downstream of the channel inlet using laser-Doppler anemometers with measuring control volumes of various sizes. Two different LDA optical systems were employed to ensure the correct measurement of near wall time averaged flow properties. Since the limiting values of the measured flow properties for $y^+ \rightarrow 0$ were of interest, reliable measurements for $y^+ \leq 4$ were needed. The parameters of the LDA systems used for this purpose are shown in Table 1. A TSI-

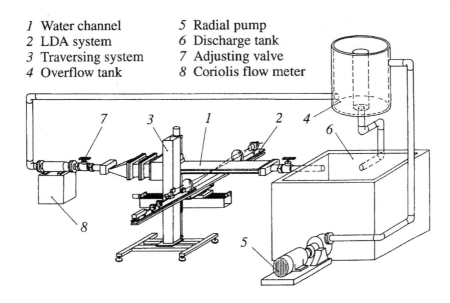

1 Water channel   5 Radial pump
2 LDA system   6 Discharge tank
3 Traversing system   7 Adjusting valve
4 Overflow tank   8 Coriolis flow meter

Figure 1: Sketch of channel flow facility showing major parts of the test-rig.

1990 counter was employed for measurement of the Doppler frequency. It was operated in the total burst mode with at least 32 cycles per burst being measured. The mean velocity and velocity fluctuations were computed using arrival time averages to correct for velocity bias at the very low data rates.

To allow measurements very close to the wall and to reduce intercepting reflected light, the transmitting and receiving optics were tilted at small angles towards the channel wall. In this way, the measuring control volume could be located with an positioning accuracy of about 5 μm, with the closest location being about half a measuring volume diameter away from the wall. The wall shear stress was determined directly from the near wall mean velocity measurements using the polynomial fit procedure described by Durst et al. (1996).

4 Measuring Control Volume Effects

Due to the spatial velocity distribution of the mean flow and the spatial distribution of the moments of the turbulent velocity fluctuations, the actual velocity measured for each scattering particle does not correspond to the velocity value at the centre of the measuring control volume but rather to time and volume integrated information. In the following derivations, the ellipsoidal shape of the control volume and the discreteness of particle detection will be taken into account.

Table 1: Computation of wall-next measuring positions for the different LDA setups and for a few exemplaric Reynolds numbers. (The measuring volume size is based on the e^{-2} Gaussian intensity cut-off point)

Laser	beam expansion factor	R_e	d	d^+	y_{min}^+
HeNe (15 mW)	2.5	3500	80μ	1.7	0.85
HeNe (15 mW)	2.5	10000	80μ	4.3	2.15
HeNe (15 mW)	2.5	21000	80μ	8.0	4.0
Nd-YAG (100mW)	10	10000	30μ	1.7	0.85
Nd-YAG (100mW)	10	21000	30μ	3.3	1.65

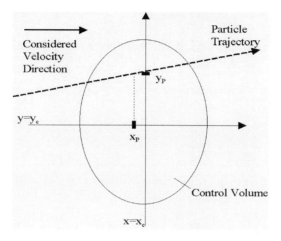

Figure 2: Control volume sketch

The time averaged as well as the fluctuating parts of the velocity may be expanded in Taylor series.

$$\underline{U}(\underline{x}, t) = \overline{\underline{U}}(\underline{x}) + \underline{u}(\underline{x}, t) \tag{9}$$

$$\overline{\underline{U}}(\underline{x}) = \overline{\underline{U}}(\underline{x}_c) + \sum_{n=1}^{\infty} \frac{1}{n!} \left(\frac{\partial^n \overline{\underline{U}}(\underline{x})}{\partial x_i^n} \right)_{\underline{x}=\underline{x}_c} x_i^n \tag{10}$$

$$\underline{u}(\underline{x}, t) = \underline{u}(\underline{x}_c, t) + \sum_{n=1}^{\infty} \frac{1}{n!} \left(\frac{\partial^n \underline{u}(\underline{x}, t)}{\partial x_i^n} \right)_{\underline{x}=\underline{x}_c} x_i^n \tag{11}$$

If one now decomposes the instantaneous velocity for one event into mean and fluctuating parts at the position of the particle trace and alternatively in expressions of the whole control volume (that correspond to the information of the measurement), a connection between measured time averaged quantities and those in the center of the control volume arises:

$$\overline{U}(x_c, y, z) + u(x_c, y, z, t) = \overline{U}_{c_v} + u_{c_v}(t) \tag{12}$$

Volume integration results in the following correction equations for mean velocity and turbulent intensity.

$$\overline{U}^+_{c_v} \approx \overline{U}^+(y_c^+) + \frac{1}{8} \left(\frac{d_y^+}{2} \right)^2 \left(\frac{\partial^2 \overline{U}^+}{\partial y^{+2}} \right)_c \tag{13}$$

$$u_{c_v}^{+2} \approx \overline{u^{+2}(x_c)} + \frac{1}{8} \left(\frac{d_y^+}{2} \right)^2 \left[2 \left(\frac{\partial \overline{U^+}}{\partial y^+} \right)_c^2 + \left(\frac{\partial^2 \overline{u^{+2}}}{\partial y^{+2}} \right)_c \right] \tag{14}$$

The measured mean velocity is very close to the time averaged value at the centre of the measuring control volume since the correction depends only on the second derivative of the mean velocity profile (which is very small in the vincinity of the wall). The correction for turbulent intensities consists of a term due to the mean velocity gradient accross the measuring control volume and a term due to the curvature of the RMS-value profile of the longitudinal velocity fluctuations. The sum of both terms enlarges the measured rms-value fluctuations in comparison to the centre value to be measured. Analysis of the wall limiting behaviour of equation (14) shows that the second correction term can contribute up to about 16% of the total correction in Eq. (14). Similar to the above derivations, correction equations can also be derived for higher order moments.

DNS data of Gilbert and Kleiser (1991) ($Re_m = 6,700, Re_\tau = 210.7$) were evaluated in order to simulate, for different control volume sizes, the expected effects of the derived corrections on the mean velocity U, the turbulence level u'/U, the Skewness factor $S = \dfrac{\overline{u_{c_v}^{+3}(t)}}{(\overline{u_{c_v}^{+2}(t)})^{3/2}}$, and the Flatness factor $F = \dfrac{\overline{u_{c_v}^{+4}(t)}}{(\overline{u_{c_v}^{+2}(t)})^2}$. In Fig. 3 (top) it can be seen that the corrections raise the measured turbulence level of the velocity fluctuations by more than one hundred percent near the wall. Computations were made only for the locations for which the control volume does not contact the wall (i.e. for y^+ values $y^+ > d^+/2$). To illustrate the difference between the two correction terms of Eq. (14) the terms are plotted in Fig. 3 (Bottom) for one size of the control volume ($d^+ = 1$) as an example. It is remarkable that the second correction term decreases the correction in the buffer layer.

The control volume size, d_y, enters the correction terms in Eqs. (13) and (14) as a weighting factor. However, use of the effective optical diameter (computed from the Gaussian light intensity distribution in the volume) for size corrections neglects the fact that each velocity measurement is dependent on the size distribution of scattering particles, on the scattered light detection system, on the LDA signal amplification, and on the discrimination level

216

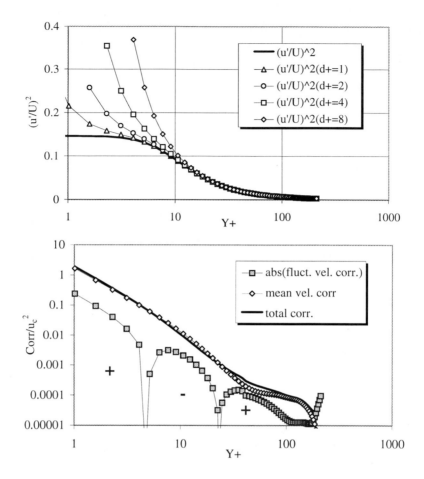

Figure 3: Expected influence on measured tubulence level using DNS-data and illustration of the contibution of the different terms in eq. (14).

employed in the signal processing electronics. To obtain the real effective measuring control volume size of a particular LDA system, it is suggested to apply the system to a laminar flow and to measure the mean velocity and variance. Applying Eq. (14) to such measurements yields that the only unknown quantity in this equation is the effective control volume size. Besides a small noise contribution, only the term defined by the mean velocity gradient remains. Thus the measured fluctuations are expected to obey the

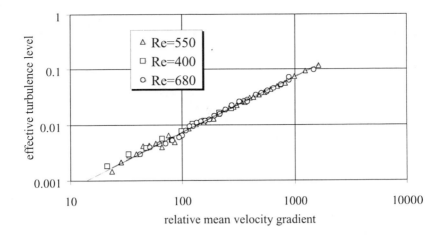

Figure 4: Effective turbulence level plottet against the relative mean velocity gradient

following expression:

$$\underbrace{\sqrt{\frac{\overline{u_{c_v}'^2}}{\overline{U}^2} - \frac{\overline{u_{noise}'^2}}{\overline{U}^2}}}_{\text{effective turbulence level}} = d_y \quad \underbrace{\frac{1}{4\overline{U}}\left(\frac{\partial\overline{U}}{\partial y}\right)_c}_{\text{relative mean gradient}} \tag{15}$$

Analysis of the slope of the curve in Fig. 4 results directly in the effective control volume size. After the determination of d_y, the same optical and electronical configuration can then be applied to the turbulent boundary layer.

5 Experimental Results

In Fig. 5 the mean velocity distributions for various Reynolds numbers are shown. A systematic variation in the core region of the flow with increasing Reynolds number can be reported. Of particular interest was the behaviour very close to the wall. Analysis of data from DNS have shown that at the outer edge of the viscous sublayer, the sum of the pressure drop term and the Reynolds stress term are nearly independent of Reynolds number and therefore cannot be resolved experimentally. According to direct numerical simulations of turbulent channel flow, the streamwise velocity component accounts for about 75% of the dissipation rate at the wall (Kim et al., 1987; Antonia et al., 1992): $(\epsilon^+)_w \approx 1.3\left(\frac{\overline{u'^2}}{U^{+2}}\right)_w$. Therefore the measurement of

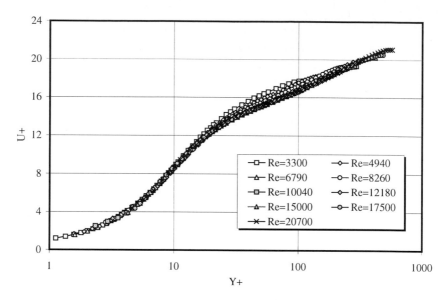

Figure 5: Mean velocity profiles non-dimensionalized with inner variables

streamwise turbulent fluctuations gives an insight into the influence of Reynolds number on $(\epsilon^+)_w$. In Fig. 6 the RMS-values of the uncorrected data are compared with the corrected ones. They reveal the influence of control volume effects on position and intensity of the measured RMS-maximum. Figure 7 shows that a Reynolds number dependence in the viscous sublayer can be seen clearly, leading to increasing wall limiting values with increasing Reynolds number. For the evaluation of these values, the near wall data in the region $y^+ < 10$ were extrapolated to the wall position. In Fig. 8 the resulting values extracted from the measurements are plotted against Reynolds number. Comparison between experiment and DNS results reveals the same trend but a remarkable deviation in the absolute values of u'/U at higher Reynolds numbers. The measured turbulent fluctuations can be approximated quite well by an analytical expression that contains a limiting value for very high Reynolds numbers and and a subtractive part inversely proportional to the Reynolds number.

$$(u'/U)_w = (u'/U)_{w;Re\to\infty} - A/Re_\tau \tag{16}$$

Fitting of Eq. (16) yields for the constants:
$(u'/U)_{w;Re\to\infty} = 0.40 \pm 0.01 \quad A = 7.0 \pm 0.5$. Extrapolating this result, one would expect the Reynolds number effects to vanish asymptotically above $Re_m \approx 35000$.

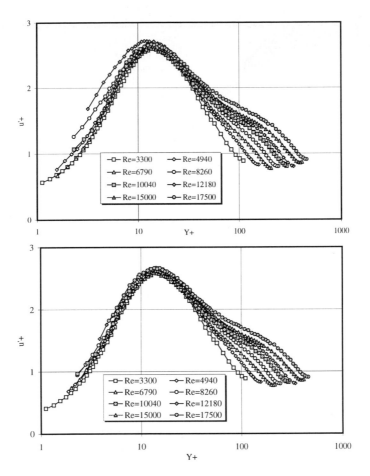

Figure 6: Profiles of the RMS values of turbulent fluctuation measured (top) and corrected (bottom)

6 Conclusions and Final Remarks

Near-wall laser-Doppler measurements were performed in a turbulent channel flow in order to examine the Reynolds number dependence of time-averaged turbulence quantities in the vincinity of the wall. A new method for the statistical interpretation of LDA-signals was proposed and realized in order to get acurate local flow information in regions of high shear rate. Deviations from the 'Law of the Wall' for the mean velocity down to the inner edge of the buffer layer could be experimentally quantified. The wall limits of fluctuations have been experimentally determined and put into the frame of an analytical treatment. The turbulent fluctuations exhibit a Reynolds number dependence in the near wall region. The experimental finding have

220

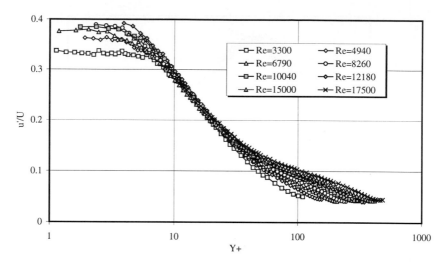

Figure 7: Streamwise turbulent fluctuation normalized with mean velocity

implications for the modelling of low Reynolds number flows.

7 Acknowledgements

This research received financial support through the Deutsche Forschungsgemeinschaft within the project Du 101/49-1.

8 References

ANTONIA, R.A; TEITEL, M.; KIM, J.; BROWNE, L.W.B., 1992, Low-Reynolds number effects in a fully developed channel flow, *J. Fluid Mech.* **236**: 579-605

DURST, F.; KIKURA,H.; LEKAKIS, I.; JOVANOVIĆ, J.; YE, Q.-Y., 1996, Wall shear stress determination from near-wall mean velocity data in turbulent pipe and channel flows, *Exp. Fluids* **67**: 257-271

DURST, F.; FISCHER, M.; JOVANOVIĆ, J.; KIKURA, H., 1998, Methods to set up and investigate low Reynols number, fully developed turbulent plane channel flows, *J. Fluid Eng.* **120**, 496-503

GILBERT, N. and KLEISER, L., 1991, Turbulence model testing with the aid of direct numerical simulation results, *8th Symp. on Turbulent Shear Flows*, Sept. 9-11, TU Munich, pp. 26.1.1-26.1.6

JOVANOVIĆ, J.; YE, Q.-Y. and DURST, F., 1995, Statistical interpretation of the turbulent dissipation rate in wall-bounded flows, *J. Fluid Mech.* **293**: 321-347

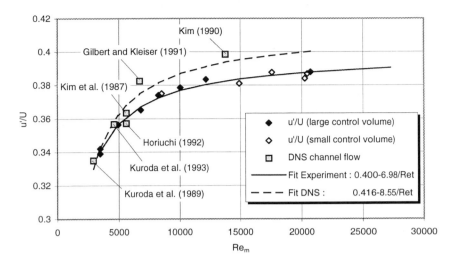

Figure 8: Plot of determined wall limiting values for u'/U. Comparison with DNS-data

JOVANOVIĆ, J.; YE, Q.-Y.; JAKIRLIĆ, S. and DURST, F., 1996, Turbulence closure for dissipation rate correlations, submitted to *J. Fluid Mech.*

KIM, J.; MOIN, P. and MOSER, R., 1987, Turbulence statistics in fully developed channel flow at low Reynolds number, *J. Fluid Mech.* **177**: 133–166

KIM, J., 1990, Collaborative Testing of Turbulence Models (Organized by P. Bradshaw), Data Disk No.4.

KURODA, A.; KASAGI, N. and HIRATA, M., 1989, A direct numerical simulation of fully developed turbulent channel flow, *Int. Symp. on Comput. Fluid Dynamics*, Nagoya, pp 1174-1179

KURODA, A.; KASAGI, N. and HIRATA, M., 1993, Direct numerical simulation of the turbulent plane Couette-Poiseulle flows: Effect of mean shear on the near wall turbulence structures, *9th Symp. on Turbulent Shear Flows*, Kyoto, Japan, Aug. 16-18, pp. 8.4.1-8.4.6

II.7. Weakly Interacting Two Parallel Plane Jets
J.C.S. Lai[1] and A. Nasr[2]

[1] School of Aerospace & Mechanical Engineering
University College, The University of New South Wales
Australian Defence Force Academy
Canberra, ACT 2600, AUSTRALIA.
[2] Department of Physics
Shahed University
Tehran, IRAN.

Abstract. The mean velocity field of unventilated two parallel plane jets for large nozzle spacing ratios has been studied using a two-component laser Doppler anemometer. Results show that by comparison with small s/w (less than 5), the interactions between the inner shear layer and the recirculation zone for large s/w are weaker. The initial peaks of turbulence intensities and Reynolds shear stress corresponding to the nozzle inner and outer edges tend to persist downstream from the location of the vortex centre without being overwhelmed by the influences caused by the recirculating flow in the recirculation zone. The lateral turbulence intensities and Reynolds shear stress in the outer shear layer spread and decay more rapidly than in the inner shear layer. The attraction of the two individual jets towards each other is clearly illustrated by the mean velocity vector field and supported by static pressure measurements. Comparisons of the results of the unventilated jets with those of ventilated jets in the literature indicate there is no recirculation zone in the ventilated jets and the unventilated jets combine to develop into a single free jet much earlier than ventilated jets.

Keywords. Two parallel plane jets, LDA.

1   Introduction

Two parallel plane jets play an important role in numerous technological applications such as entrainment and mixing processes in boiler and gas turbine combustion chambers and injection systems and thrust-augmenting ejectors for V/STOL aircraft. The flow patterns of unventilated two parallel plane jets are shown schematically in Fig. 1. Here, two identical plane nozzles each of width w are separated in the lateral (y) direction by a distance s, giving a nozzle spacing ratio of s/w. The origin of the Cartesian coordinate system (x-y) is located at the nozzle plate and the x-axis is in the plane of symmetry which bisects the two nozzles.   Owing to the mutual entrainment between the two jets, a sub-atmospheric pressure zone is formed close to the nozzle plate.   The subatmospheric pressure causes individual jets to curve towards each other in a region close to the nozzle plate known as the converging region. In the recirculation zone, the mean streamwise velocity on the x-axis (U_c) is negative. The two inner shear layers merge at the merging point (mp) where U_c is zero. Downstream from the merging point, in the merging region, the two jets continue to interact with each other and U_c increases until it reaches a maximum value (U_{cmax}) at the combined point (cp). The streamwise distances from the nozzle plate to the merging and combined points are referred to as the merging length (x_{mp}) and the combined length (x_{cp}) respectively. Downstream from the combined point is the combined region where the two jets combine together to resemble a single free jet flow.

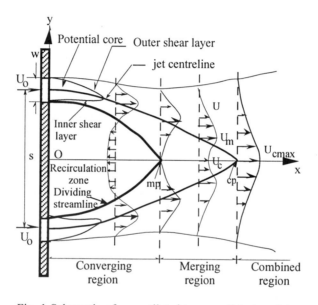

Fig. 1 Schematic of unventilated two parallel plane jets.

Although two parallel plane jets have been experimentally investigated by many researchers such as Miller and Comings (1960), Tanaka (1970, 1974), Murai et al. (1976). Militzer (1977), Ko and Lau (1989) and Lin and Sheu (1990, 1991), these studies were conducted using either hot-wire or Pitot tube which are subject to severe errors in the converging and merging regions. Recently, two component LDA measurements of unventilated two parallel plane jets were reported by Nasr & Lai (1997a, b, c) for small s/w (less than 5). By examining the strength of the interactions between the recirculating flow and the inner shear layer in terms of the turbulence intensities and Reynolds shear stress distributions for various s/w, Nasr & Lai (1997b) classified unventilated two parallel plane jets into weakly interacting and strongly interacting jets according to whether s/w exceeds 5.

Details of strongly interacting two parallel plane jets for s/w less than 5 are given by Nasr & Lai (1997b). The primary objective of this paper is, therefore, to present the velocity field of weakly interacting unventilated two parallel plane jets with large nozzle spacing ratio (namely, s/w=7.5 and 11.25) measured with a two-component laser Doppler anemometer (LDA) for a nozzle exit Reynolds number of 11,000. These data are hitherto not available in the literature and will provide a useful database for validating turbulence models for complex shear flows. Comparisons are also made with ventilated two parallel plane jets.

2   Apparatus And Experimental Conditions

Experiments were conducted for two-dimensional nozzles each with width 6 mm and two different nozzle spacing ratios, namely, s/w=7.5 and 11.25. The nozzle exit Reynolds number was 11,000 and the exit streamwise turbulence intensity was 2% at the centre-line. Velocity measurements were made with a two-component Dantech LDA and were corrected with the residence time weighting function. The uncertainty in mean velocities and turbulence intensities is within ± 0.08 m/s and that of Reynolds shear stress is within ± 0.4 m^2/s^2. In order to measure the static pressure distributions in the jets, a disc type static pressure probe was designed and validated against a standard Pitot-static tube in the test section of a low turbulence subsonic wind tunnel. Detailed experimental procedures and experimental set-up were given by Nasr & Lai (1996, 1997a).

3.   Results

3.1 Effect of Side plates

For unventilated parallel jets, there could be significant entrainment of surrounding air through the top (ceiling) and bottom (floor) sides of the jets if side plates were not installed. The effect of side plates was investigated by measuring the static pressure with a disc type pressure probe along the x-axis for s/w=4.25 and w=10 mm.

As shown in Fig. 2(a), with or without the installation of side plates, the static pressure along the x-axis decreases from subatmospheric at the nozzle plate to a minimum in the recirculation flow region and then increases to a maximum above atmospheric near the merging point due to the collision of the inner shear layers. Downstream of the merging point, the static pressure decreases to subatmospheric value due to flow acceleration followed by a slow recovery to atmospheric pressure far downstream. It can be seen that when side plates were not installed, the streamwise pressure gradient near the nozzle exit was significantly reduced due to the entrainment of the surrounding fluid through the sides of the nozzles, thereby slowing the merging of the two jets. This is supported by the streamwise velocity distribution in Fig. 2 (b) which shows that both the merging point and combined point occur further downstream when side plates were not installed. These results highlight the importance of side plates to enhance two-dimensionality of two parallel jets. Thus all the experiments reported here were conducted with side plates installed.

226

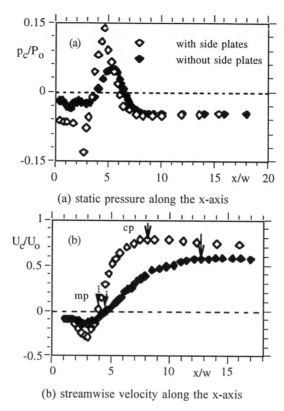

(a) static pressure along the x-axis

(b) streamwise velocity along the x-axis

Fig. 2 Effect of side plates (s/w=4.25, w=10mm).

3.2 Classification of two parallel jets

In reattaching shear layers, Bradshaw & Wong (1972) and Eaton & Johnston (1981) found that the profiles of turbulence intensities and their correlations can be altered weakly or strongly depending on the strength of the influences caused by the recirculating flow. For two parallel plane jets, the development of the inner shear layers is affected in similar way as in the reattaching flows except that the collision of the two jets and the merging process produces a much more complex flow field. Profiles of streamwise turbulence intensities and Reynolds shear stress at the streamwise location of the standing vortex centre for s/w=4.25 and 7.5 are displayed in Figs. 3(a) and (b) respectively. The stronger interaction between the recirculating flow and the inner shear layer for s/w=4.25 than for s/w=7.5 is indicated by the presence of an additional peak in the u' profile near the x-axis with a higher magnitude than that of the inner shear layer peak (Fig. 3a) and the larger negative Reynolds shear stress (Fig. 3b). Based on these results, two parallel plane jets with s/w equal to or greater than 7.5 are considered as weakly interacting jets.

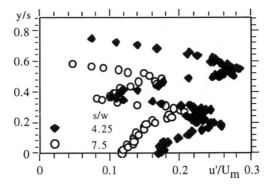

Fig. 3 (a) Effect of s/w on streamwise turbulence intensity.

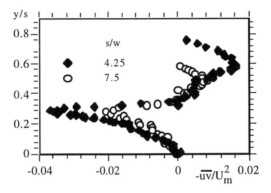

Fig. 3(b) Effect of s/w on Reynolds shear stress.

3.3 Mean flow characteristics at the nozzle exit

Figure 4 displays the nozzle exit conditions for unventilated two parallel plane jets with s/w = 7.5 and 11.25. Here, the origin of the coordinate system is located at the nozzle inner edge. The streamwise velocity profiles (Fig. 4a) have a top-hat shape similar to that of a single free jet flow, thus indicating the existence of a potential core in the main flow near the nozzle plate. The exit streamwise velocity near the nozzle inner edge is slightly higher for s/w = 7.5 than for s/w = 11.25, indicating higher flow acceleration due to the sub-atmospheric pressure zone near the nozzle plate. The lateral velocity profiles at the nozzle exit in Fig. 4(b) show that the individual jet deflects towards the x-axis even at the exit plane. Higher flow deflection can be observed for s/w = 7.5 as the lateral velocity is negative across over 90% of the nozzle width at the nozzle exit. For s/w= 11.25, both streamwise and lateral velocity profiles at the nozzle exit appear to be similar to those of a single free jet. This is deduced from the observation of both positive and negative mean lateral velocity at the nozzle exit in the outer and inner edges of the jet. The distributions of the streamwise and lateral turbulence intensities at the nozzle exit in Figs. 4(c) and (d) show that

228

their values at the central streamline and at the nozzle exit are quite low, being about 2 %. The peaks of u'/U$_o$ and v'/U$_o$ in these figures identify the inner and outer shear layers at the nozzle exit.

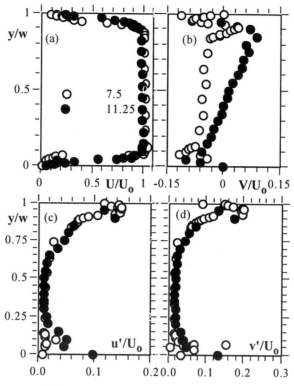

Fig.4 Nozzle exit conditions for various s/w
(a) U/U$_0$ (b) v/ U$_0$ (c) u'/ U$_0$ (d) v'/ U$_0$.

3.4 Mean velocity vectors

The mean velocity vector field of two parallel jets for s/w = 7.5 and 11.25 is shown in Fig. 5. These vectors have been determined from the simultaneous measurements of mean streamwise and lateral velocity components. Reversed flow and streamline deflection towards the x-axis are evident. The dividing streamline, standing vortex centre and merging point can be easily identified. The merging length is found to be about 8.3w for s/w=7.5 compared with 12.5w for s/w=11.25. The absolute magnitude of the reversed flow along the x-axis increases downstream from the nozzle plate until it reaches a maximum value of about 0.17U$_o$ at a streamwise distance corresponding to the location of the standing vortex centre (x$_{vc}$/w = 5.25) for s/w=7.5. These compare with 0.23U$_o$ at (x$_{vc}$/w = 9, y$_{vc}$/w =2) for s/w=11.25. U$_o$ then decreases as the flow proceeds downstream and becomes zero at the merging point. The mean streamwise velocity along the x-axis then increases downstream from the merging point until it becomes maximum at the combined point.

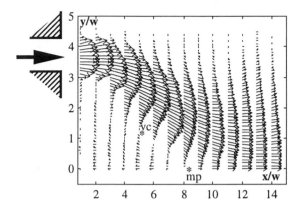

Fig. 5 (a) Mean velocity vectors for s/w=7.5.

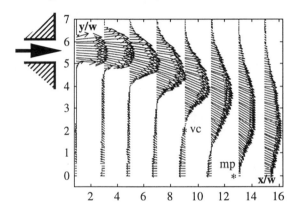

Fig. 5 (b) Mean velocity vectors for s/w=11.25.

3.5 Mean streamwise velocity distributions along the x-axis

The distributions of the mean streamwise velocity along the x-axis, U_c/U_o, are shown in Fig. 6 for s/w = 7.5 and 11.25. The negative mean streamwise velocity close to the nozzle plate for both jet configurations indicates the existence of contra-rotating standing vortices which recirculate air on the concave side of the jets. The magnitude of the maximum negative velocity along the x-axis for s/w = 7.5 is less than that for s/w = 11.25. The merging point for each s/w, where U_c/U_o is zero, can be identified by the intersection of the zero velocity dashed line with the corresponding velocity profile. Due to the interaction of the two jets, the velocity along the x-axis then increases from the merging point and becomes maximum at the combined point, where the two jets combine together to resemble a single jet flow. It can be seen from Fig. 6 that, the maximum mean streamwise velocity along the x-axis at the combined point, U_{cmax}/U_o, is less for s/w = 11.25. This is because the jet width is greater at the combined point for larger s/w. However, the velocity decay rate along the x-axis downstream from the combined point seems to be nearly the same for both s/w.

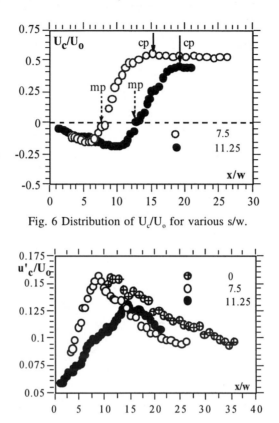

Fig. 6 Distribution of U_c/U_o for various s/w.

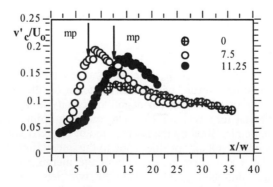

Fig. 7(a) Streamwise turbulence intensity distribution on the x-axis for various s/w.

Fig. 7(b) Lateral turbulence intensity distribution on the x-axis for various s/w.

3.6 Turbulence intensity distributions along the x-axis

Distributions of streamwise and lateral turbulence intensity on the x-axis, (u'_c/U_o and v'_c/U_o) are shown in Figs. 7(a) and (b) respectively for s/w = 7.5, 11.25 and 0 (single jet). It can be seen that, u'_c/U_o increases with x/w up to a point immediately downstream from the merging point where high velocity streamlines collide. The streamwise turbulence intensity then decreases as the flow proceeds further downstream and its decay approaches that of a single free jet. The reduction of the streamwise turbulence intensity in the merging region close to the combined point is primarily due to the relaxation effect of the merging processes on the turbulence activities. Fig. 7(a) also indicates that the maximum value of u'_c/U_o is significantly greater for s/w = 7.5 than for 11.25. Similar to u'_c/U_o, v'_c/U_o increases with x/w and reaches a maximum value immediately downstream of the merging point where the flow collision is maximum. The value of the maximum lateral turbulence intensity is slightly less for s/w = 11.25 than 7.5. As the flow proceeds downstream, the decay of v'_c/U_o approaches that of a single free jet. Comparison between Figs. 7(a) and (b) reveals that, v'_c/U_o is considerably higher than u'_c/U_o for both s/w because of higher momentum transfer in the lateral direction in the merging region.

3.7 Spatial distribution of streamwise turbulence intensity

Contours of normalised streamwise turbulence intensity, u'/U_o, in the converging and merging regions are depicted in Figs. 8(a) and (b) for s/w = 7.5 and 11.25 respectively. The inner and outer shear layers can be clearly identified by the u'/U_o peaks. Unlike the two parallel jets of s/w = 2.5 and 4.25, discussed in Nasr & Lai (1997b) where the inner shear layer initial peaks were overwhelmed by the strong effects of the recirculating flow, they persist downstream beyond the standing vortex centre (x_{vc}/w = 5.25 and 9 for s/w = 7.5 and 11.25 respectively). An additional small peak in u'/U_o within the recirculating flow region can be detected at x/w = 6 (Fig. 8a). Due to the mutual interaction between the inner and outer shear layers, their peaks begin to merge by x/w = 8.

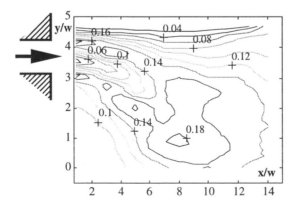

Fig. 8(a) Contours of streamwise turbulence intensity for s/w=7.5.

232

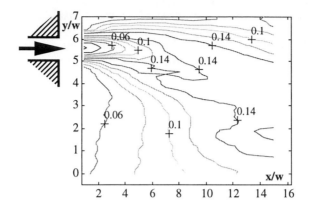

Fig. 8(b) Contours of streamwise turbulence intensity for s/w=11.25.

It can be observed that further downstream from the merging point (x_{mp}/w = 8.3 and 12.5 for s/w = 7.5 and 11.25 respectively), the turbulence in the central region of the flow decreases owing to the relaxation effect of the merging process.Downstream from the combined point, in the combined region, the streamwise turbulence intensities normalised by the local maximum mean streamwise velocity follow those of a single free jet. Fig. 8 also shows that u'/U_o in the inner shear layer peak is slightly higher than that in the outer shear layer. As the flow proceeds downstream, the magnitude of the inner and outer shear layer peaks approaches each other.

3.8 Spatial distribution of lateral turbulence intensity

Figures 9(a) and (b) display the contours of the lateral turbulence intensity v'/U_o for s/w = 7.5 and 11.25 respectively. The initial peak of v'/U_o corresponding to the lip of the nozzle outlet is smaller in the outer shear layer than in the inner shear layer. This is due to the initial convergence of the jet towards the x-axis close to the nozzle plate as a result of the lateral pressure gradient.

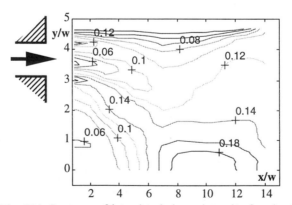

Fig. 9(a) Contours of lateral turbulence intensity for s/w=7.5.

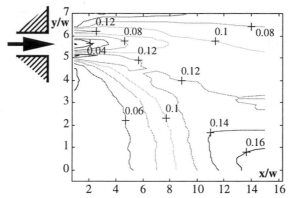

Fig. 9(b) Contours of lateral turbulence intensity for s/w=11.25.

As the jets curve towards the x-axis, the two peaks slowly disappear and by $x/w \approx 8$ and 12 for s/w = 7.5 and 11.25 respectively, v'/U_o becomes maximum due to the collision of the two inner shear layers on both sides of the x-axis. Comparison of u'/U_o and v'/U_o in Figs. 8 and 9 reveals that, downstream from merging point, v'/U_o is larger than u'/U_o in the vicinity of the x-axis. However u'/U_o is considerably higher than v'/U_o in the outer part of the flow. Hence, strictly speaking, the turbulence field is not isotropic like a single free jet.

3.9 Spatial distribution of Reynolds shear stress

Contours of the normalised Reynolds shear stress, $-\overline{uv}/U_o^2$ are shown in Figs. 10(a) and (b) for s/w = 7.5 and 11.25 respectively. The positive and negative Reynolds shear stresses indicate the outward and inward momentum transfer in the outer and inner shear layers respectively. As the jet curves towards the x-axis, the outer shear layer peak of $-\overline{uv}/U_o^2$ becomes less distinct.

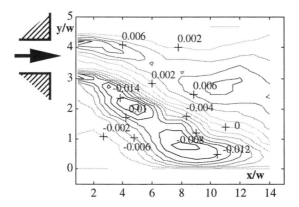

Fig. 10(a) Contours of Reynolds shear stress for s/w=7.5.

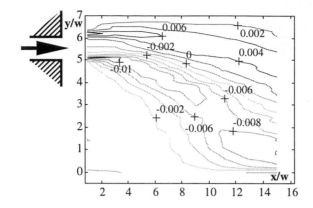

Fig. 10(b) Contours of Reynolds shear stress for s/w=11.25

On the other hand, as the inner shear layer develops, its peak first decreases with downstream distance and then increases to a maximum value around the merging point where the momentum transfer is maximum, owing to the collision of the two inner shear layers on both sides of the x-axis.

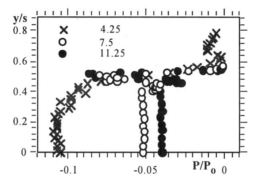

Fig. 11 Static pressure distribution at x/w=0.5 for various s/w.

3.10 Static pressure measurements

Figure 11 shows the lateral distributions of the mean static pressure P/P_o measured with a disc type static pressure probe at x/w = 0.5 for s/w = 7.5 and 11.25. Compared with strongly interacting two parallel jets of s/w = 2.5 and 4.25 reported by Nasr & Lai (1997b), the static pressure in the recirculation zone close to the nozzle plate is considerably higher for weakly interacting jets than for strongly interacting jets. Unlike strongly interacting two parallel jets where the static pressure increases continuously from the x-axis to the outer edge of the jet, P/P_o is nearly constant along the lateral direction in the recirculation zone close to the nozzle plate. For both s/w = 7.5 and 11.25, in the region close to the inner edge of the jet, the static pressure increases with the lateral distance (y) from the x-axis and then decreases again in the main flow before recovering to atmospheric pressure at the outer edge of the jet.

Distributions of the normalised mean static pressure along the x-axis P_c/P_o are presented in Fig. 12 for s/w = 7.5 and 11.25. For s/w=7.5 and 11.25, the static pressure along the x-axis behaves quite similar to that of strongly interacting jets (s/w = 4.25). However, the positive and negative peaks of P_c/P_o are relatively smaller than the corresponding peaks observed for s/w = 4.25.

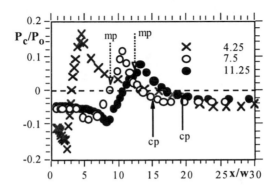

Fig. 12 Static pressure distribution on the x-axis for various s/w.

3.11 Comparison of unventilated and ventilated two parallel plane jets

Unlike unventilated two parallel jets (Fig. 1) where the surrounding air entrainment between the two jets is prevented by the nozzle plate, the inter-jet entrainment in ventilated jets (secondary flow) issuing from free-standing nozzles eliminates the formation of standing vortices (recirculating flow region) upstream of the merging point. Owing to a higher static pressure in the zone between the two jets in the converging region, the flow curvature in the ventilated two parallel jets is less than that in the unventilated jets. It is generally accepted that, in the combined region, both unventilated and ventilated two parallel jets resemble a single free jet. In the near field, however, the flow properties of unventilated and ventilated jets are quite different from each other. This has not been quantitatively examined by previous researchers and will, therefore, be considered here.

Figure 13 depicts the mean streamwise velocity profiles U/U_o for unventilated two parallel jets with s/w = 11.25 and those of ventilated jets with s/w = 12.5 measured by Elbanna et al. (1983) using constant temperature hot-wire anemometry. Here, the ordinate is normalised by the lateral distance (s) between the nozzles. Except in the region between the two jets where U/U_o is higher for ventilated parallel jets, downstream from the nozzle exit plane up to x/w = 7 the profiles of the mean streamwise velocity in the main jet are quite similar for both jet configurations. However, the effect of a higher flow curvature for unventilated jets is evident at x/w = 10 and 15 as the velocity profiles are deflected more towards the x-axis. It can be observed from Fig. 13 that, near the nozzle exit, the decay of the local maximum streamwise velocity U_m/U_o is similar for both unventilated and ventilated jets and as the flow approaches the merging point, it becomes greater for the former.

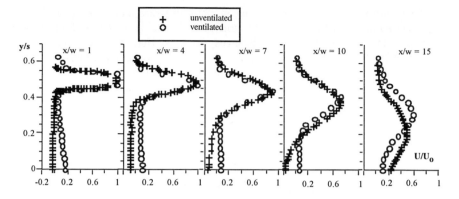

Fig. 13 Comparisons of mean streamwise velocity profiles between ventilated and unventilated jets.

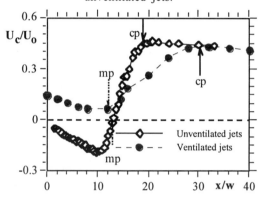

Fig. 14 Distribution of mean streamwise velocity on the x-axis.

Figure 14 compares the LDA results of the mean streamwise velocity along the x-axis, U_c/U_o, for unventilated two parallel jets with $s/w = 11.25$ with the hot-wire results of Elbanna et al. (1983) for ventilated two parallel plane jets with $s/w = 12.5$. It can be seen that, for ventilated jets, the normalised secondary flow velocity along the x-axis, U_c/U_o, is about 0.14 at the nozzle exit and decreases with x/w down to a value of about 0.06 at the merging point ($x_{mp}/w = 12$). Downstream from the merging point, U_c/U_o begins to increase with x/w up to a maximum value of about $U_{cmax}/U_o = 0.42$ at the combined point ($x_{cp}/w = 30$). On the other hand, for the unventilated two parallel jets, the maximum magnitude of the reversed flow is about -0.2 which is about 30% higher than that for ventilated jets. In the merging region, U_c/U_o increases considerably faster (owing to a higher flow collision and momentum transfer) and reaches a maximum value ($U_{cmax}/U_o = 0.46$) slightly higher than that for ventilated jets. The measured combined length of $x_{cp}/w = 30$ for ventilated two parallel jets with $s/w = 12.5$, is significantly higher than that of $x_{cp}/w = 19$ for unventilated jets with $s/w = 11.25$. This 63% longer combined length agrees with the results of Lin and Sheu (1991), Elbanna et al. (1983) and Marsters (1977). However, the merging length of

$x_{mp}/w = 12$ for ventilated jets of Elbanna (1983) measured by hot-wire is shorter than the merging length $x_{mp}/w = 12.5$ of the unventilated jets measured by LDA in the present study. This could be due to the errors in hot-wire measurements as the flow angle, turbulence intensity, and probe interference are high near the merging point.

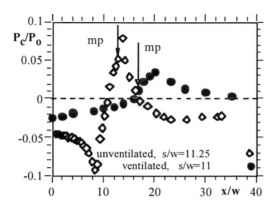

Fig. 15 Static pressure distribution on the x-axis.

The distributions of the mean static pressure along the x-axis, P_c/P_o, for unventilated two parallel jets of $s/w = 11.25$ of the present study and for ventilated two parallel jets of Marsters (1977) with $s/w = 11$ are compared in Fig. 15. For ventilated jets, P_c/P_o is minimum at the nozzle exit plane and it increases with x/w to a maximum at a location downstream from the merging point. The minimum normalised static pressure of about -0.027 for ventilated parallel jets is significantly greater than -0.098 for unventilated jets. The maximum static pressure along the x-axis, P_{cmax}/P_o, of about 0.038 is also considerably lower for ventilated jets than 0.08 for unventilated jets which is due to the higher flow collision for the latter. Examination of the wall static pressure data for ventilated two parallel plane jets obtained by Marsters (1977) suggests that the ratio of the streamwise distance from the nozzle exit at which the maximum static pressure occurs to the merging length is about 1.3 which is higher than that of 1.15 for unventilated jets reported by Tanaka (1970). This is due to the higher tendency of the two jets to deflect towards each other in unventilated jets.

4. Conclusions

Mean velocity vectors determined from simultaneous LDA measurements of streamwise and lateral velocities have been presented for two parallel plane jets with $s/w = 7.5$ and 11.25. The converging and merging regions, recirculation zone, merging and combined lengths, location of the vortex centre and magnitudes of the maximum and minimum velocities along the symmetry axis (x-axis) have been determined.

Distributions of turbulence intensities and Reynolds shear stress in the converging and merging regions and recirculation zone of two parallel jets show that the initial peaks of these turbulence properties corresponding to the nozzle inner and outer edges tend to persist downstream from the location of the vortex centre without being overwhelmed by the influences caused by the recirculating flow in the recirculation zone. The peaks of lateral turbulence intensities and Reynolds shear stress in the outer shear layer spread and decay more rapidly than in the inner shear layer.

Lateral distribution of the mean static pressure P/P_o measured using a disc type static pressure probe at $x/w = 0.5$ shows that P/P_o in the recirculation zone is considerably higher for weakly interacting jets of $s/w = 7.5$ and 11.25 than for strongly interacting jets of $s/w = 2.5$ and 4.25 examined by Nasr & Lai (1977b). It has been shown that P/P_o at $x/w = 0.5$ is nearly constant along the lateral direction in the recirculation zone close to the nozzle plate and reaches its minimum value in the main flow followed by recovery to atmospheric pressure at the outer edge of the jet. The distributions of the static pressure along the x-axis show that the positive and negative peaks of P_c/P_o are also relatively smaller for $s/w = 7.5$ and 11.25 than those observed for $s/w = 2.5$ and 4.25.

LDA results of unventilated two parallel jets with $s/w = 11.25$ of the present study and hot-wire data of Elbanna et al. (1983) for ventilated two parallel plane jets with $s/w = 12.5$ have been compared. It has been shown that the maximum normalised secondary flow velocity of about 0.15 for ventilated jets is relatively smaller than the magnitude of the maximum reversed flow velocity of 0.2 for unventilated jets. The maximum mean streamwise velocity at the combined point, U_{cmax}/U_o, has been found to be 0.42 and 0.44 for ventilated and unventilated jets respectively. Comparisons of the static pressure along the x-axis P_c/P_o for unventilated two parallel jets of $s/w = 11.25$ of the present study and for ventilated two parallel jets of Marsters (1977) for $s/w = 11$ show that for ventilated jets, P_c/P_o has a minimum of about -0.027 which is significantly greater than -0.098 for unventilated jets. The maximum normalised static pressure on the x-axis of 0.038 is also considerably lower for ventilated jets than 0.08 for unventilated jets.

Acknowledgement

This study has been partially supported by the University College Rector Special Research Grant. The second author (AN) acknowledges receipt of an Iranian Government scholarship in pursuing this study.

References

Bradshaw, P. and Wong, PY.F. (1972) The reattachment and relaxation of turbulent shear layer. *J. Fluid Mech.* **52**(1), 113-135.

Eaton, J.K. and Johnston, J.P. (1981) A review of research on subsonic turbulent flow reattachment. *AIAA J.* **19**(9), 1093-1100.

Elbanna H. and Gahin S. (1983) Investigation of Two Plane Parallel Jets. *AIAA J.*, **21** (7), 986-991.

Ko N. W. M. and Lau K. K. (1989) Flow Structures in Initial Region of Two Interacting Parallel plane Jets. *Experimental and Fluid Science*, **2**, 431-449.

Lin Y. F. and Sheu M. J. (1990) Investigation of two plane parallel unventilated jets. *Experiments in Fluids* **10**, 17-22.

Lin Y. F. and Sheu M. J. (1991) Interaction of Parallel Jets. *AIAA J* **29**(9), 1372-1373.

Marsters G. F., (1977) Interaction of Two Plane Parallel Jets *AIAA J.*, **15**(12), 1756 - 1762.

Militzer J.(1977) Dual plane parallel jets. Ph. D. Thesis, University of Waterloo, Canada.

Miller D. R. and Comings E. W., (1960) Force-momentum field in a dual-jet flow. *J. Fluid Mech.* **7**, 237.

Murai, K., Taga, M. & Akagawa, K. (1976) An Experimental Study on Confluence of Two- Dimensional Jets. *Bull. JSME*, **19**, 956 - 964.

Nasr, A. and Lai, J.C.S. (1996) The effect of offset ratio on the mean flow characteristics of turbulent offset jets. *Proceedings of ICAS96*, Sorrento, Napoli, Italy, Sept. 8-13, vol. 2, 1968-1978.

Nasr A. and Lai J. C. S. (1997a) Two parallel plane jets: mean flow and effects of acoustic excitation. *Experiments in Fluids* **22**, 251-260.

Nasr A. and Lai J. C. S. (1997b) Strongly interacting two parallel jets. *Proceedings of JSME Centennial Grand Congress, Int. Conf. on Fluid Eng.*, Tokyo, Japan, July 13-16, 1997, I, 117-122.

Nasr A. and Lai J. C. S. (1997c) Comparison of flow characteristics in the near field of two parallel plane jets and an offset plane jet. *Physics of Fluids*, **9**(10), 2919-2931.

Tanaka E. (1970) The Interference of Two - Dimensional Parallel Jets. (1st Report, Experiments on Dual Jet), *Bull. JSME*, **13**, 272-280.

Tanaka E. (1974) The Interference of Two - Dimensional Parallel Jets. (2nd Report, Experiments on the Combined flow of Dual Jet), *Bull. JSME*, **17**, No.109, 920-927.

CHAPTER III.   AERODYNAMICS

III.1. Instantaneous Doppler Global Velocimetry Measurements of a Rotor Wake: Lessons Learned

J.F. Meyers[1], G.A. Fleming[1], S.A. Gorton[2], and J.D.Berry[2]

[1]National Aeronautics and Space Administration, Langley Research Center, Hampton, Virginia 23681 USA
[2]Aeroflightdynamics Directorate (AVRDEC), U. S. Army Aviation and Missile Command, Langley Research Center, Hampton, Virginia 23681 USA

Abstract.   A combined Doppler Global Velocimetry (DGV) and Projection Moiré Interferometry (PMI) investigation of a helicopter rotor wake flow field and rotor blade deformation is presented.   The three-component DGV system uses a single-frequency, frequency-doubled Nd:YAG laser to obtain instantaneous velocity measurements in the flow. The PMI system uses a pulsed laser-diode bar to obtain blade bending and twist measurements at the same instant that DGV measured the flow. The application of pulse lasers to DGV and PMI in large-scale wind tunnel applications represents a major step forward in the development of these technologies. As such, a great deal was learned about the difficulties of using these instruments to obtain instantaneous measurements in large facilities. Laser speckle and other image noise in the DGV data images were found to be traceable to the Nd:YAG laser. Although image processing techniques were used to virtually eliminate laser speckle noise, the source of low-frequency image noise is still under investigation. The PMI results agreed well with theoretical predictions of blade bending and twist.

Keywords.   Doppler Global Velocimetry, helicopter, rotor wake

1. Introduction

The precise prediction of the helicopter main rotor wake has been cited as the driving factor for accurately predicting rotor loads, vibration, performance, and noise. It is widely accepted throughout the rotorcraft industry that correctly modeling the wake geometry and the tip vortex formation, size, strength, and position are essential to a numerical solution of the wake problem. Although there have been many efforts to characterize the rotor wake using various methods of flow visualization, these qualitative efforts are insufficient to validate and improve the numerical models of the rotor wake. Detailed, quantitative flowfield information is required.

The acquisition of the requisite flowfield measurements has been impeded by the harsh environment in which a rotor blade operates. Since the blades rotate, the flow environment is constantly changing, and each blade is affected by the wake structure shed from the preceding blades. Thus, investigations of this unsteady, vortex-dominated flow require measurement techniques that are: 1) nonintrusive,   2) instantaneous,   3) simultaneous three-component, and 4) correlated with rotor azimuth. Although fringe-type laser velocimetry satisfies the measurement requirements, the excessive acquisition time needed to obtain statistically-significant, azimuth-dependent results makes the technique impractical for surveying large areas within the rotor flowfield. Particle Image Velocimetry is limited to two dimensional measurements over relatively small measurement planes in large wind tunnel applications.

For the past several years, the NASA Langley Research Center has been developing Doppler Global Velocimetry (DGV) for use in large wind tunnels. DGV is a planar measurement technique capable of obtaining three-component flow field velocity data. Previous applications of DGV used a continuous-wave (CW) Argon-ion laser and produced flow velocity measurements that were temporally integrated over the 16.7 ms CCD camera exposure. Demonstrations of the instrument operating in this manner included the detailed measurement of wing tip vortices at focal distances of 18 m (Meyers (1996)). While these integration times were satisfactory for the measurement of stationary flows, they were unacceptable for the measurement of the unsteady flows found in rotorcraft applications. In anticipation of this limitation, the Northrop Research and Technology Center was contracted by the NASA Langley Research Center in 1990 (Komine *et al* (1994)) to conduct laboratory investigations to determine if a single-frequency, frequency-doubled Nd:YAG laser could be used in DGV applications to obtain instantaneous velocity measurements. This successful research program culminated with the first instantaneous DGV measurements, as demonstrated by one-component velocity measurements in a 10-x 10-cm free jet.

The subject of this investigation was the rotor wake developed by an isolated rotor system consisting of a Mach-scaled, four bladed rotor with a rotor disk diameter of 1.7 m. A scaled, generic helicopter fuselage shell, independent of the rotor drive system and hub, could be raised from the tunnel floor to investigate its effect on the flow and nonlinear interactions with the rotor wake. The primary objective of the current investigation was to integrate the three-component DGV technology developed by NASA with the experience derived from the Northrop effort to create an instrumentation system capable of measuring the unsteady rotor wake flow field. Complete diagnostics of the rotor wake must include blade deformation and position data since the rotor wake geometry is largely affected by these parameters. A second objective was to use the newly advanced capabilities of a laser diode-based Projection Moiré Interferometer to obtain blade position, bending, and twist data at the same instant as the velocity field was measured with the DGV system.

2. Experimental Facilities

The experiment was performed in the Langley 14-by 22-Foot Subsonic Tunnel shown in Figure1 (Gentry *et al* (1990)). This atmospheric, closed-circuit low-speed wind tunnel can be operated with a closed test section, or by raising the walls and ceiling, in an open test section mode. For this study, the tunnel was operated as a modified open test section with the walls raised for maximum optical access to the rotor wake, and the ceiling lowered to serve as a mounting platform for the isolated rotor drive system. The wind tunnel is equipped with propylene glycol vaporization/ condensation smoke generators that were mounted on a traversing mechanism located in the tunnel settling chamber.

The isolated rotor test system (IRTS) is a general-purpose rotor testing system. The fully articulated hub holds a Mach-scaled, four-bladed, 1.7-meter diameter rotor. The rotor blades have a rectangular planform and an NACA 0012 airfoil section with a chord of 6.6 cm and a linear twist of -8 degrees, nose down. Note that the rotor blades are very stiff torsionally when compared with a full-scale rotor system. A digital 1024 pulse per revolution encoder was attached to the rotor shaft to monitor rotor speed and provide an azimuthal record for conditionally sampling the instrumentation systems. A helicopter fuselage model was mounted on a vertical strut below the rotor. The strut could be raised and lowered to locate the fuselage in proper position under the rotor, Figure 2, or fully lowered to be out of the influence of the rotor wake, Figure 3. A conventional, 3-D Laser Velocimeter was also used during the test to acquire a limited number of velocity measurements in the rotor wake.

Figure 1: The Langley 14-by 22-Foot Subsonic Tunnel.

Figure 2: The 1.7-meter isolated rotor system with the fuselage placed below the rotor mounted in the 14-by 22-Foot Subsonic Tunnel.

3. Doppler Global Velocimeter

The Doppler Global Velocimeter is an Iodine vapor cell based, three-component system utilizing a pulsed, single-frequency, frequency-doubled Nd:YAG laser to obtain instantaneous (10 ns) measurements. The description of DGV technology is given by Komine (1990) and Meyers (1995). The system consists of the Nd:YAG laser, light sheet forming optics, laser

246

frequency monitoring system, and three receiver optical systems. The output laser beam was directed to the light sheet forming optics located on the wind tunnel test section floor, Figures 2 and 3. The receiver optical systems and the laser frequency monitoring system consisted of a linear polarizer, beamsplitter, an encased and insulated Iodine vapor cell, mirror, and two electronically shuttered CCD video cameras, Figure 4. Each system was enclosed to protect the optics from the wind tunnel environment and flow buffeting.

Figure 3:   The 1.7-meter isolated rotor system with the fuselage placed out of the flow.

Figure 4:   Doppler Global Velocimeter receiver optical system.

All three receiver optical systems were placed on the advancing side of the rotor to have an unobstructed view of the laser light sheet when the fuselage was in place, Figure 5. Their placement yielded an angle between the vector measured by component A and the vector measured by component B of 52 degrees, 47 degrees between the vectors measured by components A and C, and 41 degrees between the vectors measured by components B and C. These separation angles were sufficient to minimize trigonometric errors when translating the measurements to the streamwise, vertical, and crossflow velocity components. The common field of view was an area 1.02-by 1.14 meters.

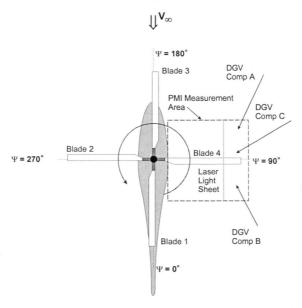

Figure 5: Planview of the model, laser light sheet, DGV receiver optical systems, and PMI measurement area.

The data acquisition system consists of a network of five PC compatible computers with one computer dedicated to each measurement component, the fourth obtaining data from the laser frequency monitor, and the fifth serving as the operator terminal and system controller. Each component computer contains two 10-bit frame grabbers to acquire single-field (512x256 pixels) images from the signal and reference cameras, respectively. The industry standard RS-170 cameras were electronically shuttered (0.1 ms) in synchronization with the Nd:YAG laser. While the cameras and data acquisition system were set to free run, images were only acquired when the rotor shaft encoder aligned with one of the selected azimuth angles: 0- to 90-degrees, in 10 ±0.7 degrees (±2 encoder steps) increments.

4. Projection Moiré Interferometer

Conventional projection moiré interferometry begins with the projection of a grid of equispaced, parallel lines onto the surface of a test object. With the object in the reference condition, a CCD video camera is used to acquire an image of the object illuminated by the lines. When the object is deformed, the grid line spatial arrangement changes. A second image is then acquired. Subtracting the reference image from the data image yields a pattern of moiré fringes tracing the contours of constant deformation amplitude. The measured deformation direction is along the bisector of the projection and viewing vector directions. The deformation amplitude between adjacent moiré fringes is proportional to the projected grid line spacing and the angle between the projection and viewing vectors.

The Projection Moiré Interferometry (PMI) system used in this study was a single-component, laser diode based system capable of obtaining instantaneous (0.1 ms) measurements. This system is described in detail by Fleming and Gorton (1998). The system consists of a pulsed 15 W laser diode bar (10 discrete emitters) operating at 800 nm, a Ronchi ruling, projecting optics, and an electronically shuttered CCD video camera. The laser diode bar was chosen as the light source because of its small size, high output power, single pulse operation, and infrared wavelength. The laser light passes through the Ronchi ruling (a diffracting grading with a square wave cross section) to a lens system that projected the resulting lines onto the underside of the rotor blades, Figure 6. The PMI optical system was placed below the test section floor to view a 1.2-x 1.2-meter area in the rotor disk plane covering 50 degrees of rotor azimuth, Figure 5. The data acquisition system, a one-camera version of the DGV component acquisition system, and the laser diode bar were triggered by the DGV synchronization signal. Thus the DGV and PMI systems acquired data at the same instant with the same measurement window.

Figure 6: Projection Moiré Interferometer laser and transmission optical system.

The standard PMI reference image could not be obtained since the rotor blades continuously moved through the viewing area. An aluminum honeycomb plate was placed in the rotor disk plane at the rotor hub height to obtain a reference image. Subtraction of this image from the data images produced moiré fringe patterns representing height changes of the rotor blade above this reference plane. Additional processing, described by Fleming and Gorton (1998), yielded the location, bending, and twist of the imaged rotor blade at the instant of the DGV velocity measurement.

5. Calibration Procedures

As with most optical instrumentation systems, DGV and PMI required daily alignment and system calibration. Their use of video cameras as the measurement sensors eased alignment procedures, but increased the number of calibration techniques needed. The systems were aligned using a simple target of equispaced dots shown in Figure 7. Each DGV component receiver optical system was aligned using a custom electronics system to normalize the signal camera image by the reference camera image in real time. Adjusting the signal camera position and viewing angle to the proper alignment would extinguish the dots from the normalized image. Aligning the PMI was even simpler: align the viewing camera to keep the dot pattern as square as possible, then focus the projected grid lines to fill the viewing area and obtain optimal contrast.

Figure 7:   Alignment and spatial calibration target for DGV and PMI.

Daily calibration of the instruments more extensive system alignment was necessary. Both techniques required optical calibrations to remove optical and perspective distortions from the acquired data images.   Spatial distortions were determined using the alignment target, Figure 7, to establish a centroid map for each camera. The maps, which embodied the observed optical and perspective distortions, were used to compute the piecewise bilinear warping coefficients necessary to remove the distortions from the data images (Meyers (1992) and Meyers (1995)).   Corrected DGV signal images could then be normalized by their respective corrected reference images to yield the component velocity data images.   In addition, the warping procedures were sufficiently accurate to allow the determination of the orthogonal U, V, and W velocity components from the three measured components. Likewise, similar procedures were used to remove spatial distortions from the PMI data images.

Since DGV measurements are based on the ratio of image amplitudes, additional calibrations were required that were not necessary for the spatially-based measurements of the PMI. The amplitude response of each DGV receiver system was flattened by determining the sensitivity of each camera pixel, and measuring any spatially dependent transmission losses through the optics.   This process generated a pixel sensitivity correction image for each camera in the DGV system, and an optical transmission correction image for each receiver. When multiplying data images by their respective pixel sensitivity correction image, the overall camera response would be held constant.   Likewise, by multiplying the normalized signal images by their respective optical transmission correction image, differences in the transmission of light through the signal and reference optical paths were removed (Meyers (1995)).

The final calibration for DGV determined the Iodine cell transfer function.   The Nd:YAG laser output beam was redirected and split into three beams that were then directed to points on the tunnel structure, each within the field of view of their respective receiver optical system.   The sampling of the output beam by the laser frequency monitor was also maintained.   Tuning the laser frequency through the Iodine absorption line produced the simultaneous calibration of the four cells. This procedure could be performed at any time, even during tunnel operation.

The PMI reference plane was established by replacing the dot target with an aluminum honeycomb plate painted flat white. The laser bar generated the grid of projected lines on the plate. A comparison of the camera image of the projected lines with the images of the dot target yielded the grid line spacing, and thus the sensitivity factor for the moiré fringes (Fleming and Gorton (1998)).

6. The Test

The Rotor Wake / Configuration Aerodynamics Test was conducted to investigate the three-component velocity flow field within the rotor wake with and without a fuselage placed below the rotor. The measurement of the trajectory and velocity of the rotor tip vortices as they moved downstream was of primary interest. Several instrumentation issues were in question since this was the first attempt at using global instrumentation to quantify the rotor wake. These included the integration of a pulsed, Nd:YAG laser into the DGV system, the ability of PMI to accurately measure instantaneous blade position, bending, and twist, and the logistics behind the simultaneous operation of the DGV and PMI systems with synchronized lasers (2), video cameras (9), and data acquisition systems (6 computers).

Velocity measurements of the rotor wake were desired at the 30-, 80-, 97-, 99-, 101-, and 103-percent span locations. Planar measurements at these spanwise locations would provide enough information to accurately determine the blade tip vortex structure and trajectory in the near rotor wake. The DGV laser light sheet was oriented vertically and directed upstream, Figure 2. An automated beam steering mechanism was devised to position the light sheet within the 97- to 103-percent span locations so that light sheet alignment changes could be made while the tunnel was operating. The 30- and 80-percent locations required realignment of the DGV optics.

Each day of testing began with the alignment of the DGV and PMI optical systems, followed by the spatial calibration of both systems at the desired span position. A rotor blade was removed during this process to allow unobstructed optical access to the measurement plane. The laser light sheet was then aligned to the span position. Once aligned, the laser beam was redirected to the tunnel structure to conduct the Iodine vapor cell calibrations. The rotor blade was then replaced and the tunnel test section sealed. The rotor was spun to 2,000 rpm and the laser beam redirected to form the light sheet. Note that the rotor was always spinning when the pulsed light sheet was crossing the blades; this technique protected the composite rotor blades from damage by the high-power light sheet. The tunnel speed was then set to 9.1 m/s and the smoke plume positioned to pass through the light sheet at the rotor wake.

Data acquisition began when the rotor system, wind tunnel velocity, and smoke plume position were stable. A portion of the Nd:YAG laser beam was sampled by a fast photodiode whose output was monitored with a high-speed digital oscilloscope. The shape of the photodiode amplitude vs. time trace was visually inspected to determine that the laser operated in single-frequency mode. If the photodiode output signal had a Gaussian amplitude profile, the laser was operating at a single frequency. A series of 100 conditionally sampled DGV and PMI image sets was acquired when single frequency operation was obtained. The laser beam was redirected to the tunnel walls and a sample of ten images was acquired. This provided a measurement of the Iodine vapor calibration stability. The light sheet was reformed and a second acquisition of 100 image sets obtained. Another calibration check was made followed by the acquisition of the third and final data set. The 300 image sets would yield approximately 30 image sets at each of the selected azimuth angles. This process was repeated at tunnel speeds of 27.7 m/s and 42.0 m/s with the fuselage in both high and low positions (Figures 2 and 3).

7. Effects of Using the ND:YAG Laser in DGV

Changing the DGV laser source from an Argon-ion laser to a pulsed, single-frequency, frequency-doubled Nd:YAG laser provided the capability to obtain conditionally-sampled rotor azimuth-dependent data. It also simplified Iodine vapor cell calibrations since the optical frequency was continuously tunable. Unfortunately, the change in laser also affected both image quality and frequency stability. Laser speckle was a far greater problem with the Nd:YAG laser than the Argon-ion laser. Although increased speckle might be expected from the narrower linewidth Argon-ion laser (10 MHz vs. 80 MHz for the Nd:YAG), the reverse was found. The level of laser speckle noise expected from the Argon-ion laser was reduced by temporal averaging the collected particle-scattered light during the 16.7 ms CCD camera integration time.

Classically laser speckle has been removed using low pass filtering techniques. These techniques include temporal averaging, spatial averaging, low frequency camera Modulation Transfer Functions (MTF) (Smith (1998)), data binning (McKenzie (1997)), and image convolution with a filtering kernel (Meyers (1995)). Temporal averaging of conditionally sampled instantaneous data requires an instrument satiability greater than that achieved in the present study. The other techniques have varying potential for removing laser speckle noise. They all low-pass filter the image data. Complete removal of laser speckle noise would require such a low filter bandwidth that the characteristics of flow structures would be masked. Since the data processing software developed for the Argon-ion based system used the image convolution technique, a new method was developed that would remove laser speckle noise without modifying the flow structure data such as for rotor vortices.

A nonlinear filtering technique developed to remove impulse noise without affecting the underlying image integrity is the median filter (Astola and Kuosmanen (1997)). Basically, the technique sorts pixel amplitudes within the processing kernel, e.g., 5x5 pixels, then selects the median amplitude as the filtered result. A median filter removes impulsive noise while the kernel based low pass filter passes an impulse, albeit wider with a lower amplitude. The effectiveness of this filter can be seen by comparing an original reference camera image with the image after filtering, Figure 8. Using this filtering method, laser speckle noise was virtually eliminated from the data images, revealing the rotor wake structures. It also revealed a low spatial-frequency modulation in the normalized signal image not found in the Argon-ion laser based system.

Laser frequency stability greatly deteriorated during the wind tunnel investigation though it was not identified as a potential concern in the laboratory. This can be illustrated by comparing a series of ratio measurements obtained by the laser frequency monitoring system during the rotor wake investigation with a similar series obtained during a subsequent investigation in the Langley Unitary Plan Wind Tunnel, Figure 9. A comparison of the two samples clearly shows a far more stable output frequency from the Nd:YAG laser during the UPWT investigation than the rotor wake test. The difference was traced to the sensitivity of the laser to vibration. The area adjacent to the UPWT test section was equivalent to a laboratory environment. However, the area adjacent to the open test section in the 14-by 22-Foot Subsonic Tunnel proved to be a very hostile. Floor vibration and wind tunnel shear layer buffeting caused the laser to lose frequency stability. Also, temperature variations, up to $30°$ C, during tunnel operation affected the Iodine vapor cell temperature stability, thus modifying its absorption characteristics.

8. Iodine Vapor Cell Stability

The inability of the Iodine vapor cell heating system to compensate for large changes in environmental temperature was discovered during a previous DGV entry in the 14-by 22-Foot Subsonic Tunnel. In preparation for the current investigation, the cells were insulated and encased with a controlled heat sink on the stem to maintain a small, but constant positive heat transfer. Also the cell in the laser frequency monitor was modified to have a limited vapor pressure by reducing the amount of Iodine placed in the cell. In addition, calibrations were conducted before the tunnel run, once during the run between changes in tunnel velocity, and after the tunnel run, along with laser frequency measurements made between the acquisition of data sets. An overlay of these three Iodine cell calibrations for component C from a typical tunnel run is shown in Figure 10. The mismatch in calibration was a result of temperature variations in the cell from 47.4° C, to 48.0° C, to 53.0° C respectively. Since the environmental effects were not eliminated by insulating and encasing the cells, bias errors were present in the velocity measurements. Although the limited-vapor cell temperature also changed, 50.2° C, to 49.8° C, to 52.4° C with the corresponding calibrations shown in Figure 11, the affects on cell calibration were reduced.

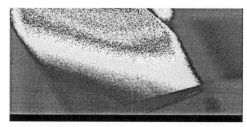

Original reference camera image

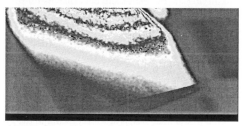

Median filtered image, 5x5 kernel

Figure 8:   Removal of laser speckle noise with a median filter.

9. Smoke System Seeding

The seeding system for the DGV measurements used a superheated propylene glycol and water mixture. The smoke was injected using the standard flow visualization smoke generator for the tunnel and three additional portable smoke generators all located on a positioning system in the settling chamber of the tunnel. The smoke generators were mounted on an array as shown in Figure 12 which was upstream of the flow straightening honeycomb and four anti-turbulence screens, 26 m upstream from the test section. The smoke output for DGV in this wide field of view application was marginal. Although each of the generators was producing the maximum volume of smoke, the smoke dispersion and distribution in the test section was unsatisfactory to seed the large planar viewing area. In addition, the heated smoke contained its own convective flow patterns at the lowest tunnel speeds. It was also difficult to locate the smoke plume in the vortex region since the laser could not be synchronized with rotor azimuth.

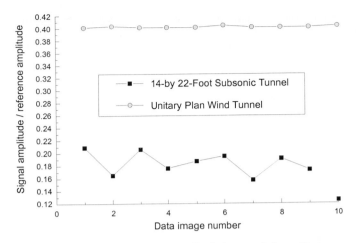

Figure 9: Comparison of Nd:YAG laser frequency stability during two wind tunnel tests.

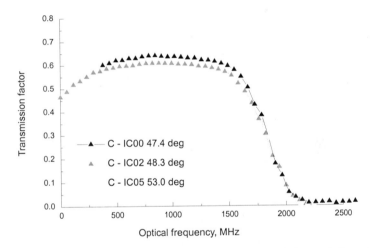

Figure 10: Iodine vapor cell calibrations for component C before, during, and after a typical tunnel run.

10. Rotor Wake Measurements

The objective of the rotor wake measurements was to obtain measurements of the rotor tip vortex trajectory and velocity for several flight conditions, with and without the presence of a fuselage beneath the rotor. The test conditions were chosen to match the conditions for which flow visualization data were reported by Ghee *et al* (1996) as well as to match the conditions for which rotor inflow data were obtained by Elliott *et al* (1988).

Although the bias error limits the usefulness of the DGV data, patterns appear to correlate to the expected behavior of the rotor wake system. The rotor wake can be described as a cylindrical column of accelerated flow which is skewed from the freestream direction by an angle determined from the downward velocity within the wake and the freestream velocity magnitude. The resulting angle from the vertical for the average wake trajectory is known as

254

the skew angle, X. Using the analysis of Stepniewski and Keys (1984) a derivation of skew angle for a rotor in forward flight can be found as:

$$X = \tan^{-1}(V\cos(\alpha)/(v_f - V\sin(\alpha)))$$

(1)

where V is the freestream velocity, α is the rotor shaft angle of attack, and v_f is a function of rotor thrust coefficient and freestream velocity.

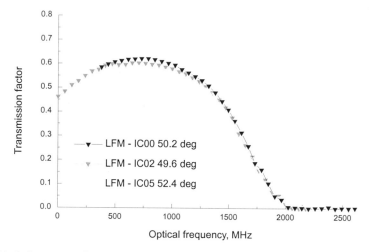

Figure 11: Iodine vapor cell calibrations for the laser frequency monitor obtained simultaneously with Figure 10.

Figure 12: Vaporization/condensation smoke generators attached to the laser velocimeter particle generation nozzle array in the tunnel settling chamber.

For this rotor system, operating at a rotor thrust coefficient of 0.0064 and a rotational speed of 209 rad/s, the following table shows the skew angle calculated from the above reference for a rotor in forward flight. Also included in table 1 are the skew angles determined from the experimental data found by Ghee *et al* (1996) as well as the skew angle measured in the DGV data and discussed below the table.

Table 1: Experimental and theoretical skew angles.

Freestream Velocity, V,	9	27.7	42
α, deg	0	-3	-3
v_f, m/s	10.6	3.7	2.5
X, deg Theoretical	40	79	84
X, deg Experimental Flow Vis (Ghee, et al (1996))	NA	77	79
X, deg DGV Experimental Images	57	78	83

Figures 13 and 14 show a DGV image of the freestream velocity component for a rotor position of zero degrees azimuth for the 27.7 and 42.0 m/s test conditions, respectively. The view in the image is from the right side of the model looking inboard at the 80-percent radial location. The flow is from right to left in the figures, and the rotor blade is rotating from left to right. The grayscale in the image indicates where the smoke was observed in the field of view. The black areas are the absence of the seeding smoke. In each image, the center of the hub is located by a small crosshair and the blades are overlaid at their forward and aft positions. The overlaid dot card pattern gives a sense of scale with 6.35 cm spacing between the dots.

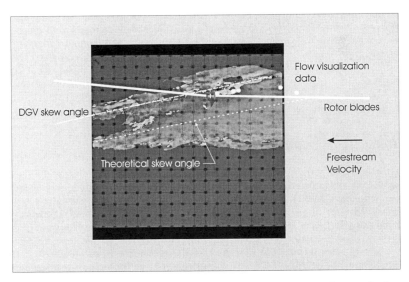

Figure 13: DGV measurement of the streamwise velocity with overlays of theoretical rotor wake skew angle and vortex positions obtained from flow visualization, freestream velocity = 27.7 m/s.

The long dashed white line in each of the images highlights the skew angle seen in the DGV measurements. The boundary of the wake should convect downstream at the wake skew angle and should theoretically be along the short dashed lines. However, vortex

256

positions obtained from the flow visualization data (symbols), (Ghee *et al* (1996)), show the location of the wake boundary to match the DGV results for these test conditions. Both theoretical and experimental data indicate good agreement in skew angle.

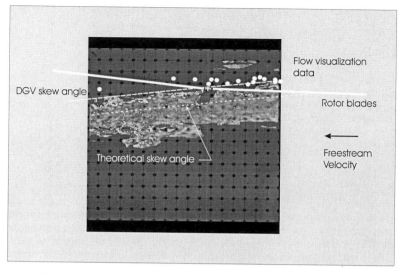

Figure 14: DGV measurement of the streamwise velocity with overlays of theoretical rotor wake skew angle and vortex positions obtained from flow visualization, freestream velocity = 42.0 m/s.

While the good agreement of skew angle is a positive indication for the DGV measurement, further comparisons with wind tunnel freestream data and with independent, three-component laser velocimetry data have been limited by the thermal variations in the Iodine calibrations and low frequency variations in the scattered light. Calibration corrections based on cell temperature measurements obtained for every image are underway at this time to improve the DGV data for later comparisons. Also, laboratory investigations are underway to determine the source of the low frequency variations in the normalized signal image not found in the Argon-ion laser based system.

11. Rotor Blade Deformation Measurements

PMI was successfully used to measure the rotor blade deformation. Figure 15 shows the contours of the blade deformation for several test conditions, and differences in the blade shape are clearly visible. The blade shape deforms as would be expected for a rotor system undergoing changes in test condition; the most dramatic tip deformation relative to the horizontal reference plane occurred when the rotor shaft angle was tilted forward. Figure 16 shows the change in measured blade height with rotor azimuth for the rotor tip path plane tilted 3 degrees nose down. The measured 26 mm difference in the blade tip deflection between 70 and 110 degrees of azimuth corresponds well to the expected deflection due to the geometry of 32 mm.

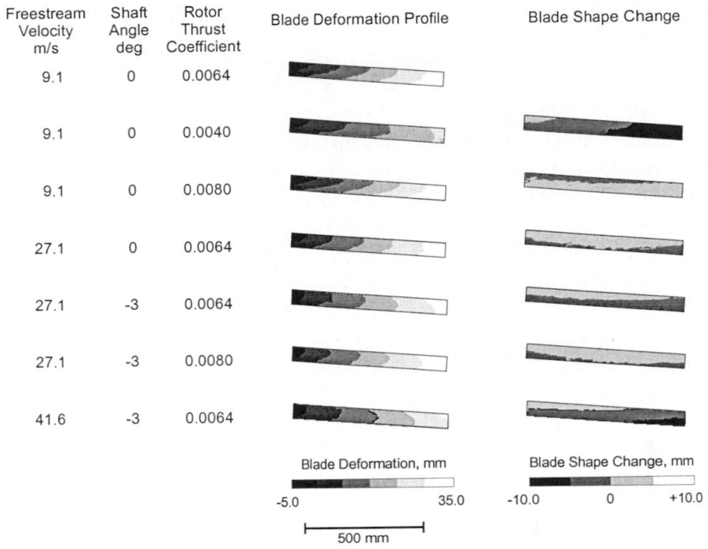

Freestream Velocity m/s	Shaft Angle deg	Rotor Thrust Coefficient	Blade Deformation Profile	Blade Shape Change
9.1	0	0.0064		
9.1	0	0.0040		
9.1	0	0.0080		
27.1	0	0.0064		
27.1	-3	0.0064		
27.1	-3	0.0080		
41.6	-3	0.0064		

Blade Deformation, mm
-5.0   35.0

Blade Shape Change, mm
-10.0   0   +10.0

500 mm

Figure 15: PMI measured blade deflection profiles for seven different flight conditions, blade #4.

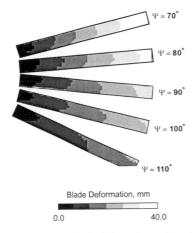

$\Psi = 70°$
$\Psi = 80°$
$\Psi = 90°$
$\Psi = 100°$
$\Psi = 110°$

Blade Deformation, mm
0.0   40.0

Figure 16: Azimuth dependent PMI measured blade deformation profiles, fuselage down, rotor shaft angle set to -3 degrees.

12. Lessons Learned

Three-component flow field investigations using DGV to conditionally sample an unsteady flow field in a large wind tunnel facility illustrates the vast differences between laboratory and large wind tunnel environments. The tasks where difficulties were anticipated such as combining the DGV and PMI systems, synchronizing the lasers, cameras, and data acquisition computers, and conditionally sampling the measurements as a function of rotor azimuth, were very successful. Several areas have been identified where previous experience was insufficient to predict system characteristics. These included insufficient isolation of the

Iodine vapor cells from the tunnel environment, laser speckle, and other noise sources originating from the Nd:YAG laser.

From these experiences, the following modifications to the system and procedures are recommended:

1. Procedures must be developed to prevent laser damage to wind tunnel models.
2. Vapor pressure limited Iodine vapor cells completely isolated from the wind tunnel environment need to be developed.
3. Vibration isolate the Nd:YAG laser and place it totally out of the flow field.
4. Develop a smoke generating system that will yield smoke plumes at least 1.5 m in diameter, preferably using cold injection.
5. Use the rotor synchronized Argon-ion based laser light sheet system to set the smoke plume position.

13. Summary

The first attempt to use pulsed, three-component Doppler Global Velocimetry to measure unsteady rotor wake flow fields was presented. These conditionally sampled results were accompanied by the simultaneously acquired pulsed laser Projection Moiré Interferometry deformation and twist measurements of the moving rotor blades. While the PMI results were very successful, the DGV results suffered from environmental problems and image noise originating by the Nd:YAG laser. The vapor-limited Iodine cell was found to be less sensitive to the environmental changes than the standard cells. Median filtering techniques virtually eliminated laser speckle noise while preserving details of the vortex structures. Work continues on reducing noise sources and correcting the data for temperature induced calibration changes.

References

Astola, J. and Kuosmanen, P., 1997, Fundamentals of Nonlinear Digital Filtering. CRC Press, New York.

Elliott, J.W., Althoff, S.L., and, Sailey, R.H., 1988, Inflow Measurement Made with a Laser Velocimeter on a Helicopter Model in Forward Flight. Volumes I-III, NASA TM 100541-100543, AVSCOM TM 88-B-004-006.

Fleming, G.A. and Gorton, S.A., 1998, Measurement of Rotorcraft Blade Deformation using Projection Moiré Interferometry. Proc. of the 3rd Int. Conf. on Vibration Measurements by Laser Techniques, Ancona, Italy.

Gentry, G.L., Jr., Quinto, P.F., Gatlin, G.M., and, Applin, Z.T., 1990, The Langley 14-by 22-Foot Subsonic Tunnel — Description, Flow Characteristics, and Guide for Users. NASA TP-3008.

Ghee, T.A., Berry, J.D., Zori, L.A.J., and, Elliott, J.W., 1996, Wake Geometry Measurements and Analytical Calculations on a Small-Scale Rotor Model. NASA TP 3584, ATCOM TR-96-A-007.

Komine, H., 1990, System for Measuring Velocity Field of Fluid Flow Utilizing a Laser-Doppler Spectral Image Converter, US Patent 4 919 536.

Komine, H., Brosnan, S.J., Long, W.H., and Stappaerts, E.A., 1994, Doppler Global Velocimetry Development of a Flight Research Instrumentation System for Application to Non-intrusive Measurements of the Flow Field. NASA Report CR-191490.

McKenzie, R.L., 1997, Planar Doppler Velocimetry for Large-Scale Wind Tunnel Applications. AGARD Fluid Dynamics Panel 81st Meeting and Symposium on Advanced Aerodynamic Measurement Technology, paper 9, Seattle, WA.

Meyers, J.F., 1992, Doppler Global Velocimetry — The Next Generation? AIAA 17th Aerospace Ground Testing Conf., paper AIAA-92-3897, Nashville, TN.

Meyers, J.F., 1995, Development of Doppler Global Velocimetry as a Flow Diagnostics Tool. Measurement in Fluids and Combustion Systems, Special Issue, Measurement Science and Technology, vol 6, no. 6, pp. 769-783.

Meyers, J.F., 1996, Evolution of Doppler Global Velocimetry Data Processing. Eighth Int. Symp. on Applications of Laser Techniques to Fluid Mechanics, paper 11.1, Lisbon, Portugal.

Smith, M.W., 1998, Application of a Planar Doppler Velocimetry System to a High Reynolds Number Compressible Jet. AIAA 36th Aerospace Sciences Meeting & Exhibit, paper AIAA 98-0428, Reno, NV.

Stepniewski, W.Z., and, Keys, C.N., 1984, Rotary-Wing Aerodynamics. Dover Publications, Inc., New York.

III.2. LDV-Measurements on a High-Lift Configuration with Separation Control

F. Tinapp and W. Nitsche

Department of Aeronautics and Astronautics
Technische Universität Berlin, Germany

ABSTRACT

The flow field around a 2D-high-lift configuration with active separation control has been investigated by using LDV. The test model, a two element high-lift configuration, was equipped with an excitation mechanism inside the trailing edge flap to generate periodical flow excitation near the flap leading edge. A reattachment of the flow was achieved at various excitation frequencies and amplitudes.

Recent investigations showed clearly that periodic excitation of the separated shear layer results in a partial reattachment and therefore in an increase of lift and decrease of drag (Hsiao et a.1990, Dovgal 1993 and Seifert et a.1996). The present paper investigates this problem and deals with experiments aimed on the separation control via excitation of the separating boundary layer on the flap.

1. INTRODUCTION

Modern transport aircraft wings have to provide very high lift-coefficients in low speed flight during take-off and landing. This leads to good payload/range capabilities for a given field length and a reduction of the noise footprint in the airport area. Therefore high-lift systems are of complex mechanics, generally consisting of a combination of leading-edge slats and multiple trailing-edge flaps. At high angles of attack the flow over high lift wings may separate, resulting in a lift reduction and in an increase of drag. If the onset of separation could be delayed towards a higher angle of attack, it will either be possible to achieve a higher lift or to reduce the complexity of the high-lift system (Fig. 1).

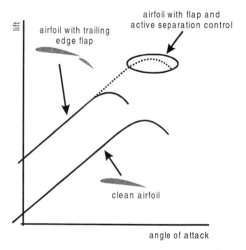

Fig 1: Schematic behaviour of the lift against angle of attack with and without active separation control

2. EXPERIMENTAL APPARATUS

The test model was a generic two element high-lift configuration, consisting of a 100 mm chord length NACA 4412 main airfoil and a NACA 4415 flap with 40 mm chord length. The flap was mounted at a fixed position underneath the trailing edge of the main airfoil, thus forming a gap of 3.5 mm height with an overhang of 2.8 mm (Fig. 1a). In accordance with the experiments made by Adair & Horne 1989, the angle of attack of the main airfoil was chosen to be fixed at 8° while the flap-angle α_{Flap} could be varied from 20° to 50°. With this configuration the flow over the main airfoil was still completely attached but separated over the flap at higher flap angles. To ensure turbulent separation of the flow, turbulators were placed close to the leading edge of the main airfoil and the flap.

For dynamic excitation of the separated shear layer over the flap, a pressure chamber was installed inside, with a 0.3 mm wide slot at 3.5% chord length (Fig. 1b).

A water cooled rotating valve was propelled by a electrical drive and was connected to a high-pressure pump. With this excitation apparatus it was possible to generate pressure pulses of varied frequency and amplitude. These externally generated pressure pulses were fed into the flap of the test model by pressure tubes, resulting in flow

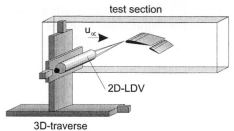

test section

u∞

2D-LDV

3D-traverse

Fig. 3: 2D-Laser-Doppler-Velocimeter mounted on a 3D-traverse system

oscillations emanating vertically from the narrow slot near the leading edge of the flap.

All experiments were carried out in a closed water tunnel, equipped with a test section made of glass with the dimensions (l x h x w) 1250 x 330 x 255 mm. The maximal achievable flow speed was 6 m/s and the mean level of turbulence was about 4%. Flow field measurements were made by means of a two component laser doppler system (Polytec LDV-580), connected with two fast fourier real-time analysers (Aerometrics RSA-1000). To allow automatic measurements of the complete flowfield, the laser optic was mounted on a 3D computer controlled traverse system (Fig. 3).

The flow speed of the present test was 1.5 m/s, resulting in a chord Reynolds number of 150.000, calculated with the flap chord of 100 mm as characteristical length.

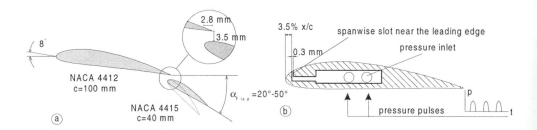

Fig.2: a) Two element high-lift configuration with fixed gap geometry
b) Flap, equipped with a pressure chamber and a narrow slot near the leading-edge

3. EXPERIMENTS

3.1 Mean Flow

For a general investigation of the flow field around the high-lift configuration at different flap deflections, extensive laser doppler measurements were undertaken. The flap deflection was varied between 20° and 50° in order to proof the separation behaviour of the configuration. Here the flow separated at α_{Flap} of about 30° forming a very small recirculation area near the trailing edge of the flap. With an increase of the flap angle, the separation point moved upstream and the flow separated at the leading edge of the flap, resulting in a big recirculation area. As an example of these measurements the velocity vector plots are shown in figure 4. In fig 4a and 4b the velocity field around the complete configuration can be seen while the fig 4c and d present the obtained results near the flap. In case of the α_{Flap}=27° (fig 4a, 4c) the flow over the flap remained attached over the flap while the flow near the trailing edge of the main airfoil tended to separate. The Increase of the flap deflection angle to α_{Flap}=32° (fig 4b, 4d), resulted in a flow separation near the leading edge of the flap. The strong shear layer between the free jet and the recirculation area can clearly be seen.

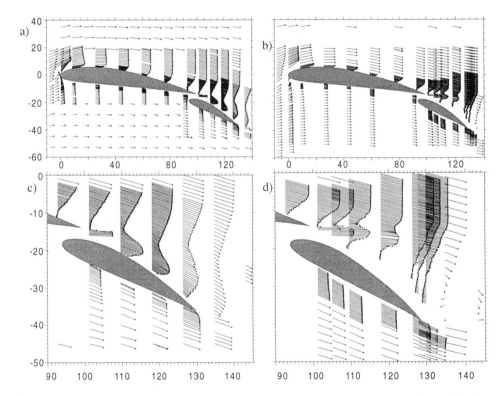

Fig 4: Mean flow field at two different flap angles around the high-lift configuration and in the vicinity of the flap: a)&c) 27°, flow remained completely attached, b)&d) 31°, flow over the flap separated at the leading edge forming a closed area of recirculation (LDA results)

3.2 The Excitation

As seen in figure 4d, the flow over the flap for $\alpha_{Flap}>30°$ separated near the leading edge. To obtain the reattachment of the flow which leads to higher lift coefficients, a periodical excitation of the free shear layer between the main airfoil and the flap was realised. This was done using a pulsed jet which was emanating perpendicularly to the flap surface next to the leading edge.

To verify the jet characteristics, velocity measurements were made in the region of the jet-slot (figure 5). Using phase-averaged laser doppler measurements, the time dependent characteristic of the jet was investigated. As can be seen in figure 5 the jet velocity was pulsating and in good agreement with the corresponding state of the rotating valve.

Phase averaged velocity measurements of the jet flow field were made for three different excitation angles Φ (fig 6). These excitation angles correspond to the three valve positions: closed, opening and opened.

For three different exciting angles the corresponding velocity distributions over the

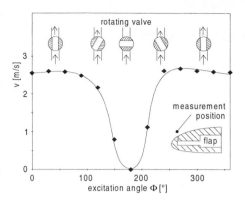

Fig 5: Vertical jet velocity at the slot exit in relation to the valve position

slot are plotted in figure 6a-c. At $\Phi=175°$ (a: closed valve) a very weak velocity field, which was caused by the residual impulse, can be detected, while at $225°$ (b: opening valve) a strong jet occurred, emanating vertically from the slot. Figure 6d shows the velocity field around the exciting slot in the moment of maximum excitation (valve open). Downstream and upstream of the jet two strong vortices, induced by the jet can be recognised. Because of the relatively long measurement time window ($260°-290°$

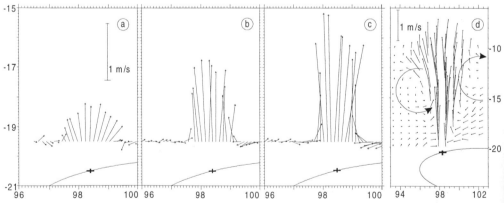

Fig 6: Velocity distribution over the slot at different exciting angles, F=25 Hz, u_μ=0 m/s (LDA results), (a) closed valve, F=175°, (b) opening valve, F=225°, (c) open valve, F=275°, (d) velocity field around the slot at open valve, F=275° (different scale) The cross marks the slot outlet

3.3 Reattachment Due to Flow Excitation

Due to the used excitation mechanism, the net momentum introduced into the flow was not zero. However steady blowing did not achieve reattachment.

Exciting the separated shear layer by a periodic jet may cause a reattachment of the flow over the flap. As an example, the velocity field over the flap at $\alpha_{Flap}=34°$ without excitation is plotted in fig 7a. The flow separated at the leading edge of the flap and formed a recirculation area. By activating the excitation mechanism a complete reattachment of the flow occurred, eliminating the recirculation area over the flap (fig 7b).

To get detailed information about the influence of the pulsed jet on the separated shear layer, detailed laser doppler measurements in the region over the flap were made. It was found, that at relatively high flap angles ($\alpha_{Flap}=30°$-$36°$) the flow reattached completely, while at angles higher than $36°$, the reattachment was only partly achieved. Several measurements were made, investigating the dependence of reattachment on the excitation parameters such as frequency and excitation amplitudes. Reattachment occurred in a frequency band around Strouhal number St=1 (St=F*c_{Flap}/u_∞). Fig.8 depicts the development of reattachment in

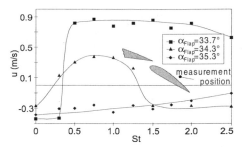

Fig. 8: Influence of the excitation frequency on reattachment at different angles of flap deflection (fixed excitation amplitude)

dependence of the excitation frequency. To illustrate this, the x-component of the flow velocity at a fixed position over the flap is plotted over the excitation frequency for different flap angels. The positive x-component of the flow velocity for the case of $\alpha_{Flap}=34.3°$, at an excitation frequency between St=0.5 and St=1.5 indicates the transition from the separated to the reattached flow. A reseparation was found when increasing the excitation frequency to a value higher than St=1.5.

In order to investigate the dependency of reattachment on the excitation amplitude, figure 9 shows some of the results at the same fixed measurement position near the

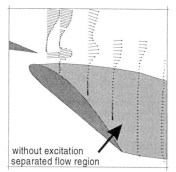

Fig. 7a: Separated flow at 34° flap angle with no excitation (LDA results)

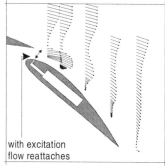

Fig. 7b: Reattachment of the flow at 34° flap angle by excitation (LDA results)

flap, as described in fig 8. The x-component of the flow velocity is plotted versus the excitation frequency for two different flap angles and two different excitation intensities. The lines marked by triangles symbolise the higher flap angle at different excitation intensities (open symbols = weak excitation, closed symbols = strong excitation). As can be seen in the graphic, the reattachment occurred in a frequency band around St=1 for all excitation intensities, but the width of the usable frequency band depended strongly on the intensity of excitation. The more intensive the excitation was, the more frequencies achieved a reattachment of the flow. Increasing the excitation intensity also resulted in a better reattachment, that means higher velocity components in the vicinity of the flap surface.

For three different excitation intensities c_μ the velocity profiles in a vertical slice at 75% flap chord length are plotted in figure 10 ($c_\mu = 2 \cdot H/c \cdot (u'/u_\infty)^2$, H=slotwidth, c=chordlength, u'=RMS excitation velocity,

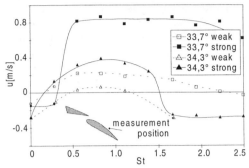

Fig 9: Dependency of reattachment on excitation intensity

u_∞=freestream velocity). The dotted line represents the velocity profile of the separated flow for the case of no excitation. It can be seen that a weak excitation (a) was not able to completely suppress the recirculation area close to the flap surface. A stronger excitation (b) resulted in a complete reattachment with totally absence of negative velocity components. Increasing the excitation intensity (c) did not change the

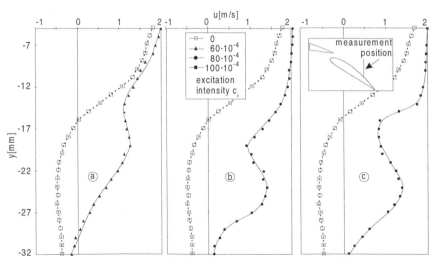

Fig 10: Mean velocity profiles over the flap at 75% x/c_{Flap} for different excitation intensities, α_{Flap}=34,5°, excitation frequency = 20 Hz $\cong$ 0,75 St, a) c_μ=60·10⁻⁴, b) c_μ=80·10⁻⁴, c) c_μ=100·10⁻⁴

fluid behaviour remarkably but still enhanced the velocity profile close to the flap surface.

Phase averaged flow field measurements in the flap region were undertaken to understand the mechanism of reattachment. In fig. 11a the time averaged separated flow at α_{Flap}=35.5° is plotted. Graphic 11b shows the results of the phase averaged measurements with activated excitation. The flow reattached, forming a small separation bubble downstream of the flap leading edge. To visualise the changes in the flow field due to excitation, the velocities of graphic b are subtracted from those of graphic a and plotted in graphic 11c. Here a transport of mo-

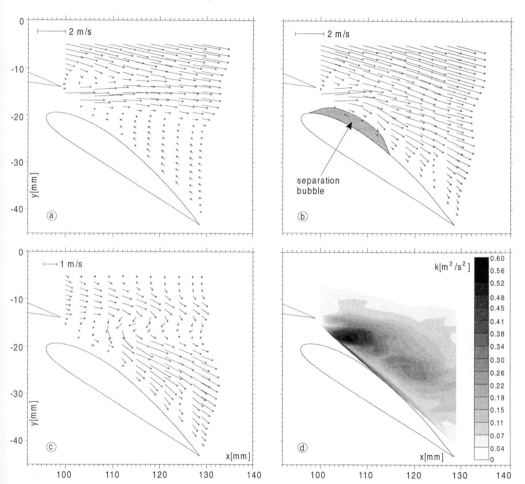

Fig 11: a) separated flow at α_{Flap}=35.5° without excitation, b) reattachment due to excitation (St=0.75, cμ=80·10^{-4}) at F=225° (opening valve) with formation of separation bubble near the leading edge, c) subtraction of both flow fields (b-a) showing a vortice structure in the shear layer induced by periodic excitation, d) turbulent kinetic energy shows a maximum near the excitation point and further downstream in the reattached flow (LDA Results)

mentum from the outer flow into the former separated flow region can be seen. To depict the amount of turbulence in the flow field due to the excitation, the turbulent kinetic energy $k=1/2\,(u'^2+v'^2)$ is plotted as isolines in graphic 11d. The dark colours symbolise high degree of turbulence. It can be recognised that the maximum peaks of turbulent energy were located close to the excitation point and further downstream in the shear layer of the reattached flow. This emphasises the fact of enhanced momentum transport from the outer flow into the separated flow area due to periodic excitation, which results in the reattachment of the flow.

4. CONCLUSIONS

The reattachment of separated flow on a simple high-lift configuration by internal periodic excitation near the leading edge of the flap was investigated by using phase averaged LDV. It was shown that it is possible to achieve a reattachment of a separated flow by exciting the separated shear layer. It also exists a strong influence of excitation frequency and intensity, showing clearly a favourite frequency of $St \cong 1$. The increase of the excitation intensity results in a widening of the usable frequency band. Phase averaged LDV measurements showed a enhanced transport of momentum from the outer flow into the separated region near the flap surface.

REFERENCES

ADAIR, D.& HORNE, W.C. 1989 Turbulent Separating Flow Over and Downstream of a Two-Element Airfoil. *Experiments in Fluids*, **7**, 531

DOVGAL, A. 1993 Control of Leading-Edge Separation on an Airfoil by Localized Excitation. *DLR-Forschungsbericht* **DLR-FB-93-16**

HSIAO, F.B., LIU, C.F.& SHYU, L.Y. 1990 Control of Wall-Separated Flow by Internal Acoustic Excitation. *AIAA Journal*, **28**, No.8, 1440

SEIFERT, A., DARABI, A. & WYGNANSKI, I. 1996 On the Delay of Airfoil Stall by Periodic Excitation. *Journal of Aircraft* **33**, No.4, 691

III.3. Flow Field in the Vicinity of a Thick Cambered Trailing Edge

G. Pailhas[1], P. Savage[2], Y. Touvet[1], and E. Coustols[1]

[1] ONERA/DMAE, Department of Modelling for Aerodynamics and Energetics, Toulouse, FRANCE
[2] presently AEROSPATIALE A/SW/SF, Toulouse, France

ABSTRACT

The aim of this experimental study is to scrutinize the flow in the vicinity of a non zero thickness trailing edge, i.e. just upstream and downstream of the base, using Laser Doppler Anemometry. Indeed, boundary layer surveys as well as wake surveys have been performed for different thicknesses and shapes of trailing edges. Experiments have been conducted with two-dimensional trailing edges in both two- and three-dimensional flows. Much of the discussion presented in this paper is relevant to flow field modification induced by thick cambered trailing edges.

1. INTRODUCTION

The aerodynamical characteristics of an airfoil are strongly dependent on the geometry of the trailing edge vicinity. Rather recent interest has suggested that trailing edge thickness could be a key feature for aerodynamical purpose [1]. For transonic applications, the interest of the namely D.T.E. (Divergent Trailing Edge) concept has arisen because when considering relatively high values of the lift coefficient of supercritical airfoils, the total drag could then be reduced [2]. Such a "DTE" concept has been mainly developed through computational tools, but there is very little published experimental results. As the physical aspects induced by such unconventional trailing edge geometries are identical whatever the flow regime, a complete experimental program has been elaborated at ONERA/CERT for several years.

Thus, a study has been launched successively at the hydrodynamic tunnel of ONERA/CERT for two two-dimensional (2D) thick trailing edges set in a 2D incompressible flow and at the F2 atmospheric wind tunnel of ONERA/Fauga-Mauzac for a single 2D thick trailing edge set in a three-dimensional (3D) incompressible flow. Apart from static pressure measurements, a 2D or 3D Laser Doppler Anemometry system has provided velocity measurements in the vicinity of the trailing edge, for the 2D and 3D configurations, respectively.

Results for two thick cambered trailing edges are mainly discussed in the present paper.

A Navier-Stokes solver has been recently applied to the airfoil equipped with these different geometries of trailing edges, for the afore-mentioned experimental conditions [3,4].

2. EXPERIMENTAL SET-UP

2.1 THALES Water Tunnel

The experimental study has been carried out in the T.H.A.L.E.S. water tunnel; it is essentially of a recirculating type and resembles a conventional closed circuit wind tunnel. Water runs into the loop by the way of a motor driving a simple impeller situated in the low corner of the tunnel. Water velocities in the working section are infinitely variable from 0.1 ms^{-1} to 7.0 ms^{-1}. The velocity variations at the test section entrance are lower than 1% of the mean velocity, whatever the value of the free-stream velocity, after an effective contraction ratio of 10:1.

The horizontal test section is 0.5 m large, 0.3 m high and 3 m long; the four walls are made of glass or plexiglass, which provides good optical access for flow visualization.

A heat exchanger, located in the lower part of the circuit, allows the temperature of the water to vary. The pressure inside the loop can be adjusted in order to avoid eventual cavitation problems. The main parameters (velocity, pressure and temperature) governing the flow are constantly controlled and adjusted by a micro computer via an automaton.

The tunnel is fitted with a displacement mechanism controlled by a computer allowing linear displacements of the optical probe of the Laser Doppler Anemometry system and measurements at any location in the test section volume. The traversing unit contains step by step motors which permit remote movement of the probe in the streamwise and vertical directions. The probe location is determined with the aid of encoders allowing position resolution of 0.01 mm per encoder pulse.

Model. The reference model for the present investigation is an OAT15A airfoil having a maximum thickness ratio of 12.3% and a chord length C (or c on some figures) of 400 mm; the upper surface contour is constrained to be the same as the original OAT15A whereas the airfoil geometry is modified on the lower surface. The airfoil is manufactured in such a way that the rear part (last 19% and 15% chord length of the upper and lower sides, respectively) be removable allowing to mount rather easily the three following 2D trailing edges (figure 1):
- the reference trailing edge, referred to as RTE, having a base thickness of 0.5% C;
- two increased cambered trailing edges with the same trailing edge angle, but different incremental base thicknesses, either 0.2%C (CTE 1) or 0.5%C (CTE 2).

These geometries were defined from Euler computations performed at transonic conditions.

practically constant water temperature of 293K. This leads to a free stream Reynolds number based on the model chord length of about 800,000.

Transition is tripped on the lower and upper sides of the model by a 0.2 mm diameter trip wire stuck parallel to the leading edge, at about 3% of chord length downstream of it.

Measurements in the wake and in the boundary layer are performed in a plane located at one third of the span airfoil from the upper test section wall.

Measurement technique. In the present experimental study, measurements include essentially pressure distribution and flow scrutinizing in the vicinity of the trailing edge airfoil through Laser Doppler Anemometry system. The sketch below shows the system of coordinates used for the present study: the X(or x)-axis is along the mean flow direction, Z is parallel to the airfoil trailing edge and Y is in the transverse direction, perpendicular to both X- and Z-axis. X=0 refers to the airfoil leading edge, Z=0 corresponds to the test section centre-line, while Y=0 is the point located at the intersection line between the suction side of the airfoil and its base.

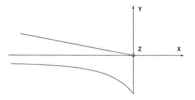

Co-ordinate system for wake surveys.

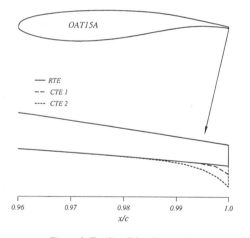

Figure 1. Trailing Edge Geometries

The model, with the reference trailing edge (RTE), is fitted with 64 static pressure taps located on its suction and pressure sides, with one on the base. The others geometries involve one or two supplementary pressure taps on their base, according to their thickness.

The wing model is mounted vertically between the top and the bottom walls of the test section; using a turntable, it could be manually rotated through a ±5° range in angle of attack by step of 0.25°.

Test conditions. The experiments are conducted for a constant tunnel water speed of 2 ms^{-1} and a

A two-dimensional component Laser Doppler Anemometer is used to measure the magnitude and direction of the velocity vector and elements of the Reynolds stress tensor in the X-Y plane. The L.D.A. system uses a coherent argon ion laser operating with a power of about 1.5W.

The light beam from the laser is split into two components: one green (λ=5.14 nm) and the other blue (λ=488 nm). Each color component of the laser is further split into two beams, one relating to the green and the other one relating to the blue passed through a Bragg cell containing a 40MHz transducer.

The L.D.A. system is used in a backward scattering mode; the four beams probe is located in such a way that its axis is perpendicular to the upper wall of the test section. So, the probe produces convergent beams that enter the water tunnel through one of its upper glass wall. The front lens of the optical system has a focal length of 480 mm in air and the laser beam pairs are focused down to a diameter of about 180 µm at the measuring point which has a length of 4 mm at an intersection angle of 5.4°. The two beams are inclined at 45° with respect to the main flow direction.

The seeding of the flow is made by means of iriodine powder (particles with a mean size of 2 μm) added to the circulating water in the tunnel. Light scattered by particles travelling through the intersection volume is collected by the probe's lens and then focuses into a receiver fiber optic cable which carries it to the output which feeds the light to the photomultiplier.

Data acquisition is done by means of Burst Spectrum Analyzers used in master-master configuration; frequencies and arrival times of particles in the B.S.A. output buffers are transferred to a Macintosh computer. A post processing of these data information validates the measured values related to the two B.S.A. counters for which the arrival time differences are within a certain preset coincidence time window.

Positioning of the measuring point in the X and Y directions is done by steps motors under computer control. Time for traversing is used for displaying mean value, standard deviation and correlation of the two velocity components at the point measured just before. Mean and fluctuating quantities are obtained from average over 2000 samples and standard statistic treatment, respectively.

The task of the LABVIEW data acquisition system is also to provide continually on a graphical display the main elements of the data processing (instantaneous velocity signals, velocity histograms, arrival times comparison, number of validated data) allowing the permanent control of the validity of the L.D.A. measurements.

Generally speaking, boundary layer measurements have been performed as close as 0.2 mm to the wall; however, the measurement volume has not been approached to the model base closer than its diameter thickness.

Measurement uncertainties, mainly attributable to the optical system, include the uncertainty in the measurement of the cross beam angles of laser beam orientation and in the misalignement of laser beams at probe volume. If so, the fringe pattern is either diverging or contracting along the optical axis and its spacing variation yields different signals frequencies for the same velocity depending where the particle passes through the sample volume. Bias errors in instantaneous velocity component may also arise as a result of a velocity lag between particles and flow, noise in signals... and so forth.

The optical probe makes an angle of 1.5° with the Z-axis direction to insure that there is no interference of the beams (at the measurement volume) with the model when measurements are being made close to the model surface in its trailing edge region. Bias error is unavoidably introduced when the laser probe axis is not perpendicular to the wall glasswindow; this bias error is increased when the environment through which the laser beam is passing changes which is the case with the present experiment. However, the measurement method with the inclined laser beam did not point out a

significant deviation in data as compared with the reference ones obtained when the probe axis is normal to the window.

A systematic analysis of the bias errors has not been performed. Nevertheless, some previous experiments conducted by letting vary the laser beam orientation and the preset coincidence time window gave us some useful information about errors occurring in velocity measurement.

2.2 F2 Wind Tunnel

Experiments concerning the same as above reference and increased cambered trailing edges set in 3D turbulent flows have been conducted in the F2 atmospheric wind tunnel of ONERA-Fauga. The test section ($1.80 \times 1.40 m^2$) is fitted with a mechanism allowing the displacement of the optical probes of the three-dimensional L.D.A. system at any location in the test section volume.

Model. The model is a cylindrical wing 400 mm chordlength and 1505 mm span; the aspect ratio being about 7. In fact, the model consist of three elements: the profile of the central element is the one previously used for the 2D analysis in the water tunnel. Measurements are made only in the wake of this central element. However, the trailing edge has the same geometry in any spanwise section of the wing.

The model is fixed with a sweep angle of 20° on a turntable (inserted in the test section floor) allowing its rotation.

Test conditions. Experiments are conducted for a constant velocity of 32 ms^{-1} leading to the same free stream Reynolds number (based on the model chord and on the velocity normal to the leading edge direction) of 800 000 obtained in 2D configuration at the water tunnel. Transition tripping is exactly the same as the one considered in 2D flows.

Measurement technique. Laser measurements are performed in a forward scattering mode. The emission heads are located in such a way that each plane formed with two beams is parallel to the model trailing edge. The focal distance of the optic laser heads is inclined at about 20° related to the horizontal reference is 1500 mm. The measurement volume has a diameter of 0.30 mm and a fringe separation of 15 μm.

From photomultipliers, signals are directed on three Dantec counters through analogic pass-band filters. A preset coïncidence time (1ms) forced the counter to detect the same particle.

3. ANALYSIS AND RESULTS

3.1 Two-dimensional Flow

Pressure Distributions. The streamwise pressure distributions obtained for the RTE, CTE 1, CTE 2 trailing edges, for the same value of the lift coefficient (nominal value: Cl=0.47), are plotted in figure 2.

The pressure coefficient has been corrected from wall effects using empirical formulae [4]. As expected, the important effect attributed to the cambered lower side and divergence angle is clearly evidenced: the airfoil loading near the trailing edge is substantially modified.

The lower and upper surfaces appear to be uncoupled, that is to say that the pressure level on the upper side differs tremendously from the one recorded on the lower side. This uncoupling effect is rather more important for the CTE 2 trailing edge than for the CTE 1 one.

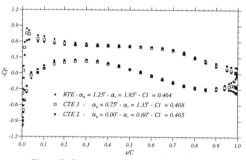

Figure 2. Streamwise Pressure Distribution

Another important feature of these increased trailing edges results is a rather large pressure gradient modification on the last 10% of chord length (figure 3):

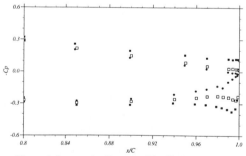

Figure 3. Streamwise Pressure Distribution (Zoom)

- on the upper surface the existing adverse pressure gradient for the standard trailing edge is reduced when considering the CTE 1 geometry; the decrease is even larger with the thickest base (CTE 2).

- on the lower surface, the pressure gradient is always negative for the OAT15A airfoil; on the other hand, this gradient is slightly positive down to x/C=0.99 and then negative again. Thus, there will be a rather small surface exposed to great acceleration.

An interesting feature to point out is that the rear loading effect does not seem to evolve too much with the angle of attack of the model, at least for the considered incidence variation.

Flow field in the vicinity of the trailing edge. One objective of the experimental programme was to scrutinize the flow just downstream of the increased trailing edges. The idea being to verify the hypothesized flow field pattern suggested by Henne et al. for the flow closure downstream of a Divergent Trailing Edge airfoil [2]. For that specific purpose, the 2D L.D.A. system has been considered.

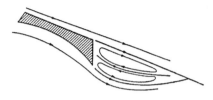

Hypothesized Flow Closure for Divergent Trailing Edge Airfoil

For all trailing edges, boundary layer surveys in the last 6.25% of chord length of the upper and lower sides as well as surveys in the near wake (towards 6.5% chord length downstream of the trailing edge) have been performed, allowing to catch a very precise description of the flow. This represents approximately an amount of 2,500 survey points.

It has to be recalled that boundary layer and wake surveys are performed with and without inclination angle of the laser beams, respectively. In any case, it allows to go as close as 0.2 mm to the wall.

Thus, for the RTE, CTE 1, CTE 2 trailing edge airfoils, set at the right incidence which corresponds to the nominal value of the lift coefficient (Cl=0.47), a very complete data base has been generated; it comprises the two components of the mean velocity, U and V, as well as the turbulent intensity and shear stress components in the close vicinity of the trailing edge.

Velocity profiles in the boundary layer are expressed in the (X,Y) cartesian frame linked to the tunnel test section. Then, knowing the geometry of the trailing edge airfoil, projection has been carried out in order to express the velocity components in directions parallel to the wall and normal to the wall, U and V respectively. However, in the last percent of chord length along the lower side, where a divergence angle is present, such a projection cannot be correctly applied.

Illustration of some boundary layer surveys is only detailed at a given location: x/C=96.25%, i.e. 15mm upstream of the trailing edge base; indeed, such a streamwise location is rather well representative of the flow modification.

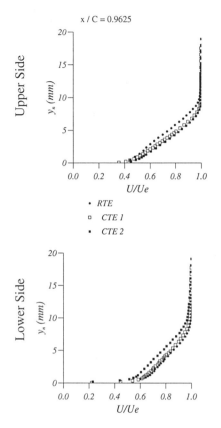

Figure 4. Mean Streamwise Velocity Profiles

For both upper and lower surfaces, mean velocity profiles, non-dimensionalized by the outer edge velocity U_e, are plotted versus y_n in figure 4; y_n is the distance normal to the wall. Being able to go as close as 0.2 mm to the wall allows to start with values of U/U_e as low as 0.4.

Considering the upper side of any airfoil model, the RTE trailing edge presents a stronger adverse pressure gradient, which explains a greater thickening of the viscous layer than that obtained from the CTE 1 and CTE 2 geometries (figure 4).

On the lower side of the airfoil, the behaviours are completely inverted: the RTE trailing edge has the most favourable pressure gradient at this measurement station compared to the CTE 1 and CTE 2 shapes and, as a consequence, the thickening of the boundary layer is

less important. Whatever geometry of the trailing edge, the V-component is either negative or almost zero due to the acceleration of the flow.

The normal and cross components of the Reynolds stress tensor are plotted in figure 5, at the same station (x/C=96.25%), for both the upper and lower sides of the three trailing edge geometries: RTE, CTE 1 and CTE 2. The profiles of u'^2, v'^2 and $u'v'$ exhibit a classic evolution versus y_n. Indeed, the streamwise turbulent intensity profile presents a peak close to the wall, the level of which increases for positive pressure gradients. On the other hand, the profile of velocity fluctuations normal to the wall has a maximum at about 20-25% of the boundary layer thickness, which moves away from the wall as the pressure gradient increases.

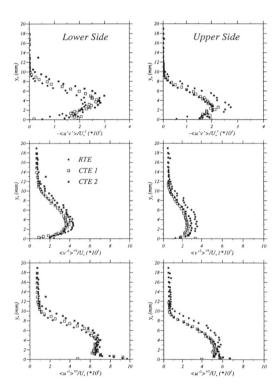

Figure 5. Components of the Reynolds Stress Tensor

One of the most evident feature concerns the shear stress profile, for which the maximum value for the upper surface (or lower one) increases (or decreases) with the adverse (or favourable) pressure gradient. That maximum shear stress is larger than the wall shear stress.

At the edge of the boundary layer the profiles of the velocity fluctuations tend towards a very weak turbulent value close to 1%; it represents the free stream turbulence rate that has been measured previously in the water tunnel for this free stream velocity.

The knowledge of the velocity field in the vicinity of the various trailing edges allows to plot the streamline pattern; in order to compare the flow field downstream of these thick cambered trailing edges, an enlargement has been produced downstream of any trailing edge, over about 3% of chord length in the streamwise as well as transverse directions.

models provide a correct prediction of the measured viscous flow behind such cambered trailing edges, but others do not [4]. An example of rather good agreement between experiments and calculation is given in figure 7, using the Spalart-Allmaras model [6].

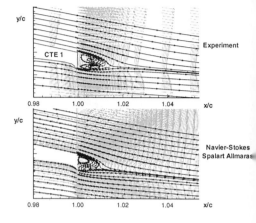

Figure 7. Computed and measured near-wake flows

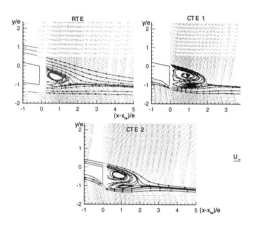

Figure 6. Streamline Pattern

Figure 6 refers to the streamline pattern of the OAT15A-RTE, CTE 1 and CTE 2 trailing edges, respectively. For every configuration, two vortices are recorded in a confined area, downstream of the base. The dimensions of the recirculating area is tightly related to the base height or thickness, e: indeed, it is about (2e x e) in the streamwise x transverse directions. The grid refinement is even such that the location of the re-attachment point can be estimated.

However, when looking at these streamline patterns downstream of such airfoils, one could question the two-dimensionality of the vortex flow, since some streamlines seem to wrap around one point while others do not. In order to try to analyse such an observation, the dW/dZ component has been derived from the continuity equation and plotted in the near wake. Although, no drawing is given in the present paper, there has been no sufficient variation in the modulus of that derivative to explain that the flow was moving inwards or outwards the (X,Y) plane.

A first numerical approach has been recently conducted, aiming at testing the behaviour of different turbulence models for such boundary and wake flows [3,4]. Several one-equation and two-equation models have been implemented in the Navier-Stokes solver, originally developed at ONERA [5]. Some turbulence

3.2 Three-dimensional Flow

Pressure measurement. Pressure measurements obtained for the model equipped with the thicker cambered trailing edge (CTE 2) have given, as expected, comparable distributions with the ones obtained for 2D flow conditions.

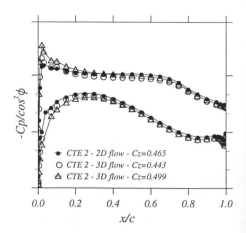

Figure 8. Pressure Distribution

With the infinite swept wing hypothesis, the pressure coefficient can be expressed in a direction normal to the leading edge. Figure 8 gives a comparison of the pressure distribution for CTE 2 (for three values of Cl surrounding the nominal value of 0.47) obtained in 2D and 3D flows. Some deviations are notable on the upper and lower side in the vicinity of the leading edge whereas on the rear part of the model, the evolutions of the pressure coefficient are very close to each other.

Flow field in the vicinity of the trailing edge. Near wake surveys with the 3D Laser Anemometer have been made in two planes P1 and P2, aligned with the free-stream direction; the distance in the transverse direction (Z) between these two planes is equal to 100mm. Results obtained then would allow to verify the invariance of mean and turbulent quantities in the Z-direction in accordance with the uniform repartition of the local lift coefficient in spanwise direction, resulting from Euler computations.

From a good refinement of the survey grid, it has been possible to catch a very precise description of the flow. Figure 9 shows a comparison of the streamline pattern in the two above mentioned planes. This comparison reveals a good agreement on the topologic structure of the flow downstream of the CTE 2 trailing edge. It should be noticed that the flow pattern just downstream of the base is correctly reproduced from P1 to P2. The two-dimensionality of the flow does not seem to be questionable regarding the location and the size of the vortices in the two measurement planes.

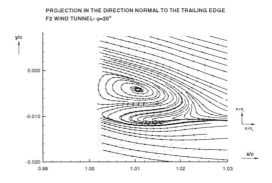

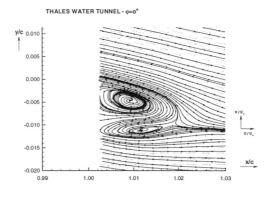

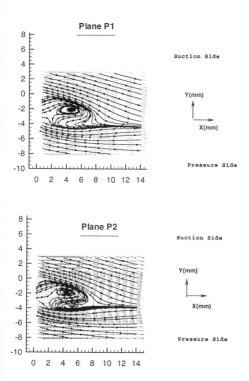

Figure 10. Comparisons of the flow field pattern downstream of the CTE 2 in 2D and 3D flows

Projecting the velocities in a plane perpendicular to the trailing edge, the flow field pattern in the near wake can be compared to the one observed in two-dimensional case (figure 10).

In a general way, behind 2D thick cambered trailing edges, 3D flow does not exhibit a topological structure far away from the one evidenced in 2D flow.

Figure 9. Streamline pattern downstream of the CTE 2 trailing edge; Check for two-dimensionality

Nevertheless, the rolling motion of the fluid in the lower side of the wake has disappeared. Setting the model at a given sweep angle of 20° seems to slightly modify the streamwise position of the upper vortex and of the reattachment point.

4. CONCLUSION

Some interest has been recently devoted to the Divergent Trailing Edge concept, following up research that has been undertaken at McDonnell Douglas Corporation. When looking at literature, there is very little published work from either an experimental or numerical point of view, as regards flow modification induced by such non-conventional trailing edge geometries.

An experimental programme has thus been carried out in order to characterize the potential and viscous flows in the vicinity of different thick cambered trailing edges in both two- and three-dimensional flows.

The two increased cambered trailing edges have induced a rather important pressure gradient modification on the last 10% chord length of both upper and lower sides of the airfoil. The effect of the cambered lower side and of the divergence angle of the trailing edge results in an increase of rear airfoil loading.

In two-dimensional configuration, a lot of boundary layer and wake surveys have been performed on both sides of the trailing edges geometries, providing with a very precise description of the flow. A consistent data base, for mean and fluctuating quantities, has been obtained even though the laser beam orientation has been changed in some areas in order to go as close as possible to the wall.

Just downstream of each increased base thickness, a recirculation area has been captured, including two contra-rotating vortices, the dimensions of which are tightly related to the base height.

The behaviour of non-conventional two-dimensional shapes of trailing edges has thus been looked at in three-dimensional turbulent flows putting the model with 20° sweep angle. Near wake surveys have also revealed a recirculation area corresponding to negative streamwise velocities. However, the topological structure is slightly modified from the one obtained for two-dimensional flow conditions.

Thus, a very complete data base has been generated and will be (hopefully) used for subsequent code validation. Indeed, boundary layer and wake surveys in the immediate vicinity of the trailing edge have provided mean and fluctuating quantities as well as integral parameters, for various geometries of trailing edges. The recorded flow field pattern downstream of such base shapes should help for validating and/or improving the existing Navier-Stokes codes for future optimisation of thick cambered trailing edges.

Acknowledgements
GP and EC are very grateful to the whole team from the ONERA/Fauga F2 wind tunnel (MM. D. Afchain, M. Deluc, R. Johannel and Ph. Loiret).
The authors wish also to acknowledge Airbus Industrie and the Service des Programmes Aéronautiques (SPAé) for supporting these studies.

REFERENCES

[1] Liebeck R.H. 1978, Design of subsonic airfoils for high lift. Journal of Aircraft, vol. 15, N°9, pp. 547-561.
[2] Henne P.A. and Gregg R. D. 1991, A new airfoil design concept. Journal of Aircraft, vol 28, N°5, pp. 333-345.
[3] Sauvage Ph, Pailhas G., Coustols E. 1997, Detailed flow pattern around thick cambered trailing edges. Seventh Asian Congress of Fluid Mechanics, Chennai Madras.
[4] Sauvage Ph. 1998, Etude expérimentale et numérique des écoulements potentiels et visqueux dans le voisinage d'un bord de fuite épais cambré, Ph. D Thesis, ENSAE, Toulouse, France.
[5] Vuillot A.M., Couailler V. and Liamis N. 1993, 3D turbomachinery Euler and Navier-Stokes calculations with a multi-domain cell-centered approach, AIAA-93-2576.
[6] Spalart P.R. and Allmaras S.R. 1994, A one-equation turbulence model for aerodynamics flows. La Recherche Aerospatiale, pp. 5-21.

III.4. Instantaneous Flow Field Measurements for Propeller Aircraft and Rotorcraft Research

M. Raffel[1], F. de Gregorio[2], K. Pengel[3], C.E. Willert[4], C. Kähler[1], T. Dewhirst[1], K. Ehrenfeld[1], and J. Kompenhans[1]

[1] DLR, Bunsenstraße 10, 37073 Göttingen, Germany.
[2] CIRA, Via Maiorise, 81043 Capua (CE), Italy.
[3] DNW, Voorsterweg 31, 8316 PR Marknesse, the Netherlands.
[4] DLR, Linder Höhe, 51147 Köln, Germany.

Abstract. This paper illustrates results of instantaneous flow field measurements which were performed by the PIV group of DLR Göttingen during test campaigns in three different wind tunnel facilities of the Dutch-German Wind Tunnel foundation (DNW). The test results described first were obtained in the low speed tunnel LST in order to obtain detailed flow field information about the wake of a new high speed propeller model. The second test was performed in the transonic wind tunnel TWG in the frame of a German helicopter research project. Here the flow field above a pitching airfoil equipped with a flap driven by an integrated piezoelectric actuator has been observed with high spatial resolution. During the third test, stereoscopic flow field measurements were performed in the wake of a rotorcraft model in order to characterize the blade tip vortices at different flight conditions. The test was conducted in the 6 x 8 m^2 open jet test section of the large low speed facility LLF which required observation distances exceeding 9 meters between camera and light sheet.

Keywords. particle image velocimetry, propeller aircraft, rotorcraft, wind tunnel, seeding, pitching airfoil, stereoscopic recording

1 Introduction

The increasing demand of air transport, the strong competition between the different airline companies and the competition with other transport systems like high speed trains - especially in Europe where the travel time between the centers of the cities can be comparable - as well as a strong interest in the protection of the environment led the propeller aircraft and rotorcraft manufactures to intensify their investigations on the efficiency and noise emission of their products.

Current testing techniques on airframe/propeller engine integration and rotorcraft research for performance enhancement and noise reduction deliver mainly test data on overall aerodynamic forces, moments, and surface pressure distributions. More detailed investigations are required for an improved CFD modeling for a better understanding of the aerodynamic phenomena in general. The unsteady complex flow structures, especially the vortices above and behind rotor blades require detailed instantaneous flow field mapping to get insight into the physics. The improvement and reliability of particle image velocimetry (PIV) applied to high speed air flows in large industrial wind tunnel facilities play a key role in today's aircraft development.

2 PIV utilities for wind tunnel tests

In large scale industrial wind tunnels one of the main requirements is to obtain a relatively large observation area. This requires a recording system with a high spatial resolution as well as a high light sensitivity. Through the use of single-exposure photo cameras or high resolution digital cameras large acquisition areas can be achieved but usually require complex additional hardware to resolve e.g. the directional ambiguity. The rapid development of full-frame interline transfer solid state sensor architectures during the past years made a new generation of cameras available which are ideally suited for PIV applications in industrial wind tunnels. The limited present resolution of approximately one million pixels is partly compensated by the fact that cross-correlation of separately-recorded single-exposed images with full frame resolution shows a principally superior signal-to-noise ratio compared to other correlation techniques, and therefore allows smaller interrogation areas that can be used for a reliable velocity estimation (Keane and Adrian 1992).

Another set of requirements arises from the fact that nearly all our applications of PIV take place in air, from moderate to high speed velocities, and sometimes with high centrifugal forces: small seeding particles are needed to accurately follow the flow which in turn requires the use of high powered lasers in conjunction with high quality imaging equipment. This holds in particular in large facilities where distances between observation area and recording equipment can be considerable. In order to fulfill all these requirements we chose the following subsystems for our wind tunnel tests.

2.1 Recording

In recent years cameras incorporating progressive scan, full-frame interline CCD technology have become available which, contrary to the more common interline transfer CCD sensors, are capable of shuttering (exposing) and storing the entire array of pixels, not just every other line. Thus, these sensors immediately offer the full vertical resolution when the CCD is used in the shuttered mode. A standard (TV) resolution camera of this type was first used for high-speed PIV measurements in air at the California Institute of Technology as part of the collaboration in the Center for Quantitative Visualization (Vogt et al. 1996).

Following this general approach, the present PIV camera systems additionally feature a non-standard, high resolution, digital video format consisting of 1008 by 1018 pixels and 1024 by 1280 pixels respectively. Using a 32^2 pixel interrogation window, this translates to a spatial resolution of up to 32 by 40 discrete vectors. Standard 35 mm photographic film in comparison, digitized at 100 pixels per millimeter, yields about 56 by 37 discrete vectors at a comparable measurement uncertainty (a 64^2 pixel interrogation window is required (Willert 1996)). As the digital video signals of these cameras cannot be viewed using a standard video monitor a PC-interface card is necessary which transfers the digital data directly into the computer's memory (RAM) and thereby allows a continuous sequence of frames to be captured and viewed. An enormous advantage of these cameras is that they are offered with both high resolution and sensor cooling which reduces the black current and increases the dynamic range to 12 bit. A fiber optic transmission link allows for very long distances between camera and frame grabber.

One major drawback of these cameras and subject of investigations (Westerweel 1998, Ronneberger et al. 1998) arises through the regular arrangement and response characteristics of picture elements on the sensor. It was found that a decrease of the particle image size to cover only a few pixels was associated with an increased systematic bias error towards integer displacement values, generally referred to as "peak locking". Since air flow PIV applications require diffraction limited imaging of very small particles this effect can cause severe problems and requires detailed studies to be able to make reliable statements about the obtained measurement accuracy. In the case of the test results presented here a significantly enhanced data quality could be achieved by applying improved correlation algorithms as presented by Ronneberger et al. (1998).

2.2 Flow Seeding

The most common seeding particles for PIV investigation of gaseous flows are oil droplets which are generated by means of Laskin nozzles. Pressurized air, injected in olive oil, leads to the formation of small oil droplets. The aerodynamic diameter of the olive oil particles is about 1μm. In wind tunnel flows the supply of tracers is often difficult. The particles, which are mostly used, are not easy to handle because many droplets formed from liquids tend to evaporate rather quickly and solid particles are difficult to disperse and very often agglomerate. Therefore, the particles cannot simply be fed preceding the measurement, but must be injected during the test into the flow upstream of the test section. The injection has to be done without disturbing the flow significantly, but also in a way and at a location that ensures homogeneous distribution of the tracers. Since the existing turbulence in many test facilities is not strong enough to sufficiently mix the fluid with particles, the particles have to be supplied through a large number of openings. Distributors like rakes consisting of many small pipes with a large number of small holes are often used. This requires particles which can be transported inside small pipes without agglomeration. Figure 1a shows the sketch of an atomizer which generates suitable particles and has been used for most of our PIV measurements in air flows. The particles generated by this device are non-toxic, stay in air at rest for hours, and do not significantly change in size under various conditions.

In closed circuit wind tunnels these micron-sized oil droplets can be used for a global seeding of the complete volume, as done in the TWG. Alternatively a local seeding of a stream tube can be achieved by a seeding rake with a few hundred small holes such as the 2.5 x 2 m^2 rake which was mounted in the settling chambers upstream the honeycombs and screens of the low speed wind tunnels LST and LLF (Fig. 1b).

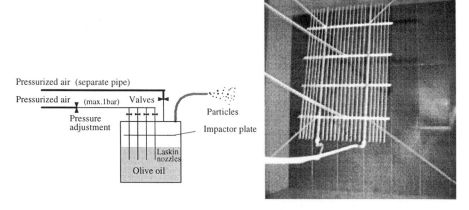

Fig. 1a Seeding generator

Fig. 1b Seeding rake in the settling chamber of the wind tunnel

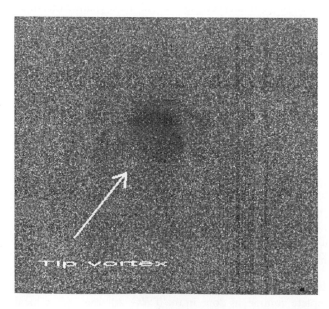

Fig. 2 Typical PIV recording of a propeller tip vortex (20 x 20 cm^2)

The goal is to obtain a high uniform seeding density in the region of interest (at least 15 pairs of particle images per interrogation window (Keane and Adrian 1990)) in order to apply statistical evaluation methods also in the region where strong recirculation or velocity gradients are present.

Figure 2 shows a PIV image portion with reduced seeding density due to the velocity lag (integrated from the blade tip to the observation area) and due to the reduced air density inside the vortex core. Because of the size of the rake, three aerosol generators, each containing 40 Laskin nozzles, are used. Each generator consists of a closed cylindrical container with five air inlets and one aerosol outlet. Several Laskin nozzles, 1 mm in diameter, are radially spaced in the inlet pipes. A horizontal circular impactor plate is placed inside the container, so that a small gap of about 2 mm is formed by the plate and the inner wall of the container. Compressed air with 0.5 to 1.5 bar pressure difference against the outlet pressure is applied to the Laskin nozzles and creates air bubbles within the liquid. Due to the shear stress induced by the tiny sonic jets, small droplets are generated and carried inside the bubbles towards the oil surface. Big particles are retained by the impactor plate; small particles escape through the gap and reach the aerosol outlet. The amount of particles can be controlled by switching four valves at the nozzle inlets. The particle concentration can be decreased by an additional air supply via the second air inlet. The mean size of the particles generally depends on the type of liquids being atomized but is only slightly dependent on the operating pressure of the nozzles. Vegetable oil and Di-2-Ethylhexyl-Sebacat (DEHS) are the most commonly used liquids since they are said to be less unhealthy than many other liquids. However, any kind of seeding particles, which can not be dissolved in water, should not be inhaled. DEHS and most vegetable oils lead to polydisperse distributions with mean particle diameters of approximately 1 µm or less and therefore can agglomerate within the lungs.

2.3 Laser

The light sources used were pulsed Nd:YAG lasers with two independent oscillators. They were driven at repetition rates of 10 Hz, the pulse energy at $\lambda = 532$ nm was 2 x 320 mJ and 2 x 150 mJ in case of the TWG test respectively. PIV measurements over long distances do not only require powerful lasers but also excellent characteristics of spatial intensity distribution of the laser beam (hole-free without hot spots especially in the midfield where it is most difficult to get), excellent co-linearity, beam pointing stability (< 100 µrad) and energy stability (< 5 %). To obtain a thin light sheet, typically with a thickness in the order of 1 mm, the laser beam passed a set of coated spherical and cylindrical lenses optimized for a homogeneous light intensity distribution versus the light sheet thickness and height. A small "misalignment" of the combination optics has partly been used in order to obtain light sheet positions which were slightly displaced in mean flow direction to compensate for the out-of-plane motion of tracer particles.

3 Propeller Investigations

The request to reduce the fuel consumption and the pollution emission of the propulsion systems has forced the aircraft manufactures to investigate and install new ultra high by-pass ratio engines, and, in case of smaller aircraft, high speed open rotors. Furthermore, these requirements have led the manufacturers to renew their interest towards propeller driven aircraft in order to develop a new generation of commuter aircraft with the same operational capacities and comfort as regional jets with reduced emission, cost and noise.

Fig. 3 The isolated test rig of the APIAN propeller in the LST wind tunnel

To develop such a propeller, it is mandatory to study the slip stream behavior and the propeller and airframe integration in order to come to higher efficiency and performance. The highly unsteady phenomena require new experimental tools in order to analyze the flow field characteristics. In parallel, the new measurement techniques should also allow to reduce the wind tunnel time in order to minimize the development cost.

In order to study the high speed propellers behavior an experimental campaign on an isolated propeller test rig (Fig. 3) has been carried out in the LST facility by means of PIV in the frame of the Brite-Euram project APIAN. For the first time, this allows one to obtain quantitative information about these unsteady and partly non-periodic flow fields for CFD code validation experimentally.

A large number of PIV recordings was acquired for each different test condition in order to also allow a statistical evaluation of the wake downstream of the propeller plane. The instantaneous results and the mean results obtained by averaging over a large number of PIV data sets were compared such that the difference between the time-averaged and unsteady flow could be studied. The dependence between blade position and wake flow could be investigated by simultaneously storing the azimuth angle of each PIV recording.

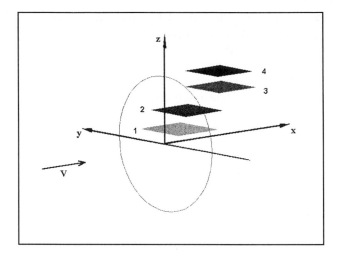

Fig. 4 Reference system and investigated areas of the APIAN propeller

The LST is a continuous atmospheric wind tunnel with closed walls, a test section of 3 m width and 2.25 m height, maximum air velocity of 80 m/s, and powered by a 700 kW fan. The measurements have been carried out downstream of an isolated propeller test rig. The propeller model, including an internal six component rotating balance, has been mounted on the external force balance of the tunnel as shown in Figure 3. The model represents a new type of high speed propeller (Mach 0.7-0.8), which is composed of six blades, and which has a diameter of 500 mm. All tests were performed at the same wind tunnel velocity (80 m/s) with the same blade angle of 40.4° but with different propeller rotational speed (8000 and 8650 rpm). This permitted the investigation of the influence of the advance ratio on the slipstream under various different incidence and jaw angles (0°, +10°,-10°) of the propeller.

The reference system is related to the model and defined as follows: the plane z = 0 is a horizontal plane containing the axis of rotation, the plane y = 0 is a vertical plane containing the axis of rotation, and the plane x = 0 is a vertical plane containing the trailing edge of the blade profile. The measurements were performed for different horizontal planes (Fig. 4), starting from the axial plane until to the z-position of 70 % of the blade radius.

The Nd:YAG pulse laser was mounted at the outer side wall of the test section above a 2D traversing system in order to be able to easily change the light sheet position. The whole laser system and light sheet optics was constructed on an optical bench using commercially available X-95 elements. This results in a stable and still very variable and compact set up. Two DC gear motors, connected to the alignment screws of the laser's second harmonic generators, allowed the phase-matching to be optimized remotely from the wind tunnel control room during the tunnel run. The light sheet formed by a 500 mm spherical and 200 mm cylindrical lens, was 300 mm wide and 2 mm thick. The flow field of the propeller slipstream presents a strong 3-D behavior. Therefore, the pulse delay time was reduced to 10 μs in order to reduce the strong out-of-plane loss of correlation.

The digital camera was located above the test section, mounted on the support of the external balance so that a realignment and a re-calibration of the camera was not necessary when changing the incidence and the jaw angles of the propeller. Calibration was done with a calibration grid inserted in the test section at the position of the light sheet as shown in Figure 4.

The measurement planes, dimension about 20 x 20 cm^2, have been acquired, by the digital camera, from a distance of 1.5 m, using a 100 mm lens with f-number of 2.8. Remotely focusing of the lens was achieved with a self made focusing system.

To correlate the propeller blade position to the acquired vector field a custom Propeller Angle Measurement device (PAM) has been designed and built. A reset signal per each propeller revolution and a Q-switch signal from the laser were used by the PAM system to evaluate the propeller position. This information was stored in the first two bytes of the image file.

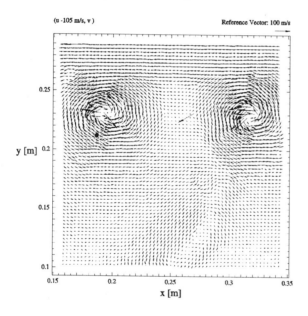

Fig. 5 Instantaneous velocity field of propeller slipstream

In total about 1700 image pairs, related to 28 different test conditions, were acquired. For each test condition 50 pairs of images were taken in order to investigate the instantaneous flow condition and, by averaging over all the instantaneous measurements, the mean flow field. In one test condition, more than 400 image pairs have been acquired so that the influence of the blade position on the slipstream could be investigated. The images have been evaluated by means of cross-correlation method using interrogation windows of 32 x 32 pixels, the interrogation windows have been overlapped 50 % providing a matrix of 61 x 62 displacement vectors. To improve the signal/noise ratio the multi pass algorithm described by Willert (1997) has been used. After that the displacement vector matrix was converted into the velocity matrix and the origin of the reference system was determined (Fig. 5). The spatial resolution of the vector matrix is better than 3 mm. A validation algorithm (Raffel et al. 1993) to remove to faulty vectors that can be seen in figure 5 and a filter pass to replace the hole with vectors evaluated from the average of the neighbor's values was applied before further analysis of the data.

PIV measurements in the plane z = 0 were performed, and 40 instantaneous vector fields are taken. The measured plane covers the area from 155 mm to 350 mm along the x-direction and from 97 mm to 292 mm along the y-direction (Area 1 in Fig. 4). In Figure 5 the instantaneous velocity field for 0 degree of incidence and jaw angles is shown. The mean component of the velocity has been subtracted in order to highlight the vortical structures. Figure 5 clearly presents two sequential blade tip vortex in the measurement plane. Furthermore the shear layer of one of the blades can clearly be seen as it rolls up in to its tip vortex.

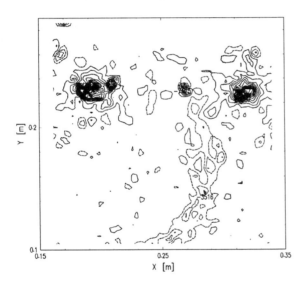

Fig. 6 Isolines of the (1d) vorticity field

These vortical structures can also be observed in Figure 6, which represents the out-of-plane component of the vorticity as isolevel lines. Continuous contours indicate positive and dashed contours negative values of vorticity. The contour levels are spaced at 1500/s.

Furthermore from our results it is possible to determine the trajectory of the tip vortex. In this case the vortex moves on a straight line nearly parallel to the free stream velocity direction of the wind tunnel at a distance of y = 227 mm from the x-axis, showing a contraction of the slip stream downstream the propeller disc. The distance between each vortex core is 120 mm.

286

For each test condition, the average of the flow field has been evaluated on the basis of 50 instantaneous velocity fields. The obtained flow field is shown in Figure 7. Also in this case the mean velocity is subtracted to highlight the vortical structures but only the shear layer is visible. These averaged flow fields are the typical output of pointwise measuring techniques like laser Doppler velocimetry, pressure probes or by hot wire anemometry, if no conditional sampling has been applied. This is also the case for most of the stationary CFD codes.

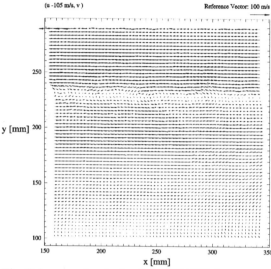

Fig. 7 Averaged velocity field (50 PIV data sets)

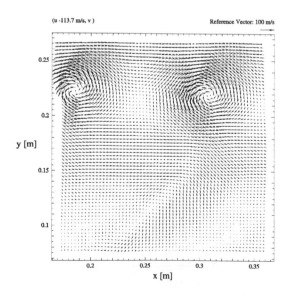

Fig. 8 Phase averaged flow field at 30° incidence angle (40 PIV data sets)

In order to investigate the influence of the blade position in more detail, more than 400 image pairs were recorded for one case. The chosen test condition is more critical than the others because the propeller had an incidence angle of 10° with respect to the wind tunnel system. In this condition the out-of-plane velocity is larger and the measurement is affected by a higher noise error. Conditional averaging was performed with reference to the position of the propeller.

In Figure 8, the conditionally averaged velocity field at the blade position of 30° is presented. The blade position is counted starting from the positive z-axis in clock-wise direction. About 40 instantaneous velocity maps, taken all at the same angular position, were averaged. Also here the mean value of the x-component of the velocity is subtracted to highlight the vortical structures.

By averaging the flow fields the noise errors can significantly be reduced, the vortical structures appear more homogeneous, and the comparison with results of other numerical or experimental techniques, which do not resolve the cycle-to-cycle variations but the periodic flow phenomena becomes easier.

This can also be seen in the vorticity map (Fig. 9), where the isolevel contour are smoother and more homogeneous with respect to the vorticity contour of an instantaneous measurement (Fig. 6).

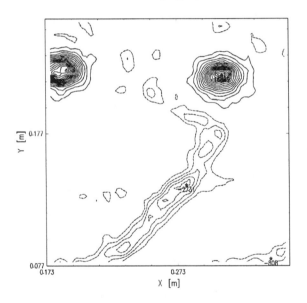

Fig. 9 Isolevel contour map of the vorticity of a phase-averaged flow field at 30° incidence angle

4 High Resolution Flow Field Mapping on An Adaptive Blade

Adaptive helicopter rotor blades are assumed to be one of the key technologies for helicopter performance enhancement for the oncoming decade. In a national project (AROSYS), the German industry (ECD and Daimler Benz Forschung) and

288

the German Aerospace Research Establishment (DLR) combined their efforts to reach reliable solutions for adaptive rotor systems.

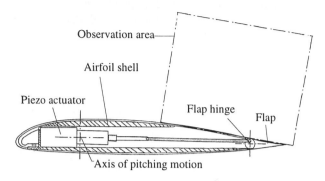

Fig. 10 Helicopter blade model and observation area.

One of the first promising results was a piezoelectric actuator small enough to be integrated into a full-scale rotor blade and strong enough to drive a servo flap at the trailing edge of the outer blade area (Fig. 10). Since performance estimations and blade profile optimization can efficiently be obtained by non-stationary Navier-Stokes simulations (see e.g. Wernert et al. 1996) there is a great interest in validating these codes and to subsequently apply them to adaptive blade geometries and concepts. For this purpose a wind tunnel test was recently performed in the transonic wind tunnel TWG in.

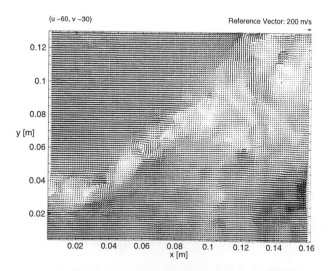

Fig. 11 Instantaneous flow velocity field above the adaptive blade at a Mach number of 0.33, pitching at 7 Hz (14.5° +/- 5°), with a flap motion of 14 Hz, +/-3°.

Forces, moments and surface pressure distributions were measured dynamically for different Mach numbers, pitching amplitudes, and frequencies. The tests were performed at different flap frequencies and phase relations by the Institute for Aeroelasticity of DLR. In order to complete this information, PIV measurements of an area above the trailing edge flap, which was pitching in phase with the airfoil frequency, were performed.

First results of these measurements at a phase angle of 320° are shown in Figure 11 and its vorticity distributions in Figure 12. A shear layer between the separated flow area and the mean flow and the influence of the dynamically pitching flap can be seen. The flap angle varied by +/- 3° at a frequency of 14 Hz and the airfoil was operated at a mean angle of attack of 14.5° with 5° amplitude at 7 Hz. Laser and camera were phase locked with the pitching motion at a pre-selected phase angle. The chord length of the blade was 0.3 m, the span 1m. The high quality schlieren windows gave optical access to the perforated transonic test section of the tunnel. Due to the high camera resolution, the high seeding density and homogeneity, and improved correlation algorithms described by Ronneberger et al. (1998) up to 125 by 100 velocity vectors could be measured with an overlap of 50 %.

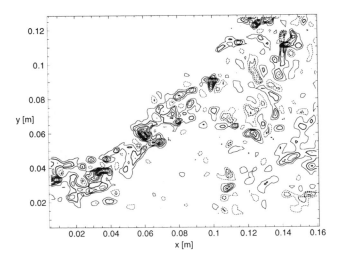

Fig. 12 Instantaneous out-of-plane component of vorticity computed from flow velocity data shown in figure 11.

5 Stereoscopic rotor measurements in a large wind tunnel

Detailed investigations employing PIV, surface pressure and acoustic measurements at an 1:4 scaled rotor were conducted in the large low-speed facility (LLF) of DNW. The successful PIV measurements on rotors in a large scale facility such as LLF are based on continual efforts over the past three years to resolve a number of special problems such as long observation distances, seeding problems, and the robustness and remote control capabilities of the systems. At this point it can be said that all technically relevant problems, even for stereoscopic measurements, can be solved and that additional efforts to employ this technique on a routine base in near future are justified. The main motivation for this is, that in spite of all its advantages, the PIV method underlies some shortcomings that make further developments on the basis of instrumentation necessary. One of these disadvantages is the fact that the 'classical' PIV method is only capable of recording the two-dimensional projection of the velocity vector. A variety of approaches capable of recovering the complete set of velocity components have been described in the literature (Hinsch 1995, Royer and Stanislas 1996). The most straightforward, but not necessarily easily implemented, method is the additional PIV recording from a different viewing axis using a second camera, which can be generally referred to as stereoscopic PIV recording (Westerweel and Nieuwstadt 1991, Prasad and Adrian 1993, Gaydon et al. 1997). Stereoscopic recording can also be achieved with a single camera by placing a set of mirrors in front of the recording lens (Arroyo and Greated 1991). To adapt the stereoscopic approach to an industrial wind tunnel environment, a number of additional developments are necessary. First of all the optical access in wind

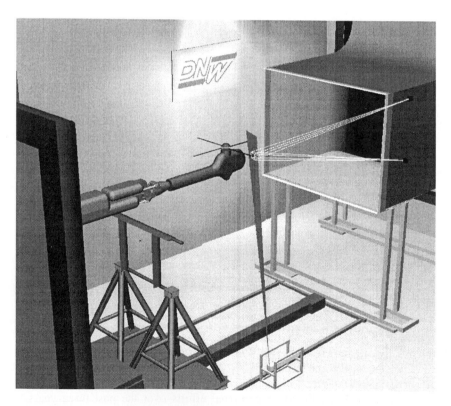

Fig 13: Experimental arrangement of the stereoscopic PIV measurement of a rotor flow in the large wind tunnel of DNW (open test section, nozzle: 6 x 8 m^2).

tunnels rarely permits the imaging configuration to be symmetric as given in all of the previous implementations. Another requirement is that the small, micron sized, seeding particles, have to be imaged over large distances at up to 10 meters. This makes the use of large focal length lenses with large light collecting capability (i.e. small $f_\#$-numbers) necessary. Since the measurement precision of the out-of-plane component increases as the opening angle between the two cameras reaches 90°, it is not always possible to mount the cameras onto a common base, much less to provide a symmetric arrangement.

In the following a description of our generalized, that is, non-symmetric, set up for stereoscopic PIV imaging is given. The feasibility of this approach was demonstrated in an actual experiment with the unsteady flow field of a rotorcraft model blade tip vortex. In the final configuration, the recording parameters were optimized such that a pulse delay of 20 μs resulted in a particle displacement of 3-4 mm normal to the light sheet (approximately 10 mm thickness) with a maximum in-plane displacement on the order of 3 mm. With an observation area of 25 by 30 cm^2 this translated to displacements in the order of 10 pixels on the digital sensor which still provides at least a factor 100 of dynamic range (the noise level in the recovered displacement data is in the order of 0.1 to 0.05 pixel). The seeding particles were illuminated by a dual oscillator, frequency doubled Nd:YAG laser with 320 mJ per pulse. The observation area was imaged by two high resolution digital cameras using two 300 mm $f_\#$ 2.8 lenses at an observation distance of 9.5 meters each. The large observation distances were chosen to reduce any blockage effects imposed by the imaging equipment.

The experimental configuration used two sensitive (12-bit converter, cooled sensor) and high spatial resolution CCD cameras to image the experimental region of interest, an area approximately 25 by 30 cm^2 over 9.5 m from each camera (see Fig. 13). The cameras were located at the side of the nozzle of the open jet test section, and were placed in a vertical plane intersecting the observation area. The angle between the camera axes and the normal of the observation area was approximately 12°. Figures 14 and 15 show the first preliminary results taken from a series of measurements performed at the DNW-LLF on a rotorcraft model. The diagrams were turned by 90° anticlockwise for this representation. The resulting vector maps, while taken at subsequent azimuth angles exhibit clear differences. The rotor blade tip, which had just passed the observation area in out-of-plane direction when the first recordings were made, had given rise to the vortices present in the velocity vector maps. The out-of-plane velocity component along a cut through the vortex center - represented by the dotted line in Figure 14b - shows strong spatial gradients that clearly exceeds the gradients of the in-plane components.

The following velocity map (Fig. 15a) was recorded at slightly higher azimuth angles when the advancing blade has moved further upstream. Therefore, the vortex roll-up and the development of the axial velocity component can be investigated in detail.

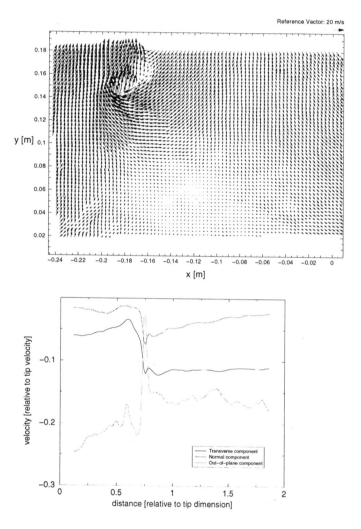

Fig. 14a (upper) and 14b (lower): In plane velocity components of the tip vortex and the wake of the blade. (The diagrams (14a and 14b) are turned by 90° anticlockwise with respect to Fig. 13.)

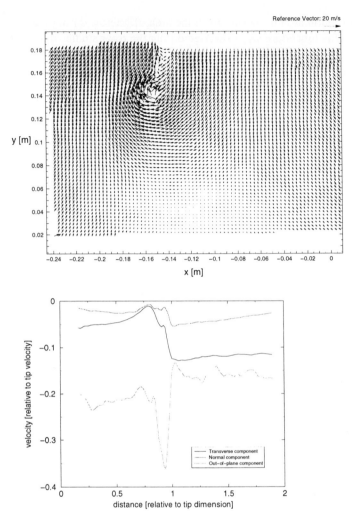

Fig. 15a (upper) and 15b (lower): In plane velocity components of the tip vortex and the wake of the blade. (The diagrams (15 and 15b) are turned by 90° anticlockwise with respect to Fig. 13.) The azimuth angle during recording was 10° larger than during the recording of data set shown in Fig. 14.

6 Conclusions and Outlook

High resolution cross-correlation cameras fill the gap in spatial resolution between the high-speed video camera and photographic 35 mm PIV recording. The high-speed data transfer (up to 40 MByte/s) allows a large number of PIV recordings to be recorded in a short time. The acquisition software was devised in such a manner that captured image pairs could be interrogated within 30 seconds allowing the camera's observation area to be shifted to the area of interest during the operation of the tunnel. This procedure was exercised during the rotor studies in mid-scale as well as in large-scale facilities at moderate and high velocities.

The seeding particles used for these investigations were oil and DEHS droplets, which were generated by means of Laskin nozzles. The aerodynamic diameter of the particles was about 1 μm or smaller. The particles have been used for a global seeding of the complete volume in the transonic wind tunnel TWG and for a local seeding of a stream tube by a seeding rake with a few hundred small holes like the rake used in the mid-scale low speed wind tunnel LST and in the large low speed facility LLF. In all cases the injection was done without significantly disturbing the flow, but also in a way and at a location that guaranteed homogeneous distribution of the tracers. The fluid mechanical properties of the seeding particles have been checked in order to avoid large discrepancies between fluid and particle motion. The particle diameter was small enough in order to ensure a good tracking of the fluid motion and a sufficient particle density inside the tip vortices to be measured.

The current development efforts regarding PIV systems at DLR Göttingen are now focused on oblique as well as stereoscopic imaging arrangements. Aside from providing image access in areas that cannot be viewed in a classical PIV imaging configuration, the oblique viewing geometry can greatly improve the sensitivity of the system especially in a forward scattering configuration. Ultimately this also permits the imaged area to be increased without a need to further increase the laser power. The large angle (30°) between the viewing axis and the normal of the light sheet in conjunction with long focal length lenses (= narrow viewing angle) no longer permits the imaging plane to be parallel to the light sheet (translation imaging geometry). Rather, the lens, image and light sheet planes are tilted with respect to each other such that they intersect in a common line (angular displacement or Scheimpflug imaging geometry). This arrangement produces images with a varying magnification factor across them, which can be accounted for by de-warping the image. By adding a second camera with a different viewing angle all three velocity components have been retrieved simultaneously. Initial experiments (Willert 1997) and our first tests in a large wind tunnel both using two digital cameras, have demonstrated the feasibility of this approach.

Acknowledgements

All three PIV tests have been carried out in major international or national projects, comprising numerous contributions from many scientists, engineers, and technicians. The investigations of the propeller flow field were performed in the frame work of a Brite-Euram Project, a research project denominated APIAN (Advanced Propulsion Integration Aerodynamics and Noise), in the collaboration between CIRA and DLR. The tests in the transonic wind tunnel in Göttingen (TWG) were performed in the frame of the national AROSYS project. Special thanks to Dr. Schewe, Dr. Wendt, Dr. Schimke, and Mr. Jänker for the fruitful cooperation during the tests. The high motivation of the wind tunnel teams in LLF, LST, and TWG is greatly appreciated.

References

Arroyo M.P., Greated C,A., 1991: Stereoscopic particle image velocimetry.
Meas. Sci. Technol., Vol 2, 1181-1186.

Gaydon M., Raffel M., Willert C., Rosengarten M. and Kompenhans J., 1997:
Hybrid stereoscopic particle image velocimetry. Exp. Fluids, Vol 23, 331-334.

Hinsch K.D. 1995: Three-dimensional particle velocimetry.
Meas. Sci. Technol., Vol 6, 742-753.

Keane R D, Adrian R J 1990: Optimization of particle image velocimeters.
Part I: Double pulsed systems. Meas. Sci. Technol. Vol. 1, 1202-1215.

Keane R D, Adrian R J 1992: Theory of cross-correlation analysis of PIV images.
Appl. Sci. Res. Vol. 49, 191-215.

Lowson M.V., 1991: Progress towards quieter civil helicopters.
Proc. 17th European Rotorcraft Forum, paper 91-59, Berlin, Germany.

Prasad A.K., Adrian R.J., 1993: Stereoscopic particle image velocimetry
applied to liquid flows. Exp. Fluids, Vol 15, 49-60.

Raffel M., Leitl B., Kompenhans K., 1993: Data validation for particle image
velocimetry, in 'Laser Techniques and Applications in Fluid Mechanics',
Eds. R.J. Adrian et al., Springer-Verlag, 210 - 226.

Raffel M., Seelhorst U., Willert C., Vollmers H., Bütefisch K.A., Kompenhans J.,
1996: Measurement of vortical structures on a helicopter rotor model in a
wind tunnel by LDV and PIV. Proc. 8th International Symposium on
Applications of Laser Techniques to Fluid Mechanics, Lisbon, paper 28.1.

Ronneberger O., Raffel M., Kompenhans J. 1998: Advanced evaluation algorithms
for standard and dual plane particle image velocimetry. Proc. 9th International
Symposium on Applications of Laser Techniques to Fluid Mechanics,
Lisbon, paper 10.1.

Royer H. and Stanislas M., 1996: Stereoscopic and holographic approaches to get
the third velocity component in PIV. von Karman Institute for Fluid Dynamics,
Lecture Series 1996-03, Particle Image Velocimetry.

Vogt A., Baumann P., Gharib M., Kompenhans J., 1996: Investigations of a wing
tip vortex in air by means of DPIV. Proc. 19th AIAA Advanced Measurement
and Ground Testing Technology, New Orleans, LA., 17-20 June 1996.

Wernert P., Geißler W., Raffel M., Kompenhans J., 1996: Experimental and numerical investigations of the dynamic stall process on a pitching NACA 0012 airfoil. AIAA Journal, Vol. 34, 982-989.

Westerweel J. and Nieuwstadt F.T.M., 1991: Performance tests on 3-dimensional velocity measurements with a two-camera digital particle-image velocimeter. Laser Anemometry Advances and Applications, Vol. 1, ed. Dybbs A and Ghorashi B (New York: ASME), 349-355.

Westerweel, J., 1998: Effect of sensor geometry on the performance of PIV interrogation. Proc. 9th International Symposium on Applications of Laser Techniques to Fluid Mechanics, Lisbon, paper 1.2.

Willert C., 1996: The fully digital evaluation of photographic PIV recordings. Appl. Sci. Res., Vol. 56, 79-102.

Willert C, 1997: Stereoscopic digital particle image velocimetry for application in wind tunnel flows. Meas. Sci. and Techn., Vol. 8, 1465-1479.

CHAPTER IV. MIXERS AND ROTATING FLOWS

IV.1. Analysis of Orderliness in Higher Instabilities Toward the Occurence of Chaos in a Taylor-Couette Flow System

N. Ohmura, K. Matsumoto, T. Aoki, and K. Kataoka

Department of Chemical Science and Engineering, Kobe University
Rokkodai-cho, Nada-ku, Kobe 657, Japan

Abstract. The purpose of the experimental work is to elucidate the mechanism for the occurrence of chaos in a Taylor-Couette flow system. The attention was focused on the higher instability region from the doubly-periodic to the weakly-turbulent wavy vortex flow. In order to take note of the effect of hysteresis, the inner cylinder rotation was accelerated at a constant rate from rest until a specified Reynolds number was reached. The test section with the radius ratio of 0.625 had an annular gap width of 22.5 mm and an aspect ratio of 4. A fiberoptic laser-Doppler velocimeter was employed to observe a time-dependent, peripheral component of velocity at the central region of vortex cells. At the same time, time-dependent peripheral component of velocity gradients on the wall of the stationary outer cylinder were measured in the neighborhood of one of the outflow cell boundaries with a hot-film shear sensor embedded flush with the inside cylindrical surface. Four vortex cells were always formed regardless of the acceleration rate of the inner cylinder rotation tested in the start-up operation, but the axial wavelength of the two central vortices was shorter than that of the remaining two end vortices. The amplitudes of the fluctuations of velocity and velocity-gradient varied irregularly whereas the first fundamental frequency was maintained in proportion to the inner cylinder rotation as long as the azimuthally-traveling waves existed. It has been concluded that chaos occurs firstly at the inflow cell boundaries as a result of instability in the viscous interaction of end vortices with the fixed bottom and top surfaces and that the chaotic turbulence is propagated along the secondary flow streamlines toward the central vortices and penetrated very slowly toward the center of vortex cells.

Keywords. Taylor-Couette flow, Taylor vortex flow, instability, bifurcation, chaos

1 Introduction

The Taylor-Couette flow system consisting of two concentric circular cylinders with the inner one rotating experiences several dynamical transitions occurring stepwisely in between laminar and turbulent flow regimes as the Reynolds number is gradually raised: (1) laminar circular Couette flow (LCF), (2) laminar cellular vortex flow (LCVF), (3) singly-periodic wavy vortex flow (SPWVF), (4) doubly-periodic wavy vortex flow (DPWVF), (5) weakly-turbulent wavy vortex flow (WTWVF), and (6) fully-turbulent vortex flow (FTVF) (for a detailed review, see DiPrima and Swinney, 1981 and Kataoka, 1986). There appears the effect of hysteresis in the bifurcation process, depending upon how to reach a steady rotation of the inner cylinder (Ohmura et al. 1995). The aspect ratio of the annular space is also a very sensitive parameter in the formation of a train of vortices (Benjamin, 1978a & b, Ohmura et al. 1995). In order to investigate the route until chaos appears, particular attention has been focused on the higher instability regions from the doubly-periodic to the weakly-turbulent wavy vortex flow. It is necessary to observe time-dependent vortex flow structures in the central region of vortex cells as well as in the near-wall region. Any probe for velocity measurement should not be inserted into the flow region of the annular space owing to the fact that an undesirable wake flow occurring behind the probe comes round back to it due to the main circulating flow. A fiberoptic laser-Doppler velocimeter was employed to observe time-dependent, peripheral component of velocities at the central region of vortex cells. Time-dependent near-wall velocity gradients were measured with a hot-film shear sensor embedded flush with the inside surface of the outer cylinder. The Reynolds number is defined based on the annular gap width d, the radius R_i and the angular velocity Ω of the inner cylinder:

2 Experiment

As shown in Fig. 1, the experimental equipment consisted of a transparent outer cylinder of acrylic resin (ID = 120 mm) and a stainless steel inner cylinder (OD = 75 mm), giving an annular gap width d of 22.5 mm and an effective height of 406 mm. The annular space was divided into three portions with two partition discs (2 mm thick). The central portion (test section) with the vertical height H = 90.0 mm had an aspect ratio $\Gamma = H/d = 4$ so as to contain 4 vortex cells in normal flow conditions. An aqueous solution of glycerin was used as the working fluid. According to the linearized instability theory (DiPrima & Swinney, 1981), the radius ratio gives the critical Reynolds number $Re/Re_c = 75.8$ for the transition from LCF to LCVF.

Figure 2 shows the location of a hot-film shear sensor and the positions of observation due to a fiberoptic laser Doppler velocimeter. The shear sensor was located approximately at the height of the outflow cell boundary of the lowest vortex in touch with the bottom surface. The velocity measurement with the LDV was made along the vertical centerline of the annular gap.

In order to take note of the effect of hysteresis, an acceleration rate of the inner cylinder rotation was controlled to be constant by a computer until a specified Reynolds number was reached. Figure 3 shows schematically the two rates of angular acceleration tested in the start-up operation: case I (0.017 s^{-2}) and case II (0.065 s^{-2}). The observation was made under the steady state operation with no axial flow after a sufficiently large time had passed.

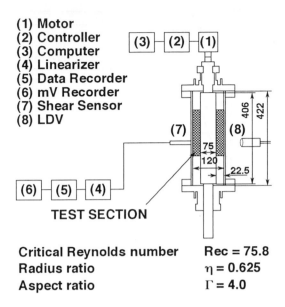

(1) Motor
(2) Controller
(3) Computer
(4) Linearizer
(5) Data Recorder
(6) mV Recorder
(7) Shear Sensor
(8) LDV

TEST SECTION

Critical Reynolds number	Re_c = 75.8
Radius ratio	η = 0.625
Aspect ratio	Γ = 4.0

Fig. 1. Schematic picture of experimental setup. Dimensions given are in mm.

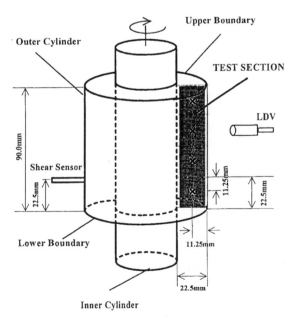

Fig. 2. Test section and measuring position.

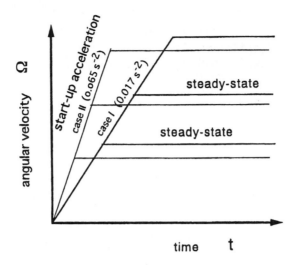

Fig. 3. History of acceleration of inner cylinder rotation until reaching a steady state.

3 Results and Discussion

Two rates of angular acceleration were tested: case I (0.017 s^{-2}) and case II (0.065 s^{-2}). However four vortex cells were always formed in the annular space, regardless of the acceleration rate tested in the start-up operation. It has been found that the two central vortices should be distinguished from the remaining two vortices with respect to the vortex structure. The two vortices being in touch with the top and bottom end surfaces were named "end vortices". The four vortices were numbered as No.1 through 4 from bottom to top. Figure 4 shows the axial distribution of radial component V_r of velocity measured along the vertical centerline of the annular space (case I). It can be seen from the figure that the two central vortices were shorter in the axial wavelength (vortex height) than the remaining two end vortices: dimensionless axial wavelength $\lambda/d \doteqdot 0.83$ (No.2 & 3 vortices), $\lambda/d \doteqdot 1.11$ (No.1 & 4 vortices).

Figures 5-1 and 5-2 show time-traces and their spectra of the fluctuations of near-wall velocity gradient detected in the neighborhood of the outflow cell boundary between the No.1 and 2 vortices by means of the hot-film shear sensor. Regarding Fig. 5-1, the upper two spectra indicate doubly-periodicity of DPWVF structure. The third spectrum indicates definitely the generation of chaotic disturbance at $Re/Re_c = 22.99$ in spite of the persistent quasiperiodic wavy structures. The lowest pair of diagrams for $Re/Re_c = 25.20$ indicate the persistent superposition of wavy motion upon the aperiodic turbulent oscillations. It can be seen from the two figures that there is not a significant difference in spectral

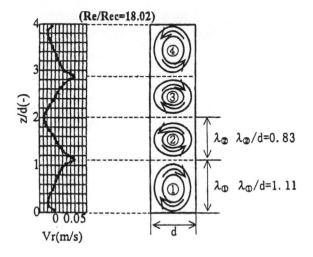

Fig. 4. Axial variation of radial component of velocity measured on the centerline of annular gap and the estimated vortex height (axial wavelength): ① end vortex, ② central vortex.

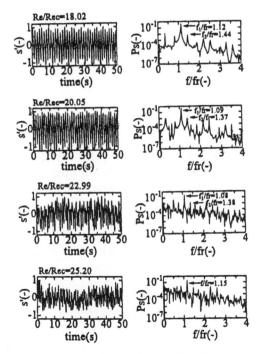

Fig. 5-1. Time-traces of local near-wall velocity gradient fluctuations observed in the neighborhood of the outflow cell boundary and their power spectra (case I).

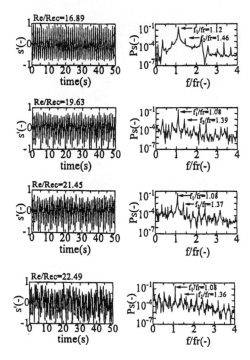

Fig. 5-2. Time-traces of local near-wall velocity gradient fluctuations observed in the neighborhood of the outflow cell boundary and their power spectra (case II).

structure between the two acceleration rates tested in the start-up operation, i.e. between case I and case II.

Figures 6-1 and 6-2 show phase space portraits constructed by embedding time-series of dimensionless near-wall velocity gradient fluctuations $s'(t)$ into two dimensional space as $\{s'(t), s'(t+\tau)\}$, where τ is a time delay (Packard *et al.* 1980, Takens 1981). Regarding Fig. 5-1, until the Reynolds number reaches $Re/Re_c = 21.09$ from 18.02, the attractors keep a shape of two-dimensional torus implying doubly-periodicity of DPWVF. When the Reynolds number goes beyond $Re/Re_c = 22.08$, the two-dimensional torus begins to get out of shape with all the orbits wrinkled. This can be interpreted as the generation of chaotic turbulence at the inflow cell boundaries of the two end vortices. As can be seen from those two figures, there is not a significant difference in variation of topological shape of attractor with the Reynolds number between case I and case II.

Figures 7 and 8 show time-traces and their spectra of circumferential component of velocity fluctuations measured at the centers of No.1 & 2 vortices by means of the LDV. The dimensionless first fundamental frequencies f_1/f_r were kept equal regardless of the vortex number as well as the Reynolds number. However in the doubly-periodic wavy vortex flow regime, the second fundamental frequency f_{22} of the central vortex (No.2) was lower than the first fundamental

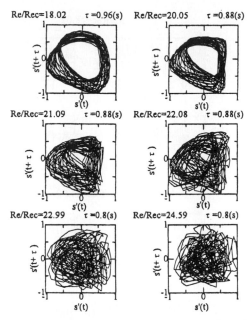

Fig. 6-1. Phase space portraits constructed by applying the Taken's embedding method to time-series of local near-wall velocity gradient signals (case I).

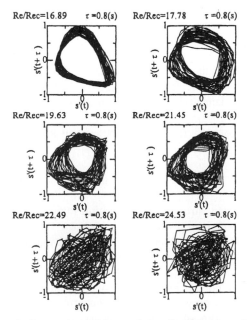

Fig. 6-2. Phase space portraits constructed by applying the Taken's embedding method to time-series of local near-wall velocity gradient signals (case II).

306

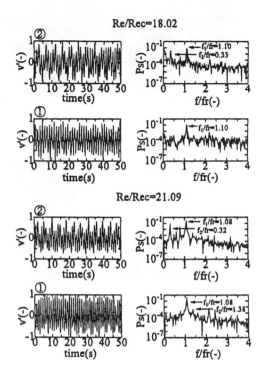

Fig. 7. Time-traces of peripheral component of velocity fluctuations observed at the center of vortex cell and their power spectra (case I): ① end vortex, ② central vortex

frequency f_{12} whereas the second fundamental frequency f_{21} of the end vortex (No.1) was higher than the first fundamental frequency f_{11}. These dimensionless first and second fundamental frequencies of the end vortex are in agreement with those of Figs. 5-1 and 5-2.

It can be seen from the figures that there is an arithmetic relation between f_{21} and f_{22}:

$$f_{11} + f_{22} = f_{21} \quad \text{and} \quad f_{11} = f_{12} \tag{1}$$

The quasiperiodic wavy structure in the central region of cellular vortices tends to be persistently maintained even when Re goes beyond 23. These pieces of evidence suggest that chaos occurs due to instability in the shear stresses at both the fixed ends, but that the chaotic disturbance is not penetrated very fast toward the vortex centers.

In addition to the Fourier analysis, a wavelet transform analysis (Walker *et al.* 1997) was utilized not only to discern the occurrence of chaos but also to extract the characteristic structure of chaotic turbulence.

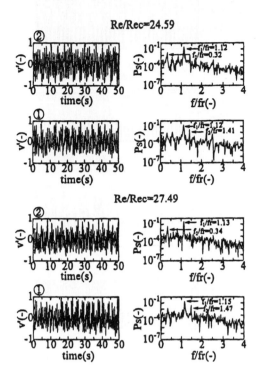

Fig. 8. Time-traces of peripheral component of velocity fluctuations observed at the center of vortex cells and their power spectra (case I): ① end vortex, ② central vortex

The wavelet transformation of continuous signal $f(t)$ is defined as

$$C(\tau,a) = \frac{1}{\sqrt{a}} \int_{-\infty}^{+\infty} g\left(\frac{t-\tau}{a}\right) f(t)\,dt \qquad (2)$$

where $g(t)$ is the wavelet mother function, a is a time scale parameter for dilatation, and t is a time-shift parameter.

The signals of the LDV and the hot-film shear sensor were analyzed by calculating Continuous Wavelet Transforms (CWTs). A Mexican-hat wavelet was used as the wavelet mother function.

$$g(t) = \left(1-t^2\right)\exp\left(\frac{-t^2}{2}\right) \qquad (3)$$

308

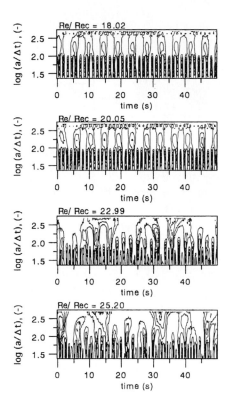

Fig. 9. Equi-correlation contour lines of Mexican-hat wavelet transforms obtained from the near-wall velocity gradient signal (case I).

The actual calculation was made in Fourier space using FFT:

$$C(\omega, a) = \sqrt{a}\, G^*(a\omega) F(\omega) \tag{4}$$

where * denotes a complex conjugate. The CWT was obtained as the inverse transform of $C(\omega, a)$.

Figures 9 through 11 show equi-correlation contour lines of CWT. As can be seen from CWT diagram of $Re/Re_c = 18.02$, in the regime of SPWVF, the point indicating maximal correlation appears regularly on the CWT diagram with a constant time-period at a constant height (time scale) of the vertical axis, i.e. at log $(a/\Delta t) = 1.67$. Those figures suggest that the instability leading to the generation of chaos appears in the end vortices earlier than in the central vortices.

In order to examine the occurrence of chaotic turbulence, a comparison was made on the diagrams of CWT between the near-wall region and the central region of cellular vortices. The CWTs of the velocity-gradient signals obtained at $Re/Re_c =$

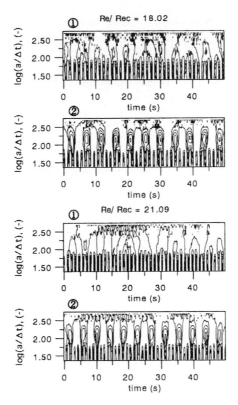

Fig. 10. Equi-correlation contour lines of CWT obtained from peripheral velocity fluctuations at the center of vortex cells (case I): ① end vortex, ② central vortex.

22.99 indicate clearly the generation of chaotic turbulence around the first fundamental component. On the other hand, the CWTs of the velocity signals obtained at $Re/Re_c = 24.59$ show the persistent existence of regularly periodical first fundamental component.

Two kinds of standard deviations as an index of irregularity were calculated respectively from the sequential data of time-scale on the vertical axis and time-interval on the horizontal axis at which every maximal correlation coefficient $C(t, a)$ max relating to the first fundamentals appears on the CWT diagram. The definition of these standard deviations can be expressed as

$$\sigma_a = \sqrt{\sum_{i=1}^{N}\left\{\log(a/\Delta t)_i - \overline{\log(a/\Delta t)}\right\}^2 / N} \tag{5}$$

$$\sigma_T = \sqrt{\sum_{i=1}^{N}\{T_i - \overline{T}\}^2 / N} \tag{6}$$

310

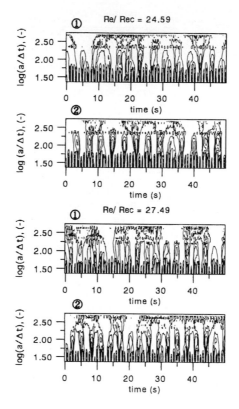

Fig. 11. Equi-correlation contour lines of CWT obtained from peripheral velocity fluctuations at the center of vortex cells (case I): ① end vortex, ② central vortex.

The variation of those standard deviations with the Reynolds number is shown in Figs. 12 and 13. Comparing between the near-wall and the central flow regions, it has been found that the dynamical transition to the occurrence of chaos is sharper in the near-wall region than in the central region of vortex cells.

Figure 12 indicates that the time-scale standard deviation shows a sharp jump at $Re/Re_c = 22$ but the time-interval standard deviation does not. This implies that the amplitudes of velocity and velocity-gradient signals vary irregularly while the dimensionless first fundamental frequency is kept constant. It can be considered that the instability of amplitude is characteristic of steadily rotating flow systems with the periodicity of time period. It can be seen from Fig. 13 that the time-scale standard deviation obtained at the center of vortices shows a dull jump around $Re/Re_c = 24$. The belated transition to chaotic turbulent flow implies the slow propagation of chaotic turbulence toward the center of vortices.

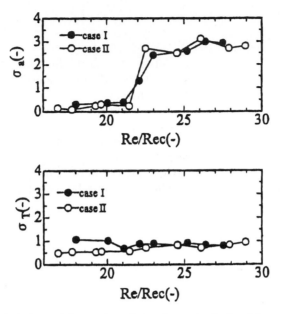

Fig. 12. Standard deviations of time-scale and time-interval calculated from maximal points of wavelet transform correlation of the near-wall velocity gradient signals.

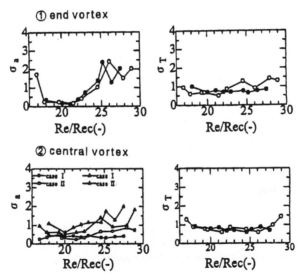

Fig. 13. Standard deviations of time-scale and time-interval calculated from maximal points of wavelet transform correlation of peripheral velocity signal observed at the center of vortex cells.

4 Conclusion

The following conclusions can be deduced:
(1) Chaos occurs firstly in amplitude of velocity fluctuations as a result of instability in the viscous interaction of the inflow cell boundaries of the two end vortices with the fixed bottom and top surfaces.
(2) The chaotic turbulence is propagated slowly along the secondary flow streamlines of the cell boundaries toward the central vortices.
(3) The chaotic turbulence is penetrated very slowly inward across the secondary flow streamlines toward the center of vortices.
(4) Chaos does not appear directly in time-period of velocity fluctuations in these kind of steadily rotating flow systems until the flow becomes fully turbulent.

5 Acknowledgement

The authors wish to thank Mr. T. Yoshimura for his experimental support. This work was supported by Grant in-Aids for Scientific Research (B)(2) (09450287) from the Ministry of Education, Science, Sports and Culture of Japan.

6 Nomenclature

a	: time-scale parameter for dilatation of wavelet mother function, s
d	: annular gap width of test section, m
f	: frequency, Hz
f_r	: frequency (revolution number) of inner cylinder rotation, 1/s
$f(t)$	: continuous signal to be analyzed by wavelet analysis
$g(t)$	: wavelet mother function
H	: vertical height of test section, m
P_s	: normalized power distribution function, -
R_i, Ro	: inner and outer cylinder radii, m
Re	: Reynolds number based on inner cylinder rotation, -
s'	: fluctuations of near-wall velocity gradient, -
T	: time-interval at which the maximal point of f_1-component appear s on CWT diagram, s
t	: time, s
v'	: fluctuations of peripheral component of velocity, -
V_r	: radial component of velocity, m/s
z	: axial coordinate along cylinder axis

Greek letters

Γ	: aspect ratio ($=H/d$), -
η	: radius ratio ($= R_i/Ro$), -
λ	: axial wavelength (vertical height of vortex), m

ν : kinematic viscosity, m^2/s
σ_a : time-scale standard deviation, -
σ_T : time-interval standard deviation, -
τ : time delay for phase space analysis or time-shift parameter for wavelet transformation, s
Ω : angular velocity of inner cylinder, 1/s
ω : angular frequency of Fourier transformation

Overline
: average

REFERENCES

BENJAMIN, T.B. 1978a Bifurcation phenomena in steady flows of a viscous liquid. I. Theory. *Proc. Roy. Soc. London* **A359**, 1-26.

BENJAMIN, T.B. 1978b Bifurcation phenomena in steady flows of a viscous liquid. II. Experiments. *Proc. Roy. Soc. London* **A359**, 27-43.

DIPRIMA, R.C. & SWINNEY, H.L. 1981 Instabilities and transition in flow between concentric rotating cylinders. In *Hydrodynamic Instabilities and Transition to Turbulence* (ed. H.L. Swinney & J.P. Gollub) pp.139-180, Springer-Verlag, Berlin.

KATAOKA, K. 1986 Taylor vortices and instabilities in circular Couette flows. in *Encyclopedia of Fluid Mechanics* (ed. N.P. Cheremisinoff) Vol.1, pp.236-274, Gulf Pub., Houston.

OHMURA, N., KATAOKA, K., MIZUMOTO, T., NAKATA, M. & MATSUMOTO, K. 1995 Effect of vortex cell structure on bifurcation properties in a Taylor vortex flow system. *J. Chem. Eng. Japan* **28**, 758-764.

PACKARD, N.H., CRUTCHFIELD, J.P., FARMER, J.D. & SHAW, R.S. 1980 Geometry from a time series. *Phys. Rev. Lett.*. **45**, 712-716.

TAKENS, F. 1981 Detecting strange attractors in turbulence. In *Lecture Notes in Mathematics Vol.989, Dynamical Systems and Turbulence* (ed. D.A. Rand & L.S. Young), pp.366-381, Springer-Verag, Heidelberg.

WALKER, S.H., GORDEYEV, S.V. & THOMAS, F.O. 1997, A wavelet transform analysis applied to unsteady aspects of supersonic jet screech resonance. *Experiments in Fluids*, **22**, 229-238.

IV.2. Experimental Investigations on Nonlinear Behaviour of Baroclinic Waves

B. Sitte and C. Egbers

ZARM, University of Bremen, Germany

(Received 11 December 1998)

Instabilities in the form of baroclinic waves occur in a rotating annulus cooled from within. Flow visualisation studies and LDV-measurements of the radial velocity component were carried out in an annulus with an aspect ratio of 4.4. The flow undergoes transitions from the laminar stable state through baroclinic waves, both stable and time-varying, to an irregular state. Based on the time series of the radial velocity at fixed point in the rotating annulus, the attractors of the flow match previous results based on temperature measurements. The slope of the dispersion relation of the baroclinic waves changes sign with increased rate of rotation. The bifurcation diagram of extrema in the radial velocity shows the existence of low dimensional chaos at the transition from the axisymmetric flow to periodic baroclinic waves.

CONTENTS

1. Introduction

In the last decades, rotating annulus experiments have been carried out to obtain laboratory simulations of the large-scale circulation of the atmosphere, with the cool polar regions at the center of rotation. Hide (1958) visualised baroclinc waves, also known as thermal Rossby waves, in a rotating cylindrical

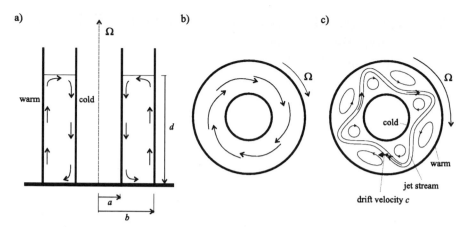

FIGURE 1. Sketch of (a) lateral view and (b) top view of the axisymmetric basic flow, (b) top view of a $m = 4$ wave pattern.

annulus cooled from within. He confined a fluid in a rotating, cylindrical gap with a free surface, cooled the inner wall and heated the outer one. The axisymmetric basic flow is characterized by an upward flow of the warm fluid at the outer wall and a downward flow at the cooler inner wall. The resulting radial flow, mainly at near surface and bottom, is deflected by the Coriolis force (see figure 1 a,b). At higher rotation rates, this azimuthal flow becomes dominant. If the rotation rate is increased, the azimuthal flow shows a wavy behaviour, the axial symmetry is broken. Hide observed waves with different wave numbers m, traveling slowly around the cylinder (see figure 1 c). Basically, with increasing rotation rates higher wave numbers are stable. At high rotation rates, the transition to turbulence takes place.

In the following years a lot of work was done to investigate the complex behaviour of the waves in various geometries and under different boundary conditions. Davies (1956) made a linear stability analysis, Fowlis, Hide & Hide (1965) investigated the effect of the Prandtl-number on the stability of the waves. In the last decade, the experiments of Hide were carried out again, this time Read, Bell, Johnson & Small(1992) used the method of time delayed coordinates to analyse their thermocouple data. By taking long temperature time series with a temperature sensor in the middle of the cylindrical gap they were able to reconstruct the phase space of the system and to analyse the nonlinear behaviour of the system. Früh & Read (1997) deepened these investigations. Different flow types of baroclinc waves were found. At the transition from the axialsymmetric basic flow to stable baroclinic waves so called 'amplitude vacillation' (AV) waves occur. The AV waves have an oscillating amplitude. Further on, 'modulated amplitude vacillation' (MAV) waves have been found, where the amplitude oscillation does not have a constant frequency. In this terminology, the structural disturbances of the waves towards the transition into turbulence is called 'structural vacillation' (SV).

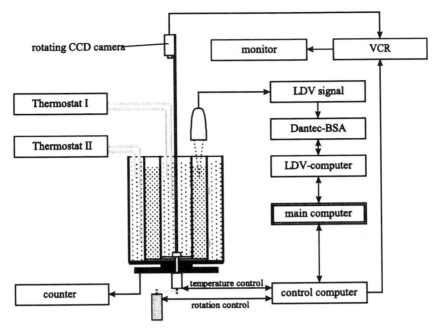

FIGURE 2. Experimental setup

Früh & Read (1997) were able to show that low dimensional chaotic states occur not only at the transition to turbulence, but also slightly above the critical point in the MAV waves.

2. The Rotating Annulus Experiment

The flow in the rotating annulus cooled from within can be characterized by the following control parameters: The geometry can be described by the radius ratio $\eta = a/b$ and the aspect ratio $\Gamma = d/(b-a)$, where a and b are the inner and the outer radii of the cylindrical gap, d is the depth. Further parameters may be defined, such as the Taylor number

$$Ta = \frac{4\,\Omega^2\,(b-a)^5}{\nu^2\,d} \tag{2.1}$$

and

$$Ro = \frac{g\,d\,\Delta\rho}{\bar{\rho}\,\Omega^2\,(b-a)^2} \tag{2.2}$$

which is known as the thermal Rossby number. Ω is the angular velocity, ν is the kinematic viscosity and ρ the density of the fluid. The Taylor number describes the rotational influence while the thermal Rossby number characterises the influence of the density gradient due to temperature differences and the influence of rotation.

Figure 2 shows the experimental setup. The cylindrical tank has three

$a = 45$ mm	Inner radius of the cylindrical gap
$b = 95$ mm	Outer radius of the cylindrical gap
$d = 220$ mm	Depth of the fluid
$\eta = a/b = 0.47$	Radius ratio
$\Gamma = \frac{d}{b-a} = 4.4$	Aspect ratio
$T_a = 19.0°C$	Temperature of the inner wall
$T_b = 22.8°C$	Temperature of the outer wall
$\nu = 1.004$ mm^2/s	Kinetic viscosity (water, 21 Celsius)
$\bar{\rho} = 0.998$ kg/l	Average density (water, 21 Celsius)
$\Delta\rho = 0.000823$ kg/l	Maximum density difference
$Pr = 8.0$	Prandtl number
0.25 rad/s $< \Omega < 6.0$ rad/s	Variation of the angular velocity
$Ro = \frac{g\,d\,\Delta\rho}{\bar{\rho}\,\Omega^2\,(b-a)^2} > 0.004$	Variation of the thermal Rossby number
$10^6 < Ta = \frac{4\,\Omega^2\,(b-a)^5}{\nu^2\,d} < 10^8$	Variation of the Taylor number

TABLE 1. Parameters of the experiment

concentric chambers. The inner and outer chambers are filled with water and connected with two thermostats. They control the wall temperatures of the cylindrical gap between these two chambers. The surfaces of all chambers are free. The outer two walls are made of acrylic glass to enable visual investigations and LDV measurements from lateral directions. The temperature is controlled by thermocouples at both walls of the gap. A co-rotating camera is mounted on top of the experiment. The whole experiment is installed on a turntable, the maximum rotation frequency is 1 Hz. As the time scales of baroclinic waves are of the order of 10^3 seconds, automation of the experiment is necessary. Therefore all components of the setup, including rotation rate, the LDV system, thermostats and VCR, are controlled by a main computer. LDV time series measurements took about 7 to 9 hours for one parameter point.

The parameters of the experiment are given in table 1. Destillated water was used as the fluid. The temperature difference between the inner and the outer walls was kept constant for all measurements. This reduces the parameter space to a line, with the rotation speed Ω left as the only free parameter.

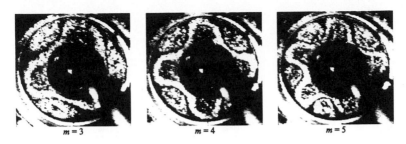

FIGURE 3. Different wave numbers $m = 3, 4, 5$ at identical Parameters
$f = 0.167$ Hz, $Ta = 6.32 \cdot 10^6$, $Ro = 0.585$

3. Flow-Visualisation

The fluid motion on the free surface of the annulus is visualised by aluminium flakes suspended in the working fluid. The reflecting flakes allow no LDV measurements during the visual investigations.

3.1. *Stability*

Beside the axialsymmetric basic flow of wave number $m = 0$, waves of different wave numbers $m \neq 0$ occur in the non-axisymmetric flow regime. However, m is limited. Hide & Mason (1970) determined the lowest and highest existing wave numbers $m_{min} \leq m \leq m_{max}$ for different geometries of the annulus and found the empirical law

$$m_{min} = 0.25 \, \frac{\pi(b+a)}{(b-a)}, \qquad m_{max} = 0.75 \, \frac{\pi(b+a)}{(b-a)}. \qquad (3.1)$$

For this geometry, $m_{min} = 2$ and $m_{max} = 6$. However, $m = 6$ did not occur if Ω is increased or decreased quasi-stationarily. $m = 6$ waves only emerge, if the rotation speed is increased or decreased rapidly. Therefore the following investigations are only focused on $m = 2, 3, 4, 5$.

The flow shows a very strong hysteresis. Figure 3 shows three different wave numbers at identical parameters. Each of the flows is stable, no spontaneous jumps between wave numbers occured within the investigated region of the parameter space.

Figure 4 shows the different flow regimes as a function of the Taylor number Ta. At the critical point $Ta = 1.76 \cdot 10^6$, $Ro = 2.11$, the axisymmetric basic flow becomes instable. This critical point is very close to the results of Hide & Mason (1970) and their measurements with water as a fluid.

At Taylor numbers slightly above the critical point the amplitude of the waves begins to oscillate. These are the AV and MAV waves described by Read *et al.* and Früh & Read. Figure 5 shows an example of an amplitude oscillation. While the wave slowly drifts with a velocity $c = 0.013$ rad/s faster than the rotating annulus, its amplitude oscillates with a period of $t \approx 1000$ s. In case of a MAV wave this period and the amplitude of the oscillation itself is not constant. At medium Taylor numbers, the waves show no fluctuations.

320

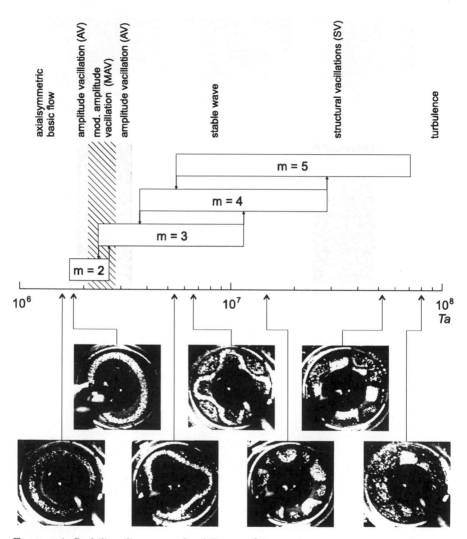

FIGURE 4. Stability diagramm for $\Delta T = 3{,}5°\mathrm{C}$, stable wave numbers at different Taylor numbers Ta and videoprints of the surface flow.

The amplitude is constant, their structure is stable. With higher Taylor numbers, structural vacillations occur. The onset of the structural vacillations at Taylor numbers $Ta > 2 \cdot 10^7$ mark the begin of the transition to turbulence. The vacillations grow stronger with increasing Ta. Finally, at Taylor numbers $Ta > 7 \cdot 10^7$, no periodic structure can be identified.

3.2. Drift velocities

The drift velocities c of the baroclinic waves have been determined by visual investigations. The drift velocities are two orders of magnitude smaller than the rotational velocity and are very sensitive to small variations of the

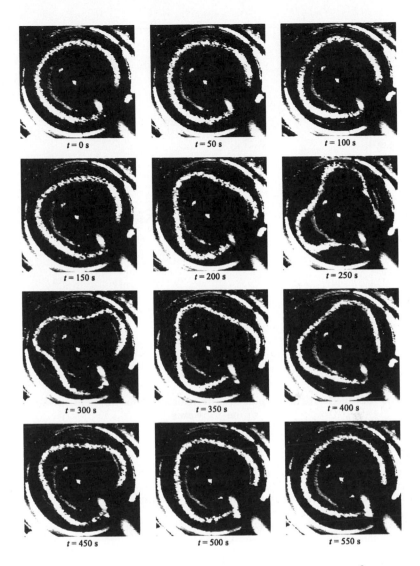

FIGURE 5. Amplitude vacillation (AV) wave, $m = 3$, $Ta = 2.86 \cdot 10^6$, $Ro = 1.29$.
The videoprints were taken at different times t.

boundary conditions. Therefore the experimental setup was located in a small seperated room. The experiment had always been running for more than 5 hours before a measurement was done. Figure 6 shows the drift rate c as a function of the Taylor number Ta, for different stable wave numbers m.

Figure 6 shows only the drift velocities of waves before the onset of strong structural vascillations. Neglecting the dispersion of the waves and taking all

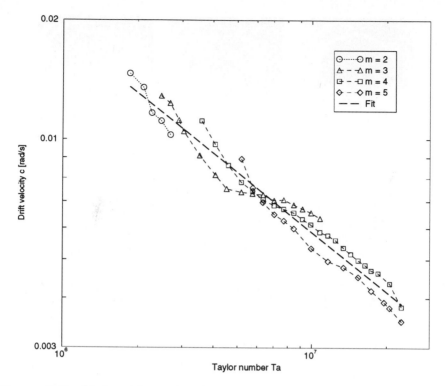

FIGURE 6. Double logarithmic plot of drift velocities c for different wave numbers m, no SV waves.

wave numbers into account, the linear fit gives

$$c \propto Ta^{-0.498 \pm 0.011} . \tag{3.2}$$

This is within the error proportional to the thermal wind relation

$$c \propto \frac{1}{\sqrt{Ta}} \propto \frac{1}{\Omega} . \tag{3.3}$$

While this result is well known (see Hide (1958)), another aspect of the drift velocities is shown in figure 6: The slope of the dispersion relation changes sign, approximately at $Ta = 6.2 \cdot 10^6$. At lower Taylor numbers, waves with higher wave numbers m travel faster than waves with lower wave number. At higher Taylor numbers the waves with high m are the slower ones.

4. LDV-Measurements

For the experimental investigations on the rotating annulus the same LDV-techniques were used as described in previous work of our group (Egbers & Rath (1996)). To detect the dynamic behaviour of the occuring waves in the rotating annulus using linear signal processing techniques, the most common

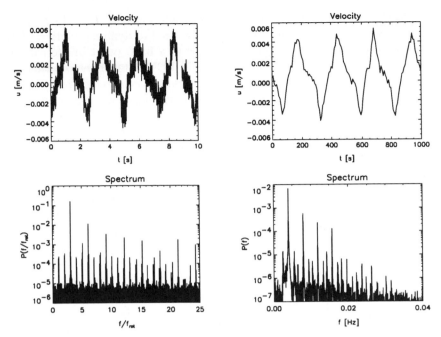

FIGURE 7. LDA-data time series and its power spectrum (left) and the and the calculated time series in the co-rotating system and its spectrum (right), $m = 3$, $Ta = 4.1 \cdot 10^6$, $Ro = 0.91$, $f_{rot} = 0.135$ Hz.

and very useful way is to analyse the time series of a representative velocity component. In this case, the radial component is chosen. The Fourier spectrum and the autocorrelation are constructed. The spectrum gives a measure of the amount of power in a given frequency band over a selected frequency range. The autocorrelation function for a periodic signal, for example, is itself periodic and can often give a less confusing representation of the data. Irregularity in the data gives rise to a decay in the autocorrelation function, and the rate of decay gives a measure of the degree of irregularity.

However, in case of chaotic dynamic behaviour of the system, these linear methods are not sufficient to describe the complex flows. Therefore, the attractor of the system is reconstructed by the method of Takens (1980). In the reconstructed phase space, the topological properties of the attractor can be calculated. The methods used for this nonlinear time series analysis are described in more detail in the work of Wulf (1997).

4.1. Measurement technique

The LDV measurement system is not co-rotating with the cylinder. Therefore, the raw data represent a circle in the rotating annulus. It is not possible to analyse the dynamic behaviour of the baroclinic waves directly from this series. However, one can determine the dominant modes of the waves from its power spectrum (see i.e. Bernardet et al. (1990)). The spectrum of the

LDV-data shown in figure 7 has the dominant peak at $f = 3 \cdot f_{rot}$, the dominant wave number is $m = 3$. To analyse the full dynamic motion of the wave, one needs to investigate a time series at a fixed point in the rotating annulus. This information is embedded in the LDV-data. If the rotation rate is known precisely, a series at a fixed point can be calculated from the raw data using only the data points at the beginning of a new rotation cycle.

Figure 7 shows the unfiltered LDV-data, the calculated time series in the co-rotating system and their Fourier spectra. One should note the different time scales, the time scale of the time series in the co-rotating system is two orders of magnitude larger than that of the raw LDV-data.

The method demonstrated above is limited in two ways. First, by the rotation frequency, because $\Delta t = 1/f_{rot}$ is the minimum time step between two points in the calculated time series. Therefore, $f_{rot}/2$ is the upper limit of frequency detection. However, the time scales of the baroclinic waves are large compared to $1/f_{rot}$ in this case, so the detectable frequency domain is sufficient. Second, the rotation frequency has to be known very precisely. A small error results in a large distortion of the calculated time series. In the experimental setup, a counter registers 1500 pulses per rotation cycle and allows to determine the mean rotation frequency f_{rot} over a longer time period with nearly arbitrary precision. But small variations of f_{rot} during the measurements broaden the frequency peaks in the power spectrum. In this setup the variation is limited to 0.05%. To take care of this effect, artificial breaks are inserted into the LDV data, marking each rotation period. These breaks can be seen in the LDV data time series of figure 7 (top, left), they are used to calculate the time series in the co-rotating system. Nevertheless, the effect of small errors in f_{rot} is not neglegible. The frequency peaks of the co-rotating time series in figure 7 are broadened compared to the spectrum of the raw LDV-data.

For the LDV analysis of the time series, as many data points as possible are needed for exact results. The LDV-data time series was taken for at least 7 hours. In the co-rotating system this time series has about 3000 up to 10000 data points, depending on the rotation rate. The radial velocity component is measured 20 mm under the surface.

4.2. Nonlinear behaviour

The phase space was reconstructed using time delayed coordinates. The fillfactor method and the integral local deformation method were used to estimate the time delay and the embedding dimension (see Buzug & Pfister (1992), Wulf (1997)). A low pass filter and a singular value filter were applied to smoothen the data.

In phase space, the correlation integrals and the correlation dimension D_2 were calculated for different embedding dimensions. In the following figures, the dimensions are calculated for the embedding dimensions 1-10 as a function of $\epsilon = R/R_{max}$, where R is the radius of a hypersphere in phase space and

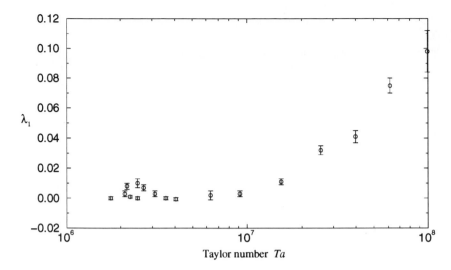

FIGURE 8. Lyapunov exponent λ_1 for different Taylor numbers Ta, $\Delta T = 3.5°C$.

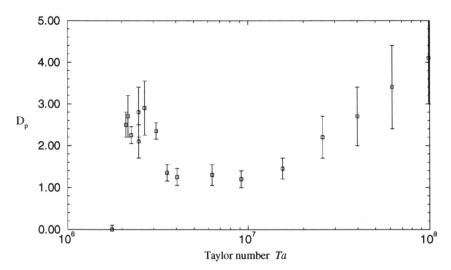

FIGURE 9. Pointwise dimension D_2 for different Taylor numbers Ta, $\Delta T = 3.5°C$.

R_{max} is the extension of the attractor. In addition, the pointwise dimension D_p and the largest Lyapunov exponent λ_1 were determined.

Figure 8 shows the development of λ_1 for different Taylor numbers at constant temperature difference between inner and outer cylinder (Compare with the stability diagram in figure 4). In the area of the modulated amplitude vacillation, small, but significantly positive Lyapunov exponents occur. For the stable baroclinic waves, represented by a limit cycle in phase space, the largest exponent decreases to zero. With the onset of the structural vacillations, λ_1 increases to relatively high values, indicating the transition to turbulence.

The pointwise dimension D_p shows the same dependency (see figure 9). At low Taylor numbers, just above the critical point in the regime of AV and MAV waves, dimensions $D_p > 2$ occur. For the limit cycle of stable waves the pointwise dimension D_p has values just above 1. At the onset of structural vacillation, the values of D_p increase significantly. Values of $D_p = 4$ mark the upper limit; higher dimensions cannot be determined, because of the restricted number of points in the velocity time series.

Figure 10 shows a $m = 3$ AV wave, where the amplitude vacillation frequency is coupled to the drift frequency. The drift frequency f_c is the main peak in the power spectrum, corresponding to a drift velocity of $c = 0.13$ rad/s, which is in good agreement with the visual measurements (figure 6). This peak is surrounded by harmonics of $1/5 \cdot f_c$. Früh & Read (1997) observed this coupling between the two frequencies in their experiments, too. However, since the coupling has not been encountered in numerical simulations, they could not rule out the possibility that their temperature sensors in the fluid or irregularities of the tank itself might be the reason for this weak coupling. This LDV measurement shows at least that the temperature sensors are not responsible for the coupling. The largest Lyapunov exponent is significantly greater than zero, indicating chaotic fluctuations in the periodic flow.

Figure 11 shows a totally different behaviour than the flow in figure 10, though the parameters are exactly the same. First, it is a $m = 2$ wave instead of $m = 3$. Second, the vacillation frequency is decoupled from the drift frequency. Therefore, the attractor is a torus and $D_p \approx 2$. The largest Lyapunov exponent is 0.

At slightly different parameters than in figures 10 and 11, figure 12 shows a more complex flow. The autocorrelation function decays. The spectrum shows broadened peaks, the Lyapunov exponent is greater than zero, indicating chaotic flow. The correlation dimension has a plateau at $D_c = 3$, indicating low dimensional chaos.

Figure 13 exhibits a stable $m = 3$ baroclinic wave, the attractor shows just the dominant drift frequency and the upper harmonics. The Lyapunov exponent is 0 within the error tolerance.

With higher Taylor numbers, the steady waves of figure 13 become unstable. The $m = 4$ wave in figure 14 shows the onset of structural vacillations. These vacillations have a small amplitude compared to the steady baroclinic wave and are not large enough to be seen in the visual investigations. They cannot be detected in the Fourier spectrum nor in the autocorrelation function, because of their very local nature. But the attractor shows the vacillations, their amplitude is significantly higher than the noise level. They cause a divergence of the correlation dimension for small $\epsilon = R/R_{max}$. The Lyapunov exponent is greater than zero, showing the chaotic nature of the fluctuations.

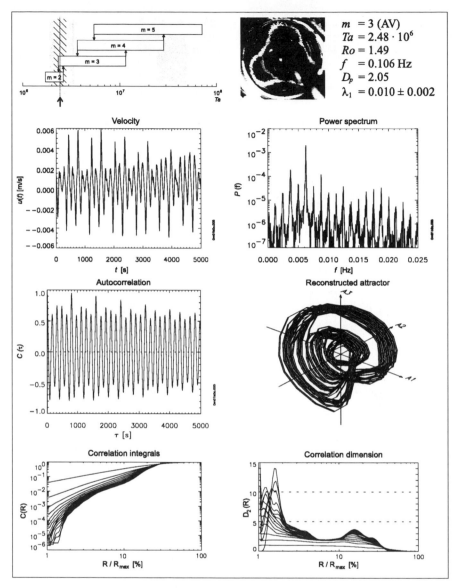

FIGURE 10. Wave with amplitude vacillation, the vacillation frequency is coupled to the drift frequency.

Figure 15 shows strong vacillations. Still, the periodic nature is evident in the Fourier spectrum. The fluctuations result in a higher 'noise' level in the spectrum, compared to a stable wave (see figure 13). The attractor is a highly disturbed limit cycle. The dimensions and the high Lyapunov exponent underline the chaotic nature of the vacillations.

With increasing Taylor numbers, the structural vacillations result in tur-

328

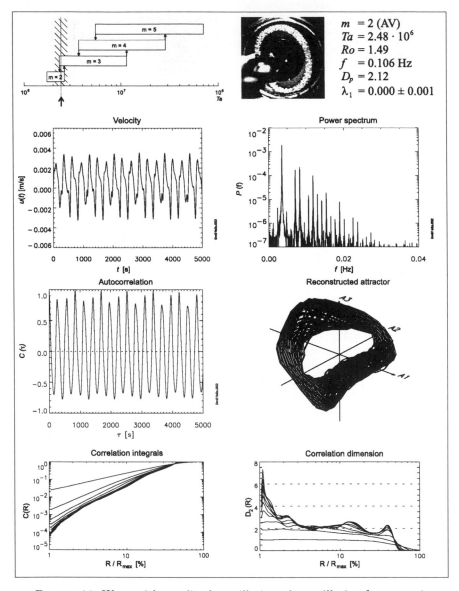

FIGURE 11. Wave with amplitude vacillation, the vacillation frequency is decoupled from the drift frequency.

bulent flow, shown in figure 16. No single significant peaks can be identified in the spectrum, the Lyapunov exponent is high, the correlation dimension does not converge.

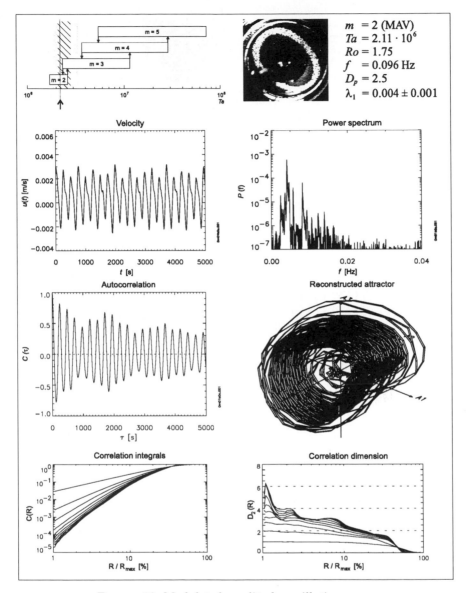

FIGURE 12. Modulated amplitude vacillation wave.

4.3. *Bifurcation scenario*

The bifurcation diagrams were taken to investigate the interesting transition from the axisymmetric basic flow (a fixed point in phase space) to stable baroclinic waves (a limit cycle). One might think of an ordinary Hopf bifurcation, but the complex behaviour of the flow at Taylor numbers just above the critical point makes this transition more complex.

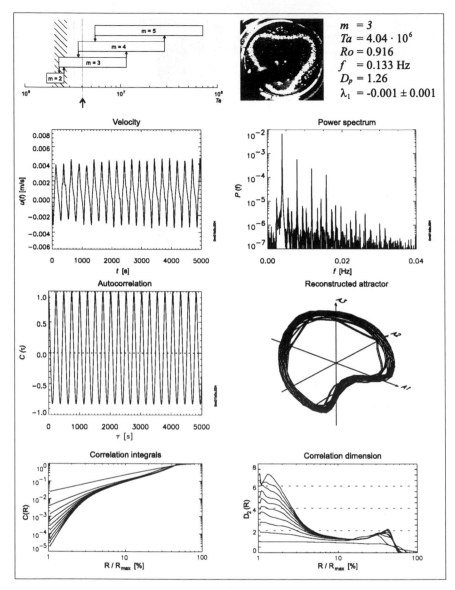

FIGURE 13. Stable baroclinc wave.

From the filtered LDV data, the local minima and maxima values of the radial velocity time series were determined. For each Taylor number, the histogram distribution of these extrema were fitted with Gauss curves. The medians of the Gauss curves are plotted in the diagramms shown in figure 17. The time series for a steady baroclinic wave has saddle points at $v(t) = 0$ (see figure 13). These saddle points are the reason for the line at $v = 0$ in the bifurcation diagrams.

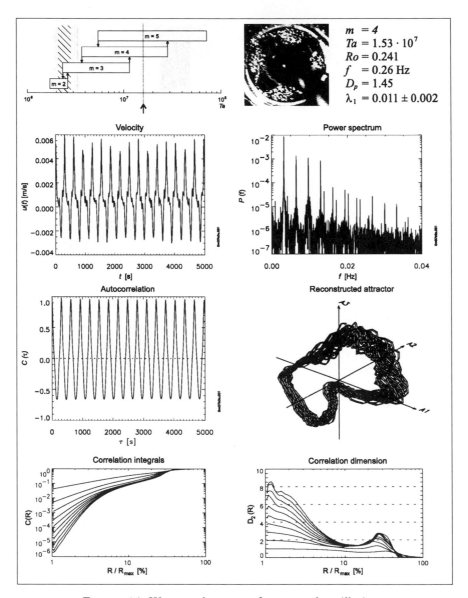

FIGURE 14. Wave at the onset of structural vacillations.

The small arrows in figure 17 denote the transition between the wave numbers $m = 2$ and $m = 3$, determined by the dominant mode in the Fourier spectra of the LDV data. The hysteresis between the diagram for increasing (figure 17, top) and decreasing Taylor numbers (figure 17, bottom) is clearly visible in the difference between the two arrows. The amplitude of the flow has a significant jump at this point, because the jetstream crosses

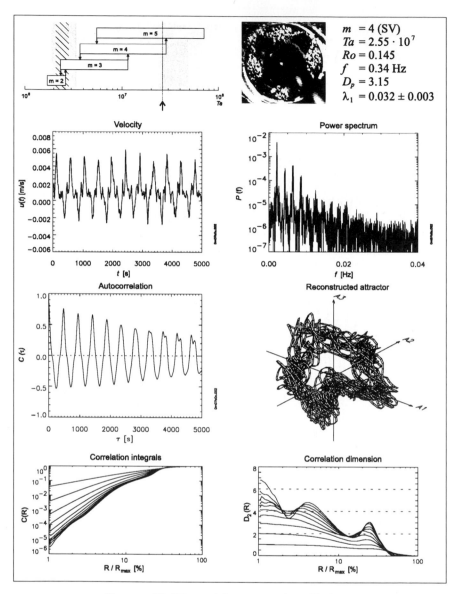

$$m = 4 \text{ (SV)}$$
$$Ta = 2.55 \cdot 10^7$$
$$Ro = 0.145$$
$$f = 0.34 \text{ Hz}$$
$$D_p = 3.15$$
$$\lambda_1 = 0.032 \pm 0.003$$

FIGURE 15. Wave with structural vacillations.

the r-direction at a different angle if the wave number jumps from $m = 2$ to $m = 3$.

The critical Taylor number is constant, it shows no hysteresis. This is a transition from a fixed point in phase space (the axisymmetric basic flow) to a limit cycle (steady waves). This seems to be a supercritical Hopf bifurcation. However, with increasing Taylor number, the amplitude of the wave begins to oscillate.

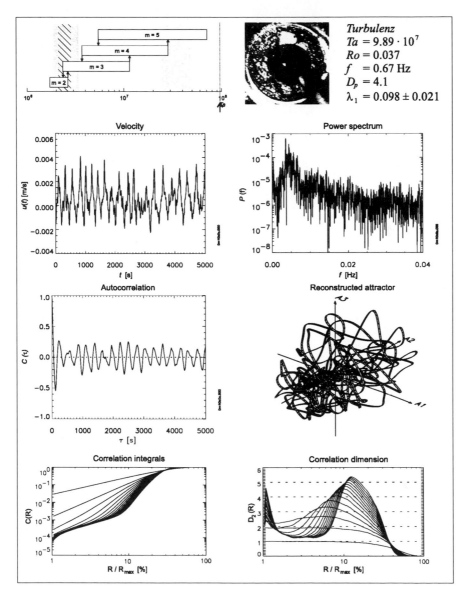

FIGURE 16. Wave at the onset of structural vacillations

At high Taylor numbers, two subcritical Hopf bifurcations occur, this is the transition from AV waves to steady waves. The dynamical characteristics of the flow in between the supercritical bifurcation on the left and the subcritical bifurcation on the right is not completely resolved by the diagrams. In this area, periodic flow (see figure 11) occurs as well as low dimensional chaotic flow (see figure 12).

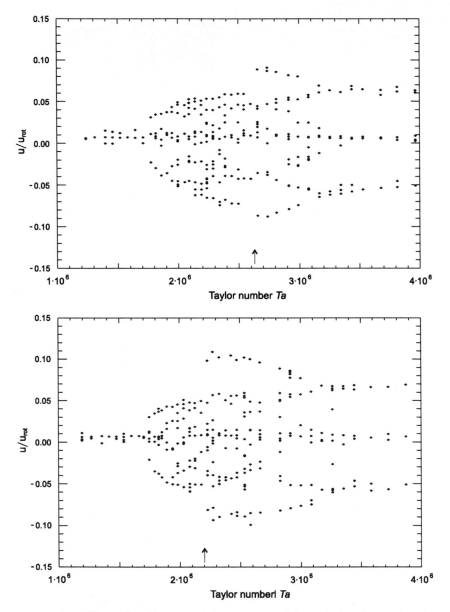

FIGURE 17. Bifurcation diagram of the extrema of the velocity time series, for increasing (top) and decreasing Taylor number (bottom).

5. Conclusions

The qualitative dispersion of the baroclinic waves was measured. The well known $1/\Omega$ dependence of the drift velocity was reproduced, as a first test of the new apparatus. Furthermore, a sign change of the slope of the dispersion relation is evident from the drift velocity data.

Supplemental to the temperature measurements of Read, Bell, Johnson & Small(1992) and Früh & Read (1997), contact free LDV measurements on baroclinic instabilities have been carried out. The LDV measurement method is good enough to apply nonlinear methods to the velocity time series. The results confirm the temperature measurements of Früh & Read (1997) and Read, Bell, Johnson & Small(1992), in particular, the coupling between the amplitude vacillation frequency and the drift frequency was observed, too. The locality of the onset of the structural vacillations is evident.

The transition from the axisymmetric basic flow to steady baroclinic waves is documented by bifurcation diagrams of the extrema of the velocity time series. The diagrams prove the complexity of this transition. A supercritical and a subcritical Hopf bifurcation occur during the transition to steady waves.

REFERENCES

BERNARDET, P., BUTET, A., DÉQUÉ, M. 1990 Low-frequency oscillations in rotating annulus with topography, *J. Atmos. Sci.*, **47**, No 24, 3023–3043

BUZUG, TH., PFISTER, G. 1992 Optimal delay time and embedding dimension by analysis of the global static and local dynamic behaviour of strange attractors, *Phys. Rev. A*, **45** (10)

DAVIES, T. V. 1956, The forced flow due to heating of a rotating liquid. *Phil. Trans. R. Soc. Lond. A* **249**, 27–64

EGBERS, C., RATH, H. J. 1996 LDV-measurements on wide gap instabilities in spherical coutte flow, in *Developments in Laser Techniques and Applications to Fluid Mechanics* (Eds. Adrian, R. J., Durao, D. F. G., Durst, F., Heitor, M. V., Maeda, M., Whitelaw, J. H.), Springer, 45–66

FOWLIS, W. W., HIDE, R. 1965 Thermal convection in a rotating annulus of liquid: Effect of viscosity on the transition between axisymmetric and non-axisymmetric flow regimes. *J. Atmos. Sci.*, **22**, 541–558

FRÜH, W. -G. , READ, P. L. 1997 Wave interactions and the transition to chaos of baroclinic waves in a thermally driven rotating annulus. *Phil. Trans. R. Soc. Lond. A* **355**, 101–153

HIDE, R. 1958 An experimental study of thermal convection in a rotating liquid. *Phil. Trans. R. Soc. Lond. A* **250**, 441–478

HIDE, R., MASON, P. J. 1970 Baroclinic waves in a rotating fluid subject to internal heating. *Phil. Trans. Roy. Soc. Lond. A* **268**, 201–232

READ, P. L., BELL, M. J., JOHNSON, D. W., SMALL, R. M. 1992 Quasi-periodic and chaotic flow regimes in a thermally driven, rotating fluid annulus. *J. Fluid Mech.*, **238**, 599–632

TAKENS, F. 1980 Detecting strange attractors in turbulence. *Dynamical Systems and Turbulence* (ed. D. Rand and L. S. Young). Lecture Notes in Mathematics. **898**, Springer, 366-381

WULF, P. 1997 Untersuchungen zum laminar-turbulenten Übergang im konzentrischen Kugelspalt,*VDI Fortschrittberichte Strömungstechnik*, Band **333**, VDI-Verlag

IV.3. Dissipation Estimation Around a Rushton Turbine Using Particle Image Velocimetry

K.V. Sharp, K.C. Kim, and R. Adrian

[1] Department of Theoretical and Applied Mechanics, University of Illinois, Urbana, IL 61801, USA

[2] School of Mechanical Engineering, Pusan National University, Pusan, Korea

Abstract. Particle Image Velocimetry (PIV) measurements have been performed in a cylindrical tank stirred by a Rushton turbine. Two datasets were acquired in the r-z plane with magnification 0.26 and 0.56. Phase-averaged velocity fields and mean square gradients were calculated. The mean velocity field averaged over all blade positions is presented. The mean square gradients are used to estimate turbulent dissipation, ε, in the measurement volumes. Two estimates of ε are presented, one based only on $\overline{u^2_{1,1}}$, and the other based on all of the available measured components. Both employ isotropic assumptions. The applicability of the isotropic assumptions is assessed by comparing the magnitudes of the mean square gradients, and the methods of dissipation estimation are compared. The normalized local dissipation, averaged over all blade positions, is calculated versus z for several values of r/R where R is the radius of the blade. The limitations of using PIV to calculate dissipation directly are addressed.

Keywords. Rushton turbine, particle image velocimetry, dissipation

1. Introduction

Mixers consisting of a Rushton turbine blade rotating in a cylindrical chamber are common in industry. The mechanisms for mixing in such systems are not well understood, nor are the relationships between mixing mechanisms and dissipation mechanisms. These deficiencies manifest themselves in the form of difficulty in determining the proper relationships for scaling up model results to full scale systems.   This study elucidates some of the fundamental structures and characteristics of this flow by performing two-dimensional field measurements in the *r-z* plane using Particle Image Velocimetry (PIV).

Previous studies using primarily single-point or photographic measurement techniques have provided significant insight into dominant flow structures and characteristics such as the vortices shed from the blade tips. Yianneskis *et al.* (1987) used laser-slit photography and Laser Doppler Velocimetry (LDV) techniques to investigate the mean flow in the tank, the formation of ring vortices in the bulk flow, and the trajectory of the tip vortices. The trajectory and structure of the tip vortices was analyzed by Van't Riet and Smith (1975) using photographic velocity measurements. Stoots and Calabrese (1995) also presented measurements of the mean velocity field using LDV.

Turbulent dissipation in a stirred mixer has been measured by Wu and Patterson (1989), Rao and Brodkey (1972), Cutter (1966), Okamoto *et al.* (1981), and Komasawa *et al.* (1974) among others. Wu and Patterson (1989), using LDV data, calculated ε locally on the basis of the turbulent macroscale and turbulent velocity. Rao and Brodkey (1972) determined the velocity microscale (dissipation length) and mean square velocity from hot-film data. These parameters were then used to estimate turbulent dissipation. Cutter (1966) used an energy balance coupled with photographic velocity data to estimate dissipation. Both Komasawa *et al.* (1974) and Okamoto *et al.* (1981) used energy spectra results to estimate the turbulence microscale, λ. From this, ε was calculated using the isotropic relationship:

$$\varepsilon = 15\nu\overline{u^2}\big/\lambda^2 .$$

In the current study, direct calculation of mean square gradients in the plane of the flow is available from the 2-D PIV data. The velocity fields can be considered as filtered fields, since the resolution of PIV data is spatially limited and may not resolve the smallest scales of velocity, which are on the order of the Kolmogorov scale. Direct measurement of the mean square gradients in phase-locked velocity fields allows for the calculation of two different dissipation estimates.   Both estimates employ isotropic assumptions for the gradient components requiring out-of-plane information.

Using information from two datasets with differing magnifications, and thus velocity spatial resolutions, it is possible to consider the effect of spatial resolution on estimates of dissipation using 2-D PIV data. It is also possible to assess the applicability of the isotropic assumptions since the magnitudes of the mean square gradients can be compared for a given dataset.

2. Experimental Set-up

The mixer and tank configuration, along with the coordinate system, is shown in Figure 1. A photo of the six blade symmetric Rushton turbine used in this experiment is shown in Figure 2. The turbine diameter (D) is 50.8 mm, and the tank was designed to have a test section diameter, T, equal to $3D$. The tank was filled such that the water depth equaled the tank diameter, and the blade was centered vertically at $T/2$. To best match computational boundary conditions, a non-baffled tank was used for the measurements.

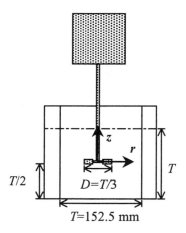

Figure 1: Dimensional relationships between blade diameter, filled tank depth, turbine clearance, and tank diameter.

Figure 2: Photo of the Rushton turbine used in this experiment.

All of the measurements were taken with an impeller speed of 100 RPM. A pulsed Nd:Yag laser was used to illuminate the particles, providing a projected lightsheet in the *r-z* plane of the flow. The camera was mounted such that the 1K by 1K CCD array was parallel to the lightsheet; thus projections of the flowfield velocities in the *r-z* planes were measured. The full experimental set-up is shown in Figure 3. Using this six-blade symmetric disk turbine, only 60 degrees of phase-locked information was required to define the entire flowfield assuming that the flow around each of the blades is the same. Phase-locked data were acquired in increments of 10 degrees over the total range of 60 degrees. Two hundred images were acquired at each radial blade position.

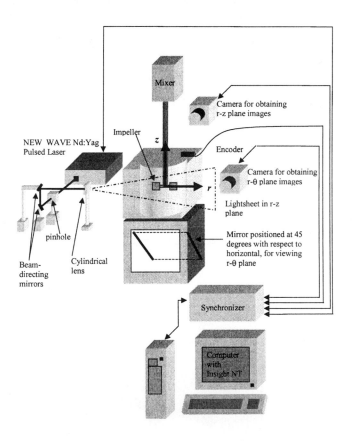

Figure 3: Experimental set-up.

PIV measurements were taken for two differing fields of view. The fields of view were 34.4 mm by 34.7 mm and 16.1 mm by 16.2 mm. An appropriate time between pulses was selected for each case, limited at the upper end by the significant out-of-plane particle motion caused by the strong tangential velocity component. Interrogation of the images was performed using a two-frame cross-correlation algorithm in INSIGHT NT Software from TSI, Inc. Thus, 100 velocity fields were obtained from two hundred images at each blade position. The interrogation window size was 24 pixels by 24 pixels in the first field of view and 32 pixels by 32 pixels in the second field of view. This provided velocity spatial resolution of 0.82 mm and 0.51 mm respectively. The magnification for the first case was 0.26, hereafter referred to as Case 1, and 0.56 for the better-resolved case, hereafter referred to as Case 2. Spurious vectors were removed in the region with high out-of-plane velocity component using CLEANVEC (Soloff and Meinhart, 1998).

3.   Results

The mean field averaged over all realizations and blade positions (600 velocity fields) is shown in Figure 4. A one-dimensional plot of radial velocity versus axial position is shown in Figure 5. The axial position, z, is normalized by blade width, W. The radial velocity in this figure is averaged over all blade positions, and the result compared with a velocity profile from Rutherford et $al.$ (1996). A series of phase-averaged velocity fields is shown in Figure 6. Each of these velocity fields represents an average over 100 velocity fields. A series of typical instantaneous (single realization) velocity fields is shown in Figure 7.

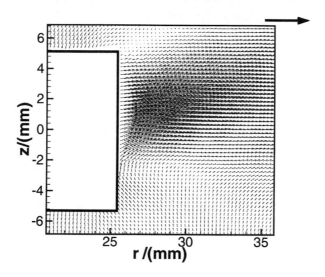

Figure 4: Mean velocity field averaged over all blade positions for Case 2 dataset, including blade outline. The reference vector has magnitude V_{tip}.

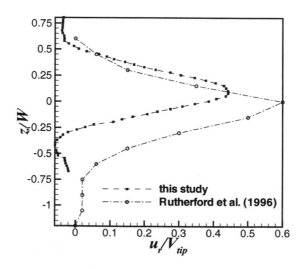

Figure 5: Comparison of mean radial velocity averaged over all blade positions for this study and Rutherford *et al.* (1996).

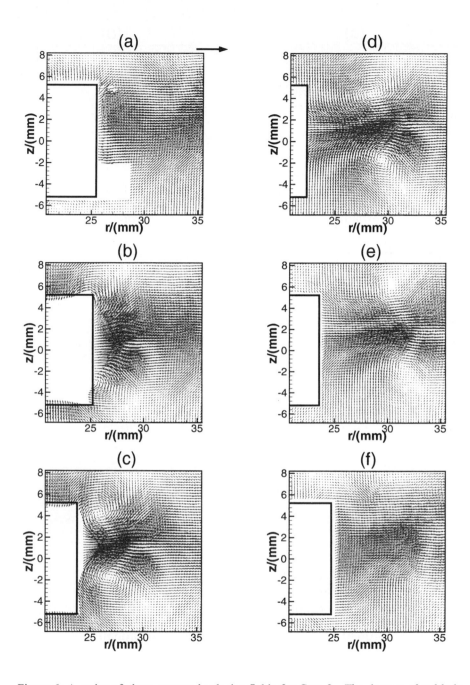

Figure 6: A series of phase-averaged velocity fields for Case 2. The degrees after blade passage are as follows: (a) 0, (b) 10, (c) 20, (d) 30, (e) 40, and (f) 50. The reference vector in frame (a) has magnitude of V_{tip}.

344

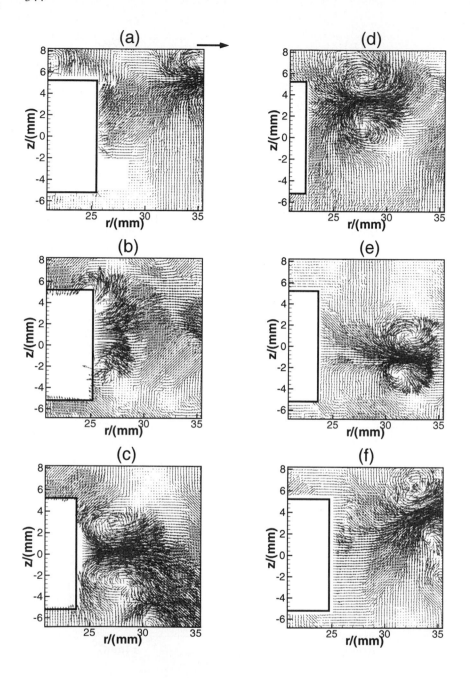

Figure 7: A series of typical instantaneous velocity fields for Case 2. The degrees after blade passage are as follows: (a) 0, (b) 10, (c) 20, (d) 30, (e) 40, and (f) 50. The reference vector in frame (a) represents the magnitude of V_{tip}.

Direct measurement of turbulent dissipation, ε, requires simultaneous knowledge of velocity gradients in all three directions. In this two-dimensional study, simultaneous velocity gradient information is available for $u_{1,1}$, $u_{1,2}$, $u_{2,1}$, $u_{2,2}$ where the "1" corresponds to the radial direction, "2" corresponds to the axial direction and u represents the Reynolds decomposed fluctuation about the unconditionally averaged mean velocity. Knowledge of these four components allows for computation of five terms in the full twelve-term turbulent dissipation equation:

$$\varepsilon = v \left\{ \begin{array}{l} 2\left(\overline{u_{1,1}^2} + \overline{u_{2,2}^2} + \overline{u_{3,3}^2}\right) + \\ \overline{u_{1,2}^2} + \overline{u_{2,1}^2} + \overline{u_{1,3}^2} + \\ \overline{u_{3,1}^2} + \overline{u_{2,3}^2} + \overline{u_{3,2}^2} + \\ 2\left(\overline{u_{1,2}u_{2,1}} + \overline{u_{1,3}u_{3,1}} + \overline{u_{2,3}u_{3,2}}\right) \end{array} \right\}.$$

The tangential direction is "3", and no information regarding this velocity component is obtainable using PIV in the r-z plane.

Data acquisition using two magnifications allows for analysis of the effect of velocity scale resolution on the estimation of dissipation from PIV data. Two *estimates* of dissipation are presented. The velocity fields used for these estimates can be considered to be filtered velocity fields, since it is likely that all of the small scales are not accurately resolved. Thus, the estimates of dissipation presented are not intended to replace a full three-dimensional simultaneous measurement of all velocity gradient components, but rather to provide insight into the different methods of estimating dissipation using PIV data.

For isotropic turbulence,

$$\varepsilon = 15v\overline{u_{1,1}^2}.$$

We use this relationship for the first estimate of dissipation, ε_{is}. It relies only on a single velocity gradient and assumes the values for all other components are statistically isotropic. The second estimate of the turbulent dissipation, ε_{iso}, uses all known components and assumes the unknown values to be statistically isotropic. For this second estimate, the average of $\overline{u_{1,1}^2}$ and $\overline{u_{2,2}^2}$ is used to approximate $\overline{u_{3,3}^2}$. Similarly, the average of $\overline{u_{1,2}^2}$ and $\overline{u_{2,1}^2}$ is used to approximate $\overline{u_{1,3}^2}$, $\overline{u_{3,1}^2}$, $\overline{u_{2,3}^2}$, and $\overline{u_{3,2}^2}$. The averages of the cross terms, $\overline{u_{1,3}u_{3,1}}$ and $\overline{u_{2,3}u_{3,2}}$, are estimated as $-\frac{1}{2}\left(\overline{u_{1,1}^2} + \overline{u_{2,2}^2}/2\right)$. This estimate is thus defined as:

$$\varepsilon_{iso} = v \left\{ \begin{array}{l} 2\left(\overline{u_{1,1}^2} + \overline{u_{2,2}^2} + \frac{1}{2}\left(\overline{u_{1,1}^2} + \overline{u_{2,2}^2} \right) \right) + \\ \overline{u_{1,2}^2} + \overline{u_{2,1}^2} + 4\left(\frac{1}{2}\left(\overline{u_{1,2}^2} + \overline{u_{2,1}^2} \right) \right) + \\ 2\left(\overline{u_{1,2}u_{2,1}} \right) - 4\left(\frac{1}{2} \right)\left(\frac{1}{2}\left(\overline{u_{1,1}^2} + \overline{u_{2,2}^2} \right) \right) \end{array} \right\}.$$

These methods of calculating the dissipation estimate were used with both of the acquired datasets, Case 1 and Case 2.

Table 1 presents area-averages of the dissipation estimates. These estimates are averaged over the same physical area (z =-7.36 mm to z = 8.70 mm and r/R = 1 to r/R = 1.43) in both cases. This physical region was selected since it represents the physical region of measurement for Case 2. Note that this is not an area average over the entire field of view for Case 1, as shown in Figure 8. Due to the significant loss of data in the near-blade region due to blade reflections in the 0-degree case, dissipation results for this position are not included in the tables.

Table 1: Comparison of area averaged dissipation estimated between Case 1 and Case 2. Results from both dissipation estimates, ε_{i15} and ε_{iso}, are shown for the two datasets.

Degrees behind blade passage	Case 1(M = 0.26)		Case 2(M = 0.56)	
	ε_{i15} (m²/s³)	ε_{iso} (m²/s³)	ε_{i15} (m²/s³)	ε_{iso} (m²/s³)
10	0.0073	0.0058	0.0277	0.0200
20	0.0108	0.0086	0.0281	0.0198
30	0.0123	0.0100	0.0356	0.0261
40	0.0099	0.0082	0.0317	0.0244
50	0.0078	0.0067	0.0263	0.0211

To assess the applicability of the isotropic estimates on dissipation estimation in a filtered field such as that acquired by PIV, the area-averaged magnitudes of the mean square gradients were computed. Table 2 presents the ratios of mean square gradients used in the isotropic assumptions. If the isotropic assumptions were perfect, all of the values in this table would be equal to 1.

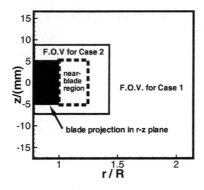

Figure 8: Schematic of regions of interest including near-blade region and fields of view for both cases. All area-averages of dissipation, unless denoted as near blade region averages, were performed over the physical region encompassed by the field of view for Case 2.

Table 2: Comparison of the area-averaged ratios of mean square gradients for Case 2.

Degrees behind blade passage	$\overline{u_{1,1}^2}\Big/\overline{u_{2,2}^2}$	$\overline{u_{1,2}^2}\Big/\overline{u_{2,1}^2}$	$\overline{u_{1,2}^2}\Big/2\overline{u_{1,1}^2}$	$\overline{u_{2,1}^2}\Big/2\overline{u_{1,1}^2}$	$\overline{u_{1,2}u_{2,1}}\Big/-\frac{1}{2}\overline{u_{1,1}^2}$
10	1.00	1.19	0.66	0.55	0.21
20	1.27	1.67	0.79	0.47	0.32
30	1.20	1.54	0.77	0.50	0.33
40	1.05	1.32	0.79	0.60	0.38
50	1.00	1.27	0.82	0.65	0.35

For Case 2, with higher magnification and thus better small scale velocity resolution, the area average dissipation estimates are shown versus estimation method and specified area in Table 3. The region close to the blade is defined as the region between the top and bottom blade edges in z and $r/R = 1$ to $r/R = 1.25$.

Table 3: Comparison of the area-averaged dissipation estimates for the near blade region and the entire measurement region.

Degrees behind blade passage	ε_{iso}	ε_{iso} near blade	ε_{i15}	ε_{i15} near blade
	(m^2/s^3)	(m^2/s^3)	(m^2/s^3)	(m^2/s^3)
10	0.0200	0.0290	0.0277	0.0441
20	0.0198	0.0346	0.0281	0.0510
30	0.0261	0.0500	0.0356	0.0671
40	0.0244	0.0327	0.0317	0.0394
50	0.0211	0.0171	0.0263	0.0212

The turbulence microscale (λ) and turbulence Reynolds number (Re_λ) are found using the following relationships:

$$\lambda^2 = \frac{\overline{u_r^2}}{\overline{u_{r,r}^2}}$$

$$Re_\lambda = \frac{\sqrt{\overline{u_r^2}}\,\lambda}{\nu}$$

where u_r is the radial velocity. These quantities, shown in Table 4, are calculated on the average over the previously defined near-blade region, on the average over the entire field of view, and at the specific point $r/R = 1.02$ on the centerline of the blade.

Table 4: Turbulence microscale and turbulence Reynolds number for Case 2.

Region or point	λ /(mm)	Re_λ
Near blade	1.1	60
Whole field of view (Case 2)	1.1	50
Blade centerline and $r/R = 1.02$	0.8	54

In this flow, N_{Re} is on the order of 4000, where $N_{Re} = \rho ND^2/\mu$, $N =$ number of revolutions per second and $D =$ blade diameter. The power number, N_p, is defined as $P/\rho N^3 D^5$ where P = the power input to the flow. Using a Reynolds number versus power number curve for the Rushton turbine (Bates *et al.*, 1966, p. 133), the power number is approximately 4.5. Therefore, the power input is calculated to be 0.007 Watts. The LIGHTNIN mixer used in this experiment was unable to measure power accurately enough to use a direct reading at this power setting.

The mean turbulent dissipation in the flow is defined as:

$$\bar{\varepsilon} = \frac{P}{\rho V}.$$

The volume of this apparatus is 0.0028 m³. Thus the mean turbulent dissipation, $\bar{\varepsilon}$, is calculated as 0.0025 m³/s². Using this value, the Kolmogorov scale is estimated as 0.14 mm. Although this estimate has its limits of applicability due to the significant non-isotropic and inhomogeneous structure of the flow, it provides at least an idea of the required resolution of velocity scales. The minimum resolution in this experiment is 0.52 mm, which is four times the Kolmogorov scale. The estimates of dissipation presented in this paper are not meant to provide final values for dissipation in a mixing flow, but instead to evaluate methods of surrogate dissipation calculation and to elucidate trends in dissipation concentrations.

Figure 9 shows a plot of $\varepsilon*$ vs. z for several r/R positions in the flow, where $\varepsilon*$ is the estimated dissipation normalized by the tank-averaged dissipation, $\bar{\varepsilon}$. The results are averaged over all blade positions for Case 2 (higher magnification). The estimation method used in this data is ε_{iso}, which relies on all the measured components plus estimates for the others.

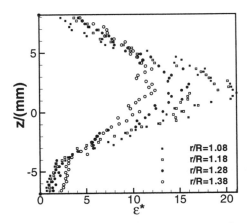

Figure 9: $\varepsilon*$, averaged over all blade positions, vs. z for varying r/R.

4. Discussion

4.1 Velocity Fields

The general shape of the radial velocity profile versus axial position near the blade is similar in this study and in Rutherford *et al.* (1996). There are some differences in experimental configurations for these studies, namely the distance from the centerline of the disk to the bottom of the tank, the impeller rotation speed, the dimensional size of the impeller and tank, and the existence of baffles. The impeller speed in Rutherford *et al.* (1996) is 200 RPM, twice that used in the current study. The maximum radial velocity in this study appears lower than in Rutherford *et al.* (1996), even when normalized by V_{tip}. The maximum radial velocity is seen above the centerline in the unbaffled tank with the clearance below the disk centerline equal to $T/2$, where T is the diameter of the tank and also the fluid depth. This slight inclination of the mean flow has also been seen in Yianneskis *et al.* (1987) for the same clearance.

The plots of the mean velocity field show well-defined tip vortices in the region close to the blade, as expected. The tip vortices appear to have similar axial convection velocities for the first 20 degrees following blade passage. Thus the phase-averaged fields for 10 and 20 degrees behind blade passage show axial and radial locations of the vortex cores approximately equal to the axial and radial locations of the vortex cores in instantaneous velocity fields. Further behind the blade, the vortices convect both positively and negatively in the axial direction. The instantaneous velocity fields shown in Figure 7 are not sequentially acquired, but rather random realizations from each blade position. In Figures 6 and 7, the blank region just outside the lower half of the blade in the 0-degree position represents a region of high laser light reflection from the blade itself and thus a region of poor data acquisition. The mean velocity field averaged over all blade positions shows evidence of a radial jet being pumped in the r-z plane by the passage of the blades.

4.2 Dissipation estimates

A comparison of the estimates of dissipation between Case 1 and Case 2 shows that the estimates are consistently higher with the Case 2 data. The lower estimates in Case 1 correspond to the data with lower resolution and it is likely that the mean square gradients are being underestimated.

For Case 2, the better-resolved case, the ratios of mean square gradients are used to assess the applicability of isotropic assumptions to dissipation estimations using PIV data. The ratio of $\overline{u_{1,1}^2}/\overline{u_{2,2}^2}$ is within 30% of 1 for all individual blade positions. Generally, $\overline{u_{1,1}^2}$ is larger than $\overline{u_{1,2}^2}$. In all cases $\overline{u_{1,2}^2}$ and $\overline{u_{2,1}^2}$ are less than $2\overline{u_{1,1}^2}$. The cross term, $\overline{u_{1,2}u_{2,1}}$, is also less than its' corresponding isotropic assumption $-\frac{1}{2}\overline{u_{1,1}^2}$. Thus the estimation of dissipation (ε_{ns}) using only multiples of $\overline{u_{1,1}^2}$ is higher than that which employs more of the directly measured components (ε_{iso}), as shown in Table 1.

In Table 3, the average of both dissipation estimates is shown for Case 2 data. The area average of the dissipation estimates is taken over the entire measurement region, and also over a smaller measurement volume close to the blade, corresponding to $-W < z < W$ and $1 < r/R < 1.25$. This table shows that, in all cases except the 50-degree position, the area average over the region close to the blade is considerably higher than that over the entire measurement region. The ratio of the area average close to the blade to that in the entire measurement region is highest at blade positions 20 and 30 degrees behind blade passage. This is the plane in which the tip vortices are strongest inside $1 < r/R < 1.25$.

The calculated turbulence microscales are on the order of 1 mm, consistent with those found by Wu and Patterson (1989) and Rao and Brodkey (1972). Using these microscales, the turbulence Reynolds number is also on the order of turbulence Reynolds numbers in previous studies, specifically the study by Wu and Patterson (1989).

Figure 9 shows that $\varepsilon*$ is highest just outside the tip of the blade ($r/R = 1.08$). The maximum value of $\varepsilon*$ occurs near $z = 0$ for all profiles, and generally decreases with axial and radial distance from the blade tip. These results are similar to those of Wu and Patterson (1989), though they found a double-tipped profile shape for r/R of 1.08 and had a much larger range of values for r/R. This double-tipped shape is not evident in Figure 9 at $r/R = 1.08$ but is slightly evident at the higher r/R values of 1.18 and 1.28. Data at larger r/R values with the current dissipation estimation technique and experimental configuration are required to definitively evaluate the dissipation magnitude trends at higher distances from the blade tip. The maximum value of $\varepsilon*$ found by Wu and Patterson (1989) is approximately 22 near the vertical centerline at $r/R = 1.29$. These results were for a rotation rate of 200 RPM, twice that in this study. The maximum value of $\varepsilon*$ in this study at $r/R = 1.28$ is 15 near $z = 0$.

The principle conclusion to be drawn from Figure 9 is that the general level of dissipation in this region is 10-20 times the volume mean dissipation, emphasizing the intense concentration of dissipation in the near-blade region.

5. Conclusions

This work describes a method for estimating turbulent dissipation using 2-D PIV data. Although the 2-D PIV measurements do not provide enough information to directly measure all the required velocity gradients, they are considerably more rich than measurements from single-point probes, and they eliminate the need to invoke Taylor's hypothesis. Two estimates of dissipation are calculated, one using only one of the velocity gradient components and the other using all known components. The isotropic estimate using only $\overline{u_{1,1}^2}$ gives higher overall values for dissipation than the estimate which relies on all known components and uses isotropic assumptions for the others. A comparison of the mean square gradients shows that the statistically isotropic assumptions are off by up to 70% when considering area-averages of phase-locked velocity fields. Although this is a significant variation, it may decrease for higher Reynolds numbers.

   Even with limited spatial resolution of velocity, the dissipation estimates calculated in this work are within the range of the dissipation calculations in previous studies. The estimate of dissipation in the near-blade region is 10-20 times higher than the volume averaged dissipation, emphasizing the high concentration of dissipation in this region.

   The spatial resolution of these measurements is limited by the camera lenses, experimental set-up, and issues of seeding density in the interrogation spots. To definitively assess the effects of limited spatial resolution on dissipation calculations, higher magnifications and higher seeding density are required. The increased magnification would shrink the size of the interrogation region in physical coordinates. If the seeding density were increased, it would be possible to use an interrogation approach combining traditional PIV and Particle Tracking Velocimetry (PTV) to obtain more velocity vectors (Takehara, 1998) and thus increase resolution of the system. The strong out-of-plane motion causes reduced particle density in the interrogation spots, especially in such regions as the tip vortices.

   For a *complete* calculation of dissipation directly from the velocity gradients, three dimensional information is required. Holographic PIV measurements have not yet been taken in a similar flowfield. Stereo PIV has been performed in this flow (Hill, 1998), though the measurements were taken in planes differing from those presented in this work. The spatial resolution of the Stereo PIV measurements was not as high as in the 2-D PIV case. Further Stereo PIV measurements at higher resolution would aid in evaluating the assumptions of isotropy.

References

Bates, R.L., Fondy, P.L & Fenic, J.G. 1966, Impeller Characteristics and Power, in Mixing: Theory and Practice, ed. by Uhl, V.W. & Gray, J.B., vol. 1, pp. 111-178, Academic Press, New York.

Cutter, L.A. 1966, Flow and Turbulence in a Stirred Tank, A.I.Ch.E.J., vol. 12, pp. 35-45.

Komasawa, I., Kuboi, R. & Otake, T. 1974, Fluid and Particle Motion in Turbulent Dispersion-I: Measurement of Turbulence of Liquid by Continual Pursuit of Tracer Particle Motion, Chem. Eng. Sci., vol. 29, pp. 641-650.

Hill, D. 1998, Private communication.

Okamoto, Y., Nishikawa, M. & Hashimoto, K. 1981, Energy Dissipation Rate Distribution in Mixing Vessels and its Effects on Liquid-liquid Dispersion and Solid-liquid Mass Transfer, Int. Chem. Eng., vol. 21, no.1, pp. 88-94.

Rao, M.A. & Brodkey, R.S. 1972, Continuous Flow Stirred Tank Turbulence Parameters in the Impeller Stream, Chem. Eng. Sci., vol. 27, pp. 137-156.

Rutherford, K., Mahmoudi, S.M.S., Lee, K.C. & Yianneskis, M. 1996, The Influence of Rushton Impeller Blade and Disk Thickness on the Mixing Characteristics of Stirred Vessels, Trans. I.Chem.E., vol. 74, part A, pp. 369-378.

Soloff, S., and Meinhart, C. 1998, CleanVec: PIV Vector Validation Software, Private Communication.

Stoots, C.M. & Calabrese, R.V. 1995, Mean Velocity Field Relative to a Rushton Turbine Blade, A.I.Ch.E.J., vol. 41, no.1, pp. 1-11.

Takehara, K. 1998, Private communication.

Van't Riet, K. & Smith, J.M. 1975, The Trailing Vortex System Produced by Rushton Turbine Agitators, Chem. Eng. Sci., vol. 30, pp. 1093-1105.

Wu, H. & Patterson, G.K. 1989, Laser-Doppler Measurements of Turbulent-Flow Parameters in a Stirred Mixer, <u>Chem. Eng. Sci.</u>, vol. 44, no. 10, pp. 2207-2221.

Yianneskis, M., Popiolek, Z. & Whitelaw, J.H. 1987, An Experimental Study of the Steady and Unsteady Flow Characteristics of Stirred Reactors, <u>J. Fluid. Mech.</u>, vol. 175, pp. 537-555.

IV.4. Visualization of the Trailing Vortex System Produced by a Pitched Blade Turbine Using a Refractive Index Matched Automated LDA-Technique

M. Schafer[1], P. Wachter[1], F. Durst[1], and M. Yianneskis[2]

[1] Institute of Fluid Mechanics, University of Erlangen-Nürnberg
[2] King's College, London

Abstract. The trailing vortex system near impeller blades has been identified as the major flow mechanism responsible for mixing and dispersion in stirred tank reactors. In the area of the trailing vortices highest values of turbulence occur and the major portion of the total energy introduced to the stirred vessel is dissipated here.

Despite the importance of trailing vortices for the mixing process, little is known about their formation and development within and in the immediate vicinity of the impeller blades. In particular, there is a lack of detailed quantitative information on the characteristics of trailing vortices, for example vorticity, vortex dimension and the distribution of turbulence kinetic energy.

This paper provides detailed information on the flow field produced by a 45° pitched blade turbine and describes a technique to visualise the formation and development of trailing vortices using laser-Doppler-anemometry. The results lead to a better understanding of trailing vortices and provide valuable information to engineers for the optimisation of mixing processes in industry. In addition, the validation of numerical computations with CFD of the flows in stirred tank reactors requires more detailed information on the flow field than is available to date so that the data gained in the present study can significantly support the ongoing developments in CFD.

Keywords. Stirred vessel, Pitched blade turbine, Trailing Vortices, LDV

1 Introduction

The flow generated in a stirred tank reactor is very complex and extensively three dimensional, especially in the impeller region. The flow conditions near the impeller blades are responsible for the formation of so called trailing vortices. Extensive studies have been carried out in the past to learn more about the characteristics of trailing vortices and their importance for several mixing processes. This is mainly due to the high levels of turbulence and local energy dissipation rates associated with the formation, development and break down of the trailing vortex system. Therefore trailing vortices can be considered as the essential flow mechanism for following mixing applications:

- Gas dispersion
- Liquid dispersion like in emulsion processes
- Solid-liquid systems in which high energy dissipation rates are needed (e.g. dissolution of agglomerates)
- Solid-liquid systems in which high energy dissipation rates must be avoided due to the formation of fine and labile products (e.g. cell damage in fermentation processes)
- chemical reactions which are controlled by micro-mixing phenomena (see for example Fournier *et al.* (1996)).

The trailing vortex system produced by a Rushton turbine is the most extensively studied in the literature. Van't Riet and Smith (1975) used photographic techniques to visualise the formation and development of trailing vortices in a gas dispersion process. They reported that high rates of turbulence exist within the trailing vortices and it was concluded that most of the energy introduced to the stirred tank is dissipated here. In other reported studies measurement techniques like Pitot tubes and hot wire anemometry were also used for the investigation of the flow field produced by a Rushton turbine (see for example Gunkel and Weber (1975)).

The development of the laser-Doppler-velocimetry (LDV)-technique enabled a more reliable quantitative study on flow fields in stirred tank reactors, since (a) LDV provides flow information even in unsteady and highly turbulent flow regions as well as in the recirculation regions in the tank and (b) it is an unobtrusive technique. Therefore the LDV-technique has been widely employed to study the characteristics of the flow in stirred tanks. Also the trailing vortex system generated by a Rushton turbine was focused in more detail (see for example Yianneskis *et al.* (1987), Wu and Patterson (1989a), Stoots and Calabrese (1995)). These investigations provided quantitative information on the velocity components and their turbulent fluctuations within the trailing vortex pair generated at each impeller blade of a Rushton turbine. In addition, it was reported that the periodic nature of the flow in the vicinity of the impeller necessitates angle-resolved LDV-measurements, in which the flow information is assigned to the corresponding angle of the impeller blades. Yianneskis and Whitelaw (1993) have shown that if the LDV-measurements are processed as 360° ensemble-averaged measurements, then the fluctuating quantities contain both periodic and turbulence contributions and this can lead to an overestimation of apparent turbulence quantities in the impeller stream of a Rushton turbine by up to 400%. Detailed quantitative flow information within the stirrer element between the rotating impeller blades where reported by Schäfer *et al.* (1997), who used a refractive index matching method to gain optical access to the inner part of the impeller, and the results gave details of the exact formation of the trailing vortices behind each stirrer blade.

Pitched blade turbines (PBTs) have been found to be more efficient than other impellers (e.g. Rushton turbine) with respect to liquid phase mixing and solid suspension processes. They are also employed in gas dispersion processes. The trailing vortex system generated by pitched blade turbines was studied by Tatterson et al. (1980) by using flow visualisation techniques. Compared with a Rushton turbine only one trailing vortex is formed behind each blade of a PBT. Ali *et. al* (1981) used a high speed, stereoscopic motion picture technique, to visualise the trailing vortex system in an oil into water dispersion and the trailing vortices have been identified as the major flow mechanism responsible for dispersion. LDV-measurements on the flow field were also reported (see for example Ranade and Joshi (1989), Kresta and Wood (1993), Hockey and Nouri (1996)), but most of these investigations have been concerned with 60° PBT and/or with the mean and fluctuating (r.m.s) velocities in the bulk flow of the vessel or at least outside the impeller swept volume. The only investigations concerned primarily with the trailing vortex structure were the flow visualisation studies of Tatterson and co-workers but the understanding and quantification of the vortex mean flow and turbulence structure is still far from complete.

The main objectives of the present investigation are to provide a better understanding of the generation and formation of the trailing vortex system by a 4/45° pitched blade turbine and to quantify the characteristic flow variables like mean and r.m.s velocities and turbulence kinetic energy within and in the vicinity of the trailing vortices. Another major objective is to provide a comprehensive and detailed experimental data set for the flow field produced by a PBT, which is suitable for the validation of numerical simulations of stirred vessel flows and the improvement of the turbulence models used in such calculations. Progress in modern computational techniques led recently to reliable calculations of single-phase flow in stirred tank reactors (see for example Wechsler *et al.*). These developments will be supported by the detailed LDV-experiments carried out during the present study.

For this purpose a LDV-system was developed at the Institute of Fluid Mechanics in Erlangen which allows for automated angle-resolved measurements within the impeller region (i. e. between the blades) through refractive index matching of the complete measuring section. The complete set-up, the measuring equipment and the flow configuration is described in more detail in section 2. For the visualisation of the results a commercial software "TECPLOT" was used and a movie was created to animate the flow around the impeller which shows the generation, development and break down of the trailing vortices. Characteristic results are presented to show the most important flow features which are discussed in section 3. As mentioned above the energy dissipation rates within the trailing vortices are an important parameter to assess the efficiency of mixing processes. Therefore an rough estimate of the local energy dissipation rate was carried out which is described and presented in section 4. The paper ends with a summary of the main findings and some conclusions which can be drawn from the results.

Some of the results of the present investigation have been reported by Schäfer *et al.* (1998). This paper presents additional results in an effort to provide a fuller characterisation of the complex three-dimensional structure of the flow around the turbine blades.

2 Experimental Set-Up and Flow Configuration

A fully automated test rig for detailed LDV-measurements in stirred tank reactors has been developed at the Institute of Fluid Mechanics within the framework of several research projects. The set-up of the test rig is shown in Figure 1. The set-up included three main parts: the measuring section, the traversing equipment for automation and the LDV measurement system consisting of a diode fibre laser-Doppler-anemometer operating in backscatter mode, a traversable probe and a frequency counter.

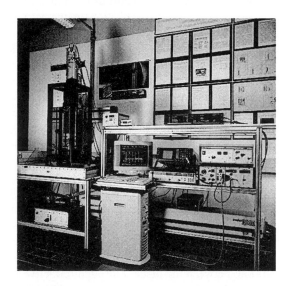

Fig. 1 Stirrer test rig at the Institute of Fluid
Mechanics in Erlangen

The measuring section consisted of a cylindrical baffled vessel of diameter $T = 152$ mm and a four-bladed pitched blade turbine of diameter 50 mm ($D = 0.329\ T$) installed at a clearance of $C = T/3$. Figure 2 shows the geometry of the mixing vessel and the coordinate system used. Figure 3 gives the details of the turbine. The liquid height was equal to the vessel diameter ($H = T$). The top of the vessel was closed with a lid to avoid air entrainment into the liquid from the free surface.

Nouri and Whitelaw (1990) have shown that the effect of the lid on the flow field has only an influence in the immediate vicinity of the lid/free surface. Four equally spaced baffles of width $B = T/10$ and thickness of 3 mm were mounted along the inner wall of the cylinder at a distance of 2.6 mm. The vessel could be rotated about its axis which facilitated the adjustment of the vessel for measurements in different vertical planes.

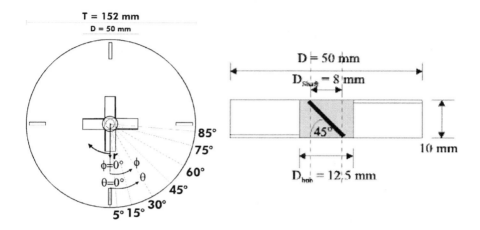

Fig. 2 Geometry and coordinate system   **Fig. 3** Geometry of pitched blade turbine

The turbine used here was a four bladed stirrer with a blade pitch of 45°. The blade thickness was 0.9 mm and the blade height was 0.264 D (13.2 mm) corresponding to a projected height of 0.2 D. The hub in which the blades were fixed had a diameter of 12.5 mm.

The entire measuring section was refractive index matched, i.e. the walls, the baffles and the impeller blades were constructed from Duran glass that has the same refractive index as the working fluid. In addition, the tank vessel was located in a rectangular trough filled with the working fluid, in order to eliminate the distorting effect of the rounded surface of the cylindrical vessel on the path of the laser beams. The refractive index depends mainly on the molecular structure of the material/fluid, the temperature and the wave length of the light (laser beam). At the operating wave length of 832 nm of the laser system and an average temperature of 21° C during the experiments Duran glass has an refractive index of $n = 1.468$. Two silicone oils, from which one has an refractive index above and one below $n = 1.468$, were mixed in that way that the fluid matches exactly the refractive index of Duran glass at a temperature of 21° C. The kinematic viscosity of the final mixture of silicone oils was measured to $v = 15.3 \times 10^{-6}$ m²/s. The temperature within the vessel and the trough was controlled with a cooling coil so that the temperature was kept constant during the measurements. The full refractive index matching offered optical access to the inner part of the impeller without any distortion of the laser beams allowing for detailed studies of the flow field in the important region, where the trailing vortices are generated.

Periodic variations in the flow field were taken into account by angle resolved measurements close to the impeller. For this purpose an optical shaft encoder was used providing 1,000 pulses and a marker pulse per revolution. The marker pulse corresponded to the angle $\phi = 0°$ of the blade and was set at the middle of the blade at the radial tip with an accuracy of $\pm 0.36°$.

For the measurements a Diode-Fiber-Laser-Doppler-Anemometer (DFLDA) was used which has been previously described in detail by Stieglmeyer and Tropea (1992). The DFLDA-system consists of a miniaturised, traversable optical probe containing transmission and receiving optics, separated mechanically from the bulky components (laser, beamsplitter, Bragg cell, fiber in coupling, avalanche photo diode (APD) for signal detection, etc.). Optical fibers connect the probe with the laser, detector and other optical components and the probe can be easily mounted on a traversing unit. The working fluid was seeded with titanium dioxide particles (3 µm mean diameter) and the scattered light was focused back to the APD. Radial and tangential velocity components were measured with the beams located in a horizontal plane and axial components were measured by rotating the optical probe 90° so that the beams are located in a vertical plane. The signal from the APD was filtered, validated and evaluated by using a frequency counter (TSI model 1980b) which was interfaced to a PC.

As the laser system used only supplies flow information at a single point, the entire flow field can only be determined rapidly if data acquisition is automated. For this purpose the optical probe was mounted on a 3-D traversing unit which was controlled via a CNC-controller by the PC. The accuracy of locating the measuring volume (i. e. the probe) was 0.1 mm.

The velocity data as well as the angular information were stored for each measuring point in a data file. Software was written to assemble all data files obtained to one input file for the postprocessor, which was the commercial software "TECPLOT".

The results which will be shown in the next section were conducted at a stirrer speed of $N = 2,672$ r.p.m. ($V_{tip} = \pi DN = 7$ m/s) corresponding to a Reynoldsnumber of $Re = ND^2/\nu = 7,300$. The stirrer speed was kept within ± 30 r.p.m (1%).

At each measuring location 10,000 single velocity data were collected for the ensemble-averaged measurements whereas this number was increased to 60,000 for the angle-resolved measurements. These numbers of velocity samples were sufficient to minimise statistical errors in the determination of mean and fluctuating flow velocities. There are further possible error sources in an LDV-system and measurement uncertainties vary with location. The accumulated errors in mean and fluctuation (rms) velocity measurements have been estimated to be, on average, 1 - 3 % and 5 - 10% of V_{tip}, respectively, with the higher errors expected in region of steep velocity gradients.

3 Results

3.1 Large-Scale Flow Field

For the purpose of obtaining a better understanding of the flow generated by the pitched blade turbine throughout the whole vessel ensemble-averaged measurements and angle-resolved measurements were carried out for the complete flow field. Angle-resolved measurements were made only in those flow regions, where periodic variations in the flow variables occurred. The boundaries of this region were determined via test measurements which indicated that periodic variations are apparent beneath the impeller down to the bottom of the vessel (from $z/T = 0$ to $z/T = 0.4$). In radial direction periodic variations were obtained only close to the impeller (from the shaft up to $r/T = 0.2$). The measuring grid for the complete flow investigations extended in axial direction from the bottom to the top (6 mm - 150 mm or $z/T = 0.039 - 0.987$) and in radial direction from the shaft to the cylinder walls (6 mm - 74 mm or $r/T = 0.039 - 0.487$) and had a fine resolution of 4 mm ($= 0.026\ T$) in each direction. In addition, seven different planes between the baffles were examined to resolve also the influence of the baffles on the bulk flow. In Figure 2 the locations of the planes of measurements within the stirred vessel are shown.

By way of example the flow field and the distribution of turbulence kinetic energy obtained for a vertical plane (r, z-plane) located half way between two baffles at $\theta = 45°$ before a baffle are shown in Figure 4 and Figure 5, respectively. The mean velocity components were normalised with the blade tip velocity V_{tip} and are denoted $\overline{U} / V_{tip}$, $\overline{V} / V_{tip}$ and $\overline{W} / V_{tip}$ for the axial, radial and tangential components respectively. The turbulence kinetic energy k were calculated from the fluctuating velocity components u', v' and w' according to

$$k = \frac{1}{2}\left(u'^2 + v'^2 + w'^2\right) \tag{1}$$

and normalised with the square of the stirrer tip velocity V_{tip}^2.

It is important to note that the velocity vectors close to the impeller (i. e. in the region, where periodic variations occur) were obtained from angle-resolved measurements, which means that they refer to a blade angle of $\phi = 45°$. The vectors shown in Figure 5 reveal the characteristic flow pattern produced by a pitched blade turbine. The fluid is entrained into the stirrer element from the top and is ejected axially at axial peak velocities of 0.45 V_{tip} (at $r/T = 0.125$). The discharge flow becomes then increasingly more radial and reaches the bottom approximately half way between the shaft and the vessel wall. A large scale ring vortex is formed, which extends only over slightly more than half of the vessel height ($z/T = 0.55 - 0.65$), therefore poor mixing exists in the upper bulk flow of the tank with almost no exchange of fluid. The flow velocities in the ring vortex reach magnitudes of 0.2 V_{tip} close to the vessel wall whereas in the uppermost area only slow velocities below 0.05 V_{tip} were obtained. A secondary vortex is formed at the bottom close to

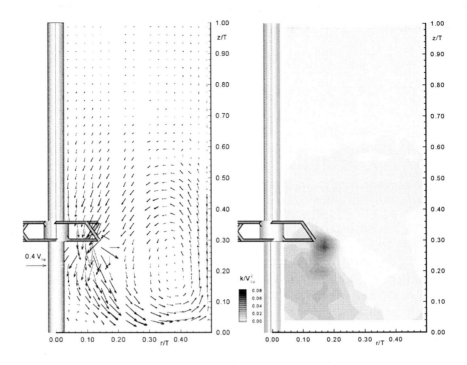

Fig. 4 Bulk flow field   **Fig. 5** Distribution of TKE at θ = 45°

the shaft with very low magnitudes of velocity. This may cause problems in this area in solid-liquid mixing tasks with the suspension of the solids off the vessel bottom.

It must be noted that the flow structure changes slightly for different measuring planes which are not shown here. In the vertical plane at θ = 5° the axial extension of the large ring vortex reaches 3/4 of the liquid height due to additional axial flow which results from the deflection of tangentially propelled fluid at the baffles. The extension in axial direction decreases with increasing distance from the baffles and at the plane at θ = 75° the vortex is only able to form over one half of the liquid height.

The flow field close to the impeller beneath the outer tip of the blade at around r/T = 0.16 reveals that a vortex is generated in the impeller region which is still present at ϕ = 45° behind the blades. As it can be seen in Figure 5 this vortex is associated with high turbulence intensity. The magnitudes of turbulence kinetic energy reach here peak values of about 0.04 - 0.085 V_{tip}^2, whereas in the bulk flow of the vessel only very little turbulence of less than 0.015 V_{tip}^2 is present.

3.2 Impeller Flow Field

Considering that for a reliable assessment of mixing processes the exact distribution of turbulence kinetic energy is required, the generation and development of the trailing vortices constitutes an important phenomenon of the flow in stirred tank reactors. In order to capture all the details of the flow in the region of the trailing vortices the resolution of the measuring grid was increased to a step width of 1 - 2 mm within and close to the impeller.

The results will be presented in form of an animation of the impeller flow field, in which planes at each degree between two blades (1° to 90°) are shown step by step leading to a rotating impeller movie. By way of example Figures 6 and 7 show the flow field and the distribution of turbulence kinetic energy for two selected planes located at $\phi = 5°$ and $\phi = 15°$ behind the blades. The blades must be considered as moving out of the page towards the reader. It must be mentioned that the blade cuts across the measurement plane and as a result some of the vectors show the flow in front and some behind the blade.

The trailing vortex is formed by the interaction of the streams issuing from the top and side of the blade and it is fully developed at $\phi = 5°$ behind the blade with an extension of over 5 mm in diameter. The trailing vortex moves further down with increasing angular distance from the blade and reaches the lower tip at approximately $\phi = 15°$. The peak values of turbulence kinetic energy exist in the region of the trailing vortex below the impeller. It is interesting to note that above the impeller where the fluid is sucked in turbulence values are low, as indicated by the magnitudes of the turbulence kinetic energy that are similar to those in the bulk flow ($k < 0.015\ V_{tip}^2$). The second circulatory motion of the flow below the impeller at approximately $z/T = 0.25$ in the $\phi = 5°$ plane indicates the presence of the trailing vortex from the preceding blade which is still in evidence at $\phi = 15°$ or $\phi = 105°$ after that blade has crossed this plane. Also higher turbulence kinetic energy values were obtained in this region.

Although the depicted flow fields clearly show the trailing vortices and the vortex centres can be determined accurately due to the high density of the measuring grid, it is more difficult to locate exactly the vortex edges, which are needed to give a full description of the trailing vortices. As vortices are characterised the intensity of vortex motion was found to be a valuable parameter for determining the extension of the trailing vortices in a more quantitative manner. For this purpose the vorticity (ζ) in ϕ-planes were calculated using the following equation:

$$\zeta = \frac{\partial \overline{V}}{\partial z} - \frac{\partial \overline{U}}{\partial r} \tag{2}.$$

The vorticity is very intensive within and very low outside trailing vortices. A limiting value of vorticity can be found that indicates the edges of the trailing vorticity. For the present study a value of $\zeta = 330$ 1/s was determined. It was then possible to visualise the trailing vortices with a contour plot in which contours lower than the limiting value of the vorticity (cut-off value) were erased.

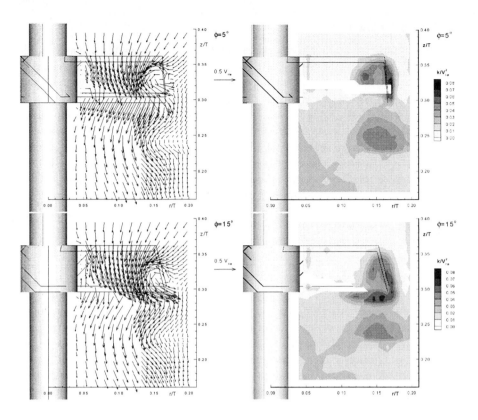

Fig. 6 Flow field at $\phi = 5°$ and $\phi = 15°$ **Fig. 7** Distribution of TKE at $\phi = 5°$ and $\phi = 15°$

From the vorticity contour plot the total extent of the trailing vortices generated behind each stirrer blade can be determined. Vorticity originates at the side and upper tips of the blade. The vortex extends to around 135° behind a blade and stays in an almost constant radial position. The axial inclination of the vortex axis to the horizontal after entering the flow around the impeller is approximately 20° which remains constant up to the total dissipation of the vortex in the bulk flow. From the impeller flow movie, which gives more details of the flow than shown in the present paper, one interesting fact has to be mentioned, that with increasing diameter of the vortex the distribution of the turbulence kinetic energy along the vortex radius becomes more uniform, which means that the difference between the peak value in the core of the vortex and the values at the vortex edges diminishes with distance from the blade.

Comparisons with the results of earlier investigations on the flow field produced by pitched blade turbines (e.g. flow visualisation results of Tatterson and co-workers) show good qualitative agreement in the formation and extent of the trailing vortices. However, it is difficult to do quantitative comparisons, since there are a number of differences in the individual experiments, such as different flow configurations (geometry of turbine, vessel, etc.) and different measuring techniques (flow visualisation, ensemble-averaged LDV/angle-resolved LDV).

4 Energy Dissipation

The distribution of the local energy dissipation rate ε within the stirred vessel is an important key variable for the design of mixing processes, in which gas and/or liquid dispersion, forces on particles or micro-mixing phenomena must be considered. In the past several methods have been used to estimate energy dissipation rates in stirred tank reactors by experimental approaches (see for example Kresta and Wood (1993)). One of the most commonly used expression for ε, which was also used in the present study, is Brodkey's (1975) recommendation based on dimensional analysis:

$$\varepsilon = A\frac{u'^3}{L} \tag{3},$$

where u' represents a characteristic turbulence velocity, L a characteristic length scale of the flow and A is a constant of proportionality. For the characteristic turbulence velocity the square root of measured turbulence kinetic energy k was chosen. The proportionality constant A can be set to 1 in jet flows according to Batchelor (1953). This was also assumed here by approximating the discharge of the impeller as a jet-type flow. For the characteristic length scale several methods have been applied in literature. The most accurate way is to determine the macro scale of turbulence from the autocorrelation coefficient. This method has been employed by Wu and Patterson (1989b) and Lee and Yianneskis (1998) for a Rushton turbine flow and Kresta and Wood (1993) for a pitched blade turbine flow. However, this method could not applied in the present study due to the absence of sufficient data rates provided by the instrument that is required to obtain the energy spectra. Kresta and Wood (1993) reported that the macro scale L is almost constant in the impeller discharge stream of a pitched blade turbine and that L is approximately equal to the width of the trailing vortices which were found to be of diameter D/10. It must be noted that the recent study of Lee and Yianneskis (1998) showed that this may not be valid in the discharge of a Rushton turbine.

In absence of more detailed information on the macro scale a constant length scale within the total vessel volume was assumed in the present study, but since the flow data was so detailed for the complete flow field this length scale could be determined via an energy balance calculation throughout the total volume V_{tot} of the stirred vessel:

$$L = \frac{\sum u_i'^3}{\varepsilon_{tot}}, \text{ with } \sum u_i'^3 = \frac{\sum V_i k_i^{3/2}}{V_{tot}} \tag{4}.$$

The total energy dissipation ε_{tot} was determined via an integral power measurement and introduced into the equation given above. Using this method a length scale of $L = 13.2$ mm $= 0.264\,D$ was calculated, which is exactly the blade height and similar to the diameter of the trailing vortex before detaching from the blade. The fact that the length scale corresponds to the diameter of the trailing vortex is in agreement with the result of Kresta and Wood, but on the other hand the magnitudes differ more than 100% from each other (0.1D compared to 0.26D).

The calculated length scale can be inserted in the above equation for calculating the local energy dissipation in each separate volume element V_i of the entire vessel. This will lead to contours which are similar to those for the turbulence kinetic energy shown in Figures 5 and 7 for the bulk flow and the impeller flow field, respectively. It must be pointed out, that although the used method for estimating the dissipation rate is very rough, it provides an approximate magnitude of local dissipation and together with the detailed flow studies presented in section 3 the regions of highest dissipation rates were resolved.

5 Concluding Remarks

The results presented here provide the most comprehensive and detailed quantitative data set that is available to date for the flow field produced by a pitched blade turbine. The trailing vortex system generated at each stirrer blade has been characterised in detail and it has been shown that the introduction of the vorticity function is a valuable tool for an accurate description of the extension of trailing vortices. The LDA-measurements confirm the findings of earlier investigations resulting from flow visualisations, but with the presented data it is now possible to give also detailed information on flow quantities within the trailing vortices, such as mean and fluctuating velocities and turbulence kinetic energy.

An estimate of the distribution of local energy dissipation rates was carried out. Although the method is rough it provides valuable information for engineers, since this parameter is essentially required for the lay-out of mixing processes like in gas and liquid dispersions.

For numerical simulations of stirred vessel flows, a data set is now available, that is required for accurate validation and improvements of the turbulence models used in such calculations. The results will therefore significantly support the on-going developments in the field of computational fluid dynamics and chemical reactor simulation.

Acknowledgements

The authors acknowledge financial support provided by the Commission of the European Union under the BRITE EURAM Programme, Contract number BRPR-CT96-0185. Further partners in this research project are Neste Oy, INVENT Umwelt- und Verfahrenstechnik GmbH & Co. KG, AEA-Technology, BHR-Group, EniChem SpA and PFD-Limited.

References

Ali, A. M., Yuan, H.-H. S., Dickey, D. S. & Tatterson, G. B. 1981, Liquid Dispersion mechanisms in Agitated Tanks: Part I. Pitched Blade Turbine, Chem. Eng. Commun., vol. 10, pp. 205 - 213.

Batchelor, G. K. 1953, The Theory of Homogenous Turbulence, Cambridge University Press, Cambridge.

Brodkey, R. S. 1975, Turbulence in Mixing Operations, Academic Press, New York.

Fournier, M.-C., Falk, L. & Villermaux, J. 1996, A New Parallel Competing Reaction System for Assessing Micromixing Efficiency - Determination of Micromixing Time by a Simple Mxing Model, Chem. Eng. Sci., vol. 51, no. 23, pp. 5187 - 5192.

Günkel, A. & Weber, M. 1975, Flow Phenomena in Stirred Tanks, Part I: The Impeller Stream, AIChE-J., vol. 21, pp. 931 - 949.

Hockey, R. M. & Nouri, J. M. 1996, Turbulent Flow in a Baffled Vessel Stirred by a 60° Pitched Blade Impeller, Chem. Eng. Sci., vol. 51, no. 19 pp. 4405 - 4421.

Lee, K. C. & Yianneskis, M. 1998, Turbulence Properties of the Impeller Stream of a Rushton Turbine, AIChE-J., vol. 44, no. 1, pp. 13 - 24.

Nouri, J. M. & Whitelaw, J. H. 1990, Effect of Size and Confinement on the Flow Characteristics in Stirred Reactors, Proc. Fifth Int. Symposium on Application of Laser Techniques to Fluid Mechanics, Lisbon, Portugal, pp. 23.2.1 - 23.2.8.

Kresta, S. M. & Wood, P. E. 1993, The Flow Field Produced by a Pitched Blade Turbine: Characterisation of the Turbulence and Estimation of the Dissipation Rate, Chem. Eng. Sci., vol. 48, pp. 1761 - 1774.

Ranade, V. V. & Joshi, J. B. 1989, Flow Generated by Pitched Blade turbines I: Measurements Using Laser Doppler Anemometer, Chem. Eng. Commun., vol. 81, pp. 197 - 224.

Schäfer, M., Höfken, M. & Durst, F. 1997, Detailed LDV Measurements for Visualization of the Flow Field within a Stirred-Tank Reactor Equipped with a Rushton Turbine, Trans. I.Chem.E., vol. 75, Part A, pp. 729 - 736.

Schäfer, M., Yianneskis, M., Wächter, P. & Durst, F., 1998, Trailing Vortices around a 45° Pitched-Blade Impeller, AIChE J., vol. 44, no. 6, pp. 1233 - 1246.

Stieglmeier, M. & Tropea, C. 1992, Mobile Fiber-Optic Laser Doppler Anemometer, Applied Optics, vol. 31, no. 21, pp. 4096 - 4105.

Stoots, C. M. & Calabrese, R. V. 1995, The Mean Velocity Field Relative to a Rushton Turbine Blade, AIChE J., vol. 41 no. 1, pp. 1 - 11.

Tatterson, G. B. Yuan, H.-H. S. & Brodkey, R. S. 1980, Stereoscopic Visualization of the Flows for Pitched Blade Turbines, Chem. Eng. Sci., vol. 35, pp. 1369 - 1375.

Van't Riet, K. & Smith, J. M. 1975, The Trailing Vortex System Produced by Rushton Turbine Agitators , Chem. Eng. Sci., vol. 30, pp. 1093 - 1105.

Wechsler, K., Breuer, M. & Durst, F. 1998, Steady and Unsteady Computations of Turbulent Flows Induced by a 4/45° Pitched Blade Impeller, submitted for publication.

Wu, H. & Patterson, G. K. 1989a, Laser-Doppler Measurements of Turbulent-Flow Parameters in a Stirred Mixer, Chem. Eng. Sci., vol. 44, no. 10, pp. 2207 - 2221.

Wu, H. & Patterson, G. K. 1989b, Distribution of Turbulence Energy Dissipation Rates in a Rushton Turbine Stirred Mixer, Exp. Fluids, vol. 8, pp. 153 - 160.

Yianneskis, M., Popiolek, Z. & Whitelaw, J. H. 1987, An Experimental Study of the Steady and Unsteady Flow Characteristics of Stirred Reactors, J. Fluid Mech., vol. 175, pp. 537 - 555.

Yianneskis, M. & Whitelaw, J. H. 1993, On the Structure of the Trailing Vortices around Rushton Turbine Blades, Trans. I.Chem.E., vol. 17, Part A, pp. 543 - 550.

IV.5. Phase-Resolved Three-Dimensional LDA Measurements in the Impeller Region of a Turbulently Stirred Tank
J.J. Derksen, M.S. Doelman, and H.E.A. van den Akker

Kramers Laboratorium voor Fysische Technologie, Delft University of Technology, Prins Bernhardlaan 6, 2628 BW Delft, The Netherlands

Abstract. Predictions on the quality of mixing in a turbulently operated stirred tank ask for a detailed knowledge of the flow field, including its turbulence characteristics. In this paper we present the results of three-dimensional, angle-resolved LDA measurements in the vicinity of a Rushton turbine in a baffled tank at $Re=29,000$. For optimal coincidence of the signals in the three velocity channels, an accurate laser beam alignment procedure was developed. The average flow field is characterized in terms of velocity and vorticity fields. In the characterization of the turbulence, the emphasis is on the anisotropy of the Reynolds stress tensor.

Keywords. Three-dimensional LDA, beam alignment, Reynolds stresses, anisotropy, mixing

1 Introduction

Understanding the fluid dynamics of turbulently operated stirred tanks is inevitable for optimizing this widely used piece of process equipment. Investigations into stirred tank flow often exploit numerical (computational fluid dynamics) as well as experimental techniques. A key role for experiments is in validating simulations. With the availability of large and fast computational resources, simulations are able to resolve more and more details of the turbulent flow field (Eggels 1996, Derksen & Van den Akker 1998). As a result, the demands on the quality and resolution of the experiments increase. Not only accurate experimental data on the average flow fields, but also on turbulence characteristics, need to be available. With respect to turbulence characteristics, we have focused on the Reynolds stress tensor, and, more specifically, on the extent to which it is (an-)isotropic.

In this paper, we report on phase-resolved, three-dimensional LDA experiments in the impeller outflow region of a stirred tank at $Re=29,000$. In a three-dimensional LDA setup, the full Reynolds stress tensor can be measured in one go. Additional advantages of a three-dimensional setup are the possibility for direct velocity bias correction, and an enhanced spatial resolution when operating the LDA equipment in an off-axis scattering mode. In a three-dimensional setup, laser beam alignment is an important and difficult issue, especially when the working fluid has an index of refraction that strongly differs from its environment. An accurate alignment procedure, based on a submerged micro-mirror system, was developed for this goal.

2 Flow Geometry

The flow geometry is depicted in Fig. 1. The baffled, flat-bottomed tank was filled with water up to a level $H=T=288$ mm. At the top level there was a free surface. The tank was entirely made of glass. To reduce refraction at the cylindrical side-wall, the tank was surrounded by a square glass box filled with water. The impeller was set to rotate with an angular velocity of $N=2\pi\Omega =3.138(\pm0.002)$ rev/s. The Reynolds number, which for this flow system is traditionally defined as $Re=ND^2/\nu$, amounted to $2.9\cdot10^4$. The velocities in the tank will all be scaled with the impeller tip speed $v_{tip}=\pi ND$.

The Froude number ($Fr=N^2D/g$, with g the gravitational acceleration) was 0.1. It was observed that, under the conditions described above, the free surface was not distorted by air entrainment.

Together with each individual velocity measurement, the angular position of the impeller was recorded. This was done by resetting every impeller revolution a 16

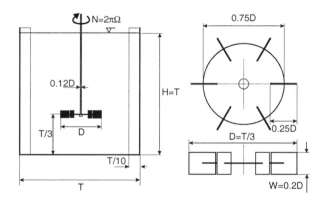

Fig. 1. The flow geometry. The tank (left) is equipped with four baffles to prevent solid body rotation of the fluid. The impeller (right) is a Rushton turbine. The diameter of the tank is $T=288$ mm. The thickness of the disk and the impeller blades is 2 mm. The impeller was mounted on the shaft without a hub.

bit clock, running at 10 kHz and writing two clock values to file. The first value represents the time since the last clock reset, the second value is the time of the last full impeller revolution. Together they determine the impeller angle at the instant of the velocity data acquisition.

3 LDA Setup

The sending optics of the LDA setup consisted of an Ar-ion laser, operating in all-lines mode, a beam splitter system (TSI ColorBurst model 9201) and an optical fiber system that guided the beams to the flow.

The ColorBurst selected the three major wavelengths of the laser and produced three pairs of beams. One beam of each pair was given a frequency pre-shift of 40 MHz. All beams were coupled into single-mode, polarization preserving optical fibers. At the flow system, they were coupled out in two laser probes. The initial spacing between the two beams in each pair was fixed to 50 mm. From probe #1 two beam pairs emerged. They were focused by a 500 mm lens, and entered the flow through the (flat) bottom of the tank (see Fig. 2a). This probe allowed for measuring the radial and tangential velocity component. The axial velocity component was measured with the single beam pair emerging from probe #2, which entered the vessel through the side-wall. This probe was equipped with a 250 mm front lens. If all beams show maximum overlap, a three-dimensional measurement volume of approximately 100^3 μm^3 has been created.

Both probes contained the receiving optics as well. A lens system projected the measurement volumes onto flat, multi-mode fiber-ends. These fibers guided scattered light to three photo multiplier tubes (one for each wavelength). The signals emerging from the photo multipliers were fed into the signal processor (an IFA 750 by TSI). The LDA system was operated in side scatter mode.

The flow was seeded with aluminum coated polystyrene particles with a mean diameter of 4 μm. They were chosen because of their good light scattering properties. A disadvantage of these particles is their relatively high density ($2.6 \cdot 10^3$ kg/m^3) which forced us to keep the impeller running during the course of the experimental sessions to prevent sedimentation.

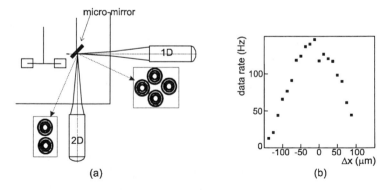

Fig. 2. (a) The alignment system: two fiber optic LDA probes (a 1D probe and a 2D probe) are aligned with help of a micro-mirror. (b) Data rate in coincidence and side-scatter mode as a function of beam misalignment Δx.

3.1 Laser Beam Alignment

Optimum overlap of the three measurement volumes is crucial for the signal quantity (in terms of the data-rate in coincidence mode) and quality (e.g. with respect to spatial resolution and geometric bias effects).

The alignment was controlled and checked by means of an aluminum micromirror with elliptical shape (long axis: 70 μm, short axis: 50 μm) deposited on a sheet of glass, that was placed in the tank at the measuring position (see Fig. 2a). Optimal alignment was achieved when all the six laser beams that were involved in the experiment, showed (in reflection) a concentric diffraction pattern. Figure 2b (the data rate in coincidence mode at a certain position in the vessel as a function of beam misalignment Δx) gives an impression of the quality and resolution of the alignment system.

3.2 Data Processing

In the vicinity of the impeller tip a measurement grid was defined in a vertical plane, midway between two baffles, see Fig. 3. In Fig. 3 the coordinate system that will be used for presenting the flow field results is defined as well. The spacing of the grid in axial and radial direction was $\Delta z = \Delta r = 2.5$ mm, i.e. $\Delta z \approx W/8$ (with W the height of an impeller blade). In tangential direction the resolution was 3° (at the inner bound of the grid this corresponds with a spacing of 2.2 mm, at the outer bound with 3.9 mm). Per grid point, i.e. per measurement position and tangential interval, $6 \cdot 10^3$ velocity samples were collected.

Burst type LDA data in turbulent flows are inflicted by velocity bias: the particle transfer rate through the measurement volume is approximately proportional to the absolute value of the velocity. Therefore, unweighted averaging will lead to biased results. The average data presented in this paper were all corrected for velocity bias with a velocity weighing scheme, as introduced by McLaughlin & Tiederman (1973). Note that, with this bias correction procedure, explicit use is made of the availability of three velocity components in the same sample, as the inverse of the absolute velocity is the weighing factor. In the impeller outstream, the bias corrected mean tangential and radial velocity

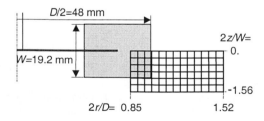

Fig. 3. The measurement grid was located in a vertical plane midway between two baffles. The grid spacing is $\Delta r = \Delta z = 2.5$ mm.

components were significantly smaller than their unweighted equivalents. The differences amounted up to $0.1 \cdot v_{tip}$.

4 Anisotropy Characterization

The major reasons for measuring Reynolds stresses can be found in the field of turbulence modeling. An ongoing debate in the field of modeling complex turbulent flows, such as the flow under investigation, is whether second-order closure models have significant added value over first-order (e.g. k-ε) models. Since the standard k-ε model is an eddy-viscosity model, it locally assumes isotropic turbulent transport. The k-ε model is known to be inappropriate in rotating and/or highly three-dimensional flows (Wilcox 1993). Assessment of the isotropy assumptions by experiment is therefore necessary.

The Reynolds stress data will be presented in terms of the anisotropy tensor a_{ij} and its invariants (Lumley 1978). The anisotropy tensor is defined as

$$a_{ij} = \frac{\overline{u_i u_j}}{k} - \frac{2}{3}\delta_{ij} \tag{1}$$

with k the turbulent kinetic energy: $k = \frac{1}{2}\overline{u_i u_i}$; u_i the fluctuations about the mean of the i-th velocity component (i.e. $U_i = \overline{U}_i + u_i$); and δ_{ij} the Kronecker delta. The anisotropy tensor has a first invariant equal to zero by definition. The second and third invariant respectively are $A_2 = a_{ij} a_{ji}$ and $A_3 = a_{ij} a_{jk} a_{ki}$. The range of physically allowed values of A_2 and A_3 is bounded in the (A_3, A_2) plane by the so-called Lumley triangle, see Fig. 4. The boundaries that appear to be most relevant in the current study are the ones associated with axisymmetric turbulence.

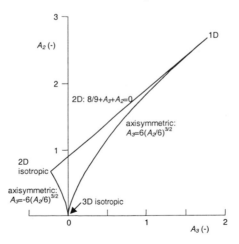

Fig. 4. The plane with coordinates A_3, A_2 (i.e. the invariants of the anisotropy tensor) was used to characterize the anisotropy of the turbulence. The physically possible states are bounded by the Lumley triangle (Lumley 1978).

Axisymmetric turbulence is defined as a situation in which the kinetic energy in two orthogonal directions is equal. Hence, a_{ij} reads

$$a_{ij} = \begin{pmatrix} -\alpha & 0 & 0 \\ 0 & \alpha/2 & 0 \\ 0 & 0 & \alpha/2 \end{pmatrix} \qquad (2)$$

(the reference frame is chosen such that the axis of symmetry of the turbulence coincides with the direction of the first coordinate). The energy contained in the first coordinate direction is $\left(\dfrac{1}{3} - \dfrac{1}{2}\alpha\right)k$, in the second and third direction the energy is $\left(\dfrac{1}{3} + \dfrac{1}{4}\alpha\right)k$. As a result of equation 2, $A_2 = \dfrac{3}{2}\alpha^2$ and $A_3 = -\dfrac{3}{4}\alpha^3$.

Consequently $A_3 = 6\left(A_2/6\right)^{3/2}$ if $\alpha < 0$, i.e. the energy contained in the symmetry direction is more than $k/3$. This is the right branch of axisymmetric turbulence (see Fig. 4). If $\alpha > 0$, then $A_3 = -6\left(A_2/6\right)^{3/2}$ This constitutes the left branch, where the energy contained in the symmetry direction is less than $k/3$. As a reference, we calculate the point on the right axisymmetric branch where the energy contained in the symmetry direction is twice the energy in the other two directions. At this point $\alpha = -1/3$, and therefore $A_3 = 1/36$ and $A_2 = 1/6$.

5 Results

5.1 Average Flow Field

The structure of the average flow field, relative to an impeller blade (see Fig. 5), qualitatively corresponds with the results presented by Stoots & Calabrese (1995) and Lee & Yianneskis (1998). The wake behind an impeller blade can be clearly observed in the top vector plot. In Lee & Yianneskis (1998) a velocity vector plot at the level $2z/W=+0.51$ (i.e. *above* the impeller disk) in the tank is presented. Their wake is somewhat larger than the wake observed in Fig. 5, which can be most likely attributed to the flow's asymmetry with respect to the impeller disk. The bottom part of Fig. 5 shows the development and advection of a strong vortex behind the impeller blade.

The strength of the vortex can be expressed in terms of the vorticity, which could be calculated because three-dimensional velocity data on a three-dimensional grid were available. The vorticity was calculated from the average flow field with a central differencing scheme. This way, vorticity data reside on a grid that is staggered with respect to the velocity measurement grid depicted in Fig. 3. The vorticity field in a single horizontal plane (Fig. 6) shows that the trailing vortex is associated with a concentrated region of high vorticity. Note that at the lower right corner of the vector field the vortex from the preceding blade can still be distinguished.

Yianneskis *et al.* (1987) proposed to characterize the curve along which the trailing vortex is swept into the tank by connecting the points in a *horizontal* plane

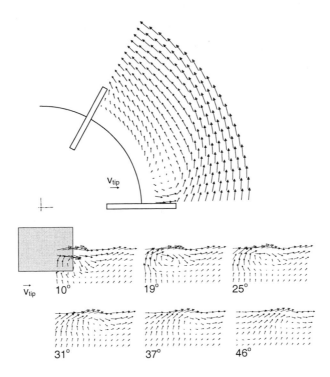

Fig. 5. The average flow field. Top: velocity relative to the impeller in a horizontal plane at the level $2z/W=-0.52$. Bottom: velocity in six vertical planes at different angles with respect to the leading impeller blade. For reference, the position of the blade at the zero-angle position is drawn in the vector plot at 10°.

with a mean axial velocity component equal to zero. An assumption was that the vortex core did not move in the vertical direction. In Fig. 7 the curves found in literature (Yianneskis *et al.* 1987; Stoots & Calabrese 1995; Lee & Yianneskis 1998; Van 't Riet & Smith 1975) are compared with the one deduced from the present flow field. The strong deviation between curve [5] and the other curves is commonly attributed to inertia of the relatively large particles (needed for a photographic technique) used by Van 't Riet and Smith. We calculated, by linear interpolation, the position of zero axial velocity in the horizontal plane at a distance $2 \cdot \Delta z$ below the impeller disk (i.e at the $2z/W=-0.52$ level). According to Fig. 5 this approximately was the plane in which the vortex core moved. Yianneskis *et al.* (1987) focused on the vortex that originated from the upper part of the impeller blade. They claim that its core moved in the $2z/W=1$ plane, which again clearly demonstrates the asymmetry of the flow with respect to the horizontal plane containing the impeller disk. The asymmetry is also reflected in the curves as given in Fig. 7. Curves [1] and [2] were taken form the upper vortex, curves [3] and [4] from the lower. It should also be noted, however, that Stoots & Calabrese (1995) used a flow geometry which differed quite strongly from the rest. Its disk had a diameter of $0.6 \cdot D$, while in the present work it was $0.75 \cdot D$; their bottom clearance was $T/2$.

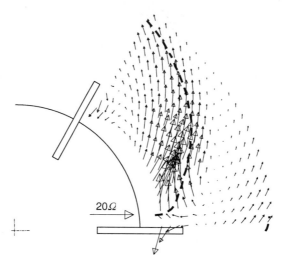

Fig. 6. The vorticity field in a horizontal plane at $2z/W=-0.39$. The vorticity is scaled with the angular velocity Ω $(N=2\pi\Omega)$. The dashed line is the position of the vortex core in the plane $2z/W=-0.52$.

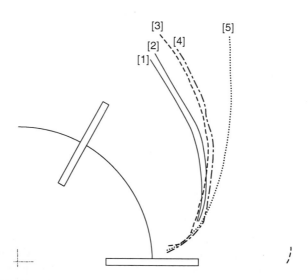

Fig. 7. The vortex core position deduced from several investigations reported in literature and from the present work. [1] Lee and Yianneskis (1998); [2] Yianneskis *et al.* (1987); [3] present work; [4] Stoots and Calabrese (1995); [5] Van 't Riet and Smith (1975).

5.2 Turbulence Characteristics

First, the regions of high turbulent activity will be identified. This is done by plotting the turbulent kinetic energy $k = \dfrac{1}{2}\overline{u_i u_i}$ as a function of the position, see

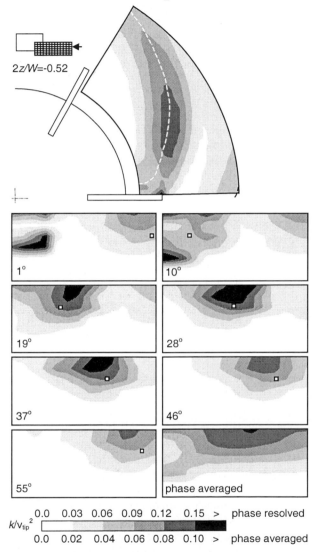

Fig. 8. The distribution of the turbulent kinetic energy $k = \dfrac{1}{2}\overline{u_i u_i}$ in the horizontal plane at $2z/W=-0.52$ (top figure); in seven r-z planes with different angles with respect to the impeller blade; and in the r-z plane averaged over all impeller angles. The vortex core position is indicated with a dashed line in the top figure. In the r-z planes it is indicated with a small white square.

Fig. 8. In the horizontal plane with $2z/W=-0.52$ it can be observed that the region with high turbulent activity more or less coincides with the trailing vortex path. The vertical sections through the flow field show a concentrated region of high k just above the vortex core. We speculate that this distribution of kinetic energy is (at least partly) due to random fluctuations in the vortex core position. Combined with the steep gradients in the flow field right above the vortex core (see Fig. 5), erratic motion of the core position will locally lead to high levels of velocity fluctuation.

The local anisotropy of the turbulence has been characterized with the second and third invariant of the anisotropy tensor (A_2 and A_3, see section 4). Both parameters define a location within the Lumley triangle (Fig. 4). A subset of all measured locations has been depicted in Fig. 9. It can be observed that the most common situation in the present flow system is close to the right-hand-side boundary of the Lumley triangle. At this boundary the turbulence is axisymmetric, with the energy contained in the direction of the axis of symmetry being highest. Many points in Fig. 9 go beyond the state of anisotropy characterized by $A_3=1/36$ and $A_2=1/6$. At this latter state, the energy contained in the symmetry direction is equal to the sum of the energies contained in the other two directions (see also section 4).

From Fig. 9, it can be concluded that in the near wake of an impeller blade the anisotropy is strongest. Also the core of the vortex can be associated with relatively high anisotropy of the turbulence. If we focus on the horizontal plane with $2z/W=-0.52$ (i.e. the plane in which the vortex moves) and the angular positions $22°$ and $40°$, we observe that the extreme locations of anisotropy are encountered at $2r/D=1.16$ and at $2r/D=1.37$ respectively, whereas the vortex core at the corresponding lines is at $2r/D=1.13$ and $2r/D=1.33$ respectively. Away from the vortex core, the trajectories of Fig. 9 indicate that the turbulence tends more and more to isotropic.

6 Conclusions

Detailed data on the mean flow and the Reynolds stresses in the complex flow field in the vicinity of a Rushton turbine have been obtained. The three-dimensional setup allowed for measuring the full Reynolds stress tensor, and, as a result, an unambiguous characterization of the anisotropy of the turbulence.

Special attention was paid to beam alignment. A submerged micro-mirror system was developed to approach optimum beam overlap. Its resolution is of the order of 10 µm (i.e. about 1/10 of the width of the measurement volume). Operation of the LDA system in side scatter mode resulted in a measurement volume with a linear dimension of approximately 0.1 mm in all three coordinate directions.

The average, phase-resolved flow field is characterized by a vortex that originates from the near wake of the impeller blade. This coherent structure could be clearly identified behind the blade over an angular range of more than $60°$. As has been observed before (Lee & Yianneskis 1998), with the trailing vortex a

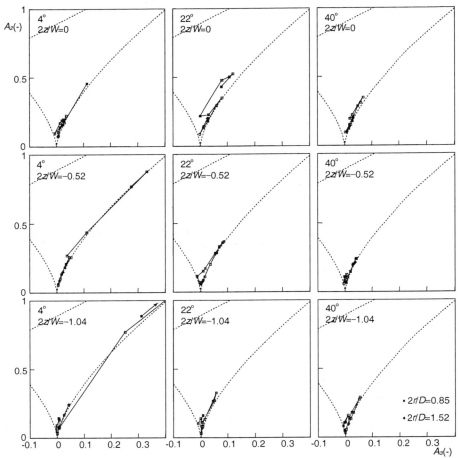

Fig. 9. The locations within the Lumley triangle (the open squares) along radial trajectories in the flow field at three levels ($2z/W=0$, $2z/W=-0.52$, and $2z/W=-1.04$), and at three different angles ($4°$, $22°$, and $40°$) with respect to an impeller blade. The dotted lines denote the boundaries of the Lumley triangle. The drawn lines connect neighboring grid points. The points on the innermost radius of the measurement grid are denoted with ■, the outermost points with ♦.

region of high turbulent activity is associated. The strongest turbulent fluctuations were measured closely above the vortex core.

The turbulence in several flow regions close to the impeller was demonstrated to be anisotropic. Especially in the near wake of an impeller blade, and near the trailing vortex core the turbulence tends to a strongly anisotropic, axisymmetric state, i.e. the turbulent kinetic energy contained in two orthogonal directions is almost equal and significantly different from the energy in the third direction.

References

DERKSEN, J.J. & VAN DEN AKKER, H.E.A. 1998 Parallel simulation of turbulent fluid flow in a mixing tank. *Lecture Notes in Computer Science* **1401**, 96.

EGGELS, J.G.M. 1996 Direct and large eddy simulation of turbulent fluid flow using the lattice-Boltzmann scheme. *Int. J. Heat & Fluid Flow* **17**, 307.

LEE, K.C. & YIANNESKIS, M. 1998 Turbulence properties of the impeller stream of a Rushton turbine. *AIChE J.* **44**, 13.

LUMLEY, J. 1978, Computational modeling of turbulent flows. *Adv. Appl. Mech.* **26**, 123.

MCLAUGHLIN, D.K. & TIEDERMAN, W.G. 1973 Bias correction methods for laser Doppler velocimeter counter processing, *Phys. Fl.* **16**. 2082.

STOOTS, C.M. & CALABRESE, R.V. 1995 Mean velocity field relative to a Rushton turbine blade. *AIChE J.* **41**, 1.

VAN 'T RIET, K. & SMITH, J.M. 1975 The trailing vortex system produced by Rushton turbine agitators. *Chem. Eng. Sci.* **30**, 1093.

WILCOX, D.C. 1993 *Turbulence modeling for CFD*. La Cañada (CA), DCW Industries.

YIANNESKIS, M., POPIOLEK, Z. & WHITELAW, J.H. 1987 An experimental study of the steady and unsteady flow characteristics of stirred reactors. *J. Fluid Mech.* **175**, 537.

CHAPTER V.   COMBUSTION AND ENGINES

V.1. On the Extension of a Laser-Doppler Velocimeter to the Analysis of Oscillating Flames

E.C. Fernandes and M.V. Heitor

Instituto Superior Técnico, Mechanical Engineering Department,
Av. Rovisco Pais, 1049 - 001 Lisboa; Portugal

ABSTRACT

Optical and probe techniques are used to analyse the coupling mechanism between pressure, velocity and temperature fluctuations typical of pulsed flames, through the combination of laser velocimetry, digitally-compensated thermocouples and a probe microphone. The system is based on a digital signal microprocessor and is shown to run up to 12.5 kHz of data ready velocity signals, allowing the time-resolved representation of vectors of turbulent heat flux along a typical cycle of oscillation. The detailed results obtained along the oscillating reacting shear layer show time sequences of zones characterised by gradient and non-gradient turbulent heat flux, which are structurally similar to previous results reported in the literature for steady recirculating flames.

1. INTRODUCTION

Pulsating flows have been the subject of many investigations covering a wide range of situations, such as: acoustics (Herzog et al., 1996), heart valves (Hirt et al., 1996), in-cylinder combustion chambers (e.g. Liou and Santavicca, 1985, Witze, 1984), unstable shear layers (Hussain and Zaman, 1980), pulsed flames (e.g. Lovett and Turns, 1993), pulse combustors (Keller and Saito, 1987) and afterburners (Heitor et al., 1984, Gutmark et al., 1991 and Sivasegaram and Whitelaw, 1987). Different diagnostic techniques have been used, depending on the purpose of the investigation and on the nature of the flow. For example, the quantification of heat release in pulsed flames has been made through PLIF of OH and CH concentrations (e.g. Gutmark et al., 1989). The velocity field has been measured with hot-wire velocimetry in isothermal flows (e.g. Hussain and Zaman, 1980 and Kya and Sasaki, 1985), and with laser velocimetry in oscillating flows (Dec et al., 1991). Temperature measurements have been reported with TLAF technique (Dec and Keller, 1990) and with fine bare-wire thermocouples (e.g. Ishino et al., 1996). High speed Schlieren cinematography has also been used (e.g. Keller et al., 1982 and Ganji and Sawyer, 1989) to record the structure of the reacting flowfield, together with phase average digital photography as a non-expensive visualization technique (Chao et al., 1991).

While most of the works mentioned above report the acquisition of single variables, other have attempted to correlate multiple signals (e.g. Lang and Vortmeyer, 1987; Keller, 1995), or to control the process of vortex shedding associated with unsteady flames (e.g. Schadow and Gutmark, 1992), but in general there is a need of reliable data on the coupling mechanisms between pressure, velocity and heat release fluctuations in order to better understand the onset of combustion induced oscillations. In this context, this paper analyzes the extension of a conventional laser velocimeter to the simultaneous measurement of the time-resolved fluctuations of velocity, pressure and temperature, in unsteady reacting flows. The system was developed in order to characterize natural unsteady flames such as those presented in figure 1 downstream of two distinct flame holders, but both of them driven by similar large-scale structures that develop along the shear layer and modulate the process of heat release. The following paragraphs describe the experimental method, including the data acquisition system, and present sample results obtained in the reacting shear layer of the flames of figure 1.

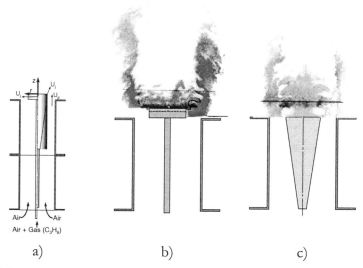

a)   b)   c)

Fig. 1. Schematic diagram of flame holders and image of flames studied.

a) *Diagram of flame holders, conical and cylindrical, with identification of principal variables*
b) *Short time film exposure (1/8000sec) of an unsteady flame- Flame A (cylindrical burner)*
c) *Short time film exposure (1/8000sec) of an unsteady flame- Flame B (conical burner)*

2. EXPERIMENTAL METHOD and PROCEDURES

2.1 The Flames considered

The flames considered throughout this paper were stabilized downstream flameholders located at the end of a 0.52m long tube (see figure 1a), which was placed on the top of a plenum chamber (for details see Fernandes, 1998). The resulting flame, burning a mixture of propane and air, is open to the atmosphere, and offers the advantage of easy access to the techniques described below.

For the range of unsteady conditions considered here, the spectrum of the pressure fluctuations in any location of the pipe wall is associated with the excitation of a predominant frequency in the range of 265-275 Hz (Fernandes, 1998), which is associated with a longitudinal standing half-wave. The flames are then located in a velocity antinode and the resulting time-resolved characteristics are described below. Both flames are also characterised by a sound pressure level of about 110dB. The analysis consider boundary conditions of $\emptyset_i = 8$, $U_j =8.8$m/s and $U_p =2.9$m/s for the Flame A presented in figure 1b and $\emptyset_i = 6$, $U_j =15$m/s and $U_p=3.4$m/s, for Flame B of figure 1c.

2.2 Experimental Techniques

Figure 2 shows schematically the various experimental techniques used throughout this work. Time-resolved velocity information was obtained with a laser-Doppler velocimeter, which comprised an Argon-Ion laser operated at a wavelength of 514.5 nm and a power of around 1W. A fiber optic (DANTEC) was used to guide the beam to an optical unit arranged with a two beam system with sensitivity to the flow direction provided by light-frequency shifting from a Bragg cell at 40MHz, a 310 mm focal length transmission lens, and forward-scattered light collected by a 300 mm focal length lens at a magnification of 1.0. The half-angle between the beams was $5.53°$ and the calculated dimensions of the measuring volume at the e^{-2} intensity locations were 2.3 and 0.219 mm. The output of the photomultiplier was mixed with a signal derived from the driving frequency of the Bragg cell and the resulting signal processed by a commercial frequency counter (DANTEC 55296) interfaced with a 16-bit DSP board. Measurements were obtained with the laser beams in the horizontal and vertical planes and by traversing the control volume along the horizontal and vertical directions to allow the determination of the axial, U, and radial, V, time-resolved velocities, respectively.

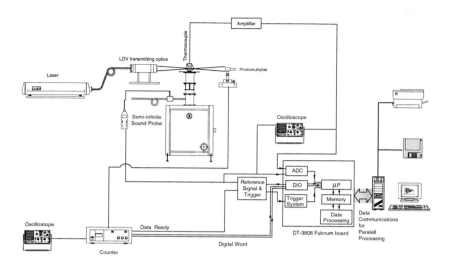

Fig. 2. *Schematic drawing of experimental apparatus with identification of the instrumentation used for simultaneous measurements of velocity-temperature-pressure in unsteady flows. Source: Fernandes (1998).*

Temperature measurements were obtained making use of fine-wires thermocouples, with 38 μm in diameter, made of Pt/Pt-13%Rh. The thermocouple output signal was digitally compensated from thermal inertia, following the procedure first outlined by Heitor et al. (1985), but optimised and used by Ferrão and Heitor (1998) and Caldeira-Pires and Heitor (1998) in premixed and non-premixed flames, respectively. The related uncertainities are quantified elsewhere (see for example Ferrão and Heitor, 1998 and Fernandes, 1998) and shown not to be higher than 60K for time-averaged values at the maximum temperature obtained in the open flames considered here, and up to 15% for the variance of the temperature fluctuations. In fact, the largest random error incurred in the values of temperature-velocity correlation, as higher as 15%, are due to the spatial separation of the measurements locations of the temperature and velocity because of the thermocouple junction must lie outside the measuring control volume of the laser velocimeter (e.g. Caldeira-Pires and Heitor, 1998, Ferrão and Heitor, 1998).

The sound intensity signal from the flame was acquired with a semi-infinite probe with a flat response up to 1kHz. The system is based on a free-field condenser microphone (B&K 4130) and a pre-amplifier (B&K 2130) with a flat response over a frequency band of 20Hz to 10kHz (Fernandes, 1998). This signal was used as a reference signal for data acquisition and post-process. The instantaneous pressure fluctuations correspond to signals measured at middle of the pipe, where the amplitude of the pressure fluctuations is maximum.

The complete measuring system was mounted in a three-dimensional traversing unit, allowing an accuracy of the measuring control volume within ±0.25mm.

2.3 Data Acquisition and System Control

The Doppler digital word and the two scalar signals, pressure and temperature, where acquired simultaneously and post-processed making use of a microprocessor, Texas Instruments-TMSC320C40, which also controls the LDV counter and the phase encoder box. The scalar signals where digitised with a sample-and-hold 16bits A/D converter at a rate of 40kHz/channel and stored in a circular memory buffer. The digital Doppler signal (8 bits mantissa and 4 bits exponent) is acquired with a dynamic digital input port.

According to the scheme presented in figure 3, the system initiates the DSP internal timer counter after a initial start flag is set to a logic value "true" on the transition of the pressure signal from positive to negative, controlled by the phase encoder box. Meanwhile, the process of analogue acquisition, which is running at a rate of 40kHz/channel is used to save temperature and pressure data into a circular memory by the action of an internal DMA co-processor. Then, for each LDV burst event, the sequence of actions is as follows:

- the time at which the burst event occurs is stored

- the DMA pointer of the exact location in the circular buffer at that time is stored

- the DSP external trigger is set to a logic value of "true", causing the counter to be stopped after 100ns

- the digital word is read and converted into velocity

- the pressure signal value at that time is read from the circular buffer

- two points before and after the temperature value at that time is also read from the circular buffer

- time, velocity, pressure and temperature are transferred to vector memory and calculations of the time derivative of temperature are made.

388

The process is concluded after a total acquisition time, T_{acq}, is concluded, following which all internal control variables are reset. The cycle repeats until a total population size of about N points (10240 or 20480) is acquired. For the measurements presented in this paper, the complete system runned up to 12.5kHz of data ready signals, with the delay of the board to the data ready signal less than 100ns, and with the window resolution between velocity and scalars less than 1/40kHz.

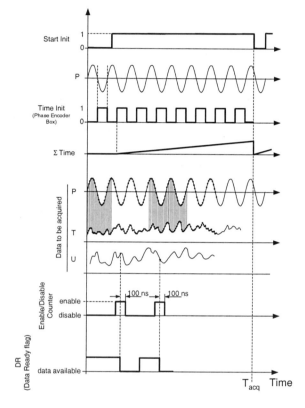

Fig. 3. Temporal diagram for simultaneous measurements of velocity (random data), pressure and temperature (continuous signals) in unsteady flows, and for hardware control

3. RESULTS AND DISCUSSION

Figure 4 show results of velocity acquired with different data rates and of temperature (non-compensated signal) acquired with a constant data rate of 40kHz. Both signals were obtained for unsteady Flame A and show a quasi-periodical variation in time. The amplitude and frequency are variable in time and, together with the random nature of the Doppler signals, imply that the detailed analysis of the time-resolved local flame characteristics has required the development of a purpose-built system for frequency analysis and phase-averaging, as discussed in the following paragraphs.

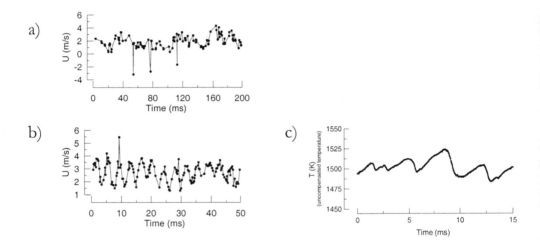

Fig. 4. Sample of time-resolved measurements of velocity and temperature in an unsteady flame oscillating with a frequency of about 275Hz±1.24Hz

   a) Axial velocity time series acquired with a mean data rate of 700Hz – (Flame A, r/R=0.92, z/R=0.53)

   b) Axial velocity time series acquired with a mean data rate of 3500Hz– (Flame A, r/R=0.92, z/R=0.53)

   c) Temperature time series acquired with a constant data rate of 40kHz– (Flame B, r/R=1.24, z/R=0.58)

3.1 Frequency Analysis

Although various weighting methods have been proposed to correct for velocity bias effects (e.g. Durst et al., 1981), no corrections were applied to the measurements reported here. The systematic errors that could have arisen were minimised by using high data acquisition rates in relation to the fundamental velocity fluctuation rate, as suggested for example by Erdman and Tropea (1981). This could be easily achieved because the rate of naturally occurring particles was sufficiently high for the flow conditions considered here.

Spectral analysis of LDV signals was carried out by resampling the time series after a linear interpolation with minimum interval time given by the mean data rate.

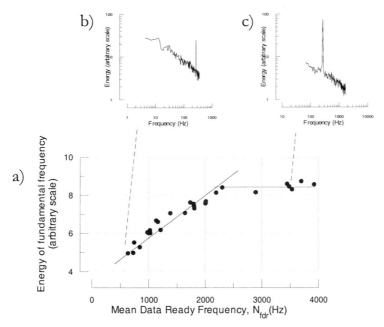

Fig. 5. Spectral analyses of LDV data as a function of mean data rate for typical unsteady flame conditions (Flame A) characterised by a fundamental frequency of $1/T_{osc} = 265 \pm 1.24 Hz$

a) *Energy of fundamental frequency as a function of mean data rate*

b) *Velocity spectrum for $N_{fdr} = 700 Hz$ ($N_{fdr}T_{osc} = 2.64$)*

c) *Velocity spectrum for $N_{fdr} = 3500 Hz$ ($N_{fdr}T_{osc} = 13.2$)*

Figure 5 show results of peak amplitude at 265Hz as a function of the data ready frequency, N_{fdr}, of the counter for a typical location in the reacting shear layer studied through this work. The measurements were obtained for conditions corresponding to data ready signals associated with each new Doppler burst. The results show that the predominant flow frequency of 265Hz could be identified in the spectra of the velocity fluctuations for any data rate, although the energy associated with this frequency (in a band of ±2.4 Hz) is independent of that rate only for $N_{fdr} >$ 2.3kHz. This agrees with the conditions given by, for example, Adrian and Yao (1989) for which the output of the counter yields a satisfactorily spectral analysis for the frequency range considered here.

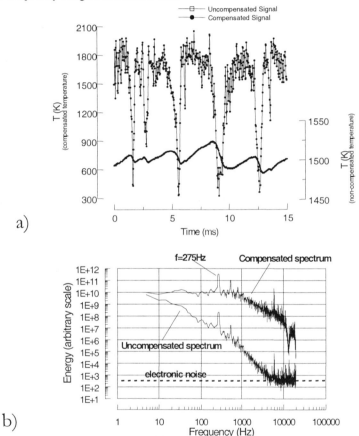

Fig. 6. Results of digital compensation of thermocouple thermal inertia obtained for Flame A at z/R=0.58, r/R=1.24.

a)   Comparison of typical non-compensated and compensated temperature time-series for a narrow time window of 15ms.

b)   Comparison of typical non-compensated and compensated temperature spectra

Spectral and temporal analysis of temperature time-series requires a compensation procedure as outlined by Ferrão and Heitor (1998) because of the thermocouple time-lag. In figure 6a the results are those of original temperature time-series, as acquired by the thermocouple and the corresponding digitally compensated time series. While there are significant differences in the amplitude domain, as also reported by the results of figure 6b, the phase compensation is quite insensitive to the numerical compensation because the dominant frequency (275Hz) is quite above 100Hz, as already pointed pout by Lovett and Turns (1993) and further discussed by Fernandes (1998).

3.2 Phase-Averaging Process

Important sources of uncertainity that may arise when measurements are taken in periodic flows include *"cycle-to-cycle" variations*, *temporal-gradient bias* and *phase bias,* and which can affect the phase-averaging process. The former are typically associated with in-cylinder flows in combustion engines, due to intake and exhaust process. To quantify this type of uncertainty, data acquisition and processing should include FFT analysis to separate the periodic from the non-periodic component of the flow (Dimopoulos et al., 1996). The *temporal-gradient bias* arises when phase-locked ensemble-averaging methods are employed to process signals with large temporal gradients, over a finite time window. The phase bias problem was addressed, for example, by Hussain and Zaman (1980) and consists essentially in a "jitter" effect, i.e. phase variations that generates virtual fluctuations. In the case reported here, this problem may arise through the pressure fluctuations, since this variable is used as a reference signal and is not strictly constant, therefore contaminates the velocity and temperature signals. In the work presented here, these uncertainities were quantified and shown not to affect the results.

A main question that arises in the detailed analysis of the present periodic flames is related with the adequacy of the pressure signal emitted by naturally pulsed flames as a reference signal for velocity measurements and for the phase-averaging process. This is because frequency variations can affect the velocity signal, creating a virtual turbulent fluctuation that is known to be a function of acquisition time (Fernandes, 1998). While pressure signals are only used as absolute reference signals, the observed spread of data points can have a significant influence on the associated velocity, affecting in this case the velocity rms. Analysis have shown that the influence of frequency uncertainties on the accuracy of the acquired data can be minimized if the acquisition time T_{acq} (see figure 3) is kept lower than 40ms (Fernandes, 1998). However, the choice of this value can have a dramatic influence on the total time spent to acquire N data points, as can be observed in the results of figure 7. This figure shows results obtained simulating a constant data rate of LDV triggers of about 6kHz, which represent the time taken by the hardware-software to perform all step functions specified in figure 2, in obtaining 10240 data points. Here a value of 40ms was chosen to allow a statistically independent sample of, at least, 10240 data points (Yanta and Smith, 1978), as shown in figure 8.

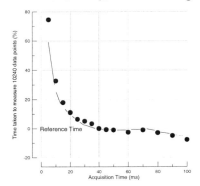

Fig. 7. The influence of maximum exposure time (before reset of the hardware) on the total time taken to acquire 10240 data points

Figure 8 shows a "partial phase-averaged" time series of pressure, axial and radial velocity and non-compensated temperature obtained with an average data rate of 200Hz for the unsteady Flame B, following the procedure outlined in previous paragraphs. The unsteadiness of the flow field is clearly visualised in the "partial" phase-averaged signal, and is associated with the temporal evolution of the turbulent or incoherent fluid motion. This is clearly identified through phase-resolved measurements of velocity, temperature and pressure obtained under periodic oscillations, figure 8b-c, which were statistically analysed following the decomposition proposed by Hussain and Reynolds (1970). For a generic variable γ:

$$\gamma(t) = \overline{\gamma} + \tilde{\gamma}(t) + \gamma'(t),$$

where $\gamma(t)$ is the instantaneous value, $\overline{\gamma}$ is the long time average mean, $\tilde{\gamma}$ is the statistical contribution of the organised wave, and $\gamma'(t)$ is the instantaneous value of turbulent fluctuations. An ensemble average over a large number of cycles yields (Hussain and Reynolds ,1970 and Tierderman et al. 1988):

$$< \gamma > (t) = \overline{\gamma} + \tilde{\gamma}(t) \quad ;$$

$$< \sqrt{\gamma'^2} > (t_i \pm \Delta t) \equiv < \gamma_{rms} > = \sqrt{\frac{\sum_{i=1}^{N} (\gamma'(t_i \pm \Delta t_i) - "< \gamma > *_{med})^2}{n-1}}$$

The phase interval, $t_i \pm \Delta t$, was chosen to be $18°/360°$, to minimise the influences of the phase averaging window size on the determination of turbulence quantities in unsteady turbulent flows, as discussed by Zhang et al. (1997). In addition, the average values at each window, $<\gamma>^*_{med}$ were determined using a best-fit polynomial of first degree to minimise the temporal bias (Fernandes, 1998).

3.3 Sample Results

Figure 9 presents the temporal evolution of $<U,V>$, $<U_{rms}>$, $>V_{rms}>$, $<T>$, $<T_{rms}>$ and velocity-temperature correlation, for the region $0.9<r/R<1.5$ at $z/R=0.58$. The ensemble-average velocity vectors are presented in figure 9a) together with mean vectors distribution. Together, they identify a central recirculation zone that is quite insensitive to the natural oscillations that are occurring in the shear layer. The large deflections of the velocity vectors in this region are due to the presence of coherent structures, as observed in the images of figure 1b-c, and confirmed by the velocity vectors of figure 9b. These velocity vectors were obtained with a *Galelian transformation* where a velocity of 7.7m/s, due to the vortex convection, was subtracted from the axial velocity component. The velocity vectors are not constant in time and identify the passage of a vortex $t/T=0.5$ and $r/R=1.2$ through the presence of a temporal "saddle point". This unsteadiness is accompanied by a singular distribution of turbulent kinetic energy, as shown in figure 9-c1, which double peaks, with a the time evolution of velocity fluctuations as in figures 9c2 and 9c3. The results confirm the high level of velocity fluctuations near the vortex core, where the radial gradient of axial velocity is positive and maximum.

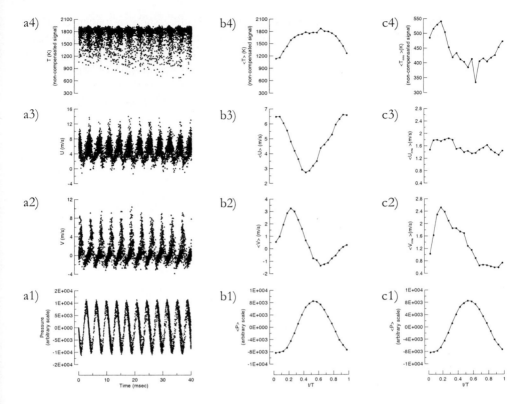

Fig. 8. Typical data reduction process from acquired signals to an evolution within a period of oscillation, for the unsteady Flame B, at z/R=0.58, r/R=1.24

a1-a4) Partial phase-averaged signals, as acquired by the system, of pressure a1), radial velocity a2), axial velocity component a3) and temperature a4).

b1-b4) Phase averaged results of mean pressure, axial and radial velocity components and compensated temperature signals

c1-c4) Phase averaged results of velocity and temperature turbulent fluctuations

The temperature characteristic profiles, figure 9d1-d3), show a rather complex nature, in that they suggest two typical high temperature regions, for r/R<1.04 and at r/R=1.2. Both maxima temperature is of about 1800K while the lowest temperature measured is of the order of 800K. The phase averaged evolution of temperature, figure 9d2, denotes the influence of large vortical structures on the heat release in the shear layer. The rms of temperature fluctuations, figure 9d3), shows values in the range of

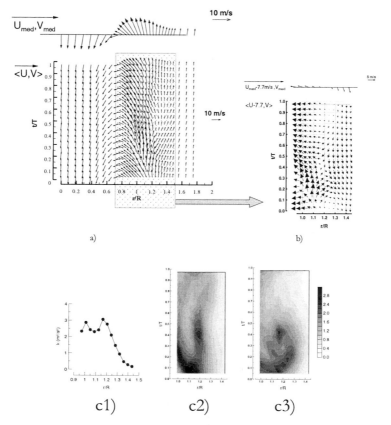

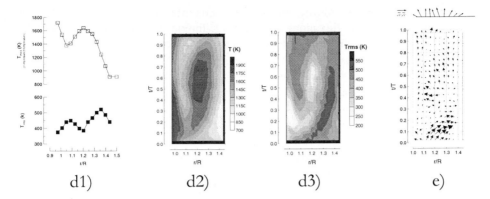

d1)   d2)   d3)   e)

Fig. 9. Time evolution of the phase averaged velocity vectors, temperature characteristics and velocity-temperature correlation at axial station z/R=0.58 for the Flame B

a)   *Time evolution of phase averaged velocity vectors*

b)   *Time evolution of phase averaged velocity vectors after the application of a "Galelian transformation" to subtract a vortex convection axial velocity of 7.7m/s*

c)   *Turbulent velocity characteristics*

c1) *Radial evolution of mean turbulent kinetic energy*

c2) *Time evolution of $<U_{rms}>$*

c3) *Time evolution of $<V_{rms}>$*

d)   *Turbulent temperature characteristics*

d1) *Radial evolution of mean and turbulent fluctuations of temperature*

d2) *Time evolution of $<T>$*

d3) *Time evolution of $<T_{rms}>$*

e)*Time evolution of $<u''t'',v''t''>$ correlation*

200K to 500K, in a way which is consistent with the findings of figure 9d2) in the sense that the maximum temperature fluctuations, at each time instant, occurs close to the regions where the instantaneous mean temperature exhibit a higher radial gradient.

Figure 9e) presents new information, in that it quantifies the temporal evolution of vectors of turbulent heat flux along a cycle of oscillation. These vectors represent the exchange rate of reactants responsible for the phenomenon of flame stabilisation and are high in regions associated with large temperature gradients. Previous results in the literature for steady recirculating flames (e.g. Fernandes et al., 1994, Hardalupas et al., 1996, Duarte et al., 1997) have shown the occurrence of zones of non- and counter-gradient diffusion of heat, which have been explained in terms of the interaction between gradients of mean pressure and density fluctuations. The present results provide new evidence of this interaction in oscillating flames, which is associated with periodic fluctuations in flame curvature. In general, the results quantify the periodic ignition of large-scale reaction zones, which drive the combustion-induced oscillations reported in this paper.

4. SUMMARY

A measuring system based on a laser-Doppler velocimeter has been extended to include time-resolved temperature and sound pressure measurements, in order to allow the analysis of unsteady reacting flows. Measurements are reported for open recirculating flames oscillating with of a predominant frequency in the range of 265-275 Hz, and have been obtained for data ready velocity signals up to 12.5 kHz, allowing the time-resolved representation of vectors of turbulent heat flux along a typical cycle of oscillation.

The results were obtained for flames stabilised in the wake of a bluff-body located on a velocity acoustic antinode. The acoustic resonance process driven by coherent structures along the reacting shear layer results in appreciable spatial and temporal deformations of the reacting flow field, which generate large periodic fluctuations in streamline curvature. This is accompanied by large fluctuations in the axial and radial velocity components of the flow and, together with the unsteady nature of the turbulent temperature field, results in a process of turbulent heat flux exhibiting a time sequence of gradient and non-gradient characteristic modes.

REFERENCES

Adrian, R.J. and Yao, C.S. (1989). "Power Spectra of Fluid Velocities Measured by Laser-Doppler Velocimetry. ASME, Winter Annual Meeting, Miami Beach, Florida, November 17-22, 1995.

Caldeira-Pires, A. and Heitor, M.V. (1998). "Temperature and Related Statistics Measurements in Turbulent Jet Flames",Experiments in Fluids, 24, pp. 118-129.

Chao, Y.C., Jeng, M.S. and Han, J.M. (1991). " Visualization of Image Processing of an Acoustically Excited Jet Flow", Experiments in Fluids, 12, 29-40

Dec, J.E. and Keller, J.A. (1990). "Time Resolved Gas Temperature in the Oscillating Turbulent Flow of a Pulse Combustor Tail Pipe", Comb. And Flame, 80, pp.358-370

Dec, J.E., Keller, J.O. and Hongo, I. (1991)." Time-Resolved Velocities and Turbulence in the Oscillating Flow of a Pulse Combustor Tail Pipe", Comb. and Flame, 83, pp.271-292

Dimopoulos, P., Boulouchos, K. and Valentino,G. (1996). " Turbulent Flow Field Characteristics in a Reciprocating Engine: Appropriate cut-off Frequencies for Cycle-Resolved Turbulence, an analysis of Co-incident 3-D LDV Data Based on Combustion-Related Dimensional Arguments", 8th Intl. Symp. On Appl. Of Laser Techniques to Fluid Mechanics, July, 8th-11th, Lisbon Portugal

Duarte, D., Ferrão,P and Heitor, M.V. (1997). "Turbulent Statistics and Scalar Transport in Highly-Sheared Premixed Flames", Proc. 11th Turbulent Shear Flows Symposium, Grenoble, 8-10 September

Durst, F., Melling, A. and Whitelaw, J.H. (1981). "Principles and Practice of laser-Doppler Anemometry", Academic Press.

Erdman, J.C. and Tropea, C.D. (1981). "Turbulence-induced Statistical in Laser Anemometers". Proc. 7th Biennal Symp. on Turbulence, Rolla, Missouri.

Fernandes, E.C. (1998). "The Onset of Combustion Driven Acoustic Oscillations", Ph.D. Thesis, Instituto Superior Técnico, Lisbon-Portugal (in English)

Fernandes, E.C., Ferrão,P., Heitor,M.V. and Moreira A.L.M (1994). "Velocity Temperature Correlation In Recirculating Flames With and Without Swril", Experimental Thermal and Fluid Science, 9, pp.241-249

Ferrão, P. and Heitor, M. V. (1998). "Probe and Optical Diagnostics for Scalar Measurements in Premixed Flames", Experiments in Fluids, 24, pp.389-398

Ganji, A.R. and Sawyer,R.F. (1980)." Experimental Study of the Flowfield of a Two-Dimensional Premixed Turbulent Flame", AIAA Journal, Vol.18,no.7

Gutmark, E., Parr, T.P., Hanson-Parr, D.M. and Schadow, K.C. (1989). " On the Role of Large and Small-Scale Structures in Combustion Control", Combust. Sci. and Tech., 66, pp.107-126

Gutmark, E., Schadow, K.C., Sivasegaram, S, and Whitelaw, J.H. (1991). " Interaction Between Fluid-Dynamic and Acoustic Instabilities in Combusting Flows Within Ducts", Combust. Sci. and Tech., vol.79, pp. 161-166

Hardalupas, Y., Tagawa, M. and Taylor, A.M.K.P (1996). "Characteristics of Counter-Gradient Heat Transfer in a Non Premixed Swirling Flame". In: Developments in Laser Techniques and Applications to Fluid Mechanics, ed. Durst et al., Springer-Verlag, pp. 159-184

Heitor, M.V., Taylor, A.M.K.P. and Whitelaw, J.H. (1984). "Influence of Confinement on Combustion Instabilities of Premixed Flames Stabilised on Axisymmetric Baffles", Combust. and Flame, 57, pp. 109-121

Heitor, M.V., Taylor, A.M.K.P. and Whitelaw, J.H. (1985). "Simultaneous Velocity and Temperature Mesurements in a Premixed Flame". Exp. in Fluids, 3, pp. 323-339.

Herzog, P., Valière, J.C., Valeau, V, and Tournois, G. (1996)."Acoustic Velocity Measurements by means of Laser Doppler Velocimetry", 8th Intl. Symp. On Appl. Of Laser Techniques to Fluid Mechanics, July, 8th-11th, Lisbon Portugal

Hirt,F., Eisele, K., Zhang, Z. and Jud,E. (1996). "Pulsatile Flow Behaviour near Cardiac Prostheses Application and Limitation of Laser and MRI Techniques", 8th Intl. Symp. On Appl. Of Laser Techniques to Fluid Mechanics, July, 8th-11th, Lisbon Portugal

Hussain, A.K.M.F. and Reynolds, W.C. (1970). "The Mechanics of an Organized Wave in Turbulent Shear Flow", J.Fluid Mech., vol.41, part2, pp.241-258

Hussain, A.K.M.F. and Zaman, K.B.M.Q. (1980). "Vortex Pairing in a Circular jet Under Controlled Excitation. Part 2. Coherent Structure Dynamics", J. Fluid Mech., vol.101, pp.493-544

Ishino, Y., Kojima, T., Oiwa, N. and Yamaguchi, S. (1996). "Acoustic Excitation of Diffusion Flames with Coherent Structure in a Plane Shear Layer", JSME International Journal, series B, Vol.39, no.1

Keller, J.O.(1995). "Thermoacoustic oscillations in combustion chambers of gas turbine", AIAA J., 33(12), pp.1125-1234.

Keller, J.O. and Saito, .K. (1987). "Measurements of the Combusting Flow in a Pulse Combustor", Combust.Sci. and Tech., 53, pp. 137-163

Keller, J.O., Vaneveld, L., Korschelt, D., Hubbard, G.L., Ghoniem, A.F., Daily, J.W. and Oppenheim, A.K. (1982)." Mechanism of Instabilities in Turbulent Combustion Leading to Flashback", AIAA Journal, vol. 20, no.2

Kiya, M. and Sasaky, K. (1985). " Structure of Large-Scale Vortices and Unsteady Reverse Flow in the Reattaching Zone of a Turbulent Separation Bubble", J. Fluid Mech., vol. 154, pp.463-491

Lang, W. and Vortmeyer, D. (1987). "Cross-correlation of sound pressure and heat release rate for oscillating flames with several frequencies excited", Combust. Science and Tech.,54. pp. 399-406.

Liou, T-M. and Santavicca, D.A. (1985). "Cycle Resolved LDV Measurements in a Motored IC Engine", Transactions of the ASME, Vol.107, pp232

Lovett, J. A. and Turns, S. (1993). "The Structure of Pulsed Turbulent Nonpremixed Jet Flames", Combust. Sci. Tech, 94, pp. 193-217

Schadow, K.C. and Gutmark, E. (19992). "Combustion instability related to vortex shedding in dump combustors and their passive control", Prog. Energy and Combust Science, 18, pp.117-132.

Sivasegaram, S. and Whitelaw, J.H. (1987). " Oscillations in Confined Disk-Stabilized Flames", Comb. and Flame, 66, pp.121-129

Tierderman, W.G., Privette, R.M. and Philipds, W.M. (1988). "Cycle-To-Cycle Variation Effects on Turbulent Shear Stress Measurements in Pulsatile Flows", Exp. in Fluids, 6, pp.265-272

Witze, P.O. (1984). " Conditionally-Sampled Velocity and Turbulence Measurements in a Spark Ignition Engine", Combust. Sci. and Tech., vol.36,pp. 301-317

Yanta, W.J. and Smith, R.A. (1978). "Measurements of Turbulent Transport Properties with a Laser Doppler velocimeter", 11th Aerospace Science Meeting, AIAA paper 73-169, Washington, USA

Zhang, Z. Eisele, K. And Hirt, F. (1997). "The Influence of Phase-Averaging Window Size on the Determination of Turbulence Quantities in Unsteady Turbulent Flows", Experiments in Fluids, 22, pp.265-267.

V.2. Liquid-Fuelled Flames with Imposed Air Oscillations

Y. Hardalupas, A. Selbach, and J.H. Whitelaw

Mechanical Engineering Department, Imperial College of Science,
Technology and Medicine Exhibition Road, London SW7 2BX, UK

Abstract. The secondary air flow of a swirl stabilised burner was oscillated at a frequency of 350 Hz for swirl numbers of 0.66, 0.7 and 0.81. Time dependent phase-Doppler velocimetry quantified the spray and the air flow and chemiluminescence of the CH radicals the shape and area of kerosene fuelled flames. The results suggest that the imposed oscillations shed a series of vortex rings at the burner exit with the oscillation frequency, increasing mixing and distorting the reaction zone, and potentially reducing the NO_x emissions. The flame lifted off for a critical amplitude, which increased with swirl number, due to the translation of the recirculation zone at a downstream position leading to increased stretch rate, evaluated from the temporal variation of the flame area, and local extinction in the near burner region.

1. Introduction

The investigation follows from previous considerations of the stability of steady liquid fuelled flames by Hardalupas et al. (1990, 1994a and b) and gaseous flames with imposed oscillations by Bhidayasiri et al. (1997), which revealed details of the variations of velocity, temperature and droplet size, including the contribution of recirculating small droplets of fuel in disk- and swirl-stabilised flames. Previous examination of the addition of fluctuating energy to the air supply of burners include that of Keller & Hongo (1990), who suggested reductions in NO_x in pulsed combustors and described a possible mechanism, the experiments of Haile et al. (1996) which showed that acoustic excitation led to increased mixing and Delabroy et al. (1996) who linked the interaction of a vortex and a flame to increased strain rate, leading to NO_x reduction. However, the understanding of the processes involved is limited and cannot provide design guidelines for imposed oscillations to the air flow to burners for reduced NO_x emissions with stable and efficient combustion. In addition, information on the interaction of imposed oscillations on the air supply and air swirl in swirl stabilised burners is not available.

The purpose of this study is to understand the way imposed oscillations on the air supply to a swirl stabilised burner influences the air flow, spray and flame characteristics and discuss the potential impact of imposed oscillations on flame stability, combustion efficiency and emissions. It presents visualisation of the effects of imposed oscillations on the shape of kerosene-fuelled flames and of their reaction regions, quantifies the mean stretch rates and focuses for the inert flow on the amplitude of the imposed oscillation in terms of velocity fluctuations at the exit of the burner and within the spray. The arrangement of the burner and the measurement methods are described in the following section and the results are described in the third section. The fourth section provides a summary of the more important conclusions.

2. Experimental Arrangement

The burner of figure 1 provided swirl with an air flow-rate through the annulus of 0.03 m^3/s resulting in a bulk velocity of 17 m/s and a potential heat release of 22.6 kW. Kerosene was delivered by a coaxial air-assist atomiser located on the axis with liquid and atomising air flow-rates of 7.7×10^{-7} m^3/s and 1.8×10^{-4} m^3/s respectively, so that droplet velocities ranged from 3 to 20 m/s and their arithmetic mean diameters from 20 to 70 μm with the radial distance from the axis. Three swirl numbers of the air flow of 0.66, 0.7 and 0.81 were considered, which resulted in lifted flames with the lower values. The overall equivalence ratio was constant at 0.25 and its local value at the exit of the atomiser was 38. Oscillations were imposed on the air flow by two acoustic drivers (figure 1), with a power of 38 W, except for the swirl number 0.66, where the power was reduced to 9 W to maintain stability, since higher power led to blow-off. The oscillation frequency was 350 Hz, which corresponded to a Helmholtz resonance at the exit of the burner and the resulting higher amplitude led to larger effect on the flame shape than frequencies at 200 and 920 Hz, for which the amplitude was low. This was confirmed by tests at frequency of 460 Hz, corresponding to second resonance, which led to a similar effect on flame shape as for 350 Hz.

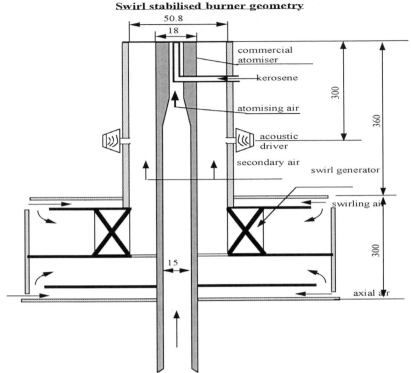

Figure 1 : The swirl stabilised burner geometry.

The velocities within the spray and the air flow at the exit plane of the burner were quantified with an isothermal flow of water with the same flow-rates and similar droplet characteristics, an oscillation power of 38 W and swirl numbers of 0.66 and 0.81. A phase-Doppler velocimeter measured velocities and droplet diameters and comprised transmission optics based on a rotating grating, and receiving optics with three photodetectors. The light refracted from droplets was observed at a forward angle of 30° on the bisector plane of the two laser beams, was focused on a 100 μm slit and passed through a mask with three evenly spaced rectangular apertures before reaching the photodetectors. The optical geometry of table 1 allowed measurement of droplet diameters up to 165 μm and droplet velocities were recorded in 30 size intervals of 5.5 μm. Results are presented for ranges centred at 13.8 and 46.8 μm with statistical uncertainties of less than 2 and 5% in mean and rms values respectively, based on average sample size of 1000. The counter processor was gated at the maximum and minimum velocities in the 2.9 ms oscillation cycle with data acquisition time of 0.5 ms.

The air velocities in the exit plane of the burner were quantified in terms of aluminium oxide particles with a nominal diameter of approximately 1.0 μm at six times within each oscillation cycle, within 0.5 ms windows overlapping by 0.02 ms.

Transmitting optics		
Laser : 1 Watt Ar-ion laser		
Operating power	0.4	W
Wavelength	514.5	nm
Beam intersection angle	3.2	deg.
Fringe spacing	9.23	μm
Number of fringes	14	
Frequency shift	6	MHz
Receiving optics		
Focal length of collimating lens	310	mm
Equivalent aperture at collimating lens: dimension of rectangular aperture 1, 2 and 3	53.3 × 10.6 mm	
separation between aperture 1 and 2	13.3	mm
separation between aperture 1 and 3	26.6	mm
Spatial filter slit width	100	μm
Observed spatial slit width	387.5	μm
Phase angle-to-diameter conversion factor for channel 1 and 3	0.468	μm/deg

Table 1: Phase-Doppler velocimeter

The reaction zone was visualised in terms of chemiluminescence of CH radicals at a wavelength of 430 nm and with a lifetime of less than 0.1 ms. The detection of chemiluminescence was achieved by installing an optical filter (centre frequency 429 nm, bandwidth 8.2 nm, peak transmission 45%) in front of an eight-bit black and white intensified CCD camera (Proxitronic HF 1). The images had a resolution of 580 × 770 pixels and the chemiluminescence intensity of each pixel was resolved on a scale from 0 to 255, with spatial resolution of 0.31 mm per pixel after considering the magnification of the lens. The CCD camera was connected to a PC with a frame grabber card (Data-Translation DT-3152) and an exposure time of 0.5 ms. The oscillations imposed with a period of 2.9 ms (350 Hz) and the images of the CH radicals in the oscillating flames were obtained by gating the camera 45° after the velocity maximum and minimum.

Figure 2a shows that a typical flame spectrum within a range of wavelengths from 200 to 900 nm was dominated by the continuum of soot emissions with two distinct peaks corresponding to emissions of C_2 and CH radicals at 516 and 430 nm respectively. The continuum of the soot emissions is equal to the theoretical emissions of a blackbody at the temperature of the soot, Solomon & Best (1991), so that the soot emission at 430 nm was estimated by linear interpolation between those measured at 420 nm and 442 nm and subtracted from the total intensity measured at 430 nm, figure 2b, to separate the emission due to the CH radicals.

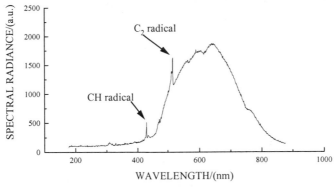

Figure 2a : Flame spectrum for swirl number 0.81

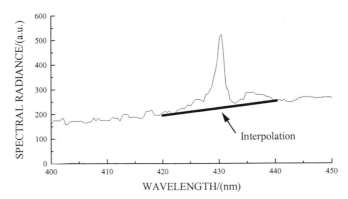

Figure 2b : Interpolation of soot emissions

The mean grey levels were obtained from the images by extracting the intensity at every tenth pixel in the vertical and horizontal directions from 200 pictures and averaging at each position. Spatial variations in the averaged intensity smaller than 3.1 mm were due to interpolation. The contribution of soot was subtracted and the CH concentration is presented as contour plots, normalised by the maximum intensity, and graded linearly from black, corresponding to a low value of CH-radical concentration, to white, corresponding to a high value, on a grey scale divided into twenty equally sized steps of 0.05.

3. Results

The results are presented in two sections. The first describes the velocity characteristics of the isothermal air flow in the exit plane of the burner and the droplet characteristics downstream of the burner exit without and with imposed oscillations. The second section describes the changes of the position and shape of the reaction zones due to imposed oscillations.

3.1 Velocity Characteristics

The axial velocities of the air flow corresponded to swirl numbers of 0.66 and 0.81 and the gated measurements with oscillations confirmed a sinusoidal variation with 350 Hz frequency. Those of figure 3 comprise mean and rms values in the absence of oscillations, the ensemble-averages of maximum and minimum velocities within oscillation cycles, and time-averaged velocities independent of the period of the oscillation cycle. Without oscillations the velocities decreased towards the axis for both swirl numbers, consistent with the findings of Ribeiro and Whitelaw (1980) and Liu et al. (1989) in their near fully-developed annular flows. With imposed oscillations, the time-averaged velocities had a similar trend with the highest swirl number and a more uniform profile with the lowest swirl number. The rms of time-averaged velocity fluctuations was increased by up to 100% with oscillations, as expected, due to the bimodal probability function of velocity at the burner exit and, therefore, did not indicate increased levels of turbulence. This is confirmed by the similarity of the ensemble-averaged rms of the velocity fluctuations without and with oscillations (figure 3). This suggests that the flow turbulence at the burner exit was not responsible for the observed effects on the flow downstream of the burner exit and on the flame.

The ensemble-averaged mean axial velocity profiles of figure 3a & c were integrated with radial distance to quantify the variation of the volume flow-rate during the cycle of the imposed oscillations. The results showed that the volume flow-rate fluctuated with the oscillation frequency and an amplitude of around 1.42 $\times$ 10^{-2} m^3/s between the minimum and maximum velocity in the cycle for the swirl number of 0.81, which was around 50% of the total air flow-rate to the burner. This remained unchanged with the reduction of the swirl number to 0.7 and reduced by around 15 % for swirl number of 0.66. The observed influence of the swirl number on the amplitude is possibly limited to values of swirl numbers below 0.7, which corresponds to the value required for the generation of a recirculation zone at the exit of swirling coaxial jets (Wall 1987). Decrease of the input power at the acoustic drivers from 38 to 9 W for the swirl of 0.66 resulted in 68 % lower amplitude.

The amplitude of the air flow-rate fluctuation during the cycle quantifies reliably the amplitude of the oscillation and takes into account the enhancement due to resonance, which is not possible either in terms of the input power to the acoustic driver or the rms of the time-averaged velocity at the burner exit. The former method depends on the characteristics of the acoustic driver, even for different units of the same model, and cannot take into account resonances. The latter method did not represent any physical quantity which can compare levels of oscillations at different burners and operating conditions.

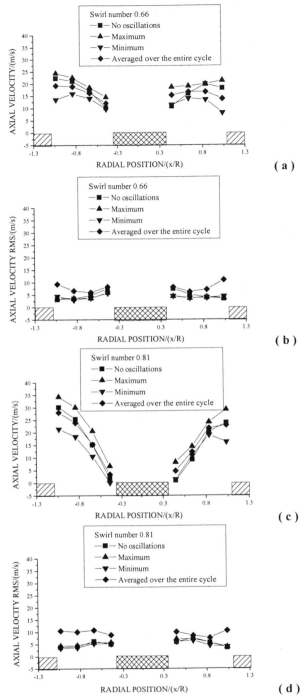

Figure 3 :   Velocity profiles for swirl numbers of 0.66 **(a)** and 0.81 **(c)** and rms
profiles for swirl numbers of 0.66 **(b)** and 0.81 **(d)** at the exit of the burner

The droplet velocities measured in the plane 40 mm downstream of the exit, near the lower boundary of the recirculation zone for the flow without imposed oscillations, correspond to droplet sizes of 13.6 and 46.8 μm with their response times of 0.6 and 6.6 ms. It is expected that the smaller droplets followed the mean and turbulent flow time scales of around 3 and 0.9 ms respectively, assuming that the characteristic length scale and velocity were the burner exit diameter, D, and the length scale of the energy containing eddies, D/10, and the area-averaged air flow velocity at the burner exit and the rms of the velocity fluctuations respectively for the mean and the turbulent flow. The small droplets could also follow the imposed oscillations with period 2.9 ms. The larger droplets could not respond to the characteristics of the flow and indicate the motion of most of the liquid fuel. Figure 4 shows that, without oscillations, the smaller droplets had negative velocities around the axis and the larger droplets did not. Thus, the air flow recirculated in a substantial region around the axis and larger droplets tended to travel through the recirculation region with near-ballistic trajectories. The two swirl numbers gave rise to similar axial velocities with a tendency for increased swirl to spread the recirculation region with small increase in the maximum negative velocities. Imposed oscillations and the larger swirl number led to averaged velocities, which were close to those of the non-oscillated flows (figure 4c & d). The difference of the axial velocity of the small droplets between the minimum and maximum point in the cycle is small within the recirculation zone and increased at the outer edge of the flow to around 4 m/s. Therefore, the imposed oscillations mainly influenced the outer part of the air flow downstream of the burner exit. With the smaller swirl number (figure 4a & b), the averaged velocities of both droplet sizes increased, the recirculation zone shifted downstream and differences between minimum and maximum of the cycle were negligible. The flow characteristics of figure 4b, for the low swirl number and imposed oscillations, led to flame extinction, probably caused by the downstream shift of the recirculation zone, which increased the local stretch rate above a critical value for extinction.

The findings of the velocity characteristics of the air flow help to understand the flow structure with imposed oscillations. The sinusoidal oscillation of the volume air flow-rate at the burner annulus, the presence of the velocity difference between the minimum and the maximum in the cycle at the outer region of the flow and the shifting of the recirculation zone downstream of the burner exit for low swirl number are consistent with the shedding of vortex rings from the burner exit with frequency of the imposed oscillations, similar to starting vortices in jet flows, in agreement with Maxworthy, 1972 and Weigand & Gharib, 1997. These remained at the outer part of the flow downstream of the burner exit, where velocity fluctuations were observed during the cycle and induced a flow field close to the burner axis with direction away from the burner exit, which could become strong enough to overcome the inverse flow of the recirculation zone generated by the swirling motion, and this occurred for low swirl number with imposed oscillations. The presence of vortex rings at the burner exit is consistent with the velocity measurements and will be further discussed below in association with the flame shape.

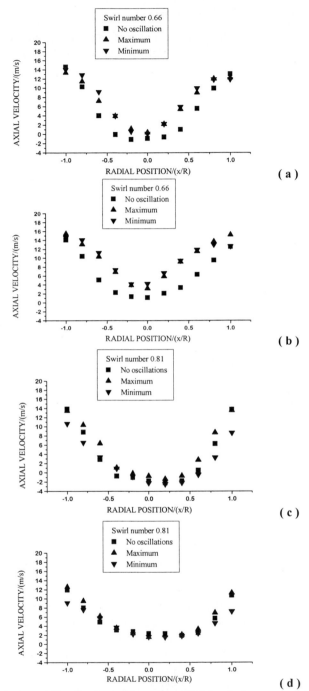

Figure 4 : Velocity profiles for the swirl number of 0.66 and size range centred
at 13.8 (**a**) and 46.8 μm (**b**) and for the swirl number of 0.81 and size range
centred at 13.8 (**c**) and 46.8 μm (**d**) at an axial distance to the exit of 40 mm

3.2 Visualisation of the Reaction Zone

Figure 5 shows CH chemiluminescence concentrations in the near burner region of flames with and without imposed oscillations and with swirl numbers of 0.66, 0.7 and 0.81. The results with imposed oscillations correspond to the time of 45° after the maximum velocity of the cycle. In the absence of imposed oscillations, the boundaries of the time-averaged images are smooth, since the random disturbances caused by turbulence were removed by averaging over 200 images, and the small scale disturbances are associated with the spatial resolution of the images. The flow field in the near burner region was dominated by the recirculation zone with the small droplets reversing their direction and evaporating to provide fuel vapour at the base of the flame. Reaction occurred in the shear layer of the secondary air stream where the fuel vapour mixed with the air stream, as in diffusion flames.

High CH emissions corresponding to the reaction zone are present in the shear layer for all swirl numbers with less intense emission close to the axis of symmetry, figure 5, where a fuel rich region is present and reaction is limited. The lower boundary of the reaction zone was some 35 mm above the burner exit with the swirl number of 0.66, figure 5a, and, although imposed oscillations (9W) led to higher CH concentrations close to the axis, the lower boundary remained at the same position but the flame was shortened, figure 5b. The amplitude of the imposed oscillation for 9W power input was 15% of the total air flow-rate to the burner and represents the maximum amplitude before flame extinction. The velocity results suggest that the shift of the recirculation zone downstream combined with the increased velocity of the large droplets for higher amplitude of oscillation, was responsible for flame extinction. With the swirl number of 0.7, the lower boundary of the reaction zone was 25 mm from the burner, figure 5c, and the flame could be stabilised with imposed oscillations of power input of 38W, corresponding to oscillation amplitude of 50% of the total air flow-rate. However, the flame was lifted at an increased distance of 40 mm, figure 5d, and the length of the flame considerably shortened. The shortening of the flames, figures 5b & d, was probably associated with the large droplets escaping the reaction zone, which did not allow them to evaporate. The total flame area, determined by the area of the normalised intensity contours of figure 5 for values higher than 0.82, for the low swirl numbers 0.66 and 0.7 increased with imposed oscillations, suggesting locally higher reaction rate. However, the reaction zone was located at a larger distance from the burner exit due to increased lift-off and with short flame length resulting in short residence times for the large droplets to evaporate and burn completely.

For swirl number of 0.81, the flame was attached to the burner exit and imposed oscillations with amplitude 50% of the total air flow-rate did not lift-off the flame (figure 5e & f). This is consistent with the negative velocities of smaller droplets and the wide recirculation zone of the air flow on the previous figure, which did not change with imposed oscillations (figure 4c & d), and, therefore, the flame remained attached to the burner. However, further increase of the amplitude of the oscillations led to flame lift-off and extinction, which was similar to observations for lower swirl numbers. Therefore, there is a critical value of the amplitude of the

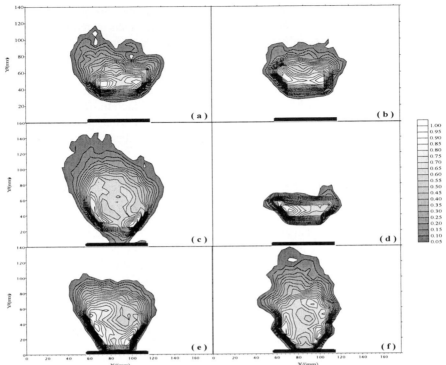

Figure 5 : Distributions of the CH radical concentration of the flame for swirl numbers
of 0.66 **(a)**, 0.7 **(c)** and 0.81 **(e)** without imposed oscillations and with
imposed oscillations for flames with swirl numbers of 0.66 **(b)**, 0.7 **(d)** and
0.81 **(f)** an exposure time of 0.5 ms averaged over 200 images, normalised
by the maximum values

imposed oscillations for each swirl number, which led to flame lift-off and
extinction and, since the amplitude of the oscillations should be large to increase
mixing and potentially reduce NO_x emissions, the imposed oscillations should be
applied to burners operating at high swirl numbers.

The contours of flames with imposed oscillations had large scale wrinkles, which
were not evident by visual observation, and the attached flame with the highest
swirl number had a reaction zone with two axi-symmetric wrinkles which were
convected downstream from the burner exit. The shape of the wrinkles suggests
that they were caused by a sequence of vortex rings shed at the burner exit with the
frequency of the oscillation and with initial length scale of the order of the annular
width, in agreement with the velocity measurements, as observed in the flames of
Delabroy et al. (1996).

Chemiluminescence contours were also observed as a function of time within the
oscillation period of 2.9 ms (not presented here) and comparison of the
downstream position of wrinkles at two times within the cycle suggested a velocity
in direction of the main flow of 15.5 m/s which was lower than the bulk velocity at

414

the exit of the burner, 17 m/s. This is expected since the vortex ring caused the wrinkles on the flame and entrained gases, which will lead to reduction of their propagation velocity according to the conservation of momentum and this suggests that 10% of the initial flow-rate of the vortex was entrained within one cycle. Thus, the vortex rings increased mixing in the near burner region, leading to an increase of the time dependent stretch rate. The increased stretch rate can give rise to local extinction and a lifted flame and, in turn, to reduction of thermal NO_x by quenching, Drake and Blint (1989). Therefore, the increased mixing and stretch rate combined with an attached flame for high swirl numbers could lead to lower NO_x emissions.

The mean stretch rates due to imposed oscillations were calculated by estimating the area of the reaction zone in the near burner region at two observation times during the cycle corresponding to 45° after the maximum and minimum of the sinusoidal velocity fluctuation at the burner exit. Thus, the term $[A^{-1} dA/dt]$ is proportional to the propagation velocity normal to the reaction zone over the radius of curvature and the velocity gradient of the flow normal to the reaction zone, Bradley et al. (1992), so that the stretch rates with the swirl numbers of 0.7 and 0.81 were 700 and 250 s^{-1} respectively, around three times higher for the lifted flame. Therefore, flame lift-off occurred due to local extinction by the increased stretch rate.

It is interesting to consider the differences between the gas flames of Milosavljevic (1993) and Bhidayasiri et al. (1997) and those of the present investigation. The first two authors observed that imposed oscillations stabilised lifted flames and are in contrast to the increase in flame lift-off determined with the present liquid-spray fuelling system. Thus, the smaller droplets required to stabilise the flame were downstream of the burner exit for the two lower swirl numbers and oscillations could not modify the ballistic trajectory of the droplets associated with most of the liquid fuel. Bhidayasiri also found a small reduction in NO_x emissions, and similar reductions will be more difficult to determine in the present open flame in which it is more difficult to ensure complete combustion within a well defined region. However, the current study provided understanding on the influence of imposed oscillations on the flame and flow structure. It is anticipated that in a confined flame, the vortex rings will entrain surrounding hot combustion gases and mix them with fuel vapour and air and droplets to enhance evaporation, improve combustion efficiency and reduce NO_x emissions.

4. Summary - Conclusions

Oscillations were imposed on the secondary air flow of a swirl-stabilised burner and the air flow, the droplets within the isothermal spray and the kerosene-fuelled flames were examined for swirl numbers of 0.66, 0.7 and 0.81, overall equivalence ratio of 0.25 and frequency of 350 Hz, which had the largest effect within a range from 200 Hz to 920 Hz. A method to visualise quantitatively the reaction zone in liquid fuelled flames, based on the detection of the chemiluminescence from CH radicals emitted at wavelength of 430 nm was also developed. The more important results are summarised below.

1.   The acoustic excitation caused the formation of a sequence of vortex rings at the burner exit with dimensions close to the burner annulus, propagating at the outer region and in the direction of the main flow with the frequency of the imposed oscillation, increasing mixing between incoming air and surrounding gases, leading to increased stretch rates and potential reduction of NO_x emissions. Increase of the amplitude of the oscillations above critical values, which increased with swirl number, led to flame lift-off and extinction. Therefore, imposed oscillations should be used to burners with high swirl numbers to achieve high combustion efficiency and potentially lower NO_x emissions.

2.   Ensemble-averaged measurements of the isothermal spray showed that flame lift-off occurred when the oscillation caused a transformation of the recirculating zone to a downstream position. Two competing mechanisms were identified, the first, the swirling flow leading to a reverse flow region and, the second, the induced flow on the burner axis away from the exit by the vortex rings. When the latter mechanism dominated, the recirculation zone shifted downstream and the stretch rate increased, leading to local extinction and flame lifted off.

3.   Ensemble-averaged measurements of the rms of air velocity fluctuations in the cycle at the burner exit showed that air flow turbulence remained unchanged. Therefore, increased mixing occurred due to interaction between the vortex rings and the flow downstream of the burner exit.

4.   Ensemble-averaged measurements of the air flow at the burner exit quantified the amplitude of imposed oscillations, in terms of the magnitude of the temporal fluctuation of the air flow-rate to the burner normalised by the total air flow-rate. This quantity allows comparisons between different burners and operating conditions, in the presence of resonance.

Acknowledgements

The authors gratefully acknowledge support from the EPSRC, Grant G/K 97097 and US Navy Contract N68171-97-C-9009. Y. Hardalupas was supported by an EPSRC Advanced Fellowship.

References

BHIDAYASIRI, R., SIVASEGARAM, S. & WHITELAW, J.H. 1997 The effect of flow boundary conditions on the stability of quarl-stabilised flames. *Combust. Sci. and Tech.*, **123**, 185-205.

BRADLEY, D., LAU, A.K.C. & LAWES, M. 1992 Flame stretch rate as a determinant of turbulent Burning Velocity. *Phil. Trans. R. Soc. Lond.*, **A 338**, 359-387.

DELABROY, O., HAILE, E., LACAS, F. & CANDEL, S. 1996 Controlled pulsed combustion for NO_x reduction in domestic oil burners. *Proc. 1st European Conf. on small burner Technology, Zurich*, **1**, 55-64.

DRAKE, M.C. & BLINT, R.J. 1989 Thermal NO_x in stretched laminar opposed flow diffusion flames with $CO/H_2/N_2$ fuel. *Combust. Flame*, **76**, 151-167.

HAILE, E., DELABROY, O., LACAS, F., VEYNANTE, D. & CANDEL, S. 1996 Structure of acoustically forced turbulent spray flame. *Proc. 26th Symp. (Int) on Combustion*, 1663-1670.

HARDALUPAS, Y., LIU, C.H. & WHITELAW, J.H. 1994b Experiments with disk stabilised kerosene-fuelled flames. *Combust. Sci. and Tech.*, **97**, 157-191.

HARDALUPAS, Y., TAYLOR, A.M.K.P. & WHITELAW, J.H. 1990 Velocity and size characteristics of liquid fuelled flames stabilised by swirl burners. *Proc. Roy. Soc. Lond.*, **A 428**, 129-155.

HARDALUPAS, Y., TAYLOR, A.M.K.P. & WHITELAW, J.H. 1994a Mass flux fraction and concentration of liquid fuel in a swirl-stabilised flame. *Int. J. of Multiphase Flow*, **20**, 233-259.

KELLER, J.O. & HONGO, I. 1990 Pulse Combustion: The mechanism of NO_x production. *Combust. Flame*, **80**, 219-237.

LIU, C.H., NOURI, J.M. & WHITELAW, J.H. 1989 Particle velocities in a swirling, confined flow. *Combust. Sci. and Tech.*, **68**, 131-145.

MAXWORTHY, T. 1972 The structure and stability of vortex rings. *J. Fluid Mech.*, **51**, 15-32.

MILOSAVLJEVIC, V.D. 1993 Natural gas, kerosene and pulverised fuel fired swirl burners. Ph.D. thesis, University of London.

RIBEIRO, M.M. & WHITELAW, J.H. 1980 Coaxial jets with and without swirl. *J. Fluid Mech.*, **96**, 769-795.

SOLOMON, P.R. & BEST, P.E. 1991 Fourier transform infrared emission/transmission spectroscopy in flames. In *Combustion Measurements* (ed. N. Chigier), pp. 385-444, Hemisphere Publishing Corporation.

WALL, T.F., 1987 The combustion of coal as pulverised fuel through swirl burners. In *Principles of Combustion Engineering for Boilers* (ed. C.J. Lawn) pp. 197-335, Academic Press, London.

WEIGAND, A. & GHARIB, M. 1997 On the evolution of laminar vortex rings. *Experiments in Fluids*, **22**, 447-457.

V.3. Application of New Light Collection Probe with High Spatial Resolution to Spark-Ignited Spherical Flames

S. Tsushima, F. Akamatsu, and M. Katsuki

Department of Mechanical Engineering, Osaka University, 2-1 Yamadaoka, Suita, Osaka 565-0871, Japan

Abstract. A newly developed light collecting probe named the Multi-colored Integrated Cassegrain Receiving Optics (MICRO) is applied to spark-ignited spherical spray flames to obtain the flame propagation speed in freely falling droplet suspension produced by an ultrasonic atomizer. Two MICRO probes are used to monitor time-series signals of OH chemiluminescence from two different locations in the flame. By detecting the arrival time difference of the propagating flame front, the flame propagation speed is calculated with a two-point delay-time method. In addition, time-series images of OH chemiluminescence are simultaneously obtained by an intensified high-speed digital CCD camera to ensure the validity of the two-point delay-time method by the MICRO system. Furthermore, relationship between the spray properties measured by phase-Doppler technique and the flame propagation speed are discussed with three different experimental conditions by changing the fuel injection rate. It is confirmed that the two-point delay-time method with two MICRO probes is very useful and convenient to obtain the flame propagation speed and that the flame propagation speed is different depending on the spray properties.

Keywords. spray combustion, flame propagation, flame speed, detection unit, chemiluminescence

1. INTRODUCTION

Since liquid fuel spray flames are heterogeneous turbulent reacting two-phase flows, they have inherently complicated transient structures. Inhomogeneity of spray properties, such as number density, droplet size and velocity, is further complicated as the result of turbulent interactions, which produce various combustion phenomena in the same flow. It was reported that not only diffusion-like phenomena but also premixed-like behaviour were observed in spray flames by Continillo and Sirignano

(1990), Cessou *et al.* (1996), Greenberg *et al.* (1996), Li and Williams (1996), Chiu and Lin (1996), Gutheil and Sirignano (1998), and so on.   Tsushima *et al.* (1998) concluded that some portions of the fuel spray disappear rapidly in a premixed-combustion-mode due to preferential flame propagation through easy-to-burn regions, whereas the nonflammable spray regions consist of droplet clusters surrounded by diffusion flames.   The results suggests that spray flames do not behave as a conglomeration of single droplets but as a group or cluster of droplets with premixed and diffusion characteristics.   Furthermore, to resolve the premixed-combustion-like behaviour in spray flames, clarification of the flame propagation mechanism and quantitative measurement of the flame propagation speed are strongly needed.

Akamatsu *et al.* (1994) observed a spark-ignited spherical flame propagating through a droplet suspension, which is freely falling and entraining surrounding air, in order to observe the flame propagation mechanism and the detailed flame structure in sprays under the minimal influences of atomization and fluid motion.   In the study, a pair of short-exposure images of OH-radical chemiluminescence and either of flame luminosity in C_2-radical band or of Mie-scattering from droplet clusters were taken simultaneously to clarify the spatial relation between the nonluminous and luminous flames and unburned droplet clusters.   Furthermore, this observation was compared with local continuous measurements, in which OH chemiluminescence, CH bands emission and droplet Mie scattering were simultaneously monitored together with the droplet sizes and velocities by phase-Doppler technique.   It was found that a nonluminous flame first propagated continuously through coexisting regions of small droplets and gas-phase mixture and that a number of small-scaled droplet clusters burned randomly associated with discontinuous luminous flames behind the nonluminous flame front.

In the present study, a pair of newly developed light collecting probe named the Multi-color Integrated Cassegrain Receiving Optics (MICRO) is applied to the spark-ignited spherical spray flame for monitoring the time-series signals of OH chemiluminescence from two different locations separated at 6 mm in the vertical direction.   The flame propagation speed is calculated by a two-point delay-time method, in which the time difference between onsets of two chemiluminescence signals detected by each MICRO probe can be measured.   In addition, time-series images of OH chemiluminescence are simultaneously obtained by a high-speed digital CCD camera to ensure the validity of the two-point delay-time method with two MICRO probes.   Furthermore, the relationship between the spray properties measured by phase-Doppler technique and the flame propagation speed is investigated using three different fuel injection rates.   It is confirmed that the two-point delay-time method using the two MICRO probes is very useful and accurate to obtain the flame propagation speed, and that the flame propagation speed are different depending on the spray properties.

2. EXPERIMENTAL APPARATUS

Figure 1 shows the experimental apparatus. An magnetostriction type ultrasonic atomizer (resonance frequency 18.5 kHz) located at 400 mm above a spark gap (4.0 mm) produces kerosene spray. The fuel injection rate is adjusted by an electrically-controlled microsyringe pump. A vertical square duct ($280 \times 280 \times 1325$ mm) is used to cover the whole test section in order to shield the droplet suspension from surrounding disturbances. The atomized droplets freely fall entraining surrounding air and are ignited by intermittent electric sparks at the gap.

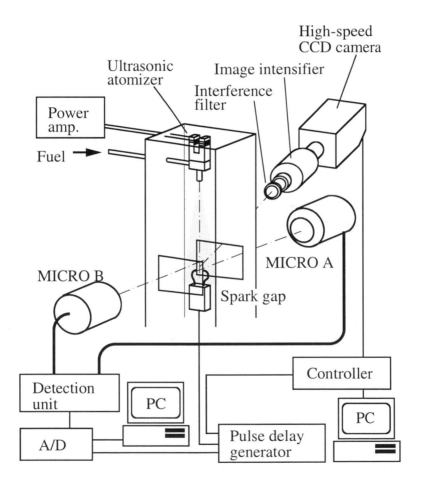

Fig. 1 . Experimental apparatus

After the ignition, a spherical flame emerges near the spark gap and kept on growing with spherical shape for several tens of millisecond. Finally, the spherical flame deforms flowing upward due to the buoyancy effect, and is extinguished by CO_2 injection just below the ultrasonic atomizer. Therefore, the observations are concentrated in the period of the spherical flame propagation. The experiments are conducted at three different fuel injection rates, 4.4, 6.3, and 11.0 cm^3/min. The ignition discharge duration time is 20 ms.

To detect OH radical chemiluminescence locally in the flame, two sets of newly developed light collecting probes named Multi-colored Integrated Cassegrain Receiving Optics (MICRO) are used. Kauranen *et al.* (1990) and Nguyen and Paul (1996) have previously applied Cassegrain-type optics to combustion flows to monitor multi-colour images. Figure 2 shows the MICRO probe system used in this investigation, which consists of an optimized pair of concave and convex mirrors combined with an optical fiber cable (Mitsubishi Densen Co. Ltd., ST-U200D-SY,

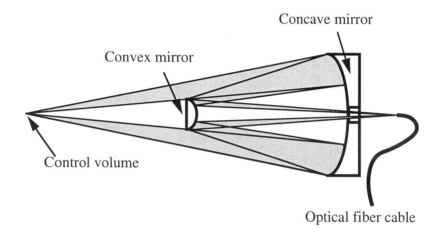

Fig. 2. Configuration of the Multi-color Integrated Cassegrain Receiving Optics (MICRO)

NA = 0.2, core diameter = 200 μm). The MICRO probe system has no chromatic aberration and minimized spherical aberration. In addition, the measurement volume center can be easily visualized using visible laser light, such as a small He-Ne laser or a solid state diode laser, guided into an optical fiber through a collimating lens (see Fig. 3) in the reverse direction. This feature enables precise optical alignment, which is necessary when dealing with UV signals, such as OH chemiluminescence. The light collection efficiency distribution of the system is evaluated using a ray-tracing method by Wakabayashi *et al* (1998), and the effective control volume size is numerically estimated at 1.6 mm long and 200 μm in diameter.

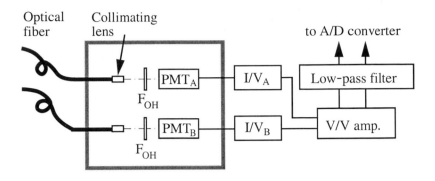

Fig. 3. Configuration of detection unit of the MICRO system

The light collected by each MICRO probe is directed to a detection unit through the optical fiber cable, as shown in Fig. 3. The light emitted from the other end surface of the optical fiber cable expands according to the numerical aperture (NA). To collimate the expanding light, a collimating lens is attached at the end surface. The OH chemiluminescence signal in the collimated light is detected by photomultiplier tubes (Hamamatsu, R106UH) after passing through an narrow-band optical interference filter (peak wavelength = 308.5 nm, half-value width = 18 nm) to reject background noise. The current signals from the individual photomultiplier tubes (PMTs) are converted to voltage signals by I-V converters (NF Electronic Instrument, Model LI-76), amplified, and then filtered by an electrical low-pass frequency filter (NF Electronic, FV-665, cut-off-frequency = 5 kHz). The time-series signals are digitized into 12 bits, *i.e.* 0 to 4096, by an A/D converter (Elmec, EC-2390) and recorded by a computer with digital sampling time of 10 μs (100 kHz).

Figure 4 shows the measurement location of two MICRO probes (MICRO A and MICRO

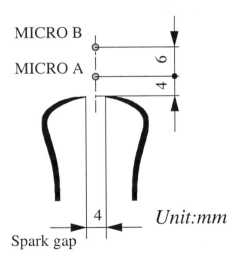

Fig. 4. Arrangement of measurement locations of two MICRO probes and the spark gap

B) and the spark gap. The MICRO A and B probes are aligned at 4 and 10 mm above the spark gap, respectively. The distance between the two measurement locations is 6 mm.

As shown in Fig. 3, to observe the evolving processes of the spherical spray flame, time-series images of OH chemiluminescence are recorded by a high-speed digital CCD camera (Kodak, EKTAPRO HS4540) simultaneously with the measurement of the two MICRO probes. The OH chemiluminescence from the flame is imaged onto a CCD array of 256×256 pixels in the high speed camera through an optical interference filter (peak wavelength = 308.5 nm, half-value width = 18 nm) and an image intensifier (Hamamatsu, C4273). Images are recorded in the transient memory of the image processing system and digitized into 8 bit, $i.e.$ intensities between 0 and 255. The frame rate is 4,500 frames/s and the maximum number of frames obtained continuously in one run is limited to 1,024, which corresponds to the actual time of 228 ms due to the memory capacity of the image processor. A pulse-delay-generator (Stanford Research Systems, WC Model DG535) is used to control the triggering timing of the spark discharge and each measurement device.

3. RESULTS AND DISCUSSION

3.1 Performance Evaluation of the MICRO Probe System

Before the MICRO probe system could be applied to the spherical spray flame, experimental verification of the MICRO system was conducted. Although the effective control volume size of the MICRO system was estimated at 1.6 mm long and 200 μm in diameter using the ray-tracing method, a fraction of the light emitted from outside the effective control volume does reach the detectors since chemiluminescence is spatially dispersed in flames and the MICRO system is designed for high sensitivity inside the effective control volume and not for completely blocking external signals. Therefore, simultaneous

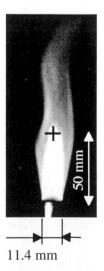

Fig. 5. A direct photograph of the premixed Bunsen flame with the measurement point of both the MICRO system and the electro-static probe

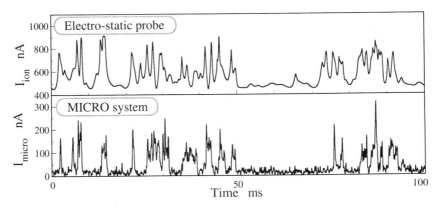

Fig. 6. Time-series signals of OH chemiluminescence by the MICRO probe system and ion current detected by the electro-static probe

measurements of OH-radical chemi-luminescence by the MICRO probe system and the ion current by an electro-static probe (Langmuir-probe), whose sensor tip is a Pt-Pt13%Rh thin wire of 100 μm in diameter and 1 mm long, were conducted in a premixed natural gas Bunsen flame. The inner diameter of the burner port is 11.4 mm. The flame appearance and the measurement point, which was 50 mm above from the burner port, are shown in Fig. 5. The measurement point is located at the fluctuating flame tip on the central axis of the flame. Figure 6 shows the simultaneous time-series signals of the ion current captured by the electro-static and OH radical chemiluminescence obtained by the MICRO system. The OH signals obtained by the MICRO probe show highly spatially resolved features equivalent to the ion current signals detected by the thin tip of the Langmuir probe. These results confirmed that the fraction of light collected from outside the effective control volume is negligible compared to the light collected inside the effective control volume. The MICRO probe system thus exhibited sufficiently high spatial resolution to observe local combustion phenomena without perturbing the flow.

3.2 Observation of Growing Processes of the Spherical Flame

Figure 7 shows time-series signals of OH-chemiluminescence, I_{OH}, obtained by the two MICRO probes together with the corresponding images of OH-chemiluminescence observed by the high-speed digital CCD camera. In image A at t = 0 ms, when the spark ignition is initiated, the spark gap is superimposed. The white cross in each image shows measurement locations of each MICRO probe (see Fig. 4 for the detailed arrangement). As seen in these images, after ignition, the flame front with low brightness is first propagating through the droplet cloud, and thereafter bright luminous lumps appear randomly inside the flame ball. The abrupt

424

increase in time-series signal intensities of the MICRO A probe implies that the flame front passes through measurement location A at t = 10.4 ms, while the MICRO B probe, focused at 6 mm above location A, detects the flame front at t = 47.7 ms. Local chemiluminescence signals detected by the two MICRO probes are in good agreement with time-series OH chemiluminescence images observed by the high-speed digital CCD camera. Hence, the two-point delay-time method is adopted to deduce the flame propagation speed, V, which is defined as

$V = L / \Delta t$ (1),

where L denotes the distance between the two measurement locations, which corresponds to 6 mm in this experiment, and Δt is the time difference between the onset of the chemiluminescence signal detected by each MICRO probe. In the case

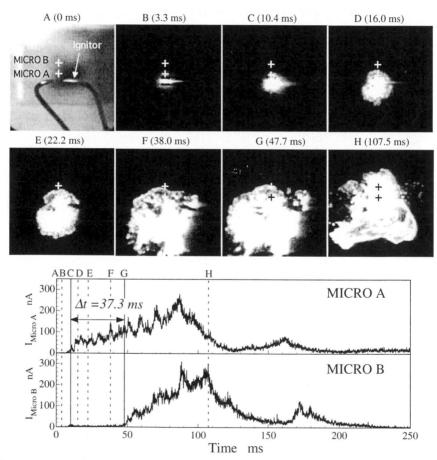

Fig. 7. Time-series data of OH chemiluminescence images recorded by the intensified high speed CCD camera and local OH chemiluminescence monitored by the two MICRO probe systems

shown in Fig. 7, the Δt is 37.3 ms. The flame propagation speed in this case is calculated at 0.16 m/s.

As reported by numerous researchers, such as Edwards and Marx (1992), sprays have very inhomogeneous spatial and temporal structures such that statistical analysis is required for understanding spray flame behavior. In this study, ignition tests were conducted approximately 100 times at three different fuel injection rates, 4.4, 6.3, and 11.0 cm^3/min. Figure 8 shows the probability distribution of flame propagation speed obtained in each experimental condition. The average flame propagation speed, $\overline{V}$, measured in the three different conditions is almost the same, ranging from 0.26 m/s to 0.29 m/s, while the RMS value, V', increases slightly with the fuel injection rate. This is explained by the increase in spray inhomogeneity with increased fuel injection rate, resulting in increased, non-uniform spatial and temporal flame propagation.

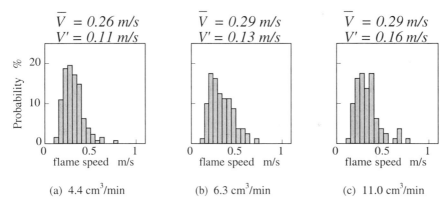

Fig. 8. Probability distribution of the flame propagation speed

Since the droplets are freely falling with some velocity, the actual corrected flame propagation speed, V_c, in which downward droplet velocity is taken into account, should be deduced. Furthermore, spray characteristics are directly connected with flame structure. Hence, it is necessary to know spray properties such as mean diameter, size distribution, number density, slip velocities between gaseous phase and liquid phase and so on. In this study, phase-Doppler technique is applied to the freely falling droplets under isothermal conditions. The control volume of the phase-Doppler system is located in the middle between the two measurement locations of the MICRO probes (7 mm above the spark gap). The droplet size distribution for the three different fuel injection rates are shown in Fig. 9, together with the Sauter mean diameter, D_{32}, mean velocity, $\overline{U}$, and RMS velocity, U', of the freely falling droplets. It is shown that the Sauter mean diameter, the mean and

426

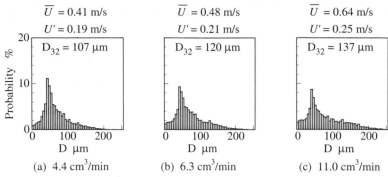

(a) 4.4 cm³/min   (b) 6.3 cm³/min   (c) 11.0 cm³/min

Fig. 9. Size distribution of the droplet suspension at the three different fuel injection rate

RMS velocities increase with the fuel injection rate. The corrected actual flame propagation speed, V_c, is defined as follows,

$$V_c = V + \overline{U} \qquad (2).$$

Figure 10 shows the variation of the corrected flame propagation speed, $\overline{V_c}$, in the three different fuel injection rates, together with the nominal flame propagation speed, $\overline{V}$, shown in Fig. 8, and the mean downward velocity of the freely falling

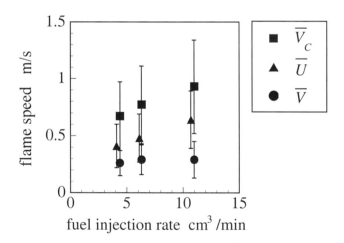

Fig. 10. Variation of actual flame propagation speed at the three different fuel injection rate

droplets, $\overline{U}$, shown in Fig. 9. The solid circle, ●, indicates the average value, and the length of a line segment corresponds to the RMS value. It is shown that the actual corrected flame propagation speed, $\overline{V_c}$, gradually increases with the fuel injection rate. Increasing the fuel injection rate increases the mass of small droplets, which plays important role in flame propagation due to the increased evaporation rate of small droplets compared to larger droplets. Mizutani and Nakajima (1973) reported that flame propagation speed in their spray flame was enhanced by adding gaseous fuel at the optimized condition. Future work for this experiment will identify the optimized fuel injection rate to enhance the flame propagation speed by conducting a number of experiments in various conditions.

4. CONCLUSIONS

In order to measure flame propagation speed in combusting spray systems, a pair of newly developed light collecting probes named the Multi-color Integrated Receiving Optics (MICRO) was applied to a spark-ignited spherical spray flame, propagation in a droplet suspension to monitor the time-series signals of OH chemiluminescence from two different locations. Firstly, the performance of the MICRO probe was experimentally investigated in comparison with an electro-static probe (Langmuir-probe), with thin sensor tip. The results showed the MICRO probe has sufficiently high spatial resolution to observe local combustion reaction in flames without perturbing the flow. The flame propagation speed was calculated by a two-point delay-time method, in which the time difference between onsets of two chemiluminescence signals detected by each MICRO system was measured. In addition, time-series images of OH chemiluminescence were simultaneously obtained by a high-speed digital CCD camera to ensure the validity of the two-point delay-time method with two MICRO probes. It was demonstrated that the newly developed MICRO probe system was useful to observe time dependent phenomena such as flame propagation. Furthermore, the relationship between the spray properties measured by phase-Doppler technique and the flame propagation speed was investigated for three different fuel injection rates. As a result, the flame propagation speed was different depending on the spray properties.

ACKNOWLEDGMENTS

This research was partially supported by the Grant-in-Aid for Scientific Research, the Ministry of Education, Science and Culture, Japan. The authors wish to express their gratitude to Mr. Yasuhide Okazaki of the Hitachi Zosen Corporation for the courtesy of special arrangement on the measurement equipment.

REFERENCE

Akamatsu, F., Nakabe, K., Katsuki, M., Mizutani, Y. and Tabata, T., 1994, Structure of Spark-Ignited Spherical Flames Propagating in a Droplet Cloud, in Developments in Laser Techniques and Applications to Fluid Mechanics, Ed. R. J. Adrian, D. F. G. Durao, F. Durst, M. V. Heitor, M. Maeda, J. H. Whitelaw, pp. 212-223, Springer-Verlag.

Cessou, A. and Stepowski, D., 1996, Planar Laser Induced Fluorescence Measurement of [OH] in the Stabilization Stage of a Spray Jet Flame, Combust. Sci. and Tech., Vol. 118, pp. 361-381.

Chiu, H. H. and Lin, C. L., 1996, Anomalous Group Combustion of Premixed Clusters, Proc., 26th Symp. (Int.) on Combust, The Combustion Institute, Pittsburgh, pp. 1653-1661.

Continillo, G. and Sirignano, W. A., 1990, Counterflow Spray Combustion Modeling, Combust. Flame, Vol. 81, pp. 325-340.

Edwards, C. F. and Marx, K. D., 1992, Analysis of the Ideal Phase-Doppler System: Limitations Imposed by the Single-Particle Constraint, Atomization and Sprays, Vol. 2, pp. 319-366.

Greenberg, J. B., Silverman, I. and Tambour, Y., 1996, A New Heterogeneous Burning Velocity Formula for the Propagation of a Laminar Flame Front Through a Polydisperse Spray of Droplets, Combust. Flame, Vol. 104, pp. 358-368.

Gutheil, E. and Sirignano, W. A., 1998, Counterflow Spray Combustion Modeling with Detailed Transport and Detailed Chemistry, Combust. Flame, Vol. 113, pp. 92-105.

Kauranen, P. Anderrsson-Engles, S. and Svanberg, S., 1990, Spatial Mapping of Flame Radical Emission Using a Spectroscopic Multi-Colour Imaging System, Appl. Phys., B53: pp. 260-264.

Li, S. C. and Williams, F. A., 1996, Experimental and Numerical Studies of Two-Stage Methanol Flames, Proc. 26th Symp. (Int.) on Combust, The Combustion Institute, Pittsburgh, pp. 1017-1024.

Mizutani, Y. and Nakajima, A., 1973, Combustion of fuel vapor-drop-air systems, Combust. Flame, Vol. 21, pp. 343-357.

Nguyen, Q. V. and Paul, P. H., 1996, The Time Evolution of a Vortex-Flame Interaction Observed via Planar Imaging of CH and OH, Proc. 26th Symposium (International) on Combust., The Combustion Institute, Pittsburgh, PA, pp. 357-364.

Tsushima, S., Saitoh, H., Akamatsu, F. and Katsuki, M., 1998, Observation of Combustion Characteristics of Droplet Clusters in a Premixed-Spray Flame by Simultaneous Monitoring of Planar Spray Images and Local Chemiluminescence, Proc. 27th Symp. (Int.) on Combust., The Combustion Institute, Pittsburgh, (in Press).

Wakabayashi, T., Akamatsu, F., Katsuki, M., Mizutani, Y., Ikeda, Y., Kawahara, N. and Nakajima, T., 1998, Development of a Multi-Color Light Collection Probe with High Spatial Resolution; Part 1 Evaluation of Spatial Resolution by Ray-Tracing Method, Transactions of JSME, Vol. 64, No. 619B, pp. 277-282 (in Japanese).

V.4. **Laser Ignition of Single Magnesium Particles**

J.F. Zevenbergen, A.E. Dahoe, A.A. Pekalski, and B. Scarlett

Delft University of Technology, Particle Technology Group,
Julianalaan 136, 2628 BL Delft, The Netherlands

Abstract

The minimum ignition temperature and minimum ignition energy of single magnesium particles was determined as a function of particle diameter. The particle was levitated ultrasonically and was ignited by a short laser pulse. The temperature transient of the particle was captured using a fast, optical fiber thermometer. It is shown that the minimum ignition temperature and the normalized minimum ignition energy are material constants of magnesium and are independent of particle diameter in the size range measured.

Keywords. Laser ignition, Ignition energy, Ignition temperature, Optical fiber thermometry

1 Introduction

The minimum ignition energy (MIE) and the minimum ignition temperature (MIT) are two key parameters in the assessment of the ignitability of combustible mixtures, like dust clouds. The MIT of a powder is determined by means of a furnace [1], which is maintained at a variable but uniform temperature. It consists of a tube, 23 cm in length and with a diameter of 3.9 cm. The dust/air mixture is introduced into the furnace by an air blast from one end of the tube. Effects of fouling, dust concentration and residence time of the particles in the furnace play a significant role in the determination of the MIT. Therefore, it is generally accepted that a furnace will not give reliable and reproducible values of the MIT, even though it is used as an international standard.

The MIE of a powder is determined with standardized equipment and involves a capacitive electric spark. By gradually lowering the voltage of the capacitor and hence the energy content of the spark, a minimum will be reached at which ignition fails to occur in ten successive attempts. The MIE is then calculated as:

$$MIE = \frac{1}{2}CV^2 \qquad (1)$$

With this equation the amount of energy stored in the capacitor can be calculated, but it does not give the amount of energy actually transferred between the electrodes. For the latter case the following equation should be used

$$MIE = \int_0^{t_{dis}} IV\,dt \qquad (2)$$

where t_{dis} is the discharge time of the capacitor. Despite the simplicity of this equation, its application is far less straightforward because of practical implications, for example impendance [2]. Even if these implications are resolved it can be questioned whether all the transferred energy between the electrodes is used to heat the particles. To answer that question, the attention should be focussed on the particles.

Currently a new method is being developed to determine simultaneously the MIE and the MIT of a dust/air mixture by means of laser ignition. A laser offers, in comparison to a spark and furnace, a far better control over the heating process of the particles. In order to be able to model these parameters for a dust cloud, it is necessary to perform experiments with single particles.

Lasers have previously been applied in ignition research. Zhang et al [3, 4] used a continuous Nd:YAG laser with a wavelength of 1.06 μm and a variable output power to heat single particles of coal placed on the end of an optical fiber. The minimum power required to ignite the coal particle was determined and related to the type of coal and to the particle size. Dependent on the volatile content of the particles, three different ignition mechanisms were identified: homogeneous (gas-phase), heterogeneous and hybrid. The higher the volatile content the more homogeneous the ignition. Wong et al [5] studied the temperature history before ignition of a char particle suspended in an electrodynamic balance and heated by a CO_2 laser. The temperature history was measured with an optical pyrometer. The effect of oxygen concentration and of the ignition delay time on the probability of ignition were studied. Qu et al [6] determined the temperature history of graphite and of several types of coal particles which were heated amd ignited on a silica plate by a CO_2 laser. The influence of the volatile matter, particle diameter, oxygen content and laser intensity on each type of carbon particle was studied. Increasing the laser intensity shortened the heating time, whilst an increase in oxygen concentration gave rise to a higher combustion temperature.

2 Experimental

The experimental setup for igniting single particles is shown in Figure 1. A CO_2-laser (Synrad) with a wavelength of 10.59 μm and a nominal output of 60 W is used as the heating source. The laser can be operated in a continuous mode as well as in a pulse mode with a variable pulse duration. The laser beam emitted (4 mm diameter) is split into two equal parts by means of a ZnSe beamsplitter. This to achieve symmetrical irradiation of the particle. Each of the two beams is focussed onto the particle which is held at the focal point of both laser beams by means of an ultrasonic levitator. The temperature of the particle is recorded by means of an optical fiber thermometer (OFT), which can capture temperatures ranging from 200 °C to 2000 °C at a rate of 10.000 measurements per second.

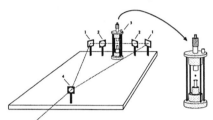

Figure 1. Schematic diagram of the setup used for single particle ignition. (1) Mirror (2) Lens (3) Ultrasonic levitator (4) Beamsplitter.

Spherical magnesium particles with diameters between 400 and 900 μm were used in these ignition experiments. The particles were not pretreated and represented a realistic sample of the material.

3 Theory

The temperature distribution within a spherical particle during laser heating is described by solving the heat conduction equation [7] in spherical coordinates. Thus, we start with:

$$\frac{\partial(\rho_s C_{p,s} T)}{\partial t} = \frac{1}{r^2} \frac{\partial}{\partial r}\left(r^2 \frac{\partial(\lambda_s T)}{\partial r}\right) \tag{3}$$

Equation (3) is solved subject to the following conditions:

(i) The initial condition.

$$\forall r, \quad t = 0, \quad T = T_\infty \tag{4}$$

(ii) The symmetry condition at the center.

$$\forall t, \quad r = 0, \quad \frac{\partial T}{\partial r} = 0 \tag{5}$$

(iii) The energy delivered to the surface of the sphere.

$$\forall t, \quad r = R, \quad \frac{\partial(\lambda_s T)}{\partial r} = \frac{Q_a I_0}{4 * 2} - h(T - T_\infty) - \epsilon_m \sigma(T^4 - T_\infty^4) \tag{6}$$

For the heat transfer coefficient, h, the Nusselt number for free convection is used.

$$\frac{hd}{\lambda_g} = 2 + 0.59 \left(\frac{d^3 \rho_g^2 g(T_{]R} - T_\infty)}{\eta_g^2 T_\infty} \right)^{\frac{1}{4}} \left(\frac{C_{p,g} \eta_g}{\lambda_g} \right)^{\frac{1}{4}} \tag{7}$$

The subscript "g" denotes the properties of the surrounding gas. Equations (3) to (7) were solved numerically using an implicit Crank-Nicholson scheme. It should be noticed that the particle was assumed to be irradiated from two sides, thus the factor 2. The absorption efficiency, Q_a, is cross-sectional based and thus the factor 4.

The absorption efficiency is determined by applying the $\overline{T}$ matrix method [8, 9] for a plane, monochromatic wave incident upon an arbitrarily shaped particle with a given size and optical constants. Linearity of the Maxwell equations and the boundary conditions implies that the coefficients of the scattered field are linearly related to those of the incident field. The linear transformation connecting these two sets of coefficients is called the $\overline{T}$ matrix. If, as in this case, the particle is spherical, the $\overline{T}$ matrix is diagonal. The strength of this method lies in the fact that the scattering and absorption by any arbitrarily shaped particle can be calculated and, once the $\overline{T}$ matrix is found for one scatterer, it can potentially be used to construct the solution of the scattering by many particles, which will be needed to describe dust cloud ignition.

Because of the low penetration depth of the laser light, in the order of nanometers, the possiblity of the formation of so-called hot spots due to resonances of the electromagnetic field in the deep interior of the magnesium particles is not present. Even if hot spots were to occur, the high thermal conductivity would quickly smooth out temperature gradients. It was found that the temperature difference between the surface and the centre of a 900 micron magnesium particle was 3 Kelvin, or less, for all possible conditions to be encountered in this work.

A schematic representation of the evolution of the particle temperature is shown in Figure 2. Two regions may be distinguished, a first part in which the particle is slowly heated by irradiation followed by a second part in which the temperature increases rapidly due to combustion. Ignition is defined [10] as the transition from a non-reactive to a reactive state in which external stimuli lead to thermochemical runaway followed by a rapid transition to self-sustained combustion. Hence Figure 2 allows the determination of the MIT of the single particle. The MIE may also be determined by integration of the energy needed to bring the particle to this point.

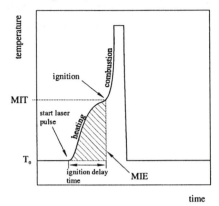

Figure 2. Determination of the MIT and MIE (T_0 denotes the initial temperature of the particle).

4 Results and Discussion

Figure 3 shows the successful irradiation to ignition of a magnesium particle with a diameter of 800 μm.

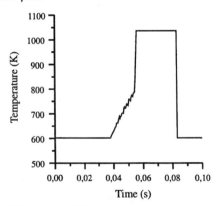

Figure 3. Temperature trace of an ignited magnesium particle (d = 800 μm),

The melting point and boiling point of magnesium are 922 K and 1363 K [11] respectively. This indicates that the particles undergo heterogeneous combustion when irradiated to ignition, since for homogeneous combustion the particles must first enter the vapour phase. As can be noticed in the figure,

436

the upper limit of the OFT was set at 1050 K, since the part of the temperature trace beyond this temperature is irrelevant for the determination of the MIE and MIT. In seperate experiments it was found that the temperature of the combusting particles reaches values well above 2000 °C , out of range of the OFT. This is in agreement with data from literature on the adiabatic flame temperature for magnesium which varies between 3250 and 3500 K, depending on the equivalence ratio [12]. Note that sometimes, during the heating process, the particle vibrates with small amplitude, causing the oscillation in the temperature trace shown in Figure 3.

In Figure 4 the values of the MIT are shown as a function of particle size.

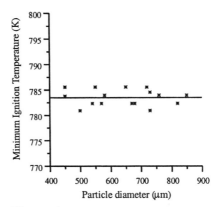

Figure 4. Minimum Ignition Temperature of magnesium particles as function of particle size.

As can be seen from this figure, the MIT is a material constant of magnesium. The MIT of the magnesium particles is 783.48 ± 1.49 K, as determined with the emissivity coefficient taken from the literature [13]. The MIT is in agreement with seperate results obtained with a thermogravic analyzer. According to the standard procedure, using a furnace, the MIT of magnesium is 1033 K [1] for a cloud of magnesium particles with a median particle size of 240 μm, and is 833 K for a cloud of magnesium particles with a median particle size of 28 μm [14]. This discrepancy is primarily due to the residence time of the particles in the furnace, which, in turn, is directly related to the settling velocity and hence to the particle size. The smaller the particles, the lower this settling velocity and the longer the residence time. Thus smaller particles require a lower furnace temperature in order to ignite before reaching the end of the furnace.

In Figure 5 the MIE is plotted as a function of particle size.

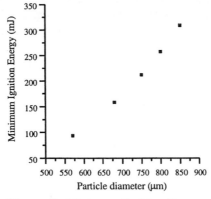

Figure 5. Minimum Ignition Energy of magnesium particles as function of particle size.

The MIE is here taken as the amount of energy accumulated by the particle before ignition and is determined by integration of Equation 3. This is in contrast to the electric spark method where the MIE is taken as the amount of energy stored in the capacitor that just fails to ignite the cloud. Figure 6 shows an overlay of a measured temperature trace and the model, supporting the validity of this procedure for determing the MIE.

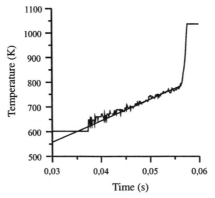

Figure 6. Overlay of the model (Equation 3) on an experimentally measured temperature trace (d = 850 μm).

As expected the MIE increases with the third power of the particle size, implying that the amount of energy accumulated scales linearly with particle volume (d^3). Under the condition that the intensity is sufficiently large to heat the particles to their MIT, both the MIE and the MIT are independent of the duration of the laser pulse

When the MIE is normalized with respect to particle volume, a constant value is obtained. This can be explained as follows. For calculating the MIE the left hand side of Equation 3 should be integrated and multiplied with particle volume. Since the MIT is constant and density and heat capacity for each particle follow the same temperature trajectory from room temperature up to the MIT, only the particle volume will change the actual value of the MIE. Therefore, when normalizing with respect to particle volume, a constant value will be obtained, see Figure 7. The particle volume normalized MIE of magnesium is $9.618 \pm 0.009 * 10^8 J/m^3$. The presence of a temperature gradient in a particle (large metal particle or particle with low thermal conductivity) will cause a deviation from the constant value.

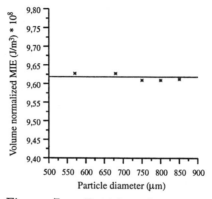

Figure 7. Particle volume normalized minimum ignition energy as function of particle size.

5 Conclusions

The minimum ignition temperature of magnesium is a material constant and equals 783.48 ± 1.49 K. For all heterogeneously igniting materials, the MIT will be a constant. In principle, for homogeneously or hybridly igniting materials, the MIT will not exhibit a constant MIT as function of particle size, since the heating time will determine the formation of gaseous products that can subsequently ignite and thus dictate the determination of the MIT. This in turn depends on the volatile content of the particle, the gaseous components that are formed and their concentration in the oxidizing medium. However, if the heating time is small compared to the time needed for devolatization then a constant value should be obtained.

The internationally recognized standard on the determination of the MIT, which utilises a furnace, lacks a fundamental basis. Measurements performed according to this standard show a decrease of the MIT as function of particle size. However, this effect is not directly related to particle size, but to the residence time of the particles in the furnace, which in turn is related to this settling velocity. The concept of determining the MIT by laser heating is more basic. The minimum ignition temperature of a cloud of magnesium particles is the same as that of a single magnesium particle. The key question now to be answered is "how many particles must achieve the MIT before the ignition of an explosion can be considered to become the propagation of an explosion?"

The minimum ignition energy of the heterogeneously igniting particles is, due to the constant MIT, a function of particle size only. When normalizing the MIE with respect to particle volume a constant value is obtained. The MIE, determined by laser heating and integration of the accumulation term in the well-known heat conduction equation will give the minimum amount of energy that is needed to heat the particles to their MIT. This is contrary to the spark method, where a part of the energy stored in the capacitor will be dissipated in other forms than only the energy used to heat the particles. The ignition energy determined with a spark will not therefore be the minimum amount of energy, contrary to what the name suggests. Only in the case of gases, can the spark be used with reasonable confidence for the determination of the actual minimum ignition energy of these gases since all parameters involved can be quantified. For dust/air mixtures the spark should be replaced by another technique, for instance a laser. In the latter case the minimum ignition energy of the cloud can be calculated if the required number and size of the particles that are needed to propagate the ignition into the rest of the dust cloud are known. The minimum ignition energy of the cloud is then the product of the sum of these particle volumes and the normalized single particle minimum ignition energy.

List of Symbols

C	Capacitance	$A^2 s^4 kg^{-1} m^{-2}$
$C_{p,g}$	Heat capacity of bulk gas	$m^2 s^{-2} K^{-1}$
$C_{p,s}$	Heat capacity of particle	$m^2 s^{-2} K^{-1}$
d	Particle size	m
g	Gravitational accelaration	ms^{-2}
h	Heat transfer coefficient	$kgs^{-3} K^{-1}$
I	Current	A
I_0	Initial intensity laser light	kgs^{-3}
Q_a	Absorption efficiency	$-$
R	Particle radius	m
r	Radial coordinate	m
T	Temperature	K
T_∞	Temperature of gas bulk	K
t	Time	s
t_{dis}	Discharge time	s

440

V	Voltage	$kgm^2 A^{-1} s^{-3}$
ϵ_m	Emissivity of particle	–
η_g	Dynamic viscosity of bulk gas	$kgm^{-1} s^{-1}$
λ_g	Thermal conductivity of bulk gas	$kgms^{-3} K^{-1}$
λ_s	Thermal conductivity of particle	$kgms^{-3} K^{-1}$
ρ_g	Density of bulk gas	kgm^{-3}
ρ_s	Density of particle	kgm^{-3}
σ	Stefan-Boltzman constant	$kgs^{-3} K^{-4}$

References

[1] Eckhoff R.K. *Dust Explosions in the Process Industries*. Butterworth, 1997.

[2] Bazelyan E.M. and Y.P. Raizer. *Spark Discharge*. CRC Press, 1998.

[3] Zhang D.K., Wall T.F., and Hills P.C. The ignition of single pulverized coal particles:minimum laser power required. *Fuel*, 73:647–655, 1994.

[4] Zhang D.K. Laser-induced ignition of pulverized fuel particles. *Combustion and Flame*, 90:134–142, 1992.

[5] Wong B.A., Gavalas G.R., and Flagan R.C. Laser ignition of levitated char particles. *Energy and Fuels*, 9:484–492, 1995.

[6] Qu M., Ishigaki M., and Tokuda M. Ignition and combustion of laser-heated pulverized coal. *Fuel*, 75:1155–1160, 1996.

[7] Bird R.B., Stewart W.E., and Lightfoot E.N. *Transport Phenomena*. John Wiley & Sons, Inc., 1960.

[8] Bohren C.F. and Huffman D.R. *Absorption and Scattering of Light by Small Particles*. Wiley-Interscience, Inc., 1983.

[9] Chew W.C. *Waves and Fields in Inhomogeneous Media*. IEEE Press, 1995.

[10] Kuo K.K. *Principles of Combustion*. John Wiley & Sons, Singapore, 1986.

[11] Lide D.R. *CRC Handbook of Chemistry and Physics*. CRC Press, Boston, 1992.

[12] Glassman I. *Combustion*. Academic Press, Inc., third edition, 1996.

[13] Gubareff G.G., Janssen J.E., and Turborg R.H. *Thermal Radiation Properties Survey*. Ann Arbor, 1960.

[14] Palmer K.N. *Dust Explosions and Fires*. Chapmann and Hall Ltd, 1973.

V.5. PIV Measurements During Combustion in a Reciprocating Internal Combustion Engine

D.L. Reuss and M. Rosalik

Engine Research Department
General Motors Global R&D Operations, Warren, MI, USA

Abstract

Two-dimensional particle image velocimetry (PIV) measurements were made in the burned and unburned gas of a firing, reciprocating internal-combustion engine. Cross correlation was used to analyze the double-exposure photographs using a 1.25 mm interrogation spot on a 0.5 mm grid. Electro-optical image shifting was used to resolve the directional ambiguity. The particle properties and the seeding density are the two key features enabling simultaneous measurements in the burned- and unburned-gas regions. The flame coordinates were identified as the interface between the low and high seeding density regions on the PIV photographs. Those coordinates were used to identify the burned and unburned regions of the velocity distributions. High-pass spatial filtering is used to visualize coherent structures of spatial scales L, where 1.25 mm < L < 10 mm. Vorticity and strain-rate distributions are computed from the instantaneous velocity. PDFs of these distributions were computed, conditionally sampled on the burned gas, unburned gas and proximity to the flame front. For this paper, analysis of only one realization is presented in detail in order to maintain a manageable number of illustrations yet demonstrate flow properties observed in fifteen recorded realizations.

The high-pass filtering reveals coherent flow structures that appear to correlate with the flame wrinkling, but are otherwise not apparent in the instantaneous velocity distributions. The vorticity in the burned gas regions appears higher than that in the unburned gas regions. The normal strain-rates appear to be large and predominately positive adjacent to the flame and in the burned gas. The shear strain-rate distributions show no obvious variations ahead, behind, or adjacent to the flame. These observations were apparent both by inspection of the plotted distributions and in the conditionally sampled PDFs.

1. INTRODUCTION

The turbulent flow in reciprocating internal combustion (RIC) engines controls fuel mixing, residual mixing, heat-transfer, and the rate of combustion; consequently, it is a major factor in the engine's efficiency and emissions. Most knowledge of the in-cylinder turbulence has been acquired through single-point time-resolved measurements of motored or precombustion gases using either hot-wire anemometry (HWA) or laser-Doppler anemometry (LDA), Rask (1984), Arcoumanis and Whitelaw (1987) and Valentino et al (1997). These measurements provide the ensemble mean velocity (U_i) fluctuating velocity (u_i') Reynolds stresses, spatial-correlations, and temporal correlations. Turbulence characterization using these parameters is consistent with current computational fluid dynamics (CFD) codes that solve the time-averaged turbulent flow equations for in-cylinder flow. Few measurements of in-cylinder turbulence have been made in the burned gas regions. This is because HWA measurements are not feasible and LDA measurements in fired engines are even more difficult than in motored engines. Notwithstanding, LDA measurements in research engines have shown that there is little change in the burned-gas turbulence fluctuations between 600 and 1200 RPM, Foster and Witze (1988), but is substantially changed between 1500 and 2000 RPM, Lorenz and Prescher (1990) .

One of the most important effects of turbulence is it's effect on combustion. Advanced models of turbulent, premixed combustion include local flame-surface generation, local burning rates, and local extinction. Laboratory experiments, Mueller et al.

(1996, 1998), and flame models, Candel and Poinsot (1990), have demonstrated that these phenomena are dependent on the local fluid strain; further, baroclinic torque in the flame attenuates small eddies and generates new vorticity in the burned gas that counters the incident vorticity. One consequence of the advances in flame modeling is that the turbulence fluctuation about the mean, u', is a poor metric for characterizing the turbulence. Whereas u' provides only a random sampling of the coherent velocity structures (cf. Hussain (1986)), strain rate (e_{ij}) and vorticity (w_l) are direct measures of the fluid deformation caused by the coherent structures. Further, e_{ij} and w_l are the fluid properties that control the flame's response to the fluid. Particle Image Velocimetry, PIV, provides the ability to measure the 2-D, instantaneous velocity, out of plane vorticity (w_z) and in-plane strain rates (e_{11}, e_{22}, and e_{12}) over an extended area in a RIC engine. However, PIV measurements in RIC engines have been limited to either motored flow (Reuss et al (1989), Nino et al (1992), Valentino et al (1993), Guibert et al (1993), Sweetland and Reitz (1994) and Reeves et al (1996)) or flow in the unburned gas ahead of the flame (Reuss et al (1990), Nino et al (1993) and Reeves (1995)). The work reported here demonstrates a PIV technique for measuring the velocity, vorticity and strain-rates simultaneously in the burned and unburned gas regions of a firing RIC engine.

The ultimate goal of the work described here is to reveal the physical processes in the flame-flow interaction. With this goal in mind, the engine used in this study was configured to produce a high-swirl in-cylinder flow for three reasons. First, the instantaneous velocity from each realization is well directed and the cyclic variability of the large-scale swirl structure is small; thus, the large-scale flow during each cycle is reasonably repeatable at ignition and any changes in the flow caused by the flame are readily apparent. Second, the flame will pass through a reasonably homogeneous region of turbulence during the early part of the burn. Third, with the well-directed large-scale swirl there is a visually apparent separation between large-scale swirl and the turbulence in the intermediate scales. This property provides a conceptual link between the in-cylinder RIC-engine flow and simple flows traditionally used as starting points for understanding the principles of turbulence.

This paper describes the first step towards the study of flame-flow interaction and has three purposes. One is to describe the technique for making PIV measurements in the burned gas. Because the particle properties and the seeding density are the two key features enabling successful measurements, details of the seeding technique are provided. A second purpose is to provide a qualitative look at the PIV measurements of turbulence in the unburned and burned gases. A synopsis of the motored flow in this engine is provided as a basis of comparison. Because of the abundance of new information, the scope will be restricted to one realization (i.e. one instant of the combustion event in one cycle). The point of this focus is to demonstrate the new information that can be provided in preparation for a more comprehensive treatment. The third purpose is to demonstrate the use of conditionally sampled probability density functions (PDFs) as a means for quantitative analysis in a subsequent study. In particular, PDFs are used to quantify observations made in the distributions of the vorticity and strain rates.

2. EXPERIMENTAL HARDWARE AND METHODS

Sec. 2.1 provides a brief description of the engine and a synopsis of the motored in-cylinder flow from Reuss et al (1995), which is used for comparison with the fired results in Sec. 3. Details of the particle seeding are discussed in Sec. 2.2. In Sec. 2.3 a brief description of the PIV methodology is repeated from Reuss et al (1989, 1990, 1993), with details added where they are unique to the implementation here. In Section 2.4 the method is described for identifying the flame position on the velocity distributions and for generating the conditionally sampled PDF's.

2.1 The Engine and In-Cylinder Flow

The transparent-combustion-chamber engine used for this study is illustrated in Fig. 1. It is a research, four-stroke-cycle engine with 30 mm intake and exhaust valves (one each), a pancake-shaped combustion chamber and a centrally located spark plug. It has a 92 mm bore, 86 mm stroke, and 12.5 mm clearance at top dead center (TDC), resulting in an 8:1 compression ratio. Optical access to the combustion chamber is provided through a quartz ring, which forms the top 25 mm of the cylinder, and a 70 mm diameter quartz window in the top of the extended piston. More details of the engine are given in Reuss et al (1995). The engine was operated at 1200 rev/min, 40 kPa MAP and fueled with premixed propane at an equivalence ratio of 0.95. Fuel mixing was attained using the method described for seed mixing in Sec. 2.2. The engine was skip fired (five motored cycles and one fired cycle, repetitively) with ignition timing set at 14 ca BTDC (crank-angle degrees before top dead center) compression. Skip firing was employed so that the initial condition for combustion is nearly the same as motored cycles. The

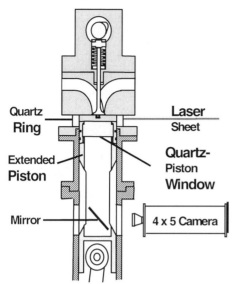

Figure 1 Illustration of the two-valve optical engine, laser sheet, and PIV camera.

here. MAP and exhaust back pressure were controlled to within ± 1 kPa and engine speed to ± 5 rev/min. The air, cooling water, and oil were controlled to 40 ± 1 C.

To create a repeatable in-cylinder engine flow, an intake valve with a 120 deg shroud was directed tangentially as illustrated in Fig. 2. The resulting high-swirl flow has been measured and computed previously by Kuo and Reuss (1995) and Reuss et al (1995) under motored conditions. Those studies have demonstrated that the largest-scale flow (on the order of the cylinder bore) is a combination of swirl and tumble with a swirl ratio of 6 and tumble ratio of 2 at intake-valve closing. The motored-engine ensemble-mean velocity distribution at the TDC, combustion-chamber mid plane (where the fired-engine measurements were made in this study) is nearly solid body rotation (see Reuss et al (1995)). At TDC the swirl ratio is less than 5 and what remains of the tumble structure is weak and relegated to the center region of the bore. The instantaneous velocity distributions, u(x,y), in Figs. 3a and b demonstrate that the shrouded valve produces large-scale-swirl flow each cycle that is nearly the same as the ensemble-average velocity, but with swirl centers that move from cycle to cycle. The instantaneous velocity distributions in Fig. 3a and b also show cyclic variability in the large swirl structure toward the cylinder wall, which manifests itself as waviness in the vector alignment compared to a nearly circular vector alignment in the ensemble-mean distribution.

PIV photographs were taken 2 BTDC compression, which captured burned and unburned regions of approximately equal areas in the field of view studied

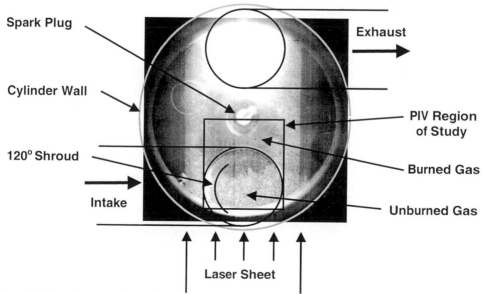

Figure 2 PIV photograph as viewed from the top, front of the engine. The overlaid illustration shows the region of study, shrouded intake valve, and ports.

444

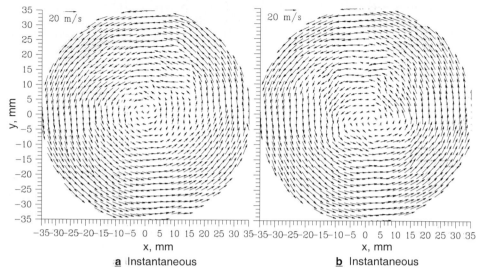

a Instantaneous   b Instantaneous

Figure 3 Instantaneous velocity distributions of two different motored cycles showing cyclic variability of the swirl center. Every fourth vector is shown.

This waviness is caused by the superposition of intermediate scale structures (1 mm < L < 10 mm) associated with the integral scales of the turbulence. These intermediate-scale structures are revealed by high-pass filtering as described in Reuss et al (1989, 1990), and will be shown in the results here. Smaller-scale structures (< 1 mm) are expected to be present in the flow, but are not resolved due to the 1.25 mm resolution of the PIV measurements.

Images of a single flame kernel 8 ca deg after ignition (6 ca BTDC) are shown in Fig. 4. These images are of the visible-light emissions taken simultaneously with two gated, image-intensified, CCD cameras, one viewing through the piston and one viewing through the quartz ring. Fig. 4 shows that the kernel is convected toward the intake valve, which was typical of all images recorded. This bias of the flame kernel is consistent with tumble causing a directed flow at the spark plug that would not otherwise exist in a pure swirling flow. The laser-sheet position shown in Fig. 4 is a reminder that the sheet shows a slice of a complex three-dimensional flame, and one must be mindful of this when interpreting the results.

2.2 Seeding

The particle seeding technique used in this study was the key to making successful measurements in the burned gas. Optimum seed density cannot be achieved simultaneously in the burned and unburned gas because there is a volumetric expansion across the flame that is greater than five. As a compromise, the

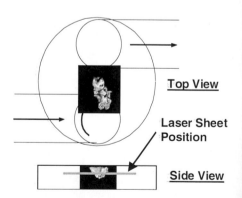

Figure 4 Bottom and side view images of the visible-light emissions recorded simultaneously 6 CA BTDC from a single flame kernel. The laser sheet position (not present for these photographs) is a reminder that the velocity measurements are a two-dimensional sample of a three-dimensional flame.

unburned gas was over seeded to achieve particle density in the burned gas that is sufficient for successful PIV interrogation. In addition, cross-correlation analysis was needed to achieve acceptable analysis of the PIV photographs in the burned-gas region.

The seeding criteria for successful measurements are as follows.
1) The particles must follow the flow.
2) The particles must scatter enough light to be visible over the background scattering and to be recorded on film.
3) The particle concentration must be high enough to yield at least 10 particles per interrogation spot in the burned gas.
4) The particle material must survive the high combustion temperatures.
5) The particle-scattered light must maintain the polarization of the illuminating laser light as required for electrooptical image shifting (to be discussed in Sec. 2.3).

The refractory particles used here were added to the engine-air flow employing a seeding technique used by Witze and Baritaud (1986) for LDV. Four disposable medical nebulizers (designed for humidifying respirators) were used to atomize a slurry of particles and water. The particle concentration in the slurry is adjusted with the expectation that each droplet will contain one particle and thus the droplet will evaporate leaving one particle rather than an agglomerate. In this study the occurrence of agglomeration was detected empirically by increasing the particle concentration in the slurry until either the photograph contained large particle images or the onset of depolarization of scattered light was observed. As will be described in Sec. 2.3, the use of a birefringent crystal for electrooptical image shifting allows depolarization to be observed as particle pairing while illuminating the particles with a single linearly-polarized laser sheet. The slurry particle concentration used for the tests here is 55 parts unpacked-dry-particle volume to 45 parts water volume.

To assure homogeneous mixing prior to entering the intake plenum, the metered engine intake air was split into two streams, recombined at the base of a "T" (thus impinging), exited through the third leg of the "T", and into a tube with fully developed turbulent flow. The seeded flow from the nebulizers (and propane) was injected through a small hole added to the centroid of the "T" fitting where mixing is expected to be high. The engine-air flow was desiccated to a dew point of less than 0 C prior to seeding in order to assure the evaporation of the particle laden droplets. The nebulizer-air flow was not metered. Rather, the metered engine air was decreased to maintain the 40 kPa in the intake plenum. The engine-air flow was 83 l/s without the seed and 45 l/s with the seed, suggesting the total seed gas flow (air and water) was 38 l/s. Based on the number of particle images in the photographs, we estimate that the seeding density was 75 and 15 particles/mm^3 in the unburned and burned gas, respectively.

Three different polishing were tried for seeding material: boron nitride, titanium dioxide, and zirconium oxide. Nipsil SS-50 White Carbon (Sanbayashi et al (1991)) was tried as well. In spite of the small particle size (all advertised to be less than 1μm) and attempts with very low concentration slurries, all four powders are unusable because the scattered light was depolarized. It is presumed that the significant depolarization is a result of the irregular (multifaceted) shape inherent in polishing particles. Hollow, spherical particles are used here to minimize the depolarization. These particles are an undisclosed silica-alumina alloy advertised to have a softening temperature of 1500K and a specific gravity of 2.2. The particles were classified and the size distribution analyzed with a Coulter counter. The mean and standard deviation of the number percent and volume percent are 1.6 ± 0.13 μm and 2.0 ± 0.12 μm, respectively, with 99.5 percent of the particles less than 4 μm by volume and weight.

The response of the particles to the fluid motion was estimated using the data of Haghgooie et al (1986). In that study it was shown that 2 μm alumina particles (at a specific gravity of 4) respond to 2 kHz fluctuations in air at STP. The lower specific gravity of the particles used in this study (2.2) and higher gas density that exist in the engine will improve (increase) the particle frequency response compared to the calculations in Haghgooie et al (1986). Further, since the spatial resolution here is on the order of 1 mm, only large-spatial-scale (low temporal frequency) fluctuations are resolved. Consequently, the better than 2 kHz particle response is considered adequate. Finally, the particles are assumed to have a negligible effect on combustion since the ratio of the particle heat capacity to gas heat capacity (assuming no particle-phase change) is estimated to be less than 10^{-5} for the 75 particles/mm^3 particle density used here.

2.3 The PIV Measurements

The particles were illuminated with a laser sheet that was formed by superimposing two 532 nm, 100 mJ, cross-polarized, 6 ns duration, Nd:YAG laser beams. The 50 mm wide sheet was made narrower than the possible 70 mm-diameter field-of-view to increase the laser power density and thereby increase the particle-scattered light intensity. The lasers used

446

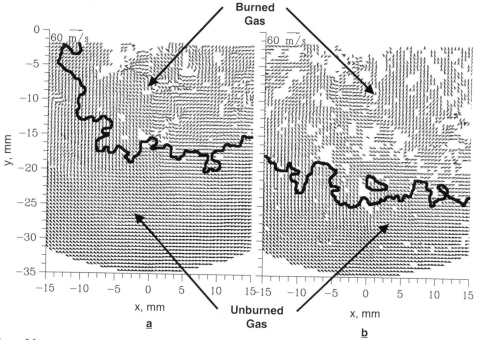

Figure 5 Instantaneous velocity distributions showing the highest (a) and lowest (b) data rates. No vectors are shown at the grid nodes where the interrogations were unsuccessful.

in this study focus near the test section but at slightly different longitudinal positions. Thus, the coincident laser sheets were approximately 0.3 mm and 0.4 mm in the test section based on Polaroid-film burns. Operating with one beam being thicker than the other improves the signal-to-noise ratio during the PIV analysis by decreasing out-of-plane pairing losses caused by (1) out-of-plane fluid velocity and (2) shot-to-shot variability in the out-of-plane beam positions. The two lasers were Q-switched 5 ± 0.01 μs apart. The laser sheets were parallel to the piston top and 6.3 mm from the head, which is the mid plane of the TDC clearance.

To avoid the directional ambiguity inherent in PIV with correlation analysis, the electrooptical image shifting technique developed by Landreth and Adrian (1988) was used and implemented as described in Reuss (1993). As implemented, the technique imparts an image displacement of approximately 0.22 mm displacement for particles with zero velocity. Consequently, positive and negative velocities add and subtract from the shift displacement and thus the

direction of the particle displacement (and therefore velocity) was known.

A large-format (100 mm by 125 mm) film camera with a 210 mm focal length Nikon APO Macor lens at 1:1 magnification is used to photograph the particle images. The photographs were recorded on Kodak 2415 Technical Pan film pushed to approximately 200 ASA for increased sensitivity and contrast. A laser-line optical filter with a 10 nm band width was used to block the flames visible-light emissions and particle incandescence. Particle-image sizes between 10 and 30 μm were observed. The particle image size is larger than the particle size due to the finite diffraction limit of the lens and aberrations caused by the filter and birefringent crystal. The PIV photograph studied here was taken at 2 ca BTDC with the ignition timing at 14 ca BTDC.

An example photograph is shown in Fig. 2, which shows laser light scattered by the cylinder head (background noise) and the particles. The background scattering reveals the valves, intake valve shroud, and spark plug. The photograph in Fig. 2 also shows that the region studied here was a subregion of the photograph near the entrance side of the laser sheet. The photograph could not be analyzed near the side of the cylinder where the laser exits due to high background light. This was caused by back-reflections (specular) off of the exit-side cylindrical surfaces of the quartz ring.

Image processing is used to analyze the photographic negatives to determine the particle displacement (and thus the fluid velocity) at each point on the 0.5 mm-spaced grid. A TSI Model 6000 interrogation system with a cross-correlation technique is used (see Keane and Adrian (1992)). The cross correlation is between a 1.25 mm square spot centered on each node and a 1.86 mm square spot centered at the position equal to the image-shift distance from the node. The larger second area significantly reduces pair losses due to in-plane displacement since almost all pairs from the first spot are captured. Thus, the signal-to-noise ratio in the correlation plane is improved.

After the interrogation of the entire region, velocity vectors were validated using the post-interrogation refinement procedure described in Reuss et al (1989). The refinement rejects those vectors whose velocity magnitude is greater than ten percent different from the nearest neighbors and thus, if retained, would indicate velocity gradients that are larger than could be reasonable. In the 15 photographs taken here, there were 85 to 96 percent valid vectors. Fig. 5a and b are instantaneous velocity distributions from two different fired-engine cycles showing only the validated vectors. These distributions have the highest and lowest data rates, respectively (4% and 14% rejection rates). It can be observed that the regions of invalid vectors occurred predominantly in the burned gas. Inspection of the correlation space (not shown here) corresponding to the invalid vectors revealed that the invalid vectors are caused by poor signal to noise. The depolarization problem identified by Reuss (1993) was present but not dominant. Inspection into the corresponding image regions indicated low particle-image number density, on the order of twenty particle images per interrogation spot. Although this is a lower number density than achieved in the consistently more successful motored PIV images in Reuss et al (1995), it is not quantifiably different from neighboring image-regions where valid vectors were present. However, at this low number density it is easy to identify the number of particle images but hard to identify image pairs. We hypothesize that regions of low data rates are caused by poor correlation due to insufficient particle pairing. This is consistent with the fact that the use of cross correlation analysis reduced the number of invalid vectors. We further speculate that, having gone to

cross-correlation, out-of-plane motion is the dominant cause of unpaired particles. This hypothesis is consistent with the fact that the 0.3 to 0.4 mm laser sheet thickness is small compared to the 1.25 mm interrogation spot thickness required to capture the in-plane motion. Decreased laser-pulse separation did improve data rates, but at the expense of velocity resolution. Attempts to further test this hypothesis by using a region of the laser sheet that is significantly thicker failed due to insufficient laser energy and due to poor laser-beam transformation. A thicker higher-energy laser pulse would have two effects. First, it would decrease the out-of-plane pair losses. Second, it would increase the total number of image pairs in an interrogation spot (due to the larger illuminated volume) without having to further increase the seeding density in the gas.

2.4 Flame Position Identification and Conditional Sampling

To identify the burned- and unburned-gas regions, it is necessary to identify and mark the flame position in each velocity distribution. It is natural to return to the corresponding PIV photographs and determine the flame-position coordinates by marking the position of the change in light scattering as was done in zur Loye and Bracco (1987). In that work, the large seeding density (many particles per imaging pixel) produced a nearly binary light-intensity distribution with a uniformly high scattered-light intensity in the unburned region and a uniformly low intensity in the burned gas. As a result, the flame position was marked unambiguously.

In this study the flame position in the photographic images is less apparent because the seeding density is optimized for PIV analysis rather than for flame imaging as in zur Loye and Bracco (1987). As a result, the PIV photograph consists of spatially separated individual "bright" particle images in a relatively dark field both in the burned and unburned regions. Thus, the image intensity in the burned and unburned regions is not uniform, but rather it differs by the relative number density of individual particles. The result is that the image contrast between the burned and unburned regions is poor compared to that in zur Loye and Bracco (1987). This can be observed in Fig. 6a, which shows an enlargement of the region of study. To acquire this image, the PIV photographic negative (at 1:1 magnification image of the object space in the engine) was scanned with a resolution of 12 pixels per mm. Also, the brightness and contrast of the scanned image were adjusted with image processing to make the flame boundary between the burned gas (low seeding density) and unburned gas (high seeding density)

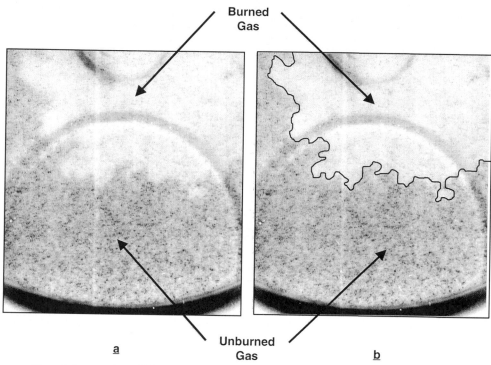

Burned Gas

Unburned Gas

<u>a</u> <u>b</u>

Figure 6 Image enhanced PIV photograph (a) without and (b) with manually drawn flame position.

more apparent to the eye as well as more suitable for publication. After this image processing, it is possible to discern the burned gas region on the largest scales. However, careful inspection of the boundary between the burned and unburned regions reveals that it is still difficult to clearly identify a well-defined position or "edge" between the two regions. An attempt to find an automatic image-processing algorithm to define the boundary between the unburned and burned regions was abandoned.

In lieu of an automatic image-processing algorithm, the flame position was traced manually using the Microsoft Powerpoint drawing tool on a bit map of the scanned negative. An overlay of a hand-drawn flame is shown in Fig. 6b. The image coordinates were carefully registered and converted to the [x,y] coordinates of the PIV interrogation grid and the flame position overlaid on the velocity-distributions as shown in Fig. 7b. The error in the flame position based on boundary ambiguity is estimated to be about 0.2 mm in most regions with occasional position errors as large as 0.5 mm. This is

considered adequate for the 1.25 mm resolution of the velocity scales and 0.5 mm grid used in this study.

In addition to marking the flame position in the velocity distributions, the image processing is used here to identify subregions of the vorticity and strain-rate distributions for conditional sampling. In particular, a flag (zero or one) is assigned to each PIV grid node to indicate if it is in the burned or unburned region. As a third condition, nodes within 1 mm of the flame are identified. PDF's are then computed from these subsamples and used to quantify visual observations.

3. RESULTS

This section presents several hypotheses concerning flame/flow interaction. The suppositions were formulated after observing the distributions of instantaneous velocity, high-pass filtered velocity, vorticity and strain-rates for all fifteen of the realizations recorded in this study. The hypotheses are presented here using distributions from a single

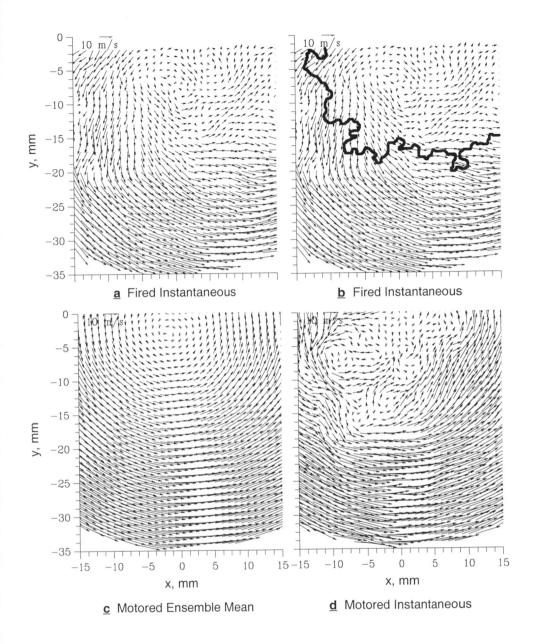

Figure 7 Instantaneous velocity measurements from photograph in Figure 6, (a) with and (b) without flame position. 7c and d show the mean and an instantaneous distribution for the same region of study from the motored tests in Reuss et al (1995). Every second vector is shown.

450

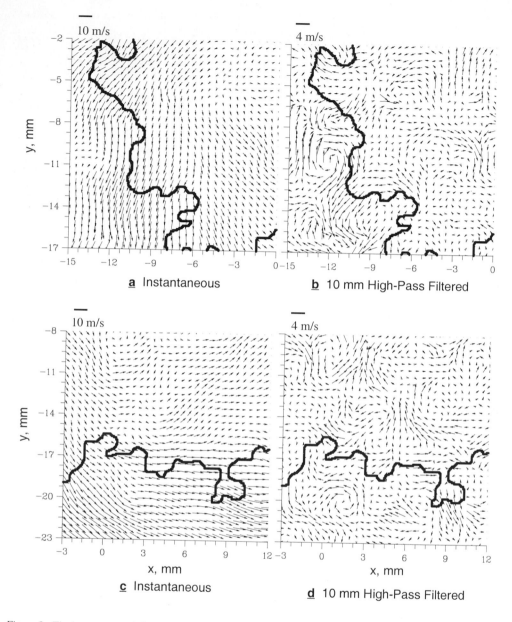

Figure 8 The instantaneous velocity (a & c) and high-pass filtered (10 mm) velocity (b & d) of two subregions in Figure 7b. All vectors are shown.

realization to avoid an unreasonably large number of figures. The hypotheses are based on qualitative observations and conditionally sampled PDF's.

3.1 Instantaneous Velocity Measurements

It is natural to first look at the instantaneous velocity ahead and behind the flame (unburned and burned gas, respectively). The instantaneous velocity distribution corresponding to the PIV photograph in Fig. 6 is given with and without the flame image in Figs. 7a and b, respectively; only every second vector is plotted for clarity. This comparison with and without the flame image is shown to demonstrate that there are no obvious flow features in the instantaneous velocity distribution to mark the flame position. However, it can be observed that the velocity magnitude in the burned gas is generally smaller than in the unburned gas. To assess if this is an effect of combustion or a feature of the motored flow, the instantaneous velocity in the burned gas can be compared with the motored ensemble-mean velocity (100 cycle mean) and randomly chosen motored, instantaneous-velocity distributions shown in Figs. 7c and d, respectively. Note that all four distributions in Fig. 7 have the same velocity scaling. As might be expected, the ensemble- and instantaneous-velocity magnitude is smaller near the center of the swirl, which is coincident with the geometric center of the cylinder ($[x,y] = [0,0]$). However, the velocity magnitude in the burned gases (Fig. 7b) near the upper right had corner of the region ($[x,y] = [5,-20]$ to $[15,0]$) has decreased considerably. This is expected since the gas expansion across the flame opposes the forced-convective (motored) flow in this region. Notwithstanding, there are remnants of the original swirling flow in the burned gas as indicated by the direction of the flow.

3.2 Turbulence Visualization

The intermediate length-scale turbulence (1.25 mm < L < 10 mm) in the instantaneous-vector distributions reveals itself as waviness in the vector alignment compared to the nearly circular vector alignment in the ensemble-mean distribution (cf. Figs. 7a and c). This turbulence is the superposition of the intermediate length-scale coherent structures with the large-scale swirl structure. It is possible to visualize the coherent structures of the turbulence using the high-pass spatial-filtering technique employed by Reuss et al (1990). This is shown in Fig. 8 for two expanded regions (every vector plotted) of Fig.7b. Figs. 8a and c show the instantaneous velocity distributions, and Figs. 8b and d show the high-pass filtered velocity distributions. The high-pass filtered velocity distributions are attained through a two step

operation. First, the local average of the instantaneous velocity centered at each grid node is computed by convolution of the instantaneous velocity distribution with a Gaussian-weighting function. Here the Gaussian weighting is 10 mm at the $1/e^2$ point. Second, this local-average velocity (low-pass filtered velocity, u_{lp}) is subtracted from the instantaneous velocity, u, leaving the high-pass filtered velocity, $u_{hp} = u - u_{lp}$. Thus, the high-pass filtered velocity is the velocity observed from a Lagrangian frame of reference where the observer is moving at the local spatial-average instantaneous velocity of the large-scale swirl flow.

Inspection of the velocity distributions in Fig. 8 reveals that the high-pass filtered distributions show coherent structures at this length scale. Using the labels of Hunt et al (1988), it is possible to identify eddies (e. g., rotational structures at $[x,y] = [-4,-10]$, $[-12,-10]$, and $[2,-20]$) and streaming flows at $[x,y] = [-12,-4]$, $[-11,-14]$, and $[10, -21]$. Note that the coherent structures are visible in both the burned- and unburned-gas regions. As noted in Reuss et al (1989), the spatial scale of these structures is consistent with the turbulence integral-length scale measured with LDA in engines of similar geometry. The appearance of these coherent structures has two implications. First, experience has demonstrated that PIV experiments configured with insufficient velocity dynamic range result in high-pass filtered velocity distributions with random vectors rather than coherent structures as observed here. Thus, the appearance of the coherent structures in the burned gas is taken as evidence that the accuracy of the measurements in the burned gas is sufficient to resolve the intermediate scale structures with approximately the same accuracy as measurements in motored and unburned flows. The second implication is that it is possible to measure and visualize the flow structures that affect the flame wrinkling. For example, in comparison to the instantaneous velocity, observations along the flame in the high-pass velocity distribution reveal streaming structures that correlate with the wrinkling. The most obvious correlation is for the streaming structures at $[x,y] = [-13,-4]$, $[-11,-14]$ and $[10,-21]$. It is not obvious that eddies directly affect the flame wrinkling since the flame wraps around but does not appear to propagate through the vortex structures. Speculating, the three dimensional vortex tubes producing the eddies observed in these two-dimensional measurements act to create the streaming flows in between the vortex tubes; the flame is then convected by the streaming flows between the eddies rather than burning through the eddies. This is supported by the streaming flow at $[-11,-14]$ that appears to be formed between the three eddies centered at $[-8,-15]$, $[-12,-11]$

452

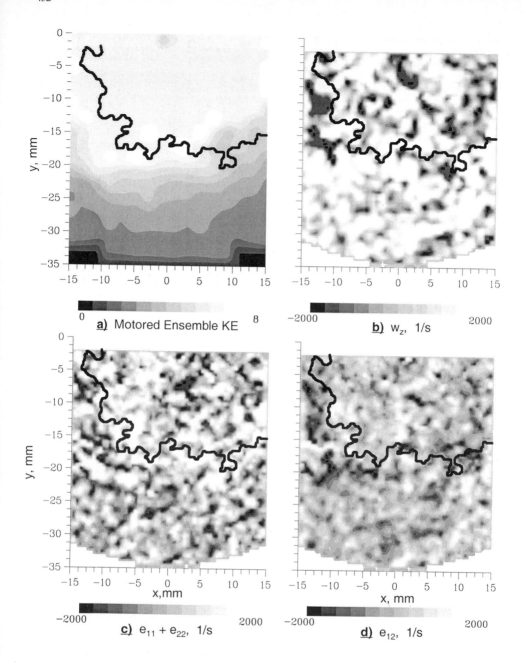

a) Motored Ensemble KE

b) w_z, 1/s

c) $e_{11} + e_{22}$, 1/s

d) e_{12}, 1/s

Figure 9 (a) The two-dimensional kinetic energy ($\frac{1}{2}\,u'^2 + \frac{1}{2}\,v'^2$) based on fluctuations about the ensemble average of 100 motored cycles. The flame studied here is superimposed to show relative position. Figures b, c, and d are the vorticity, normal- strain rate, and shear-strain rate computed from the instantaneous-velocity distribution in Figure 7a.

and [-4,-11] and, in turn, appears to be stretching the flame to be convex with respect to the unburned gas. This supposition cannot be considered universally true since it is based on observations of only 15 realizations and only one operating condition. Nonetheless, observations of the high-pass filtered distributions make two points. First, u', the traditional turbulence measure, is a random sampling of coherent structures in the burned gas as suggested in the introduction. Second, the flow-direction and spatial scale of the coherent structures appears to correspond to the flame wrinkling.

3.3 Turbulence Characterization

To determine the effect of the flame on the in-cylinder turbulence, it is of interest to compare distributions of turbulence kinetic energy, KE, vorticity, w_i, and strain-rate, e_{ij}, from both motored and fired conditions. However, at this time KE from the motored flow and w_i, and e_{ij} from fired condition are all that are available for comparison. Thus, we proceed with a qualitative comparison recognizing that KE is based on u', the random sample of velocity fluctuations about the ensemble mean, while w_i, and e_{ij} are direct measures of the fluid deformation due to turbulence from single realizations. For this comparison, the motored-engine KE distribution from Reuss et al (1995) is repeated In Fig. 9a for the region of the cylinder studied in these fired tests with the flame position superimposed. Fig. 9a shows that in the absence of combustion the turbulence is very high in the cylinder center ([x,y] = [0,0]), low near the wall, and has a relatively steep gradient between the high- and low-turbulence regions (at a radius between 20 and 25 mm). Fig. 9a shows that the flame at 2 ca BTDC has propagated just to the edge of the high turbulence KE region.

The vorticity, w_z, ahead and behind the flame is shown in Fig. 9b. It can be observed that the unburned gas has predominately positive vorticity as indicated by the predominately light regions. This trend is consistent with the dissipation of large-scale gradients in the large scale swirl structure, which has positive circulation. By comparison, the burned gas vorticity has a more equal balance of positive (light regions) and negative vorticity (dark regions) and the magnitude of the negative vorticity is much greater than that in the unburned gas. This trend was observed in all 15 realizations captured in this study. This observation provided the motivation for measuring the conditionally sampled PDF's shown in Figs. 10a and b for the unburned- and burned-gas regions, respectively. Comparison of Figs. 10a and b at $w_z \approx -2000$ 1/s demonstrates that indeed there is more negative vorticity in the burned gas. Further, the

vorticity distributions are approximately symmetric with the peak of the burned-gas PDF shifted towards zero (also compare the mean values) indicating a more equal distribution of positive and negative vorticity in the burned gas. In the absence of any knowledge about the motored flow, one might assume that the flame has indeed affected the vorticity behind the flame. However, the flame position in Fig. 9a shows that the burned gas is in the high KE region of the motored flow and the unburned gas is predominately in the low KE region. Thus, one might speculate that the flame is behaving as a passive scalar and the differences in the burned and unburned gas vorticity may be the result of the vorticity in the forced-convective motored flow. This supposition is weakly supported by the coincidence of high negative vorticity in the fired cycle and high motored KE in the region [x,y] = [-15,-20] to [-8,2] (upper left hand corner), which is in the unburned gas region.

Figures 9c and d show the distributions of the

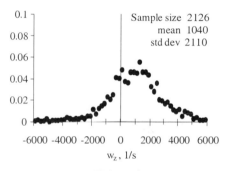

a) Unburned gas.

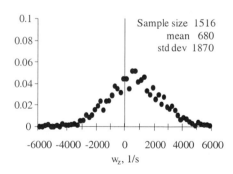

b) Burned gas.

Figure 10 Conditionally sampled PDF's of the vorticity in Fig. 9b.

sum of the normal-strain rates (e_{11} + e_{22}) and the shear-strain rate (e_{12}). The spatial scale of the structures are approaching the spatial resolution of the measurements, and thus serve as a reminder that measurements with smaller spatial resolution are desired. Unlike the vorticity distribution, visual

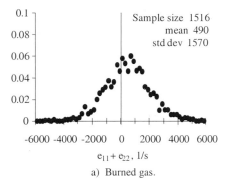

a) Burned gas.

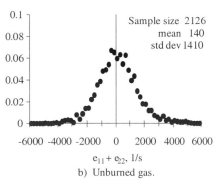

b) Unburned gas.

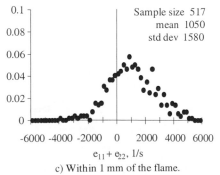

c) Within 1 mm of the flame.

Figure 11 Conditionally sampled PDF's of the normal strain in Fig. 9c

inspection of the normal- and shear-strain-rate distributions does not reveal that the structures are obviously different ahead and behind the flame. However, careful observation while following along the flame reveals that the normal-strain rates are predominately positive adjacent to the flame. The later observation is physically consistent with the high rates of heat release in the flame. In fact, normal-strain was found to be a useful flame marker for the two-dimensional, wrinkled, laminar flames studied by Mueller, et al (1996, 1998). To test the validity of these observations, PDF's sampled in the burned-gas, the unburned-gas and the region within ±1 mm of the flame are shown in Figs. 11a through c, respectively. Looking first at the unburned-gas PDF (Fig. 11b), it appears nearly symmetric with a mean of 140. By comparison, the burned-gas PDF (Fig. 11a) has shifted towards positive values with a mean of 490 apparently caused by the larger number of occurrences of positive strain rate between 0 and 4000 1/s. Thus, the PDF has shown regional differences not readily apparent from observations of the plotted distribution. The normal-strain-rate PDF sampled within a 1 mm of the flame (Fig. 11c) shows a more dramatic affect. Clearly the peak has shifted in the positive direction compared to the general population in the unburned gas. Further, the distribution has skewed significantly towards positive values with a significant increase in the number of occurrences between 1000 and 5000 1/s.

In the case of the shear strain there are no obvious differences between the unburned- and burned-gas regions, either observed in the distributions or in the PDF's. The PDF for the entire region (burned and unburned gases) in Fig. 12. However, the distribution does show a considerably narrower distribution when compared with the vorticity or normal-strain rates. (compare the standard deviations in Figs. 10 through 12).

The observations described here are insightful as a first step and raise interesting questions concerning the nature of flow and combustion in this high-swirl in-cylinder RIC engine. In particular, what is the link between u', the traditional measure of turbulence, and the direct measures of the turbulence fluid deformation (it is interesting that the highly-directed swirling flow has large-scale inhomogeneity in the vorticity and normal strains while the shear strain is relatively homogeneous)? Two more questions arise concerning the physics of the flame/flow interaction. In particular, is the flame behaving as a passive scalar leaving the turbulence relatively unaffected, and does the flame propagate preferentially around rather than through the vorticity? Investigation of these questions requires analysis and comparisons of much larger samples under a larger variety of operating conditions.

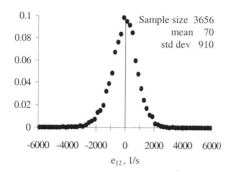

Figure 12 PDF of shear strain in the unburned and burned gas of Fig. 9d.

Even then, analysis is particularly difficult both because these PIV measurements provide only two-dimensional samples of three-dimensional structures, and because no two realizations are the same, which renders it impossible to compare motored and fired cycles directly. However, the conditionally sampled PDFs do provide a tool for comparative analysis.

4. SUMMARY

Particle image velocimetry has been used for simultaneously measuring the velocity ahead and behind a flame in a RIC engine. Electro-optical image shifting was employed to remove the directional ambiguity and a spatial resolution of 1.25 mm was achieved. The techniques required to make these measurements successful are described. We deduce that the data rates in the burned gas could be improved by using thicker higher-energy laser sheets.

Fifteen realizations of the in-cylinder turbulence were recorded at either 2 or 4 ca BTDC. However, the results from only one realization are shown in order to maintain a manageable amount of information. Visual observations of the instantaneous velocity, filtered velocity, vorticity and strain-rate distributions provide insight into the interaction of the flame and the flow. First, the high-pass filter distributions show that u', the traditional turbulence measure, is a random sampling of coherent structures in the burned gas. This is an expected result, but none the less an affirmation of what was known to be true in non-engine flows, motored-engine flows, and the unburned gas of a fired engine. Second, the gas expansion caused by the flame decreased the magnitude of the swirling-flow velocity in the burned gas region as one might expect. Third, the flow direction of coherent structures revealed in the high-pass-filtered velocity distribution show a strong correspondence with the flame wrinkling, although the flame appears to propagate around rather than through eddy structures. Fourth, the length scales of the vorticity appear to be the same size as the length-scales of the flame wrinkling.

Observations and conditionally sampled PDFs of the vorticity and strain-rate distributions showed large-scale inhomogeneity. In particular, the vorticity ahead of the flame had a positive bias compared to the burned gas region. The normal-strain rates show a positive bias behind and especially near the flame, which is consistent with the thermal expansion of heat release. Finally, the shear-strain rate appeared to be homogeneous, with a symmetric and considerably narrower PDF when compared to the vorticity and normal shear.

The restricted focus of this paper has been to demonstrate the new information that can be provided in preparation for a more comprehensive treatment. We caution that the observations can not be considered universal since they are made for only fifteen realizations at one operating condition in one engine geometry. In fact, no more than fifteen realizations were recorded due to the difficulty producing PIV photographs with acceptable data rates in the burned gas. We believe we have sufficient evidence to indicate this difficulty occurred because the laser sheets were too thin. Thus, there is a danger of bias since we might have recorded only realizations with small out-of-plane motion. None the less, this work demonstrates a technique for recording two-dimensional velocity distributions in RIC engines, suggests methods for improved experiments, and provides direction for future studies.

5. REFERENCES

Arcoumanis, C. and Whitelaw, J. H., 1987, "Fluid Mechanics of Internal Combustion Engines – A Review", Proc. I. Mech. E., 201C.

Candel, S. M., and Poinsot, T. J., 1990, "Flame Stretch and the Balance Equation for the Flame Area," Combustion, Science, and Technology, 70.

Foster, D. E. and Witze, 1988, P. O., "Two-Component Laser Velocimetry Measurements in a Spark Ignited Engine," Comb. Sci. & Tech., Vol. 59.

Guibert, P., Murat, M., Hauet, B., and Keribin, P., 1993, "Particle Image Velocimetry Measurements: Application to In-Cylinder Flow for a Two Stroke Engine", SAE Paper 932647.

Haghgooie, M., Kent, J. C., and Tabaczynski, R. J., 1986, "Verification of LDA and Seed Generator Performance", Experiments in Fluids, 4.

Hunt, J. C. R., Wray, A. A., and Moin, P., 1988, " Eddies, Streams and Convergence Zones in Turbulent Flows", In Report CTR-S88, Proceedings of the 1988 Summer Program, Center for Turbulence Research, Stanford University.

Hussain, A. K. M. F., 1986, "Coherent Structures and Turbulence," Journal Of Fluid Mechanics, 173.

Keane, R. D., and Adrian, R. J., 1992, "Theory of Cross-Correlation Analysis of PIV Images", Applied Sci. Res., 49.

Kuo, Tang-Wei, and Reuss, D. L., 1995, "Multidimensional Port and Cylinder Flow Calculations for the Transparent-Combustion-Chamber Engine," ASME, IVE-Vol. 23, Engine Modeling.

Landreth, C. C., and Adrian, R, J., 1988, "Electrooptical Image Shifting for Particle Image Velocimetry," Applied Optics, Vol. 27, No. 20.

Lorenz, M. and Prescher, K., 1990 , "Cycle Resolved Measurements on a Fired SI-Engine at High Data Rates Using a Conventional Modular LDV-System," SAE Paper 900054.

Mueller, C. J., Discroll, J. F., Reuss, D. L., and Drake, M. C., 1996, Twenty-Sixth Symposium (International) on Combustion, The Combustion Institute, Pittsburgh.

Mueller, C. J., Discroll, J. F., Reuss, D. L., Drake, M. C., and Rosalik, M. E., 1998, Combustion and Flame, Vol. 112, No. 3.

Nino, E., Gajdeczko, B. F., and Felton, P. G., 1992, "Two Color Particle Image Velocimetry Applied to a Single Cylinder Two Stroke Engine", SAE Paper 922309.

Nino, E., Gajdeczko, B. F., and Felton, P. G., 1993, "Two Color Particle Image Velocimetry in an Engine With Combustion", SAE Paper 930872.

R. B. Rask, 1984, "Laser Doppler Anemometry Measurements of Mean Velocity and Turbulence In Internal Combustion Engines," in Int. Conf. on Applications of Lasers and Electro-Optics, Boston.
Reeves, M., 1995, "Particle Image Velocimetry Applied to Internal Combustion Engine In-Cylinder

Flows", Ph. D. Thesis, Loughborough University of Technology, England, September.

Reeves, M., Garner, C. P., Dent, J. C., and Haliwell, N. A., 1996, "Particle Image Velocimetry Measurements of In-Cylinder Flow in a Multi-Valve Internal Combustion Engine", Proc. Inst. Mech. Engrs., Vol. 210.

Reuss, D. L., Adrian, R. J., Landreth, C. C., French, D. T., and Fansler, T. D., 1989, "Instantaneous Planar Measurements of Velocity and Large-Scale Vorticity and Strain Rate in an Engine Using Particle Image Velocimetry," SAE Paper 890616.

Reuss, D. L., Bardsley, M., Felton, P. G., Landreth, C. C., and Adrian, R. J., 1990, "Velocity, Vorticity, and Strain-Rate Ahead of a Flame Measured in an Engine Using Particle Image Velocimetry," SAE Paper 900053.

Reuss, D. L., 1993 , "Two-Dimensional Particle-Image Velocimetry with Electrooptical Image Shifting in an Internal Combustion Engine," in the Proceedings of the International Society for Optical Engineering, Optical Diagnostics in Fluids and Flows, Vol. 2005.

Reuss, D. L., Kuo, T-W, Khalighi, B., Haworth, D., and Rosalik, M., 1995, "Particle Image Velocimetry Measurements in a High-Swirl Engine Used for Evaluation of Computational Fluid Dynamics Calculations, SAE Paper 952181.

Sanbayashi, D., Ando, H., and Kumagai, H., 1991, "Feasibility of Using Several Powder Materials as the Seeding Particle for LDV Measurement", JSAE Review, Vol. 12, No. 2.

Sweetland, P., and Reitz, R. D., 1994, "Particle Image Velocimetry Measurements in the Piston Bowl of a DI Diesel Engine", SAE Paper 940283.

Witze, P.O., and Baritaud, T. A., 1986, "Particle Seeding for Mie-Scattering Experiments in Combusting Flows", ed. Proceedings of the Third International Symposium on Laser Techniques and Applications, Lisbon Portugal (ed.), Springer-Verlag.

Valentino, G., Kaufman, D., and Farrell, P., 1993, "Intake Valve Flow Measurements Using PIV", SAE Paper 932700.

Valentino, Gerardo, Corcione, F. E. , and Seccia, G., 1997, "Integral and Micro Time Scales Estimated in a DI Diesel Engine," SAE Paper 971678.

V.6. In-Cylinder Measurements of Mixture Composition for Investigation of Residual Gas Scavenging
P. Miles

Combustion Research Facility, Sandia National Laboratories,
Livermore, CA USA

Abstract. Laser Raman scattering with broadband signal detection is employed to simultaneously measure the mole fractions of CO_2, O_2, N_2, C_3H_8, and H_2O under conditions simulating a cold start. The engine is operated with well-premixed fuel and air, such that each of the measured species can be used to independently estimate the mole fraction of burnt residual gases. The residual gas mole fraction estimates are subsequently examined for consistency and for evidence of poor scavenging of H_2O due to condensation on the cold combustion chamber walls. The data indicate that if this mechanism inhibiting H_2O scavenging is operative, it results in H_2O number densities which are only 10-15% higher than expected. It is suggested that these higher than expected levels of H_2O may be associated with Raman scattered light from condensate films on the windows. Raman scattering techniques using backscatter collection geometries are expected to be more sensitive to light scattered from condensate films, which may explain previously measured high H_2O mole fractions.

Keywords. Raman scattering, Residual gas scavenging

1 Introduction

A number of studies have shown that a significant fraction of the unburned hydrocarbon (UHC) emissions from spark ignition engines occur during the warm-up period following a cold-start (*e.g.* Takeda, *et al.*, 1995). Directly or indirectly, the majority of these disproportionate emissions can be traced to a single common factor: cold surfaces in the combustion chamber and intake ports. Cold combustion chamber surfaces result in increased flame quenching within the combustion chamber, while cold surfaces in the intake port and valve region lead to slow or insufficient fuel vaporization. Poor fuel vaporization, in turn, results in gas phase air/fuel ratios above the flammability limit (with subsequent misfire) and in partial burning associated with mixture inhomogeneities. Dilution of the charge with burnt residuals can exacerbate each of these factors; a dilute mixture is more susceptible to flame quench and has a narrower range of inflammable air/fuel ratios.

Recently, Grünefeld, *et al.* (1995) suggested another mechanism by which cold surfaces might affect cold-start engine operation and emissions. In their study, which employed laser Raman scattering to measure the in-cylinder gas composition during a cold start simulation, high instantaneous values of residual water content were observed (between 5 and 6%), which correspond to a residual gas content of roughly 40%. Based on the simultaneously measured O_2 gas densities, a residual gas content of only half that corresponding to the water content was deemed more probable. The suggestion was made, therefore, that the residual water content may not be a reasonable marker of the total residual gas content, due to the possibility of preferential scavenging of the different residual gas species. In this preferential scavenging scenario, water vapor condenses on the cold combustion chamber walls during the expansion and exhaust stroke (while N_2 and CO_2 do not), and re-enters the gas phase through vaporization during the subsequent compression stroke. Corroborating this idea of a preferential scavenging mechanism was the observation that the scavenging of the residual water vapor when misfires occurred was 5–10 times slower than expected. It was concluded that charge dilution by residual gases, particularly the slowly scavenged H_2O, was a major cause of misfire and of increased UHC emissions during cold-start and warm-up.

The experimental system used in the above-referenced study, however, may be susceptible to interferences caused by the back-scatter collection geometry employed. The cause of this potential interference can be understood by recognizing that the incident laser beam will scatter Raman-shifted light from any films of fuel or combustion products which condense on the beam entrance or exit windows. A film thickness on the beam entrance and exit windows of roughly 1.3 μm would be sufficient to double the measured H_2O content. Furthermore, the energy required to vaporize these films can be shown to be greater than the laser pulse energy, such that window 'cleaning' by the laser pulse is not possible. Thus, in a backscatter geometry, condensate films not only attenuate the scattered light from the gas phase mixture, but may also contribute sufficient signal to invalidate measurements of the gas phase species.

In this paper the in-cylinder residual gas content under simulated cold-start conditions is revisited using a similar Raman scattering technique. An alternate experimental geometry is employed, however, in which the scattered light is collected at 90° from the incident beam. With this geometry, light scattered from films on the beam entrance and exit windows is not in the direct view of the collection optics, and does not contribute directly to the measured signal. Furthermore, in this work the engine is fueled with gaseous propane, which is well pre-mixed with the combustion air. Under these circumstances, each of the measured (non-Nitrogen) major species, CO_2, O_2, C_3H_8, and H_2O, can be used to estimate the residual gas content. The various residual gas estimates can be examined for consistency and thus for evidence of the existence of preferential scavenging of the different residual gas components.

2 Fundamental Physics

Vibrational Raman scattering is an inelastic scattering process in which there is a net exchange of energy between the incident light and the scattering molecule. In this exchange, the change in vibrational energy of the participating molecule equals the change in energy of the scattered light. Because the quantized vibrational energy states available to each molecule are unique, the change in energy of the molecule (and, thus, the change in energy, or wavelength, of the scattered light) can be used to identify the scattering molecular species. By collecting scattered light over a broad band of wavelengths, and subsequently dispersing this light onto a multi-channel detector, multiple species can be monitored simultaneously.

The energy of the scattered light from species i, $E_{scat,i}$, is given by (Eckbreth, 1996)

$$E_{scat,i} = E_{laser} \left(\frac{\partial \sigma}{\partial \Omega} \right)_i N_i \, \Omega \, \ell \, \eta_i \tag{1}$$

where E_{laser} denotes the laser pulse energy, $(\partial \sigma / \partial \Omega)_i$ the differential scattering cross-section of the ith species, N_i the molecular number density, Ω the solid angle subtended by the collection optics, ℓ the length of the beam from which scattered light is collected, and η_i the efficiency of the optical system. From Eq. 1, it is seen that the signal is directly proportional to N_i, provided the differential scattering cross-section is constant. In practice, $(\partial \sigma / \partial \Omega)_i$ is a function of temperature; however, for the limited temperature range considered here this temperature dependence is small and is neglected.

By forming a ratio of the signals collected from each species i to the weighted sum of the signals from the major species, it is possible to determine the species mole fraction independent of such parameters as instantaneous laser power or η_i (which may be affected by such factors as window fouling). The weighting factors, which represent the signal obtained per molecule of species i relative to the signal obtained per N_2 molecule, are obtained by calibration.

3 Experimental Apparatus

Measurements reported here are obtained in the Sandia side-valve engine, which is pictured in Fig. 1. The side-wall location of the valves in this engine allows unimpeded optical access to the combustion chamber through a large window in the head. The intake valve can be fitted with a shroud, which allows the introduction of various degrees of swirl to the in-cylinder flow. With an unshrouded valve, mixing of the fresh charge with the residual gases is quite rapid, while with the shrouded valve significant mixture stratification persists throughout the compression stroke (Miles and Hinze, 1998). The incident laser beam enters and exits the cylinder through two small, diametrically opposed

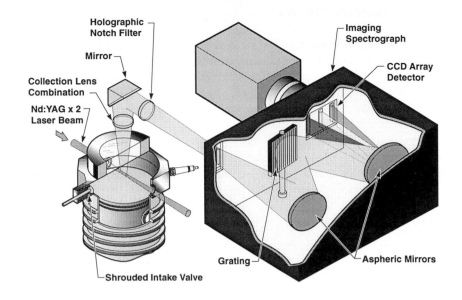

Figure 1: The side-valve research engine and layout of the optical diagnostic system

windows, and follows a path through the cylinder which approximately bisects the clearance height of the combustion chamber. The engine geometry is summarized in Table 1.

Raman scattered light is collected from a 10.9 mm length of the beam (λ=532 nm) in the central portion of the combustion chamber. The focused beam diameter within the engine, 0.49 mm as defined by the $1/e^2$ intensity contour, defines the spatial resolution in the plane normal to the beam. After passage through the engine, the beam energy is measured on an instantaneous, shot-by-shot basis. Typical pulse energies of 120 mJ were employed. The collected light is subsequently refocused to form an image of the laser beam at the entrance slit of an imaging spectrograph. As depicted in Fig. 1, the spectrograph disperses the image in the entrance slit spectrally, such that at the exit plane a two-dimensional image is formed, consisting of multiple images of the beam at the entrance slit. Each beam image at the detector plane corresponds to Raman scattered light from a particular molecular species. Thus, at each spatial location, the number density of each species can be determined from the intensity of the appropriate beam image integrated over the appropriate spatial and spectral regions. The size of the detector, and the dispersion of the spectrograph, are selected such that scattered light from all major species of combustion is imaged onto the array, from CO_2 at 571 nm to H_2O at approximately 660 nm.

The detector employed is an unintensified, cryogenically-cooled, 1024 x 1024 back-illuminated CCD array, with a (typical specification) quantum efficiency that exceeds 85% over the wavelength range of interest. Due to the lack of an image

Table 1: Engine Geometry and Operating Conditions

Bore	7.64	[cm]
Stroke	8.27	[cm]
Clearance Volume	99.0	[cm^3]
Compression Ratio	4.82	
Speed	600	[rpm]
Manifold Pressure	47.5	[kPa]

intensifier, the camera cannot be gated quickly and interference from combustion luminosity necessitates a skip-fired mode of engine operation. This additional complication is outweighed by the high quantum efficiency; typical intensified cameras are characterized by an effective quantum efficiency (accounting for the intensifier noise factor) of approximately 5% (Paul, *et al.*, 1990) in this wavelength range. Because of the limited height of the spectrograph entrance slit, only 75% of the vertical extent of the array is used. Data are obtained in two formats: high-resolution 512 x 384 images obtained using 2 x 2 charge binning on the CCD array; and low-resolution 16 x 12 images obtained binning the charge into 64 x 64 pixel "superpixels". This massive charge binning is required for instantaneous measurements, where the spatial and spectral signal integration is performed on-chip to minimize the contributions from measurement noise. Employing 64 x 64 superpixels, the combined contribution from thermally generated dark-noise, spurious charge generation, and electronic read-noise can be characterized by a normally-distributed, random noise source with a standard deviation of 11 e^- (Miles, 1998). Thus, for signals greater than approximately 120 e^-, the measurement is expected to be limited by photoelectron shot-noise. Approximately 600 signal photoelectrons are expected from each superpixel from both CO_2 and H_2O under the measurement conditions employed. Signals from the remaining species are considerably higher.

4 Experimental Procedure

The engine is run slightly lean with premixed propane fuel at an equivalence ratio of 0.96. As mentioned above, a skip-fired mode of operation is employed, wherein the engine is fired three consecutive times and the ignition disabled on the fourth cycle, during which the data are obtained at top dead center (TDC) firing. It was demonstrated that, had it been fired, the combustion performance of the fourth cycle would have been equivalent to that of steady-state operation. To investigate scavenging efficiency after a misfire, an additional motored cycle is added, and data are also obtained at TDC firing of this fifth cycle.

To simulate cold-start conditions, the engine cooling water is maintained at 21 °C. The engine is brought to speed (600 rpm) under motored operation, the ignition enabled, and data acquisition commences after speed stabilization. This start-up procedure required 33 seconds; 2/3 of which represents motored operation. Data acquisition thus began approximately 11 seconds after first-fire, or, roughly, after 55 engine cycles (3/4 of which are fired). 200 single-cycle, low-resolution images are acquired over the next 800 engine cycles, followed immediately by a high-resolution image, cycle-averaged over an additional 400 cycles (100 measurements). During this data acquisition sequence, films of condensate were observed to form on the large cylinder head window. High levels of scattered light also implied the presence of films on the beam entrance and exit windows. These films persisted through the end of the sequence, by which time the cooling water leaving the head had risen in temperature to approximately 24 °C. Finally, with the engine stopped and positioned at TDC overlap, air reference spectra for background determination were acquired while flowing air through the engine. During acquisition of the reference spectra, the films on the windows were observed to shrink but not to disappear.

After allowing the engine to cool, this procedure was repeated with data obtained every fifth cycle, as described above. To investigate the effect of the in-cylinder flow field on the scavenging performance, data were obtained both in the nominally quiescent flow generated by an unshrouded intake valve and in the highly swirling flow ($R_s \approx 10$) created by the shrouded intake valve.

5 Sample Data And Data Reduction

A sample high-resolution image, obtained in the quiescent engine flow, is shown in Fig. 2. In contrast to equivalent images obtained under fully-warmed up conditions (Miles and Hinze, 1998), the levels of background light are considerably higher. The elevated background is attributed to the formation of condensate films on the combustion chamber windows and increased elastically scattered light from these films which is insufficiently rejected. Note from the spectrum shown at the bottom of Fig. 2, generated by summing the columns of the image above, that the (spatially) integrated signals remain well defined on a variable background level. The background level can be fit quite accurately in the vicinity of the H_2O and C_3H_8 signals. In the CO_2 and O_2 region, however, the background determination is more ambiguous. To assist with background fitting in this region, the air reference spectra described above are used. Due to the variable film size during acquisition of these spectra, however, uncertainties in the appropriate form of the background are introduced. Thus, the mole fractions of CO_2 and O_2 deduced from the cycle-averaged images are less reliable than the mole fractions of C_3H_8 and H_2O.

Superposed on the image in Fig. 2 is a light-colored grid which depicts the spatial and spectral integration regions defined by the superpixels employed while acquiring the single-cycle images. Due to the extent of the spectral binning, and to

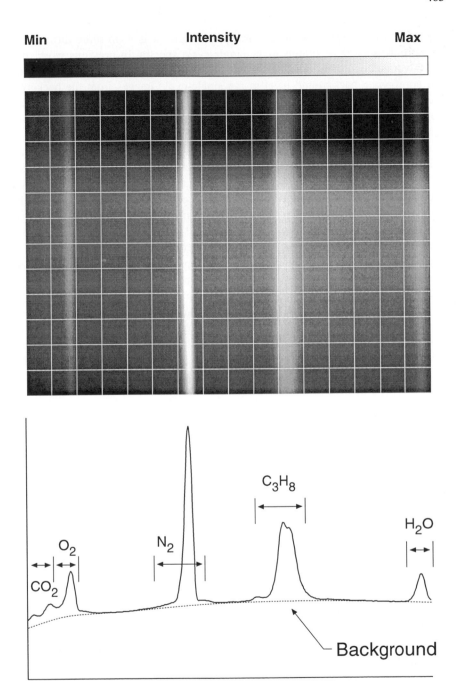

Figure 2: Cycle-averaged image and corresponding spectrum obtained with the quiescent flow

the single-cycle shot-noise on the background light, it is even more difficult to fit background spectra accurately to the single-cycle spectra. In previous work, under fully-warmed up conditions, the constancy of the background was used to deduce the appropriate background level in the single-cycle images from the background levels measured in a cycle-averaged image. This strategy is not appropriate here, however, due to the development of films on the combustion chamber surfaces during these cold start simulations, and the subsequent evolution of the background level. Quantitative species mole fractions are therefore not easily determined from this single-cycle data. Nevertheless, evidence for preferential scavenging of the combustion by-products can be obtained from the trends observed in the data. The remaining discussion of the data reduction methodology focuses on the identification of these trends within the larger trend associated with film formation and background drift.

From each single-cycle image, a spatially and spectrally integrated "signal" for each species is obtained by summing the appropriate column(s) of the image. These signal values are proportional to the molecular number density of each species, averaged over the 10.9 mm beam length, plus an offset associated with the background light. From consecutive single-cycle images (with images obtained every fourth engine cycle), data sequences are constructed for each species. To separate the background drift from the underlying trends in the data, a drift sequence $d_j = d_1 + d_2 + ... + d_{n-1} + d_n$ is defined from the measured N_2 sequence N_j by subtracting the mean signal value:

$$d_j = N_j - \langle N \rangle \tag{2}$$

The projection of any data sequence S_j onto this drift sequence can then be subtracted from the original data sequence to give a drift corrected sequence $\hat{S}_j$:

$$\hat{S}_j = S_j - \frac{(S,d)}{(d,d)} d_j \tag{3}$$

where (a,b) denotes the standard inner product between sequences a and b.

This technique for background drift removal assumes that the temporal evolution of the background is self-similar for each signal, and can be determined from the temporal evolution of the N_2 signal. Implicit here is the expectation that the true N_2 signal does not change with time. Because the fresh charge is provided to the engine at a constant mass flow rate, metered through critical orifices, the amount of fresh N_2 inducted in each cycle is expected to be approximately constant. Any changes in the charging process associated with engine warm-up will be compensated for automatically by a rising or falling manifold pressure, such that the inducted mass per cycle remains constant. There are, however, at least two mechanisms by which the N_2 signal can be envisioned to change with time: first, by changes in the residual gas mole fraction, and second, via preferential H_2O scavenging.

The first mechanism will be associated with either an increase or decrease in the trapped mass within the cylinder, with a corresponding increase or decrease in the N_2 signal levels. The drift sequence will thus have components associated with both background drift and temporal changes in the N_2 number density. The component associated with N_2 number density changes may mask changes in the true signal levels associated with H_2O and CO_2, making changes in the trapped residual mass difficult to detect. This same component, however, would introduce variations in the O_2 and C_3H_8 signals, which are expected to remain constant. By examining the corrected sequences from these fresh charge species, then, it is expected that significant temporal changes in the residual mass can be detected.

Poor scavenging of H_2O, the second mechanism, can increase the trapped mass of N_2 by effectively enriching the N_2 mole fraction in the recycled exhaust gases (because H_2O has been removed). Thus, for the same volume of trapped, gas-phase residual gas, a greater amount of N_2 is present when the proposed preferential scavenging mechanism is operative. It can be shown, however, that this effect is minimal. For sufficient H_2O condensation to double the trapped mass of H_2O, the mass of trapped N_2 is expected to increase by only 0.6%. After applying the above drift correction, this increase in N_2 signal could only mask an increase in the H_2O signal on the order of 2%.

6 Results and Discussion

Single-cycle, temporal sequences of the signals from superpixels identified with N_2, C_3H_8, and H_2O, obtained in the quiescent engine flow, are shown in Fig. 3. The equivalent sequences obtained in the swirling engine flow are shown in Fig. 4. Also shown for reference are sequences corresponding to N_2 obtained under fully-warmed up conditions, in which scattering from condensate films does not occur. Note that in both engine flows a significant temporal variation in the data sequences is observed. Furthermore, based on the N_2 sequence, it is apparent that a large portion of this trend is associated with background drift.

Focusing on Fig. 3, it can be seen that the fluctuations in each sequence are very well correlated from species-to-species. Data obtained under warmed up conditions indicate that in this nominally quiescent engine flow the in-cylinder fluid is well-mixed by TDC, and cycle-to-cycle variations in residual gas mole fraction are approximately 0.005 (Miles and Hinze, 1998). Similarly, it has been shown that under these conditions (no films on the combustion chamber windows), the fluctuations in the background levels from cycle-to-cycle are determined by the statistical shot-noise on the background (Miles, 1998). The large, correlated fluctuations seen in Fig. 3 are thus indicative of cycle-to-cycle fluctuations in scattering from films. In Fig. 4 no such clear correlation in the fluctuations is observed. This is due to two factors: the lower overall magnitude of scattering from films which occurs with this engine flow (implying smaller films); and the higher fluctuations in residual gas mole-fraction, as inferred from the data obtained under fully warmed up conditions.

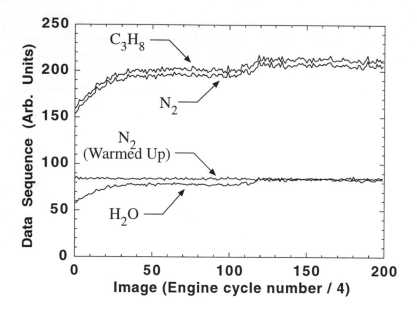

Figure 3: Single-cycle data sequences measured in the quiescent engine flow.

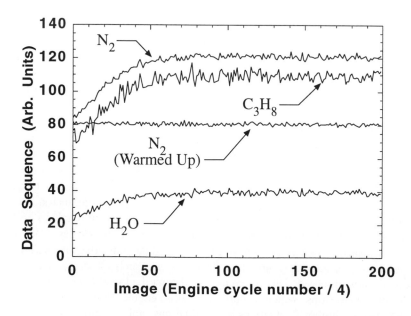

Figure 4: Single-cycle data sequences measured in the swirling engine flow.

Table 2. Residual Gas Mole Fraction Determined from Each Major Species at TDC

Species	Quiescent Flow	Swirling Flow
CO_2	0.273	0.324
O_2	0.242	0.336
C_3H_8	0.268	0.362
H_2O	0.291	0.394

Another feature apparent from Figs. 3 and 4 is the relatively steady state of the data sequences near the end of the set of single-cycle images. The asymptotic levels are very close to the levels obtained from the cycle-averaged images obtained immediately afterward[1]. For example, in the quiescent flow, the cycle-averaged signal levels in units corresponding to Fig. 3 are 210 and 80 for C_3H_8 and H_2O, respectively. In the swirling flow the respective cycle-averaged signal levels are 98 and 35 (*cf.* Fig. 4). This implies that neither the films nor the mean species number densities are evolving significantly beyond approximately image 50 in the sequences shown[2] or roughly 250 engine cycles (50 sec) after first-fire. Thus, the residual gas mole fraction, as determined from the measured species mole fractions obtained from the cycle-averaged images, is likely representative of the residual gas mole fraction fairly early in the cold-start simulation. These residual gas mole fractions, determined by assuming complete combustion of the fuel-air mixture, are presented in Table 2 for both engine flows investigated.

In Table 2, no evidence of significant levels of preferential scavenging is observed, as the residual gas mole fractions derived from the H_2O mole fractions are generally consistent in magnitude with the residual gas mole fractions derived from the remaining species. Nevertheless, residual mole fractions derived from H_2O are typically 10-15% higher than those derived from the other species. These differences, particularly the difference between the estimates from H_2O and C_3H_8, may be due in part to the precision of the cycle-averaged measurements (about 5%, at worst, under warmed-up conditions) and in part to inaccuracies in the system calibration. Temperature variation in the scattering cross-sections is insufficient to result in differences of this magnitude. The higher than expected measured H_2O content may also be due to two physical factors: a small level of H_2O

---

[1] This comparison is possible because both the single-cycle and the cycle-averaged data have been normalized by the measured laser energy delivered during the acquisition of each image.

[2] Excepting the rather anomalous change in background level observed in Fig. 3 between image 100 and image 120.

Table 3. Residual Gas Mole Fraction Determined from Each Major Species after one Scavenging Cycle

Species	Quiescent Flow	Swirling Flow
CO_2	0.203	0.185
O_2	0.216	0.264
C_3H_8	0.273	0.265
H_2O	0.235	0.245

condensation resulting in poor scavenging, or a small contribution due to Raman scattering from films on the windows.

To help illuminate the cause of the higher than expected water content, the residual gas mole fractions determined from the individual species after a non-fired scavenging cycle are presented in Table 3. There, it is observed that while the mole fractions estimated from the individual species are again generally consistent in magnitude, the residual mole fraction estimated from H_2O is no longer significantly larger than the estimates from the other species, particularly the (most reliable) C_3H_8 estimate. This behavior is consistent with *more* efficient scavenging of H_2O than the other residual gas species.

Comparing the reduction in residual H_2O after one scavenging cycle between the quiescent and the swirling flow, it is observed that the reduction is greatest for the swirling flow. This is consistent with the hypothesis that the (smaller) films formed in the swirling engine flow are more rapidly vaporized by the in-cylinder flow, leading to a reduction in the apparent H_2O mole fraction after an additional scavenging cycle. Both the apparent better scavenging of H_2O in the swirling flow, and the apparent more efficient scavenging of H_2O discussed above, suggest that the higher than expected H_2O content is likely due to Raman scattering from window films.

Comparison of Tables 2 and 3 also reveals slow scavenging of residual gases, as compared to the intuitively expected $(1 - \eta_{Scav})^n$ dependency of the residual gas mole fraction on the number of scavenging cycles n and the scavenging efficiency η_{Scav}. Slow scavenging of H_2O was also observed by Grünefeld, *et al.* (1995), who attribute it to a preferential scavenging mechanism. Here, however, slow scavenging is evidenced by all the major species. Similar slow scavenging has been reported by Galliot, *et al.* (1990).

Finally, it is interesting to note that the cycle-averaged residual gas mole fractions measured in these cold-start simulations are somewhat higher that those measured under warmed up conditions: 0.24 and 0.34 in the quiescent and swirling flow, respectively.

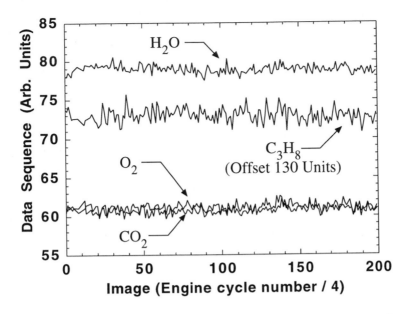

Figure 5: Drift corrected data sequences obtained in the quiescent engine flow.

Following the drift correction procedure outlined above, the data sequences corresponding to Figs. 3 and 4 have been corrected for background drift. The drift corrected sequences for the quiescent flow are shown in Fig. 5; corrected sequences obtained in the swirling flow are similar and are not shown here. The single notable observation from Fig. 5 is that the drift correction procedure has removed all obvious secular trends in the data sequences. The only data sequence which demonstrates a possible remaining trend in the data is the H_2O sequence obtained in the quiescent engine flow. This sequence is shown in expanded form by the dotted line in Fig. 6. Smoothing by convolution with an 11-point rectangular window serves to highlight the remaining trends in the data; the smoothed sequence is shown by the solid, heavy line. Although some trends remain, comparison with the scale[3] in the lower left corner of the figure indicates that these trends correspond to a rather small (about 10%) change in the H_2O number density. As noted above, the correction procedure is unlikely to mask a change in H_2O mole fraction associated with poor scavenging of H_2O. The implication from Fig. 6 is that if a mechanism resulting in poor H_2O scavenging is operative, it reaches a relatively steady-state after approximately 11 seconds of engine operation, which corresponds to the beginning of the data sequence shown in Fig. 5. Because the cycle-averaged data discussed above suggest that any

---

[3] The scale is derived from the cycle-averaged image obtained immediately afterwards.

470

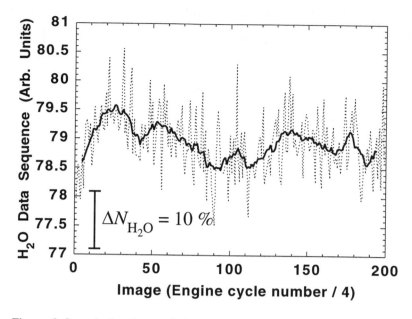

Figure 6: Smoothed and expanded version of the H_2O data sequence in Fig. 5.

preferential scavenging effects at later times are small (if they exist at all), it is reasonable to conclude that the data presented here show no strong evidence for the existence of a preferential scavenging mechanism. If such a mechanism exists, it is significant only in the first few seconds of engine operation. However, this conclusion is subject to a few implicit assumptions:

1) The low compression ratio engine used here does not inhibit a preferential scavenging mechanism due to insufficient residual gas cooling during expansion.

2) The mixture composition obtained from a fairly small volume in the center of the cylinder is typical of (or scales with) the global cylinder composition. Note that H_2O vapor which evaporates from films on the combustion chamber walls may not have time to mix in to the center of the clearance volume by the time the data is acquired, even if data obtained in a warm engine indicate that the fresh charge is well mixed with the residual gases (*i.e.* in the quiescent engine flow).

3) The evolution of the background light (scattered from films) is self-similar for each species, and can be determined from the temporal evolution of the N_2 signal.

7 Summary and Conclusions

Simultaneous, in-cylinder measurements of the major species of combustion are made using Laser Raman scattering with broadband signal detection under conditions simulating a cold start. Mean, cycle-averaged mole fractions of CO_2, O_2, N_2, C_3H_8, and H_2O are determined and compared for consistency. This comparison indicates that measured H_2O mole fractions are 10-15% higher than expected. Comparison with species mole fractions obtained after an extra scavenging cycle, as well as differences between the quiescent and swirling engine flow fields, suggest that these higher H_2O concentrations may be due to scattering from films on the window surfaces, rather than poor scavenging of H_2O due to condensation on the cold cylinder walls or other mechanisms.

Due to growth of films on the optical windows during the simulations, a time-varying signal background associated with film scattered light complicates determination of quantitative single-cycle species mole fractions. However, the temporal evolution of the signals, after correction for the varying background, can be examined for evidence of poor H_2O scavenging. Between approximately 11 sec (55 cycles) after the first fired cycle and 170 sec (855 cycles), the time period during which data was acquired, no evidence of significant variation in H_2O scavenging was observed.

Acknowledgement

This work was sponsored by the U.S. Department of Energy, Office of Defense Programs, Technology Transfer Initiative.

References

Eckbreth, A.C. 1996, Laser Diagnostics for Combustion Temperature and Species, 2nd ed., Gordon and Breach, Amsterdam.

Galliot, F., Cheng, W.K., Cheng, C-O, Sztenderowicz, M., Heywood, J.B., and Collings, N. 1990, In-Cylinder Measurements of Residual Gas Concentration in a Spark Ignition Engine, SAE Paper No. 900485.

Grünefeld, G., Knapp, M., Beushausen, V., Andresen, P., Hentschel, W., and Manz, P. 1995, In-Cylinder Measurements and Analysis on Fundamental Cold Start and Warm-up Phenomena of SI Engines, SAE Paper No. 952394.

Miles, P.C. 1998, Raman Line-imaging for Spatially- and Temporally-Resolved Mole Fraction Measurements in Internal Combustion Engines, Appl. Optics (In Press).

Miles, P.C. and Hinze, P.C. 1998, Characterization of the Mixing of Fresh Charge with Combustion Residuals Using Laser Raman Scattering with Broadband Detection, SAE Paper No. 981428.

Paul, P.H., van Cruyningen, I., Hanson, R.K., and Kychakoff, G. 1990, High Resolution Digital Flowfield Imaging of Jets, Exp. in Fluids, 9, pp. 241-251.

Takeda, K., Yaegashi, T., Sekiguchi, K, & Saito, K. 1995, Mixture Preparation and HC Emissions of a 4-Valve Engine with Port Fuel Injection During Cold Starting and Warm-Up, SAE paper No. 950074.

V.7. Laser Diagnostics of Nitric Oxide Inside a Two-Stroke DI Diesel Engine

N. Dam, W. Meerts, and J.J. ter Meulen

*Applied Physics, University of Nijmegen, NL-6525 ED Nijmegen
the Netherlands*

Keywords: Diesel engines, combustion, nitric oxide, LIF

ABSTRACT

The Nitric oxide (NO) content and distribution within the combustion chamber of an optically accessible one-cylinder two-stroke direct-injection Diesel engine have been studied by means of Laser Induced Fluorescence. Using 193 nm excitation of NO, detection of the ensuing fluorescence at 208 nm and 216 nm allows determination of the in-cylinder NO content throughout the whole combustion cycle. Images of the two-dimensional NO distribution in a plane perpendicular to the cylinder axis have been recorded for crank angles larger than 31° after Top Dead Center. The measured NO fluorescence is transformed into an in-cylinder NO content taking into account the changing in-cylinder conditions and the spectroscopic interference of Oxygen fluorescence. It is concluded that, in this engine, the bulk of the NO formation takes place relatively late in the stroke, indicating that most of the NO is formed during the diffusion burning phase.

1 INTRODUCTION

Strategies for the reduction of toxic emissions from Diesel engines focus on particulates (soot) and oxides of Nitrogen (NO_x). Although both of these components are formed during combustion, legislation is concerned only with the exhaust products. Emission control therefore aims at either catalytic exhaust gas after-treatment or at combustion optimisation, where optimisation is taken to imply reduced toxic compound formation while (at least) maintaining combustion efficiency. Combustion optimisation, arguably the more fundamental way of tackling the emission problem, poses a huge challenge both to experimental data acquisition and interpretation, and to theoretical combustion modelling. This paper intends to contribute to the former aspect. We have used non-intrusive optical diagnostics, based on Laser Induced Fluorescence, to monitor the amount as well as the distribution of Nitric oxide (NO) inside the combustion chamber of a Diesel engine.

Laser based optical diagnostics of combustion processes are appreciated for their ability to combine non-intrusiveness with selectivity for specific chemical species. As such, they have been applied both to open flames and to internal combustion engines; reviews can be found in e.g. Eckbreth (1988), Kohse-Höinghaus (1994) and Rothe and Andresen (1997). Among the manifold of optical techniques available, only Laser Induced Fluorescence (LIF) has the sensitivity to provide instantaneous, two-dimensional (2D) information on minority species distributions in combustion processes. The measurement principle of this planar LIF (PLIF) technique involves electronic excitation of the molecules of interest by a thin sheet of laser radiation, and detection of the subsequent fluorescence in a direction perpendicular to the sheet by an intensified CCD camera. PLIF has been used to demonstrate the presence of a large number of specific small molecules in a variety of combustion environments, but in general the observed data are very hard to quantify. Although in principle the LIF intensity is linearly proportional to the local number density of laser-excited molecules, the proportionality constant depends on the local physico-chemical environment, involving local temperature, density and chemical composition, and possibly spectroscopic interference by other molecules. Since these parameters are usually difficult to assess simultaneously with the (P)LIF measurements, one must have recourse to model assumptions.

The present paper reports LIF and PLIF measurements of the NO density within the combustion chamber of a small, two-stroke direct-injection (DI) Diesel engine, running on standard Diesel fuel. NO fluorescence distributions and dispersed fluorescence spectra are presented as a function of crank angle throughout the whole stroke, and ways towards their semi-quantitative interpretation are discussed.

2   ENGINE AND OPTICAL SETUP

All measurements are performed on a one-cylinder, two-stroke, DI Diesel engine (Sachs; bore 81 mm, stroke 80 mm, swept volume 412 cc) that has been made optically accessible by mounting quartz (Suprasil I) windows in the cylinder wall as well as centrally in the cylinder head (see fig. 1; W1-3). A flat piston with a shallow slot (0.5 mm depth) in its upper surface was used to provide optical access through the side windows throughout the whole engine cycle. The engine is operated steadily running (1200 rpm) on standard Diesel fuel, obtained from a local retailer, and was loaded by a water-cooled electric brake with up to 3.5 Nm (0.44 kW). Fuel is injected through a three-hole nozzle, located 6 mm above the piston surface at Top Dead Centre (TDC). Fuel injection starts at 27° before Top Dead Center (bTDC). The exhaust port opens at 105° after Top Dead Center (aTDC), the inlet ports at 121° aTDC. Inlet air is supplied under an overpressure of typically 0.2 bar to improve scavenging. With a compression ratio of 13, the

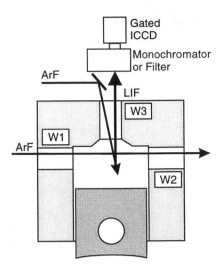

Figure 1: Schematic view of the optically accessible Diesel engine and the optical setup. The excimer laser beam (ArF) either traverses the combustion chamber (as a thin sheet parallel to the piston upper surface) through the two side windows (W1 and W2), or it is used unfocused to illuminate the combustion through the top window (W3). Fluorescence (or natural flame emission) is observed through the top window by an intensified CCD camera (ICCD) positioned behind either a narrow-band reflection filter (used for imaging) or a monochromator (used for fluorescence dispersion).

pressure P, in the engine follows the curve shown in fig. 2a reaching a peak pressure of 75 bar slightly after TDC. The pressure curve is used together with the volume to calculate the mean temperature of the gas content inside the cylinder (see Heywood (1988)). This, so-called, mean gas temperature is given in fig. 2a (solid line) and reaches a maximum of 1470 K a few degrees after TDC. Combustion is seen to start at 15° bTDC, and natural flame emission can be observed up to about 60° aTDC. Its spectrum shows no additional structure at any crank angle, indicating that it is mainly due to light emission by glowing soot particles. The natural emission spectrum can be well fitted to a Planck black body radiation curve, and can be used to derive a so-called soot temperature from the glowing soot particles as a function of crank angle (Θ). These data, showing a temperature of about 2300 K at TDC, have been included in fig. 2b (■); details will be published elsewhere by Stoffels et al. (1999). In the figure, the dashed line represents the estimated gas temperature based on adiabatic expansion ($PV^\gamma = $ constant, $\gamma = 1.36$) of an ideal gas during the later part of the stroke. It is matched to the experimental data at intermediate crank angles.

All (P)LIF measurements on NO employed a pulsed, tunable ArF excimer laser (Λ Physik Compex 350T; $\lambda = 192.9 - 193.9$ nm, bandwidth

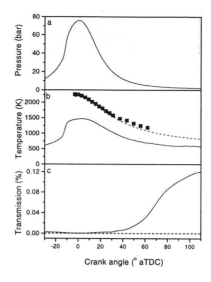

Figure 2: Parameters of the engine running steadily on standard Diesel fuel. a) The in-cylinder pressure. b) Mean gas temperature derived from the in-cylinder pressure curve (solid line) and soot temperature derived from the flame emission spectrum (■; see text). The dashed curve is based on calculations assuming adiabatic expansion of an ideal gas during the later part of the stroke. c) Transmission of the excimer laser beam (193 nm) through the firing engine, incl. window losses.

≈ 0.7 cm^{-1}, 20 nsec pulse duration) to excite NO on the R$_1$(26.5) transition of the D$^2\Sigma^+(v' = 0) \leftarrow$ X$^2\Pi(v'' = 1)$ band at 193.337 nm as determined by Versluis et al. (1991). The laser is synchronised to the engine cycle with a precision of $< 0.6°$ crank angle. It emits a beam with rectangular cross section, which, for the purpose of recording NO fluorescence distributions, is focussed down to a sheet of about 0.1 mm thickness and coupled into the engine through a side window (W1). The beam traverses the combustion chamber in a plane perpendicular to the cylinder axis (see fig. 1), and is located within the slot in the piston upper surface when the piston is at TDC. The laser illuminates the whole area beneath the top window, resulting in a measurement volume of 25 × 0.1 mm (diameter × thickness). Fluorescence of NO is recorded through the top window (25 mm diameter; fig. 1) and through a 4-plate reflection filter, tuned to 208 nm center wavelength with a bandwidth of 5 nm (FWHM), by a 576 × 384 pixels CCD camera (14 bits dynamic range) equipped with an image intensifier (Princeton Instr. ICCD-576G/RB-E) and a quartz $f/4.5$ 105 mm objective (Nikon). At this setting, the filter is centered on the NO D($v' = 0$) → X($v'' = 3$) fluorescence band. Mie scattering images were obtained in the same setup, but with the filter mirrors replaced by broad-band aluminium mirrors.

Alternatively, the unfocused laser beam can be coupled into the combustion chamber through the top window (W3). The advantage of this setup is that the laser beam enters the observation area immediately, thus avoiding attenuation of the laser radiation in the first part of the combustion chamber. (Note that the laser beam has to travel about 25 mm before entering the observation area when coupled in through the side window.) In this case NO fluorescence is detected through the top window by the camera mounted behind an imaging monochromator (Chromex 250IS) with a 1200 gr/mm holographic grating blazed at 250 nm, as part of an Optical Multichannel Analyser (OMA), to spectrally disperse the fluorescence. For all fluorescence measurements the image intensifier was used with a gate width of 50 nsec, which was sufficient to collect all LIF (no signal increase for longer gate times) while keeping the contribution of the natural flame emission down to a manageable (and usually negligible) level. The latter was, however, always measured separately and subtracted from the LIF data.

3 DATA EVALUATION

Even though qualitative data on chemical species distributions within a combustion environment are of interest in themselves, one of the goals of our present research is to quantify the LIF data as much as possible. This will allow them to be compared throughout the stroke and for different engine operating conditions, different fuels and fuel injection systems, and so on. The LIF yield will be described using a model in which the energy level structure of NO is simplified to a 3-level system coupled to a 'bath' of all other levels by collisional energy transfer processes. Rotational energy transfer in the ground state will be neglected, because of the low laser intensity within the engine. In general the fluorescence yield $S_{\mathrm{LIF}}(x,y)$ due to a local NO density $\rho_{\mathrm{NO}}(x,y)$ can be written as

$$S_{\mathrm{LIF}}(x,y) = \wp\, I_{\mathrm{L}}(x,y)\, A_{\mathrm{F}}(x,y)\, g(\nu_{\mathrm{L}},\nu_0)\quad f_{vJ}(T)\, \rho_{\mathrm{NO}}(x,y)\, \frac{A}{A+Q}\,,$$

in which x and y are the spatial coordinates within the plane illuminated by the laser and $\wp$ is a proportionality constant including experimental parameters like optics collection efficiency, camera sensitivity, absorption line strength, etc. $I_{\mathrm{L}}(x,y)$ denotes the local laser beam intensity, $A_{\mathrm{F}}(x,y)$ describes the attenuation of the induced fluorescence emitted at (x,y) and $g(\nu_{\mathrm{L}},\nu_0)$ is the overlap integral of the laser line profile with the NO absorption spectrum. The temperature-dependent fractional population of the probed rovibrational state is described by the Boltzmann fraction $f_{vJ}(T)$ and the fluorescence yield is determined by the Stern-Vollmer factor $A/(A+Q)$, in which A denotes the spontaneous emission rate on the vibronic transition that is monitored and Q the effective non-radiative decay rate. The non-radiative decay processes that reduce the fluorescence yield are a result of intermolecular collisions and include both Electronic Energy Transfer (EET; notably D $\rightarrow$ A and D $\rightarrow$ C) and quenching (D $\rightarrow$ X).

In order to extract the NO density from the LIF signal S_{LIF}, all factors in eq. 3 have to be known. Most of these factors depend on the position (x, y) and/or on the in-cylinder conditions (pressure, temperature, volume and laser beam intensity), and can therefore not be calibrated by *e.g.* an exhaust gas measurement. The evaluation of the individual factors of eq. 1 is discussed shortly below, a more extensive description is given by Stoffels et al. (1999).

Local laser intensity $I_{\mathrm{L}}(x, y)$ & Fluorescence attenuation $A_{\mathrm{F}}(x, y)$: The laser intensity suffers severe attenuation on its way through the combustion chamber. Also, the induced fluorescence is attenuated on its way to the top window. At larger crank angles (*i.e.* later in the stroke) these attenuations are expected to be caused mainly by scattering off and absorption by particulates, whereas at smaller crank angles absorption by still unburned fuel will also contribute, as reported by Sick (1997). For the purpose of imaging, an independent measure of the local laser intensity and the fluorescence attenuation is provided by Mie scattering, here understood to comprise elastic light scattering off small particles (regardless of their size). (The small difference in wavelength between the laser radiation and the induced fluorescence is neglected.) On the assumption *i*) that laser intensity and fluorescence attenuation is mainly due to scattering off and absorption by small particles (oil, soot, fuel droplets) and *ii*) that there exists a linear relationship between scattering and absorption cross sections of these particles as described in Bohren and Huffman (1983), the local laser intensity and the fluorescence attenuation can be reconstructed from measurements of the local Mie scattered intensity combined with overall transmission measurements. Details of the reconstruction method will be published elsewhere by Stoffels et al. The overall transmission of 193 nm light through the firing engine is measured by coupling the laser beam in and out through the side windows and detecting the transmitted radiation behind the outcoupling window by a CCD camera. The result is given in fig. 2c. The rise in transmission around 60° aTDC coincides with the end of the visible combustion, suggesting that unburned fuel indeed plays an important part in absorption of 193 nm laser radiation (see also Sick (1997)). In the case that dispersed fluorescence spectra are recorded, only the overall transmission is used to calculate the local laser beam intensity and the attenuation of the induced fluorescence.

Overlap integral $g(\nu_{\mathrm{L}}, \nu_0)$: The overlap integral is calculated by assuming Gaussian profiles for both the laser emission line and the NO absorption line. Lacking data on the pressure broadening and shift of transitions in the D($v' = 0$) $\leftarrow$ X($v'' = 1$) band, for the present purpose the functional pressure and temperature dependence of the A $\leftarrow$ X band as derived by Chang et al.

(1992) is taken in combination with a proportionality factor that is derived from own measurements on the D ← X band in the engine.

Boltzmann fraction $f_{vJ}(T)$: The temperature dependent fractional population of the probed state (v''=1, J=26.5) can be calculated using the well established spectroscopic data of the NO electronic ground state given in Huber and Herzberg (1979).

Stern-Vollmer factor $\mathcal{A}/(\mathcal{A} + Q)$: The spontaneous emission rate on the $D(v' = 0) \rightarrow X(v'')$ transition can be estimated from the radiative lifetime of the D-state ($\tau = 18$ nsec given in Radzig and Smirnov (1985)) and the appropriate Franck-Condon factor (0.165 for $v'' = 3$, calculated using the approximation by Nicholls (1981) and molecular data from Huber and Herzberg (1979)), yielding $\mathcal{A}_{03} = 9.2 \cdot 10^6$ sec^{-1}. The non-radiative decay rate Q poses a more serious problem. Although there is a considerable amount of data available for the NO A-state quenching (see e.g. Paul et al. (1993)), data for the D-state are all but lacking. Preliminary results of measurements on the D-state fluorescence yield under conditions of elevated temperature and pressure indicate that D → A EET by collisions with N_2 provide a very efficient decay channel (Van den Boom et al. (1999)). For the present purpose, we have simply assumed

$$Q = \bar{v}_{\text{rel}}\rho\sigma \gg \mathcal{A} = \mathcal{A}_{03} , \tag{1}$$

with $\bar{v}_{\text{rel}} = (8kT/\pi\mu)^{1/2}$ the relative velocity between collision partners, ρ the total number density and σ an effective quenching cross section taken to be independent of pressure and temperature.

In summary, using the above assumptions the local NO number density can be extracted from PLIF images and dispersed fluorescence spectra as

$$\rho_{\text{NO}}(x,y) \propto \frac{PT^{-1/2}}{g(\nu_L,\nu_0)f_{vJ}(T)} \frac{S_{\text{LIF}}(x,y)}{I_L(x,y)A_F(x,y)} , \tag{2}$$

where in the case of imaging $I_L A_F$ is reconstructed from separately recorded Mie scattering images as discussed above. Assuming the average NO density in an image or spectrum to also represent the average NO density in the whole cylinder, the total amount of NO can be written as

$$\mathcal{N}_{\text{NO}} \propto V_{\text{cyl}}\rho_{\text{NO}}(x,y) , \tag{3}$$

in which V_{cyl} is the in-cylinder volume (depends on crank angle).

4   RESULTS AND DISCUSSION

4.1   Dispersed Fluorescence Spectra

Figure 3 shows a typical series of dispersed fluorescence spectra (averaged over 100 engine cycles) for different crank angles, recorded with the OMA

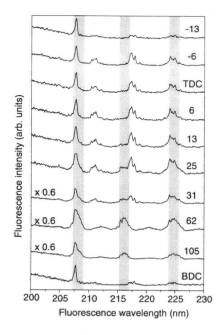

Figure 3: Dispersed fluorescence spectra of NO for different crank angles (° aTDC, indicated at the right), obtained by illuminating the combustion through the top window. All spectra are to the same scale unless indicated otherwise. The increasing intensity at the blue end of the spectra is due to directly scattered laser light. The persistent peak at 207.8 nm is an artefact due to the quartz window. The grey bars indicate the positions of the $D^2\Sigma^+(v' = 0) \rightarrow X^2\Pi(v'' = 3, 4, 5)$ bands of NO.

system (50 μm entrance slit) and illuminating the combustion through the top window (fig. 1). All these spectra show a persistent peak at 207.8 nm, which, however, is an unfortunate artefact, possibly caused by Raman scattering of the quartz observation window. The spectral structure shows a considerable qualitative change around 30° aTDC. For $\Theta \lesssim 30°$ aTDC two prominent fluorescence features are found at 211 and 217.5 nm, the latter with a persistent doublet structure even at TDC. Both features can be ascribed to fluorescence of hot Oxygen (O_2) that is also excited by the laser radiation from v''=2,3 as reported by Lee and Hanson (1986) and Shimauchi et al. (1994). For $\Theta \gtrsim 30°$ aTDC these O_2 features disappear and are replaced by somewhat broader spectral structures around 208 nm (red shoulder on the quartz peak), 216 and 225 nm. These features arise from fluorescence out of the laser-excited D(v'=0)-state to the X(v''=3,4,5)-states, respectively. The two indistinct emission signals at 212.5 and 220 nm result from fluorescence out of the C(v'=0)-state, populated by Electronic Energy Transfer (EET), to the X(v''=3,4)-states. Fluorescence bands of NO and O_2 coincide in the 225 nm region, which is the reason why the spectrum in this wavelength re-

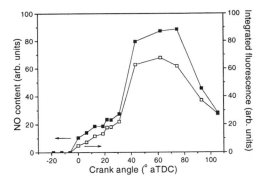

Figure 4: The integrated fluorescence signal of NO present inside the cylinder of the engine as a function of crank angle obtained from the spectra given in fig. 3 (□). The NO content curve derived from the integrated fluorescence signal processed for the change in pressure, temperature, volume and laser intensity during the stroke (■).

gion is relatively invariant. The shape of the 225 nm band does, however, change from a multiple-peaked structure (characteristic for O_2, due to strong predissociation of the upper state, described in e.g. Kruppenie (1972) and Wodtke et al. (1988)) for $\Theta \lesssim 30°$ aTDC to a structureless bump (characteristic for NO) at $\Theta \gtrsim 30°$ aTDC.

Even though the NO fluorescence is most evident in the spectra for $\Theta \gtrsim 30°$ aTDC, it should be stressed that NO fluorescence can be recognized in all spectra for $\Theta \geq$ TDC.

4.2   The in-cylinder NO content

The intensity of the NO fluorescence peaks in the dispersed fluorescence spectra provides information about the amount of NO present inside the cylinder at different crank angles. To compare the peaks of the spectra at different crank angles and engine conditions it is necessary to convert the NO fluorescence into more quantitative data. To this end the NO fluorescence signal of the $D(v' = 0) \rightarrow X(v'' = 4)$ band at 216 nm (right at the blue side of the O_2 band at 217.5 nm; central grey bar in fig. 3) is fitted to a Gaussian curve taking into account the structures due to O_2 fluorescence. The integrated fluorescence yield obtained from the spectra presented in fig. 3 is given in fig. 4 (□). To arrive at a measure for the NO density as a function of crank angle (eqs. 3 and 4) the integrated fluorescence yield has been processed according to the prescriptions discussed above (Data evaluation section). In the processing the in-cylinder pressure, mean gas temperature and transmission data given in fig. 2 are used. A curve that is proportional to the NO content within the cylinder, obtained from evaluation of the integrated fluorescence signal of the spectra in fig. 3 is also presented in fig. 4 (■). It shows a steep rise starting at $\Theta \approx 30°$ aTDC and reaches a maximum around 75° aTDC, followed by a more gradual decline during the later part of the stroke. Similar curves have been measured in several different runs, as well as for various engine operating conditions and fuels.

Because the fluorescence yield curves obtained for several engine runs reproduce well, the accuracy of the NO content curve will be determined mainly by the precision of the fluorescence yield processing method. Although the general relationship between the fluorescence yield and the NO content is well established (eq. 1) some errors will be introduced by the uncertainties involved in some of the factors of the signal processing. Another systematic error will be caused by the use of the mean gas temperature. During the actual combustion, the local temperature at the place the NO is formed is probably higher, as the cylinder contents are not in thermal equilibrium at that moment. Because the NO molecules are excited from a vibrationally excited state, the use of the mean gas temperature to calculate the population of the lower level results in too low a population around TDC, where the actual temperature is probably higher. This may lead to an overestimate of the relative NO density at smaller crank angles. However because these are systematic errors that appear in all NO content curves they do not affect the relative differences between curves and NO content curves at different engine conditions can be compared.

The curve in fig. 4 (which is typical for this engine) shows that, in this particular engine, the bulk of NO formation takes place relatively late in the stroke, indicating that the premixed combustion contributes only little and most of the NO is formed during the diffusion burning phase. This result is in agreement with theoretical predictions by the group of Pitsch et al. (1997). They also agree with recent results from Dec and Canaan (1998) and Nakagawa et al. (1997). Images presented by Dec and Canaan (1998) showed that NO formation starts around the jet periphery just after the diffusion flame forms and continues in the hot post-combustion gases, at places where the reacting fuel jet had traveled. However, it is in contrast with the models that predict that the premixed combustion has a large contribution to the NO formation and NO_x emissions correlate with the amount of fuel consumed during the initial premixed burn as described in e.g. Heywood (1988) and Warnatz et al. (1996).

A second conspicuous feature of the NO curves in fig. 4 is their decline towards larger crank angles. This can partly be explained by the fact that NO is chemically not particularly stable, and will be converted to NO_2 during the colder part of the stroke. (Note that Diesel engines operate under lean conditions, so there will be O_2 left after the combustion has subsided; see e.g. Heywood (1988).) This would lead to a reduction of the NO content in the engine, which has, in fact, been predicted by calculations of Pitsch et al. (1997), albeit not to the extent observed in fig. 4.

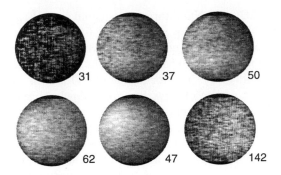

Figure 5: NO distributions, averaged over 25 engine cycles, corrected for local laser intensity variations. Crank angles are indicated below the images. The laser beam travels from right to left and fuel is injected from the bottom of the images upwards. Images are individually scaled.

4.3   Two-dimensional NO distributions

The spectra shown in fig. 3 have some consequences for PLIF measurements that aim at recording 2D NO distributions. Excitation-emission spectra recorded in the two-stroke engine at 43° aTDC have shown that all strong NO transitions (within the ArF laser tuning range) lie close to some O_2 resonance, so that efficient excitation of NO unavoidably also involves some excitation of O_2, as also reported by Lee and Hanson (1986) and Shimauchi et al. (1994). Evidently, since O_2 excitation can not completely be avoided, the NO band(s) should be filtered out of the total fluorescence. In our case, the $D(v' = 0) \rightarrow X(v'' = 3)$ NO fluorescence band at 208 nm was used for imaging, because it lies relatively isolated from O_2 fluorescence bands (nearest neighbours observed at 205 nm and 211 nm), and is also sufficiently distant from the excitation wavelength to avoid contributions of Mie scattered laser radiation.

However, the fluorescence distributions recorded this way cannot immediately be interpreted as NO density distributions, because the intensity of the laser beam is not uniform over the whole image. The local laser intensity was reconstructed from the local Mie scattered intensity by the method briefly discussed above. NO distributions (diameter 25 mm, individually scaled), averaged over 25 engine cycles and processed for the local laser intensity, are given in figure 5. To the extent that quenching and other collisional energy transfer processes are, at least on average, uniform over the field of view, the images represent NO distributions in arbitrary units (that is, the local pixel values are proportional to the local NO density). Since they are individually scaled to enhance contrast, only the distributions should be compared, not the intensities.

For $\Theta \lesssim 35°$ aTDC the NO fluorescence signal is weak, and, since at these crank angles the Mie scattering images used for reconstruction of the local laser intensity may contain a contribution from the flame luminosity, the fluorescence distribution is hard to interpret. At larger crank angles the signal level improves considerably. The formation of NO is seen to be concentrated in the leftmost part of the observation area, slowly spreading into a more uniform distribution as the crank angle increases. At 142° aTDC, when both inlet and outlet ports are open, a uniform distribution of the remaining NO is seen. (Note that the plane of observation is located high in the cylinder, whereas the scavenging ports are located near the position of the piston at BDC.) Since also the distribution of scattering particles is concentrated in the left-hand part of the observation region (data not shown), the images point to the occurrence of a reproducible flow pattern inside the cylinder of this engine.

5   CONCLUSION

The NO formation inside the combustion chamber of a DI Diesel engine operated on standard Diesel fuel has been studied throughout the whole engine cycle using ArF excimer laser induced fluorescence. LIF images visualise the two-dimensional NO distribution in a plane located 6 mm below the fuel injector. Dispersed fluorescence spectra have been used to obtain a semi-quantitative measure (up to a calibration constant) for the amount of NO present in the cylinder at any crank angle. The post-processing procedures that must be used to extract a NO content from the LIF signal strength are discussed. The results for this particular engine show a relatively late start of the NO formation with a steep rise at about 30° aTDC. The NO content reaches a maximum at about 75° aTDC, after which it gradually decreases.

Our results show the feasibility of obtaining semi-quantitative data on the NO distribution in a fairly realistic Diesel engine by means of ArF laser diagnostics. Great care must be taken to avoid spectroscopic interference of, particularly, Oxygen. In order to arrive at these data, a number of assumptions has had to be made, the substantiation of which has to be part of future research. Notably, this concerns i) information on collisional energy transfer involving both the ground state (RET) and the electronically excited $D^2\Sigma^+$-state (EET) for different collision partners, ii) data on pressure broadening and shifting of NO absorption lines and iii) methods to assess the local in-cylinder laser intensity and gas temperature.

ACKNOWLEDGEMENTS

It is a pleasure to acknowledge the expert technical assistance of especially L. Gerritsen and P. van Dijk of the workshop of the University of Nijmegen, as well as the help of E. van Leeuwen during the build-up phase of the project. This research is supported by the Technology Foundation (STW), the Netherlands Organization for Applied Scientific Research (TNO) and Esso.

REFERENCES

Bohren, C.F. & Huffman, D.R., 1983: *Absorption and scattering of light by small particles*, Wiley, New York.

Chang, A.Y., DiRosa, M.D. & Hanson, R.K., 1992: *Temperature Dependence of Collision Broadening and Shift in the NO A ← X (0,0) Band in the Presence of Argon and Nitrogen*, J. Quant. Spectrosc. Radiat. Transfer **47**, pp. 375-390.

Dec, J.E. & Canaan, R.E., 1998: *PLIF Imaging of NO Formation in a DI Diesel Engine*, SAE transactions, paper no. 980147, pp. 79-105.

Eckbreth, A.C., 1988: *Laser Diagnostics for Combustion Temperature and Species*, Abacus Press, Tunbridge Wells, UK.

Heywood, J.B., 1988: *Internal Combustion Engine Fundamentals*, McGraw-Hill, Singapore.

Huber, K.P. & Herzberg, G., 1979: *Molecular spectra and molecular structure IV. Constants of diatomic molecules*, van Nostrand Reinhold, New York.

Kohse-Höinghaus, K., 1994: *Laser techniques for the quantitative detecion of reactive intermediates in combustion systems*, Prog. Energy Comb. Sci. **20**, pp. 203-279.

Kruppenie, P.H., 1972: *The Spectrum of Molecular Oxygen*, J. Chem. Phys. Ref. Data **1**, pp. 423-534.

Lee, M.P. & Hanson, R.K., 1986: *Calculations of O_2 Absorption and Fluorescence at Elevated Temperatures for a Broadband Argon-Fluoride Laser Source at 193 nm*, J. Quant. Spectrosc. Radiat. Transfer **36**, pp. 425-440.

Nicholls, R.W., 1981: *Approximate formulas for Frank-Condon factors*, J. Chem. Phys. **74**, pp. 6980-6981.

Nakagawa, H., Endo, H., Deguchi, Y., Noda, M., Oikawa, H. & Shimada, T., 1997: *NO Measurements in Diesel Combustion Based on Laser-Sheet Imaging*, SAE transactions, paper no. 970874, pp. 187-196.

Paul, P.H., Gray, J.A., Durant Jr, J.L. & Thoman Jr, J.W., 1993: *A Model for Temperature-dependent Collisional Quenching of NO $A^2\Sigma^+$*, Appl. Phys. B **57**, pp. 249-259.

Pitsch, H., Barths, H. & Peters, N., 1997: *Modellierung der Schadstoffbildung bei der Dieselmotorischen Verbrennung*, in Berichte zur Energie- und Verfahrenstechnik, ed. Leipertz, A., Heft 97.1, pp. 139-163, ESYTEC GmbH, Erlangen, Germany.

486

Radzig, A.A. & Smirnov, B.M., 1985: *Reference data on atoms, molecules and ions*, Springer Verlag, Berlin.

Rothe, E.W. & Andresen, P., 1997: *Application of Tunable Excimer Lasers to Combustion Diagnostics: A Review*, Appl. Opt. **36**, pp. 3971-4033.

Shimauchi, M., Miura, T. & Takuma, H., 1994: *Absorption Lines of Vibrationally excited O_2 and HF in ArF Laser Spectrum*, Jpn. J. Appl. Phys. **33**, pp. 4628-4635.

Sick, V., 1997: *Mehrdimensionale Laserspektroskopische Messung von Stickoxid in der Motorischen Verbrennung*, in Berichte zur Energie- und Verfahrenstechnik, ed. Leipertz, A. Heft 97.1, pp. 177-190, ESYTEC GmbH, Erlangen, Germany.

Stoffels, G.G.M., van den Boom, E.J., Spaanjaars, C.M.I., Dam, N., Meerts, W.L., ter Meulen, J.J., Duff, J.L.C. & Rickeard, D.J., 1999: *In-Cylinder Measurements of NO Formation in a Diesel Engine*, to be published in SAE transactions.

Van den Boom, E.J. *et al.* 1999: To be published.

Versluis, M., Ebben, M., Drabbels, M. & ter Meulen, J.J., 1991: *Frecuency Calibration in the ArF Laser Tuning Range Using Laser-Induced Fluorescence of NO*, Appl. Opt. **30**, pp. 5229-5232.

Warnatz, J., Maas, U. & Dibble, R.W., 1996: *Combustion*, Springer Verlag, Berlin.

Wodtke, A.M., Huwel, L., Schlüter, H., Voges, H., Meijer, G. & Andresen, P., 1988: *Predissociation of O_2 in the B-State*, J. Chem. Phys. **89**, pp. 1929-1935.

V.8. Investigation of Oil Transport Mechanisms on the Piston Second Land of a Single Cylinder Diesel Engine, Using Two-Dimensional-Laser-Induzed Fluorescence

B. Thirouard and D.P. Hart

Sloan Automotive Laboratory
Massachusetts Institute of Technology
Cambridge, MA 02139-4307, U.S.A.

Abstract. *A two-dimensional-Laser-Induced-Fluorescence (LIF) system was developed to visualize the oil distribution and study the oil transport in the piston ring pack of a single-cylinder diesel engine through an optical window on the liner. The system provides high spatial and intensity resolutions so that detailed oil distribution on the piston as well as between the rings and the liner can be studied. This work primarily focused on investigating different oil transport mechanisms on piston second land under various engine operating conditions. Two mechanisms for the oil flow on the second land were identified, namely, inertia driven oil flow in the axial direction and oil dragging by gas flow in the circumferential direction. Finally, the effects of ring rotation were investigated.*

Keywords. Oil transport, piston, ring pack, Laser Induced Fluorescence

1. Introduction

Reduction of oil consumption is required to satisfy three factors that currently govern the development of automotive engines: utilization of hydrocarbon resources, protection of the environment, and customer satisfaction. It is estimated that more than 50% of the engine oil consumption in automotive engines is from the piston ring pack. Consequently, the investigation of oil transport mechanisms at the interface between piston and liner is of critical importance.

The ring and groove geometry along with temperature and pressure conditions at the interface between piston and liner are the governing factors in oil flow. Due to lack of experimental data, the effects of these parameters on oil transport in the piston ring-pack were poorly understood. Thus, accurate prediction of oil consumption is nearly impossible.

LIF technique have been used for measuring oil film thickness of the piston ring-pack for some time [Hoult and Takigushi (1991), Shaw *et al* (1992), Tamai

488

(1995), Casey (1998), Takigushi *et al* (1998)]. In early works, LIF was implemented to make point measurements. A great deal of knowledge on the oil film thickness between rings and the liner and on the piston was gained by using these systems. However, the obvious limitation of point LIF is that it can only display the oil film at one location and is thus not able to simultaneously show the oil distribution in the piston ring-pack. When one intends to study oil transport in piston ring-pack, it was often found to be very difficult to interpret the data from these measurements without knowing the oil film in the surrounding region [Casey (1998)].

Inagaki et al. (1995) developed a two-dimensional oil film distribution measuring system, where a flash lamp was used as the light source. Even with a narrow window, their system demonstrated the benefits of having a two-dimensional view of oil distribution on a piston. In the current study, a two-dimensional oil distribution measuring system was developed utilizing a pulsed laser. The system was implemented in a single cylinder diesel engine with a 7 mm wide window on the cylinder liner. Studies were conducted at varying engine operating conditions and utilizing different ring configurations. Several important oil transport mechanisms were identified. Theoretical models are presented to describe oil flow due to different driving forces.

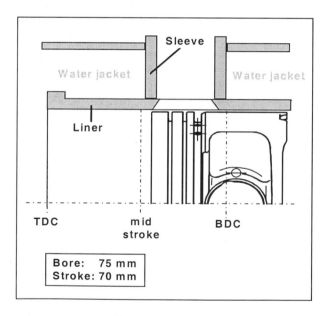

Fig. 1. Engine Setup

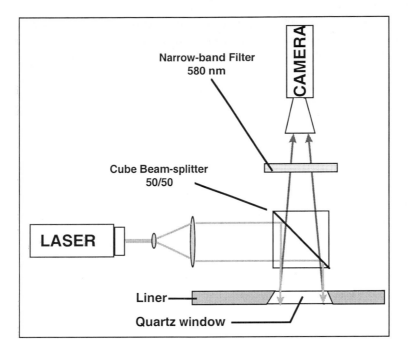

Fig. 2. Optical Setup

2. Experimental setup

2.1 Engine Setup

The engine used in this study was a 300 cc Kubota single-cylinder diesel. The liner was equipped with a quartz window on the anti-thrust side (Fig 1). The window is located between mid-stroke and the BDC so that the entire axial extent of the piston ring-pack can be observed at 88 degrees of crank angle before or after the TDC. The window was glued in the liner with Epoxy. Although, the window and the liner were honed together in order to obtain a flush surface, the window was still a few micron protruded after honing. However, ring profiles showed no damage after more than 20 hours running.

2.2 Optical Setup

Laser induced fluorescence was used to observe oil behavior in the piston ring pack. As reported by Röhrle (1995), temperature in the piston ring pack can vary from 80°C on the piston skirt up to 280°C on the crown land. Hence, to resolve all the regions in the piston ring-pack, the sensitivity of the LIF signal to oil temperature must be minimized. The dyes commonly used for this type of experiment are Coumarin 540 and 523 [Hoult and Takigushi (1991), Shaw *et al*

(1992), Inagaki *et al* (1995), Takigushi *et al* (1998)]. These dyes, however, are sensitive to temperature. In this work, Rhodamine 590 was chosen for its relatively strong stability with temperature up to 180°C (Fig 3). The implemented dye concentration was 5.10^{-4} mol/liter of oil. The wavelength of maximum absorption for Rhodamine 590 is 528 nm. Accordingly, the second harmonic of a Nd YAG laser (532 nm) was used to excite the doped oil. The fluorescence signal was acquired with an intensified CCD camera (Princeton Instrument ICCD 576 SE) through a narrow band filter centered on 580 nm (Fig. 2). The purpose of the narrow band filter was to suppress any effect of direct reflection from the quartz window.

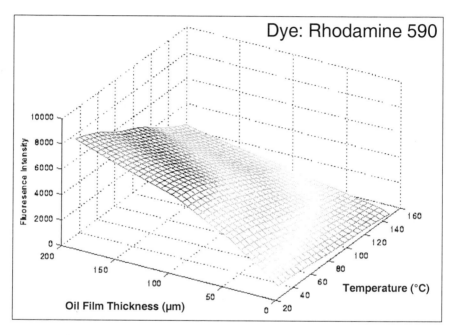

Fig. 3. Fluorescence signal versus temperature and oil film thickness (oil + Rhodamine 590).

The acquisition frequency was limited by the camera characteristics to 1 frame per second. High spatial resolution was obtained at 0.04 mm/pixel (Fig 4). In the images, the brighter the signal is, the higher the film thickness is. No calibration was attempted in this work. Therefore, observations were qualitative. The system was only able to detect accumulation of the oil in liquid form. As described by Burnett (1992), oil can also be vaporized or be transported as an oil mist in air stream. The current system is not able to detect oil in gaseous and mist forms.

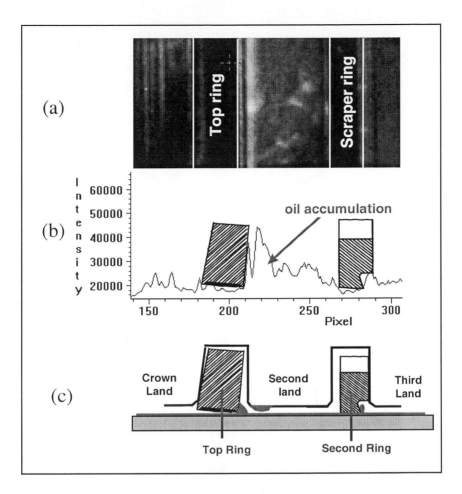

Fig. 4. (a)- Oil distribution image
(b)- Fluorescence intensity across the image
(c)- Ring pack geometry

2.3 Test Specifications

For the experimental results discussed in this paper the engine was equipped with standard rings. Three different ring configurations were tested.

- Configuration 1: standard configuration:
  non-pinned rings.
- Configuration 2: top two rings pinned with gaps located on the side opposite to the window (Fig 5).
- Configuration 3: top two rings pinned with gaps located on the window side (Fig 5).

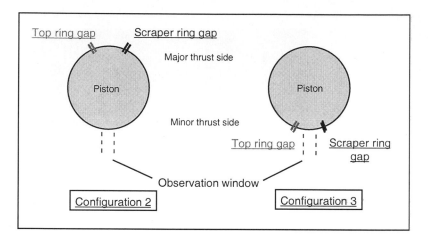

Fig. 5. Ring gap location in configurations 2 and 3.

Experiments were conducted from 1200 rpm to 2800 rpm and from no load to 60% load. Coolant temperature was kept constant at 50°C. SAE 10W30 oil was used. Due to the pollution by combustion soot, oil had to be changed every 8 hours.

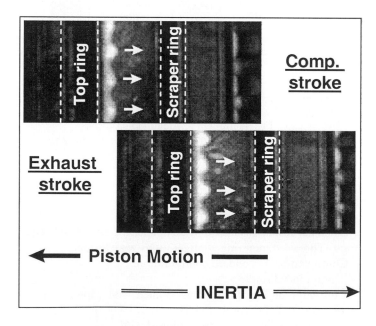

Fig. 6. Inertia driven flow on the second land
(1200 rpm - no load - Configuration 1).

3. Oil behavior analysis

Images acquired in configuration 1 during the four strokes of a engine cycle, at 1200 and 1700 rpm, and no load, have shown two important types of oil flows occurring on the piston second land:

        1- Oil flow in the cylinder axis direction,
        2- Oil flow around the piston circumference.

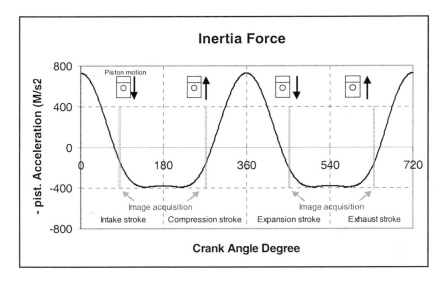

Fig. 7. Inertia Force due to piston
alternating motion (1200 rpm)

3.1 Oil Flow in the Axial Direction

Oil flow in the axial direction, driven by the inertia force due to piston acceleration/deceleration was observed. The images shown in Figure 6 were acquired around mid-stroke (Fig 7) during the compression and exhaust strokes. The direction of the inertia force switches to the downward direction after the middle of the down-strokes until the middle of the next up-stroke (Fig 7). Since the images are acquired at 88 degree before the TDC for both compression and exhaust stroke, they show the maximum effects of downward inertia force on the oil accumulation. As seen in figure 6, the pattern of the oil accumulation below the top ring clearly shows that the oil was moving toward the second ring.

    The velocity of oil moving on a flat area under inertia force can be estimated from Tian (1995).

$$V_{oil} = \frac{a.h^2}{\upsilon} \quad \text{(Eq. 1)}$$

where   V: Oil velocity.
          a : Acceleration .
          h : Thickness of the oil accumulation.
          v : Kinematic viscosity.

This equation can be established from a balance of inertia and shear stress, and neglecting the effect of surface tension. The displacement of oil accumulation during the period between a mid down-stroke and mid up-stroke (180 crank angle degrees) was calculated at 1200 rpm for two different oil accumulation thickness:

- h = 10 µm, oil displacement = 0.3 mm
- h = 20 µm, oil displacement = 4 mm.

These results show a very strong effect of oil film thickness on the oil velocity. During the experiments with non-pinned rings (configuration 1), the thickness of the oil on the piston second land was changing continuously so that it was possible to see different oil accumulation patterns at different oil film thickness. As shown in figure 8, with a thicker oil film thickness on the second land (bottom figure), indicated by a strong fluorescence signal, oil motion could be identified from the oil pattern. In the top figure where a thinner oil film thickness exists, no oil flow in the axial direction was observed.

The velocity of inertia driven flow is proportional to the piston acceleration and thus to the square of the engine rotation speed. In Figure 9, two images acquired at 1200 and 2000 rpm, during the exhaust stroke, show the effect of engine rotation speed on the second land oil accumulation. At 2000 rpm (top figure) the inertia force is almost 3 times higher than at 1200 rpm (bottom figure). The oil pattern on the second land shows that the oil accumulation, initially under the top ring, moved much further toward the second ring in the high rotation speed test (2000 rpm).

One characteristics of the inertia driven oil flow is that the net displacement of the oil in the axial direction over an engine cycle is not significant, as inertia switches direction after every half engine revolution. Unless the film thickness is large enough so that inertia can drive the oil out of the second land, oil would always move back and forth on the second land.

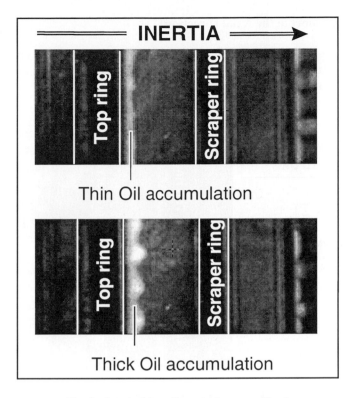

Fig. 8. Inertia driven flow on the second land
(1200 rpm - no load - Configuration 1).

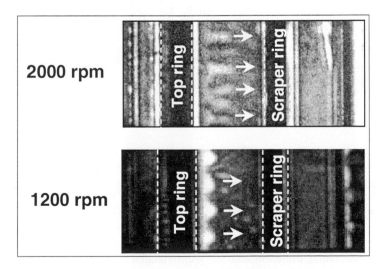

Fig. 9. Effect of engine speed on driven flow.
(1200 rpm - no load - Configuration 1).

496

3.2 Oil Transport in the Axial Direction

Strong oil motion around the piston circumference was observed on the second land. The images shown in figure 7 were acquired 10 engine cycles apart. A clear 3 mm/s circumferential flow can be identified on the second land.

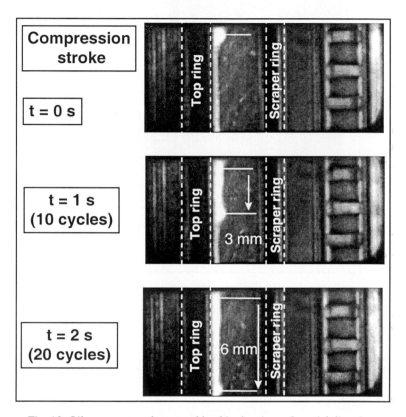

Fig. 10. Oil transport on the second land in the circumferential direction. (1200 rpm - no load - Configuration 1)

The oil flow in the circumferential direction can be driven by the gas flow. When the cylinder pressure is high, there is a gas flow on the second land from the top ring gap to the second ring gap. The shear stress at the interface of gas and oil drags the oil along with gas. Here, estimation is made to relate the velocity of the oil flow to the gas flow by balancing the shear stress at the gas-oil interface. Because the clearance between the second land and the liner is of the order of 0.2 mm and the axial width of the second land is of the order of 3mm, the gas flow in the circumferential direction can be approximated as uniform across the axial extent of the second land. Thus, the Reynolds number of the gas flow in the channel between the second land the liner can be estimated as:

$$\mathrm{Re} = \frac{\dot{V}\rho}{L\mu}$$

where $\dot{V}$: Volumetric flow rate.
ρ : Gas density.
μ_{gas} : Gas viscosity.
L : Axial width of the second land.

Using the blow-by value of this engine as the estimation of the volumetric flow rate $\dot{V}$, we found that the Reynolds number is below 1000. Consequently, the gas flow in the second land is laminar. Approximating the gas flow in the second land as a Poiseuille flow and the oil flow as a Couette flow, balancing the shear stress at the gas/oil interface gives the following relationship:

-shear stress balance: $\dfrac{dP}{dx} \cdot \dfrac{h_{gas}}{2} = \mu_{oil} \cdot \dfrac{V_{oil}}{h_{oil}}$ (Eq. 2)

where
P   : Pressure.
V_{oil} : Oil velocity at the gas/oil interface.
μ_{oil}: Oil viscosity
h_{gas} : Thickness of the gas layer between the piston
      and the liner.
h_{oil} : Thickness of the oil layer on the second land.

$$\dot{V} = \frac{L}{12 \cdot \mu_{gas}} \cdot \frac{dP}{dx} \cdot h_{gas}^3 \qquad \text{(Eq. 3)}$$

- Combining Eq. 2 and 3, we obtained:

$$V_{oil} = \frac{6 \cdot \dot{V} \cdot h_{oil} \cdot \mu_{oil}}{L \cdot h_{gas}^2 \cdot \mu_{gas}}$$

Assuming an oil film thickness of 40 µm, the calculated oil velocity is about 2 mm/s. This result is of the same order of magnitude as the observed velocity. Consequently, it can be concluded that the gas dragging effect is the main force driving oil in the circumferential direction.

There are two characteristics of this oil flow dragged by the gas flow:

- After one engine cycle, positive blow-by implies that there is more gas flow from the top ring gap to the second ring gap than the other way around despite a reversed gas flow starts at around 50 degree after the TDC of the expansion stroke (Fig 11). Thus, the mean oil transport due to the gas dragging effect is directed toward the third land.
- When gas flows down through the top ring gap during the later part of the compression stroke and early part of the expansion stroke, the pressure difference between the combustion chamber and second land is high (Fig. 11). Consequently, the downward gas flow through the top ring gap is usually choked. Thus, the gas velocity in the second land and oil velocity moving by gas dragging is constant with engine speed. The results shown in Figure 11 have been computed using MIT models developed by Tian (1996).

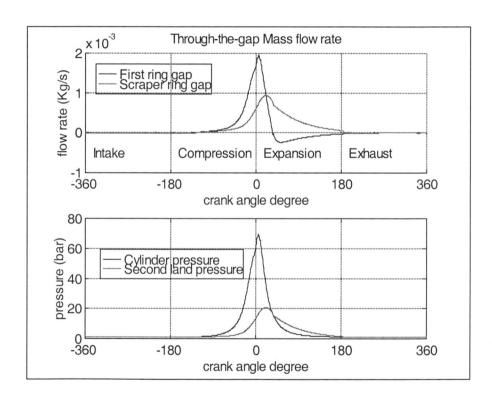

Fig. 11. Mass gas flow through the top two ring gaps and pressure difference at the top ring gap.
(1200 rpm - no load).

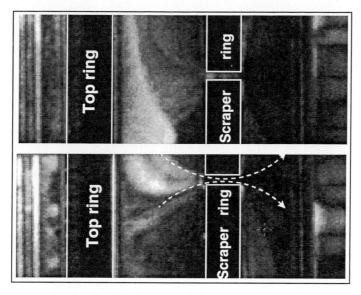

Fig. 12. Oil flow through the scraper ring gap
(1200 rpm - no load - compression stroke - Configuration 1).

Occasionally, a clear oil flow pattern just above the second ring gap is observed when the second ring gap rotates to the window location. As shown in Figure 12, oil is pulled toward the scraper ring gap by the gas flow during the compression stroke. However, no oil accumulation was detected in the third land just below the second ring gap (Fig. 12 and 13). The reason is that the high velocity gas flow in the gap is able to break up the oil and create an oil mist. Oil mist can not be observed on the third land due to low dye concentration.

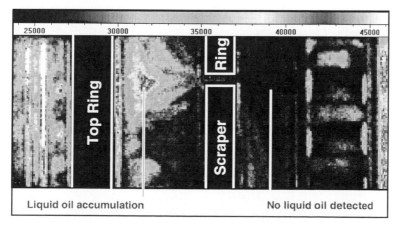

Fig. 13. Oil break up in the scraper ring gap,
(1200 rpm - no load - compression stroke - Configuration 1).

500

3.3 Comparison between Inertia and Gas Dragging Effect

Using the equation proposed to evaluate velocity of oil motion on the second land in the axial direction, the contribution of inertia and gas dragging effects was compared. The following assumptions were made:
- second land width 3 mm,
- oil accumulation lying in the middle of the second land,
- constant blow-by gas velocity on the second land.

If the alternating oil motion induced by inertia is not strong enough for the oil to reach either a ring or a groove, the mean oil axial displacement will be zero. In this case, the gas dragging effect, which slowly and continuously drives the oil toward the scraper ring gap, is the main oil transport mechanism on the second land. If inertia can push the oil in one of the two top ring grooves within half an engine revolution, however, the major flow is in the axial direction and the inertia driven transport becomes the dominant one. In figure 16, a line is drawn to separate two regions where different oil transport mechanisms dominate, based on Eq. 2. In the domain where inertia effect is dominant, oil can be moved across the second land and possibly be transported into the grooves of the top two rings. Part of the oil that flows into the top ring groove then can be transported to the crown land and consumed. In the other domain, the effect of gas dragging is the main mechanism of oil transport in the second land. Effect of the gas dragging eventually removes the oil out of the second land through the ring gaps.

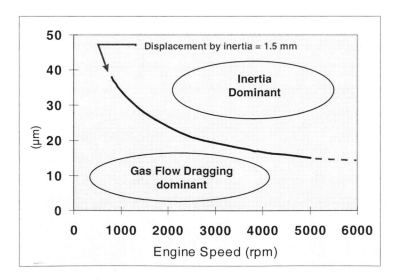

Fig. 14. Comparison of gas dragging and inertia effect on oil transport on the second land.

4. Effects of ring gap location and ring rotation on second-land oil distribution

4.1 Oil Behavior in Configuration 1

In configuration 2 (Fig. 5), the two top ring gaps were located on the thrust-side, away from the window. In this configuration, strong gas flows do not occur on the second land in front of the observation window. In this case, oil motion on the second land only occurred in the axial direction.

4.2 Oil Behavior in Configuration 3

In configuration 2 (Fig. 4), the top two ring gaps were located on each side of the window, 7 mm away from the window edges. In this configuration, the main gas flow on the second land occurs within the observation window on the anti-thrust side. When the two rings were pinned in configuration 3, no oil accumulation was detected on the piston second land. The strong gas flow which occurs between the two top ring gaps when they are close to each other carries oil into the third land and thus leaves no oil accumulation on the second land at the window location.

4.3 Ring Rotation Effect

In the experiments with non-pinned rings (configuration 1), strong unsteady oil transport was observed. The amount of oil accumulated on the second land changed significantly over several minutes whereas engine parameters were kept constant. In the experiments conducted with the two top rings pinned, however, the oil distribution on the second land was steady. Therefore, it was concluded that ring rotation was responsible for unsteady oil accumulation and oil transport on the second land.

5. Conclusions

A two-dimensional visualization system using laser-induced-fluorescence was developed to provide detailed images of oil distribution in the piston ring pack of a single cylinder diesel engine. Several mechanisms of liquid oil transport on the piston second land were identified and studied.

Oil was observed moving on the second land in both axial and cicumferential direction.

1- Oil moves in the axial direction under inertia force when the lubricant accumulation is thick. This was found to be consistent with the fact that inertia driven flows are strongly dependent on oil accumulation thickness.

2- It was observed that the blow-by gas flow drags the oil on the second land in the circumferential direction. This phenomena creates oil accumulation around the second ring gap.

Comparison of both phenomena showed that the gas dragging effect should over take inertia effects at high engine speed for thick oil accumulation.

Finally ring rotation was found to be responsible for unsteady oil behavior on the second land.

Acknowledgements

This work is sponsored by the MIT Consortium on Lubrication in Internal Combustion Engines. Current members include Dana Corporation, Mahle Gmbh, PSA Peugeot Citroën, Renault and Volvo. The authors like to thank Remi Rabute and Steven Sytsma at Dana for their input on experimental setup. The authors also would like to thank Dr. Tian Tian, Dr. Victor Wong and Prof. John Heywood for their support and discussions.

References

Hoult, D.P., Takigushi, M., 1991, "Calibration of Laser Fluorescence Technique Compared with Quantum Theory," STLE Tribology Transactions, Volume 34 (1991), 3, pp440-444.

Shaw, B.T., Hoult, D.P., Wong, V.W., 1992, "Development of Engine Lubricant Film Thickness Diagnostics Using Fiber Optics and Laser Fluoresence," SAE Paper 920651.

Tamai, G., 1995, "Experimental Study of Engine Oil Film Thickness Dependency on Liner Location, Oil Properties, and Operating Conditions," M.S. Thesis, Department of Mechanical Engineering, MIT.

Casey, S., 1998, "Analysis of Lubricant Film Thickness and Distribution along the Piston/Ring/Liner Interface in a Reciprocating Engine," M.S. Thesis, Department of Mechanical Engineering, MIT.

Takiguchi, M., Nakayama, K., Furuhama, S., Yoshida, H., 1998, "Variation of Piston Ring Oil Film Thickness in an Internal Combustion Engine," SAE Paper 980563.

Inagaki, H., Saito, A., Murakami, M., and Konomi, T., 1995, "Development of Two-Dimensional Oil Film Thickness Distribution Measuring System," SAE Paper 952346.

Röhrle, M.D., 1995, "Pistons for Internal Combustion Engines", MAHLE Gmgh, Printed in Germany 930513.

Burnett, P.J., 1992, "Relationship Between Oil Consumption, Deposit Formation and Piston Motion for Single-Cylinder Diesel Engines," SAE Paper 920089.

Tian, T., Noordzij, B., Wong, V.W., Heywood, J.B., 1996, "Modeling Piston-ring Dynamics, Blow-by, and Ring Twist Effects," ASME ICE Fall Technical Conference, October 1996.

Tian, T., 1995, "Modeling Oil-Transport and Oil-Consumption Mechanisms and the Influence of Piston and Ring Dynamics," Consortium on Lubrication in Internal Combustion Engines, Massachusetts Institute of Technology, June 13 1995, Internal Report.

V.9. Simultaneous Planar OH and Temperature Measurements for the Detection of Lifted Reaction Zones in Premixed Bluff-Body Stabilized Flames

D. Most, V. Holler, A. Soika, F. Dinkelacker, and A. Leipertz

Lehrstuhl für Technische Thermodynamik (LTT), Universität Erlangen-Nürnberg,
Am Weichselgarten 8, D-91058 Erlangen, Germany

Abstract. A bluff-body stabilized highly turbulent and premixed lean methane/air flame has been investigated with simultaneous planar laser Rayleigh scattering thermometry and planar laser-induced fluorescence of OH, systematically varying exit velocity and equivalence ratio. From that a conditional thermal flame front thickness is calculated by determining the steepest gradient while the OH-radical concentration is taken as a indicator for zones with or without a burning reaction. Thus the measured thermal fronts were distinguished in locally non-reacting and reacting flame fronts. In case of locally non-reacting flame fronts, e.g. between hot recirculating gas and fresh unburned gas, the thickness is found to be independent of equivalence ratio and exit velocity. For reacting flame fronts a relative thinning, compared to laminar unstretched flames is measured as a function of the non-dimensional stretch factor K_T, which is an expression for vortex induced local stretch. Although contact between burnt and fresh gas immediately takes place direct downstream the burner nozzle, from OH images a lifted reaction zone can be found to be at least 11 mm above the burner nozzle, which cannot be recognized with thermometry techniques. It is tried to find a unique criterion to predict conditions for this lift-off. Neither a unique local turbulent stretch scale a_T nor the non-dimensional stretch factor K_T can be found to predict this flame lift-off. Instead a unique global mean flow strain A_m seems to be a adequate criterion for most of the flame conditions. Only for very lean mixtures this does not match. Here a residential-ignition delay time model seems to fit more to the observed lift-off heights.

1 Introduction

For optimizing of combustion processes detailed knowledge about the local combustion processes and heat and mass transport mechanism are necessary.

Of special interest are "clean" flames driven with lean premixed mixtures. While for laminar and laminar-like conditions already adequate models are provided, this is not the case for highly turbulent flames. Although for this conditions models exist, e.g. the phase diagram of Peters (1986) which describes a broadening of the flame with increasing turbulence intensities, the prediction of flame behavior is still not sufficient. Recent experiments using planar laser imaging techniques under different turbulent premixed flame conditions indicate only partially modified pre-heat zone structures [Mansour et. al., 1992; Buschmann et al., 1996] and detailed flame-vortex investigations show that flames are locally quenched at these high turbulence intensities [Dinkelacker et al., 1998; Poinsot et al., 1990; Roberts et al., 1992]. These effects are not considered in the model introduced by Peters [1986]

Thus, in order to achieve temporal and spatial resolved 2-dimensional information about temperature structures of highly turbulent flames a bluff-body stabilized flame is investigated in this study. In order to achieve information about reaction the OH-radical concentration is taken as a criterion which can distinguish zones of reacting from those of non-reacting fresh fuel and those of burnt exhaust gas. The OH concentration is measured qualitatively with planar laser-induced fluorescence (LIF). Additionally the planar temperature field is measured with laser Rayleigh scattering (LRS). The combination of planar techniques allows to identify non-reacting temperature fronts and reacting flame fronts as well and to determine from that a thermal thickness distribution.

Furthermore the OH-images reveal a lifted reaction zone downstream the burner nozzle which cannot be recognized from the temperature field.

2 Experimental setup

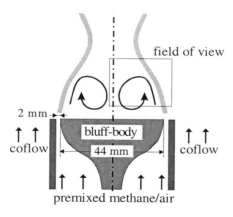

Figure 1. Schematic of the bluff-body burner and the field of view

For stabilizing a turbulent premixed flame at high exit velocities v_{exit} a bluff-body configuration has been chosen, which produces in its wake a recirculation zone through which burnt hot gases flow back, heat the fresh cold gases and ignite them. This kind of flame stabilization allows to run the burner with high turbulence intensities resulting from high shear rates over a wide parameter field.

A bluff-body of 44 mm diameter is placed concentrically in the burner tube of ∅ 48 mm. In order to get detailed information about the interaction of high turbulence with burning reactions the investigated flame conditions have been systematically varied between exit velocities from 5 to 29 m/s and lean premixed methane/air mixtures with equivalence ratios φ from 0.55 to 1.0. Flow rates of dried filtered air and methane of high purity (99.995%) are adjusted by two mass flow controllers and mixed in an external chamber. To prevent entrainment of dust particles, which cause interfering Mie-scattering, a coflow of filtered air with a mean velocity of 0.5 m/s coats the flame from the surroundings.

The velocity field is measured with a Laser Doppler Velocimetry (LDV) system, which allows simultaneous point-measurement of two velocity components of the flow field. To cover the whole field of interest measurements in several axial (up to 75 mm) and radial (up to 40 mm) positions have been performed with a positioning accuracy of ±0.3 mm. At each measured point at least 3000 samples were taken. Thus the axial, radial and tangential averaged velocities V_a, V_r and V_t and turbulent fluctuations v_a', v_r' and v_t' have been determined under isothermal flow conditions. V_t is found to be nearly zero, while v_r' and v_t' are found to be nearly equal. Thus the mean rms velocity fluctuation v'_{rms} can be expressed with the turbulent kinetic energy k from

$$v'_{rms} = (2k)^{0,5} \tag{1}$$

$$k = \frac{1}{2}(v_a{}^2 + v_r{}^2 + v_t{}^2) = \frac{1}{2}(v_r{}^2 + 2 \cdot v_r{}^2) \tag{2}$$

2.1   Planar laser techniques

With light sheet techniques 2-dimensional information about the spatial structure of the detailed combustion process can be obtained with high spatial and temporal resolution. Applying planar Laser Rayleigh Scattering (LRS) [Kampmann et al., 1993] the temperature field can be investigated. To extend information simultaneous qualitative Laser Induced Fluorescence (LIF) of OH-radicals [Krämer et. al., 1995; Roberts et. al., 1992] can be applied.

With a KrF excimer laser with wavelength $\lambda = 248$ nm and a pulse length of 15 ns a light sheet of 20 mm height and less than 200 µm thickness has been formed with cylindrical lenses. Images are taken perpendicular to the laser sheet by an intensified 14 bit CCD camera. Flame luminosity is suppressed by setting the exposure time of the image intensifier down to 100 ns, sampling only laser-induced signals.

The Rayleigh signal is proportional to the number of molecules in the gas sample weighted with their specific Rayleigh cross section factor $\sigma(\lambda)$. Therefore at isobaric conditions comparing the Rayleigh signal intensity I_{flame} of a measurement with the measured signal intensity I_{air} at known conditions (air at T_{ref}) the molecule density and from that the gas temperature T can be calculated by using the perfect gas law for every pixel.

$$T(x, y) = T_{ref} * \frac{I_{air}(x, y) - I_{back}(x, y)}{I_{flame}(x, y) - I_{back}(x, y)} * \frac{\overline{\sigma}_{flame}}{\overline{\sigma}_{air}} \qquad (3)$$

Here $\overline{\sigma}$ is the averaged Rayleigh cross section factor of the whole sample volume. The images will be corrected with the background noise (I_{back}) of the camera system. For thermometry measurements the maximum error has been estimated to be 11%, taking into account a signal to noise ratio (SNR) better than 12:1, variation of Rayleigh cross section in the gas volume of 3% and laser fluctuation of 5%.

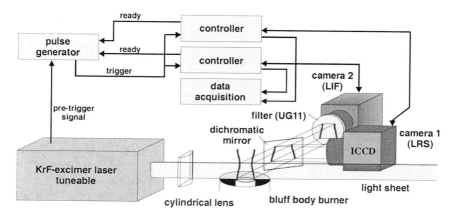

Figure 2: Schematic setup for simultaneous T- and OH-measurement

For simultaneous measurements of OH-radicals and temperature two intensified CCD cameras are imaging the same laser pulse. For OH determination the KrF excimer laser is tuned to excite the P1(8) line of the (3 - 0) transition of the A - X band at 248.17 nm. [Krämer et al.,1995]. To separate the Rayleigh and the LIF signal a dichromatic mirror has been installed, which reflects the Rayleigh signal at 248 nm, but transmits the LIF emission around 308 nm With this setup a resolutions of (58 μm)² per pixel can with both cameras. A bandpath filter (UG11) is installed in front of the LIF-ICCD camera to suppress interfering signals of other wavelengths. Both cameras are triggered by the laser pulse using suitable trigger logic to enable simultaneous measurements (see Fig 2).

3 Evaluation technique

The local flame front thickness is calculated from the extrapolation of the steepest temperature gradient to measured difference from T_{min} to T_{max}. From the received raw temperature images a 2-dimensional gradient field of the flame front is determined using a combination of a horizontal and vertical oriented Sobel filter within a 3x3 matrix [Dinkelacker et. al 1998]. For examination of the flame front only those gradients are sampled where the temperature has a value of about $(T_{max}-T_{min})/2$. This temperature level is nearly equal to the inner layer temperature of Göttgens et al. [1993]. Typically in this temperature range the steepest gradients are found. From simultaneous obtained OH and temperature images the OH radical is used to indicate the burning reaction. A point to point assignment between OH and temperature field is achieved by a transformation rule with an accuracy of 1 pixel. Regions with high OH signals (super-equilibrium) and steep OH concentration gradients indicate local reaction conditions. Thus this is used as a criterion to distinguish between local reacting or non-reacting conditions. The flame front thickness is calculated for reacting conditions from the temperature images.

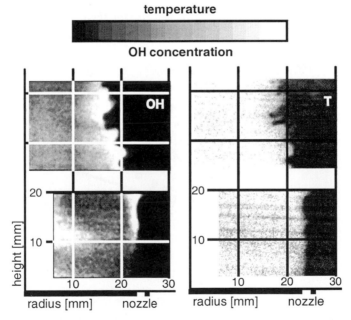

Figure 3: Simultaneously obtained image pairs of the T and OH distribution at two heights ($v_{exit} = 10.6$ m/s, $\phi = 0.6$)

510

For each investigated flame condition an ensemble of about 3000 of measured conditional thickness data from 30 single-shot images, is collected into a probability density distribution. From that the maximum value is taken as characteristic flame front thickness for this burning conditions. A systematic error is produced by occasional non-perpendicular crossing between the 3-dimensionally oriented front and the light sheet. A recent systematic comparison between 3-dimensional and 2-dimensional measurements [Soika et al., 1998] indicates, that with this light sheet setup the 3-dimensional flame front thickness can be estimated from the maximum of the 2-dimensional thickness distribution. Errors of the thickness measurement are estimated to ±50 µm.

4   Measurements

4.1   Investigated flames

The thermal flame front structure is measured at the outer flame contour in two different regions above the burner nozzle (height 1 11±8 mm, height 2 34±8 mm). As indicated in figure 4 the investigated flames are located within the *distributed* or *broadened flames* regime and the *stirred reactor* regime of the phase diagram of Peters [1986].

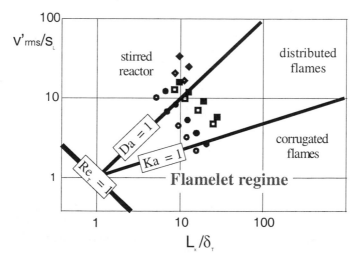

Figure 4: Phase diagram for turbulent premixed flames [Peters 1986] with exemplary investigated flame conditions. The laminar flame thickness is calculated to $\delta_L = v_i/s_L$, where v_i is the kinematic viscosity inside the flame at T ≈ 1250 K [Göttgens et al., 1993]; open (pos. 2) and filled (pos. 1) symbols indicate two different downstream positions investigated
(v_{exit} ●·5.1 m/s, ■·13.8 m/s, ◆ 28.7 m/s)

For all operation conditions the temperature images show a sharp partly wrinkled frontier between hot and cold gas (see Fig. 5). For high exit velocities single vortex structures are visible direct above the burner nozzle, looking similar to Rayleigh instability structures of corresponding flow systems. This is caused by the contact of the recirculated hot gas with fresh cold fuel. Further downstream the flame front is much more corrugated, but still a sharp front exists.

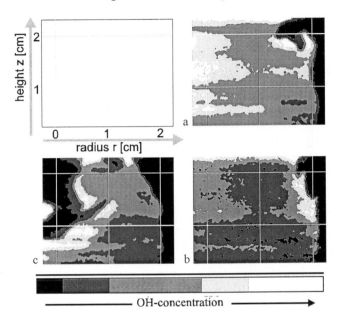

OH-concentration

Figure 5:   Exemplary images of OH-concentration at v_{exit} 21.3 m/s. The burner nozzle is located at r = 2.2-2.4 cm, z = 0 mm; image a: $\phi = 0{,}8$; b: $\phi = 0{,}7$; c: $\phi = 0{,}65$ (For case c burning occurs inside the less turbulent recirculation zone.)

The OH-concentration images are used to detect the burning reaction zone in a „tri-modal-way". First case: *fresh fuel* or air. Here no burning takes place and therefore no OH-signal can be detected. Second case: *reaction*. A relatively large OH concentration with a superequilibrium value and steep gradient indicates reaction in the early part of the reaction zone (e.g. upper part of Fig. 5b). Third case: *burnt gas*. Behind the reaction zone the high superequilibrium OH signal reduces gradually to the equilibrium value of the burnt region. Typical OH images are shown in Fig 3 and 5. Note that the burner nozzle is located at the right side (r = 2.2-2.4 cm).

As above mentioned the temperature measurements show a strong temperature gradient immediately above the burner exit. From that the reaction zone would be expected to be direct above the nozzle. But in OH images no superequilibrium-OH can be found here. Obviously the reaction zone is lifted, which is only revealed in the OH images. Up to a height of about 11 mm only the low equilibrium OH signal is visible, coming from the circulated burnt gas.

For very lean mixtures ($\phi = 0.7$) and high exit velocities (>21.3 m/s) burning occurs partially inside the less turbulent core of the recirculation zone, while no ignition occurs at the outer contour of the flame (Fig 6 c). The location of reaction zone fluctuates between the core and the contour of the flame.

5 Results

5.1 Flame front thickness

If for this flame the thermal flame thickness is determined without OH conditioning, the results are not easy to interpret.

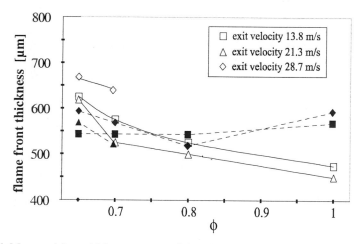

Figure 6: Measured front thickness, not conditioned with reaction criterion
(filled symbols height 1; open height 2)

Using OH concentration as reaction criteria the thermal fronts can be distinguished into flame fronts and non-reacting fronts. In height 2, usually a reaction takes place, except for exit velocities above 21,3 m/s, $\phi = 0.65$. Thus comparing only reacting cases with laminar calculated unstretched flames [Göttgens and Dinkelacker, 1997], a slightly increased flame front thickness of turbulent flames can be found for stoichiometric methane/air mixtures. For lean mixtures with equivalence ratios below 0.8 (e.g. in Fig 7) a significant decrease is visible.

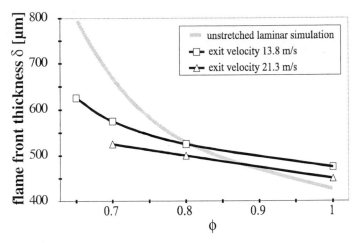

Figure 7:   Measured flame front thickness at height 2 in comparison to calculated laminar unstretched flames [Göttgens and Dinkelacker, 1992]

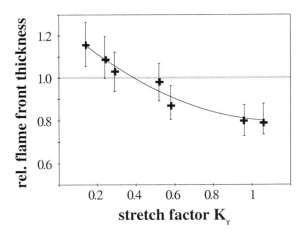

Figure 8: Flame front thickness δ normalized by thickness δ_L of a calculated laminar unstretched flame front plotted against local stretch parameter K_T

514

Plotting this flame front thickness, normalized with the thickness of the calculated laminar unstretched flame δ_L versus the local stretch factor K_T [Bradley, 1992] an influence of local stretch on the flame front is visible (Fig 8).

$$K_T = \frac{v'_{rms}/l_T}{\delta_L^{Zel}/s_L} = \frac{a_T \cdot v}{s_L^2} \qquad (4)$$

Here a_T is the turbulent strain rate, calculated with the Taylor length l_T, while v is the kinematic viscosity of the cold gas mixture and s_L the laminar burning velocity.

$$a_T = \frac{v'_{rms}}{l_T} = \frac{(2k)^{0.5}}{l_T} \qquad (5)$$

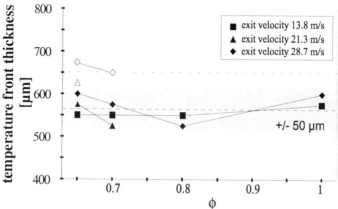

Figure 9:   Front thickness without reaction (filled height 1, open height 2)

If then the thermal flame front thickness is determined for conditions without reaction (missing superequilibrium OH) it is found to be independent of equivalence ratio and exit velocity (Fig 9). The averaged thickness at 8 mm height is determined to 560 µm and at 30 mm 650 µm. The independence of the thermal thickness from stoichiometry and the increase with height indicates that here a diffusional heat transfer without reaction takes place, since for reacting flame fronts the thermal thickness depends strongly on ϕ. If thermal diffusion is the basic heat transfer mechanism, the profile and thickness of the front should depend on time after contact with hot gases and therefore depend on the exit velocity. But no significant different profiles are visible for different exit velocities. On the other hand the heat transfer should increase with the turbulence intensity (improved mixing) which is increased with higher velocities. Obviously both effects compensate each other.

5.2 Lifted reaction zone

Lift-off heights for the different flame conditions are shown in Fig 10. The lift-off height, as the place where reaction starts indicated by a steep OH gradient, is about 11 mm for nearly all equivalence ratios and exit velocities. Only for very lean mixtures with exit velocities above 21,3 m/s the reaction starts first at a higher position. These behavior cannot be explained by a unique ignition criterion. Instead it seems that two different conditions have to be fulfilled for ignition:

(a) chemical ignition delay
(b) sufficient low turbulent strain rate

For the chemical delay a strong dependency on temperature is known [Warnatz et al., 1986]. Contact of hot burnt and fresh cold gas takes place immediately after exit. Since the temperature of the hot gas depends strongly on φ, the ignition delay time τ is significantly larger for lean mixtures with lower maximum temperatures. This will lead to a qualitative behavior as shown in Fig. 10 (dotted lines). While the chemical ignition delay time depends on the flame temperature and thus on the stoichiometry, it gets large and therefor dominant for very lean mixtures with low flame temperatures. Here an ignition height is found, which can be understood from different convection velocities.

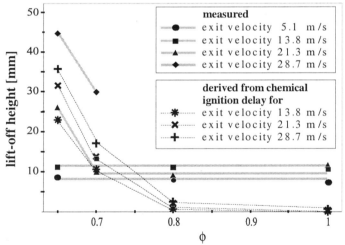

Figure 10: Height of measured (full lines) and theoretical ignition derived from chemical ignition delay (dotted lines)

516

Assuming a φ-dependent ignition delay time the "*chemical-dominated*" lift-off height is expected to increase with higher convection velocity, which fits to the observation of lean flames (e.g. v_{exit} 28.7 m/s, Fig. 10), but not for near stoichiometric flames. On the other hand high turbulent strain rates can also be the reason of quenching [Pitts et al., 1989] or in this case for the prevention of ignition. For ignition this effect should be independent on chemical mixture, while depending on turbulence conditions and therefor on the velocity field. Therefore a constant lift-off height should be found for different equivalence ratios at one exit velocity, which can be obtained for $v_{exit} \leq 13.8$ m/s and $\phi \geq 0.7$. Here the "*chemical-dominated*" lift-off is obviously below the "*turbulence-dominated*" lift-off height (see Fig. 10). A combination of both effects should lead to a lift-off behavior as found experimentally.

Thus, the combination of a turbulent dominated criterion, being independent on stoichiometry, and a chemical ignition delay criterion can explain the observed lift-off behavior in a remarkable manner.

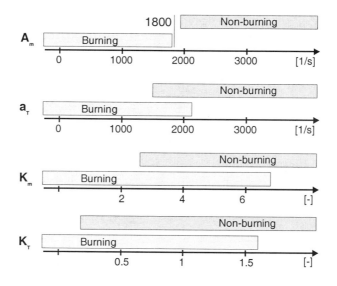

Figure 11: Comparison of different critical local strain parameters for lifted reaction zones. Marked local burning or non-burning conditions.

Furthermore we tried to find a suitable criterion for the *turbulent dominated* lift-off case (no φ-dependency, v_{exit} 13.8 m/s and 21.3 m/s with $\phi \geq 0.7$). For that we determined four different characteristic quantities from the turbulent flow field, measured with LDV and referenced to the corresponding position of the most probable local flame front location. Two quantities describe the turbulent flow field, the mean flow strain rate A_m, which is the gradient of the mean flow and the turbulent strain rate a_T, which is taken as a measure for strain induced turbulence by eddies.

$$A_m = \left| \frac{dV}{dy} \right| \qquad K_m = A_m \frac{\nu}{s_L^2} = A_m \frac{\delta_L^{Zel}}{s_L} \qquad (6)$$

$$a_T = \frac{v'_{rms}}{l_T} \qquad K_T = a_T \frac{\nu}{s_L^2} = a_T \frac{\delta_L^{Zel}}{s_L} \qquad (7)$$

For both the corresponding normalization with thermochemical data of the flame mixture is determined additionally, leading to K_m and K_T. Note that K_T correspondents to the "Karlovitz number" used by Bradley [1992].

Plotting the local burning or non-burning condition determined from the lifted reaction zone (OH concentration) as a function of these four quantities (Fig. 11) it can be found that only the mean flow strain rate $A_{m,crit} \approx 1800$ s^{-1} is a suitable criterion for ignition in these flames, while all other quantities (including Bradley's Karlovitz number K_T) show overlapping regions of burning or non-burning conditions, thus being not definite criterions.

6 Conclusion

With the investigated bluff-body burner highly turbulent flames can be produced. At this turbulence conditions a burning reaction is not always visible at the border between hot recirculated gas and fresh fuel/air mixture. Since this effect is not visible from mere temperature images, in experiments an indicator for reaction intensity, like the ascent of OH concentration has to be determined, too. Then the measured temperature fronts can be distinguished in non-reacting and reacting flame fronts.

Compared with calculated unstretched laminar flames a thinning of the local flame front is visible for the reacting case, which is contrary to the models included in the phase diagram [Peters, 1986]. This thinning is found to be dependent on the local stretch factor K_T.

Directly above the burner nozzle up to a height of 11 mm, no burning reaction can be detected. Instead a lifted reaction zone is found. While for near stoichiometric equivalence ratios (≥ 0.7) the lift-off height (the starting point of reaction) is found to be independent on exit velocity and equivalence ratio, for very lean mixtures dependency of these parameters is observed. For an explanation of this behavior two different ignition criteria seem to be necessary, taking into account both chemical ignition time and a hydrodynamic strain.

Since the chemical ignition delay time depends on the flame temperature, it is dominant for very lean mixtures with low flame temperatures. Only if this effect is shorter than a certain value, the local hydrodynamic strain seems to determine the

518

lift-off height. As a suitable lift-off criterion for the hydrodynamic case the mean flow strain rate $A_m = |dV/dy| \approx 1800$ s^{-1} distinguishes between burning and non-burning conditions. The strain rate a_T, representing the strain influence from turbulent eddies, and also the non-dimensionalized forms of the strain rates – including Bradley's stretch factor K_T – are not suitable. To predict the lift-off height of the flame both, a *turbulence-* and a *chemical-dominate*d ignition criterion have to be considered.

Acknowledgement

The authors gratefully acknowledge financial support of parts of the work by Deutsche Forschungsgemeinschaft.

References

Buschmann, A., Dinkelacker, F., Schäfer, T., Schäfer, M. and Wolfrum, J., Measurement of the Instantaneous Detailed Flame Structure in Turbulent Premixed Combustion, 26th Symp. (Int.) on Combustion, The Combustion Institute, Pittsburgh, PA, p. 437-445, 1996

Bradley, D., How Fast Can We Burn, 24th Symp. (Int.) on Combustion, The Combustion Institute, Pittsburgh, PA, p. 247-262, 1992

Dinkelacker, F., Soika, A., Most, D., Höller, V. and Leipertz, A., Measurement of Local Temperature Gradients in Turbulent Combustion Systems using Two-Dimensional Laser Diagnostics, Proc. to 11th Int. Heat Transfer Conf., Kyongju, Korea, Vol. 7, p. 373-378, 1998

Dinkelacker, F., Soika, A., Most, D., Hofmann, D. Leipertz, A., Polifke, W., and Döbbeling, K., Structure of Locally Quenched Highly Turbulent Lean Premixed Flames, 27th Symp. (Int.) on Combustion, The Combustion Institute, Pittsburgh, PA, p. 857-865, 1998

Göttgens, J., Mauss, F., Peters, N., Analytic Approximation of Burning Velocities and Flame Thickness of Lean Hydrogen, Methane, Ethylene, Ethane, Acetylene and Propane, 24th Symp. (Int.) on Combustion, The Combustion Institute, Pittsburgh, PA, p. 129-135, 1992

Göttgens, J. (data), Dinkelacker, F. (evaluation), (privat), 1997

Kampmann, S., Leipertz, A., Döbbeling, K., Haumann, J. and Sattelmayer, Th., Two-Dimensional Temperature Measurement in a Technical Combustor with Laser Rayleigh Scattering, Applied optics, 32, No. 30, p. 6167-6172, 1993

Krämer, H., Kampmann, S., Münch, K.-U. and Leipertz, A., Simultaneous Measurements of Temperature and Concentration Fields inside Technical Combustion Systems, Proc. Air pollution and Visibility Measurements, Vol. 2506, p. 85-93, Munich, 1995

Mansour, M. S., Chen, Y.-C. and Peters, N., The Reaction Zone Structure of Turbulent Premixed Methane-Helium-Air Flames near Extinction, 24th Symp. (Int.) on Combustion, The Combustion Institute, Pittsburgh, PA, p. 461-468, 1992

Peters, N., Laminar Flamelet Concepts in Turbulent Combustion, 21th Symp. (Int.) on Combustion, The Combustion Institute, Pittsburgh, PA, p. 1231-1250, 1986

Pitts, W.M., Assessment of Theories for Behavior and Blow-out of Lifted Turbulent Jet Diffusion Flames, 22th Symp. (Int.) on Combustion, The Combustion Institute, Pittsburgh, PA, p. 809-816, 1989

Poinsot, T., Veynante, D., Candel, S., Diagrams of Turbulent Combustion based on Direct Simulation, 23th Symp. (Int.) on Combustion, The Combustion Institute, Pittsburgh, PA, p. 613-619, 1990

Roberts, W. L., Driscoll, J. F., Drake, M. C., Ratcliff, J. M., OH Fluorescence Images of the Quenching of a Premixed Flame During an Interaction with a Vortex, 24th Symp. (Int.) on Combustion, The Combustion Institute, Pittsburgh, PA, p. 169-176, 1992

Soika, A., Dinkelacker, F., Most, D. and Leipertz, A., Detection of 3D Temperature Gradients in Turbulent Premixed Flames with the Dual-Sheet Laser Rayleigh Scattering Technique, Proc. to 9th Int. Symp. on Appl. of Laser Tech. to Fluid Mechanics, Lissabon, p. 25.45.1-8, 1998

Warnatz, J., 24th Symp. (Int.) on Combustion, Resolution of Gas Phase and Surface Combustion Chemistry into Elementary Reactions, The Combustion Institute, Pittsburgh, PA, p. 553-579, 1992

CHAPTER VI. TWO-PHASE FLOWS

VI.1. Molten Metal Atomization by Inert Gases: Off-Line and On-Line Metal Powder Characterization by PDA

G. Brenn[1], J. Raimann[1], G. Wolf[2], J. Domnick[3], and F. Durst[1]

[1] Lehrstuhl für Strömungsmechanik (LSTM), Universität Erlangen-Nürnberg, Cauerstr. 4, D-91058 Erlangen, FRG

[2] Applikations- und Technikzentrum für Energieverfahrens-, Umwelt und Strömungstechnik (ATZ-EVUS), Rinostr. 1, D-92249 Vilseck, FRG

[3] Fraunhofer-Institut für Produktionstechnik und Automatisierung (FhG-IPA), Nobelstr. 12, D-70569 Stuttgart, FRG

Abstract. Modern metal powder processing techniques have a continuous demand for decreasing the mean particle size of powders and increasing powder quality in terms of, e.g., particle surface structure and morphology. The industrial processes of gas or inert gas assisted atomization (IGA) of molten metals must therefore be optimized to yield the required high quality powders at competitive prices. In the present work it was attempted to reach this optimization by two steps: first, the search for a deeper understanding of the IGA process through detailed off- and on-line PDA measurements and, second, on-line process control by adjusting the dominant process parameter, the atomizing gas pressure, to an optimized level. It was found that a droplet velocity measurement at a single specific location inside the molten metal spray of IGA may be sufficient to characterize the mean droplet diameter of the whole spray and, hence, to predict the essential characteristics of the metal powder produced.

Keywords. Molten metal atomization, phase Doppler anemometry, industrial IGA process, on-line control

1   INTRODUCTION

Although metal powders cover only a small part of the worldwide market for metal raw materials, powder metallurgy shows increasing importance for many production processes. Following White (1997), powder metallurgy is the most promising way for developing new metals and creating new material characteristics.

Metal powders required by the processing industry over the last years have shown tendencies for very small mean particle diameters (Capus 1996). In order to deliver these fine powders at competitive prices, the powder production process via IGA has to be optimized.

The gas-assisted atomization of molten metals has been investigated by several groups worldwide (e.g. Anderson et al. 1991, Miller et al. 1992) in order to optimize the atomizer geometries and process parameters. The on-line measurement of particle sizes in this process was first performed using Fraunhofer diffraction-based techniques (e.g. Ridder et al. 1992). Stagnation in the dissemination of this technique to industrial application showed that this technique is only partly applicable for characterizing the dense sprays of IGA. Early investigations by Bauckhage et al. (1989) proved that PDA is more suitable for such measurements in the gas-assisted atomization of metals.

In the present paper, the steps in the application of PDA for the control of IGA are described. First a short overview of the applied PDA measurement technique is given. The off- and on-line investigations are then described in detail, together with the resulting setup of a closed control circuit for this process (Domnick et al. 1996). In addition to the PDA characterization of the metal sprays in IGA, a method to characterize the atomization process by a one-component LDA measurement is presented. The paper ends with a summary of the work and an outlook.

2   PDA MEASUREMENT TECHNIQUE

A commercial fiber-based two-component AEROMETRICS PDA system, using an ILT argon-ion laser and a DSA processor, was used to investigate the metal sprays and powders produced by IGA. The system offers a sufficient velocity bandwidth and the robustness of optical and electronic components required in the hostile environment near a metal atomization

Table 1: Optical setup of the used PDA system

Transmitting optics	Symbol	Value
laser power in probe volume	P_L	40 mW
focal length	f_{tr}	500 mm
beam separation	d_{sp}	78 mm
beam diameter	$D_{e^{-2}}$	7.8 mm
wavelength	λ	514.5 nm
Receiving optics	Symbol	Value
focal length	f_r	500 mm
slit width	s	150 μm
detector separation	$(d_1\text{-}d_2) / (d_1\text{+}d_3)$	14.71 mm / 38.01 mm

Table 2: Resulting PDA characteristics

Characteristic dimensions of the probe volume	Value
diameter	42.0 μm
length	554.0 μm
distance of the fringes	3.3 μm
number of fringes	13
Measurement range	
velocity	-66.0 - +260.5 m/s
size (reflection)	4.8 - 216.9 μm

chamber. The optical setup of the system and the resulting measurement ranges are summarized in Tables 1 and 2. The light scattering mechanism used for the PDA measurements is reflection.

3   OFF-LINE CHARACTERIZATION OF POWDERS BY PDA

There are two major reasons for performing an off-line characterization of metal powders before applying a PDA system on-line in an atomization process: first, the capability of the PDA to measure solid powder particles as well as liquid droplets has to be verified, and second, these tests allow a direct comparison of PDA and Fraunhofer diffraction measurements of the powders. According to Bauckhage et al. (1989), the surface structure of solid particles makes PDA measurements of the particle size more difficult and uncertain than drop size measurements. In an application, this might affect the small particles most strongly, since the smallest droplets solidify most quickly. Any correlation between the rejection rate of the PDA and the size of the particles would lead to a systematic difference between on-line PDA results and the real particle size distribution, which is usually measured using the Fraunhofer technique as a reference.

3.1   Laboratory setup

These reference measurements were performed with a SYMPATEC HELOS particle size analyzer. Metal powders produced by IGA are highly spherical, and therefore the SYMPATEC results agree reasonably well with sieving results for the powders tested. Additionally, it must be noted that the SYMPATEC particle sizer is widely accepted by the consumers of metal powders as a reference measurement technique. In general, typical mean diameters of metal powders, as the volume median $D_{V0.5}$, are accepted by customers only if this measurement technique has been used. Based on a scanning electron microscope (SEM) analysis of the powders produced, approximately 95% of the mass is represented by spherical particles.

In a typical SYMPATEC setup, the powder sample is passed through the probe volume by a quasi two-dimensional gas flow. The probe volume has a diameter of approximately 2 cm. The setup is chosen to allow for the analysis of the powders with minimized effects of multiple scattering of the incident laser light (Vielhaber 1989).

It is evident that PDA results comparable to the SYMPATEC data can only be obtained if a representative fraction of the powder passes the small probe volume of the PDA. Here it must be kept in mind that the PDA

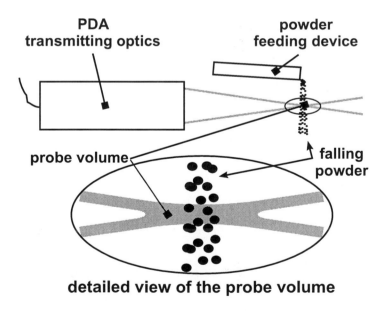

Fig. 1: Experimental setup for the off-line characterization of metal powders by PDA

measures a temporal mean size (flux weighting), while the Fraunhofer diffraction result is a spatial mean (concentration weighting). Good results were achieved by simply feeding the powder to the PDA probe volume in a stream driven by gravitation (Fig. 1). The distance between the tip of the feeding device and the probe volume location was kept very short in order to prevent aerodynamic separation of the powders and size dependence of the measured velocities. It was found that 10000 single particle measurements were necessary to ensure statistical stability of the measured mean values.

In Figs. 2 and 3 the measurement results obtained from the two techniques for a sieving fraction of copper powder and iron-nickel alloy powder of nominally the same sizes are presented. In general, the results obtained with the two techniques agree quite well. In the case of the copper powder (Fig. 2), the PDA yields the same shape of the cumulative distribution as the SYMPATEC, but slightly shifted towards smaller sizes. For the FeNi powder (Fig. 3), however, the deviations between PDA and SYMPATEC increase for particle sizes above 50 µm. This difference may be due to slightly inaccurate sieving and differences in the surface structure.

According to Göbel et al. (1997), surface irregularities of the particles lead to broadened size distributions in PDA measurements. Hence it is worth examining the surface structure of the powder particles. Examples

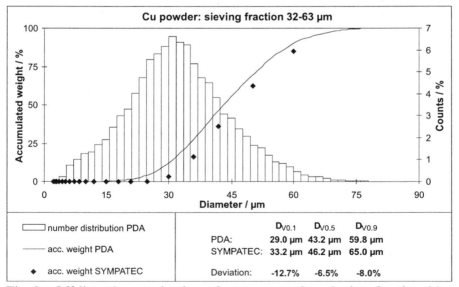

Fig. 2: Off-line characterization of copper powder, sieving fraction 32 - 63 µm: comparison of PDA and SYMPATEC results

528

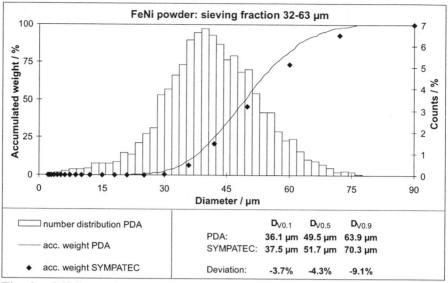

Fig. 3: Off-line characterization of iron-nickel alloy powder, sieving
fraction 32 - 63 µm: comparison of PDA and SYMPATEC results

of surface structures on different metal particles produced by IGA are
presented in Figs. 4 - 7. In Fig. 4 the surface of a copper powder particle
with a diameter of 36 µm is shown. Apparently, the surface is very
smooth. In general, IGA atomized copper particles of any diameter show
practically no surface irregularities with length scales larger than 1/5000
of the particle size. The structures visible on the iron-nickel particle are
clearly larger than 300 nm (Fig. 5). However, these disturbances are
smaller than the wavelength of the laser light applied for the PDA measurements (Table 1). It can be assumed that such surface irregularities of
the particles have no effect on the PDA measurements.

The SEM photographs of a copper-tin alloy and a tool steel powder particle in Figs. 6 and 7 reveal larger surface irregularities with length scales
clearly greater than the incident laser light wavelength. However, these
structures are smaller than the fringe spacing in the PDA probe volume.
Owing to the integration effect of the photodetectors used for the investigations, no systematic influence of the surface structures on the measurement results was found.

Figure 8 depicts the results of PDA and Fraunhofer analyses for different
sieving fractions of the four powder materials corresponding to Figs. 4 - 7.
The data given are the mean diameters $D_{V0.5}$ from the PDA measurements
versus $D_{V0.5}$ from the SYMPATEC analysis. The double dotted line represents the exact agreement of the two techniques. The results for $D_{V0.5}$ cal-

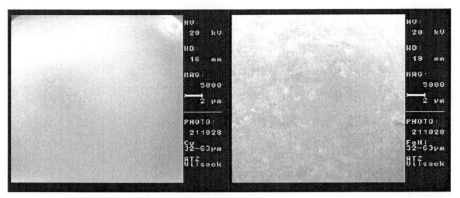

Fig. 4: SEM picture of a copper
   particle (D = 36 μm)

Fig. 5: Iron-nickel alloy particle
   (D = 42 μm)

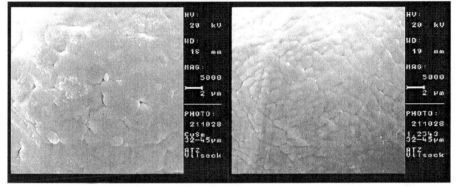

Fig. 6: Copper-tin alloy particle
   (D = 38 μm)

Fig. 7: Tool steel particle
   (D = 40 μm)

culated from the PDA measurements compared with the SYMPATEC
results for all four powders show no systematic effect of the different surface structures on the PDA measurements. All materials show a linear
relationship of the measurement results.

   Although the surface structures of the powders seem to have only slight
influence on the size measurements, the detected light intensities are, in
fact, affected by these structures. In Fig. 9 the correlation between the
PDA measured diameter and the detected signal intensity of the copper
and iron-nickel powders already discussed in Figs. 2 and 3 is depicted.
The curves show the well-known relationship between particle size and
scattered light intensities: larger particles yield higher intensities. The
fluctuations of the detected intensities are depicted as error bars on the
mean values. The larger fluctuations in the light intensities from the FeNi

particles, which even increase with the particle size, give rise to the assumption that this kind of powder may lead to slightly larger measurement uncertainties with optical techniques than the copper powder. In the present case this leads to a slight underestimation for the coarse particles.

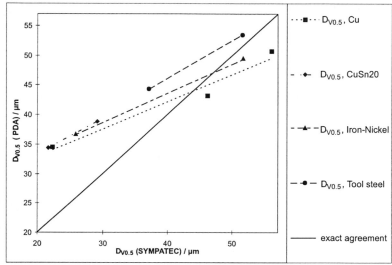

Fig. 8: Comparison of mean diameters of different alloy and pure metal powders: $D_{V0.5}$ measured by PDA versus $D_{V0.5}$ measured by SYMPATEC

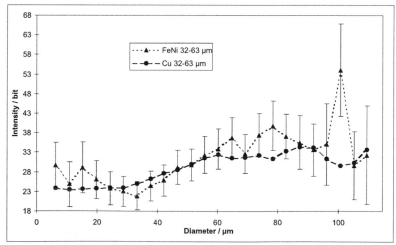

Fig. 9: Correlation between PDA measured diameter and intensity of the signals for the powders of copper and iron-nickel melts

4   ON-LINE CHARACTERIZATION OF METAL SPRAYS

The on-line characterization of the IGA process should be understood as an experimental investigation of the fluid dynamics of the metal spray inside the atomization chamber. In addition to on-line measurements in the melt atomization process, the present contribution also presents results of experiments performed in a model flow field of the IGA. These model experiments were performed by adding metal powder particles to a gas flow in order to obtain information about the correlation between locally measured particle sizes and velocities with the overall powder size distribution.

In this section, both liquid melt droplets and solidified particles are called "particles", basically because it is impossible to distinguish between the two disperse fractions in the metal spray.

4.1   Laboratory setup

For adapting the PDA to the IGA system, detailed Mie calculations were carried out (Domnick et al., 1996). In addition to the construction of windows, which was necessary to allow optical access to the measurement region in the spray about 12 cm below the atomizing nozzle, a set of baffles was introduced into the atomizing chamber.

In Fig. 10 the PDA system is shown together with the cross-section of the atomization chamber and the windows. The off-axis angle of 60° was chosen since this angle is appropriate for all planned measurements, including melt spray experiments with liquefied gases and water, where extended phase-Doppler anemometry has to be applied (Domnick et al. 1995, Raimann et al. 1997). In Fig. 11 the effect of the baffles is explained schematically. Without the baffles, the recirculation zones in the atomization chamber cover the regions of the windows and block parts of the incident and scattered laser light, which causes problems for the PDA measurements. By an appropriate layout of the geometry of the baffles, the recirculation of small powder particles was minimized. It was found that the application of such baffles is necessary for the production of powders with a $D_{V0.5}$ of less than 20 µm. In the production of coarser powders, the baffles show no effect on the measurement results.

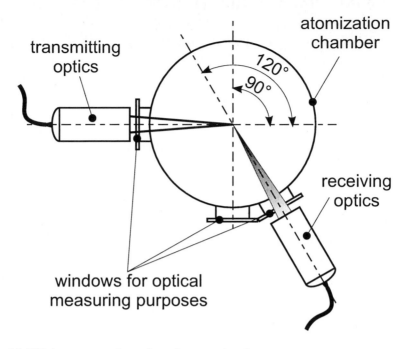

transmitting optics

atomization chamber

120°

90°

receiving optics

windows for optical measuring purposes

Fig. 10: PDA system adapted to the atomization system

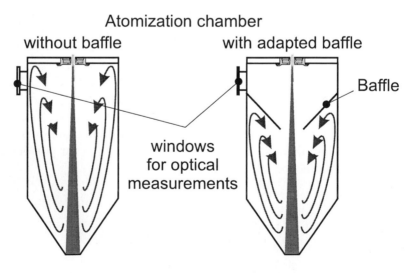

Atomization chamber

without baffle

with adapted baffle

Baffle

windows for optical measurements

Fig. 11: Effect of the baffle adapted to the atomization system

4.2   On-line measurements in melt atomization

On-line PDA measurements were performed in a melt spray facility equipped with a close-coupled atomization nozzle as drawn schematically in Fig. 12. The melt is fed through a tube surrounded symmetrically by individual gas nozzles. The gas pressure ratio between the gas ring chamber and the atomization chamber lies beyond the critical pressure ratio, so that the gas jets leave the gas nozzles underexpanded.

A view of the melt atomization process by inert gases is given in Fig. 13. The bright glooming atomized melt jet indicates the high particle density and the high temperature in the metal spray. The PDA measurement volume is formed by the two laser beams transmitted from the right-hand side.

The high-speed gas flow outside the melt tube causes a low static pressure at the edges of the tube. Owing to this pressure field, the melt flows as a thin film from the center towards the edge of the melt tube tip, where the primary atomization takes place. The disintegration process of the melt at the edge of close-coupled atomizers was visualized using the PCO technique, confirming previous results of Anderson et al. (1991), Miller et al. (1992), and Kuntz et al. (1995).

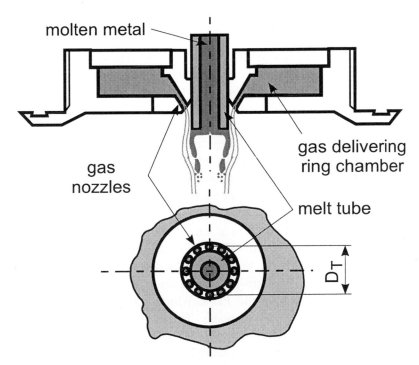

Fig. 12: Close-coupled atomizer applied to IGA

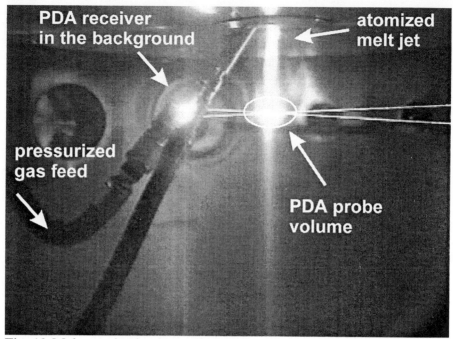

Fig. 13: Melt atomization by an inert gas

The PDA measurements at a location 12 cm below the atomizer exit were performed in a number of experiments under constant process conditions. These experiments with a typical run time of 5 - 10 min allowed complete radial profiles of mean particle size and velocity in the spray cone to be measured.

In Figs. 14 and 15, results of three experiments are depicted as radial profiles of the volume median $D_{V0.5}$ and the mean axial particle velocity. Comparing cases a) and b), it is remarkable that experiments under nominally identical test conditions result in different melt flow rates MF, but in similar $D_{V0.5}$ values, as determined by means of the SYMPATEC technique. The radial profiles of $D_{V0.5}$ as measured by PDA correspond to the global $D_{V0.5}$ measured by SYMPATEC. The application of a smaller melt outlet tube and a higher gas pressure (experiment c)) results in a finer powder. Owing to the high drop concentration in the spray, many fine particles are not detected by the PDA, which leads to too high $D_{V0.5}$ as compared with the SYMPATEC result. The measured velocity, however, corresponds to the gas pressures applied (Fig. 15). The importance of this result will be discussed in the next section.

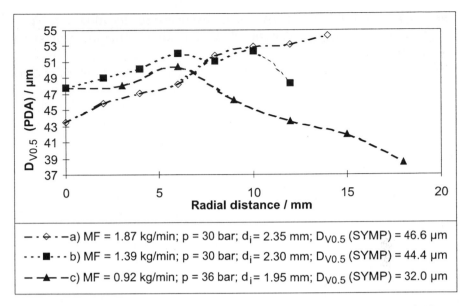

Fig. 14: Particle mean diameter profiles in the spray at a distance of 12 cm from the nozzle (legend also for Fig. 15)

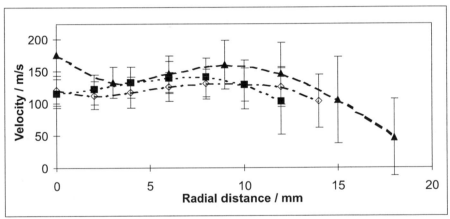

Fig. 15: Axial particle velocity profiles in the spray. Error bars represent the RMS velocities (legend see Fig. 14)

Using these measurement results, the mass flow rate of the melt was calculated. In Fig. 16 the melt flow rate computed from the PDA measurement data is plotted versus the melt flow rate determined after the experiment by dividing the mass of the powder obtained by the process run time.

The results show that the PDA results overestimate the flow rate. This may be caused by the fact that the PDA overestimates the mean diameter used for the flow rate calculation.

Finally, experiments with variations of the dominant process parameters were performed. It was found that changes of the melt temperature in the relevant range between 100 and 300 K melt superheat, which are believed to affect the fluid parameters, seem to be irrelevant for the atomization process. A stepwise variation of the gas pressure had a much stronger effect. In Figs. 17 and 18, the volume median $D_{V0.5}$ and the axial mean and RMS velocities measured by PDA during such an experiment are shown. RMS velocities are represented by error bars. The mean particle size shows a tendency to decrease with increasing gas pressure, whereas the mean particle velocity increases, as expected.

It can be seen in Figs. 17 and 18 that, before the process time 13 min 30 s, the measurements do not reveal the described spray behavior: despite the slightly increasing gas pressure, the particle size increases and the particle velocity remains constant. This must be interpreted as a measurement artifact caused by a blockage of the laser light which was removed around the time 13 min 30 s. Thereafter, the spray behaviour is repre-

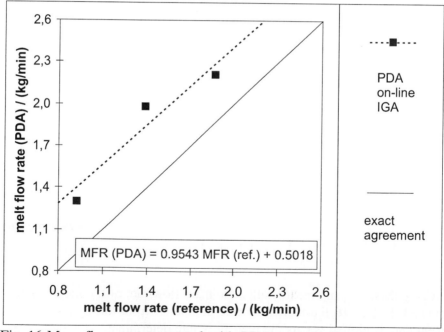

Fig. 16: Mass flow rate measured with PDA versus reference mass flow rate

sented correctly by the measurements. In combination with the obtained velocity information, the PDA measured drop size $D_{V0.5}$ can be used to define a calibration function for the measurement of the global mean power size $D_{V0.5}$ by a single PDA measurement of 10 s duration. The definition of this calibration function is discussed in the next section.

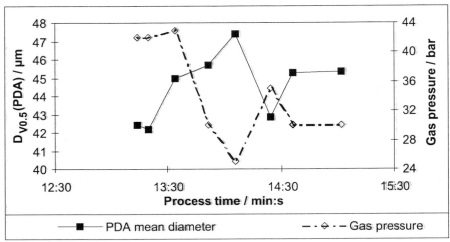

Fig. 17: Particle mean diameter measured by PDA as a function of process time at different gas pressures during an IGA process

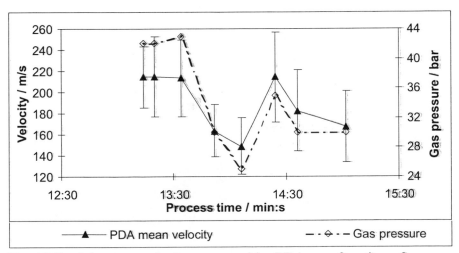

Fig. 18: Particle mean velocity measured by PDA as a function of process time at different gas pressures during an IGA process

538

4.3   Model Experiments

It was indicated above that the velocity of the particles measured at a given location in the spray is correlated with the atomization process itself. However, both the results of the off-line characterization by PDA and the on-line measurements during the powder production process revealed certain differences between the PDA particle size measurements and the results of the reference measurements by SYMPATEC. It is therefore obviously necessary to incorporate additional particle information to enable a unique determination of the mean particle size of the powder by PDA. Information useful for this purpose is the particle velocity.

In order to establish the use of the particle velocity, model experiments were performed, combining the technique of off-line PDA characterization (Fig. 1) with the measurement in the flow field of the atomizing nozzle in the chamber. Powders of well-defined mean particle diameters and size distributions were fed through the melt feeding tube of the atomizer and entrained into the gas flow.

The PDA measurements in this two-phase flow field were performed at a position 12 cm below the nozzle at an arbitrary radial distance of 4 mm from the symmetry axis of the atomizer. The gas pressure applied varied between 0 bar (free falling powder particles) and 40 bar, resulting in particle velocities around 200 m/s. The correlation of the arithmetic mean di-

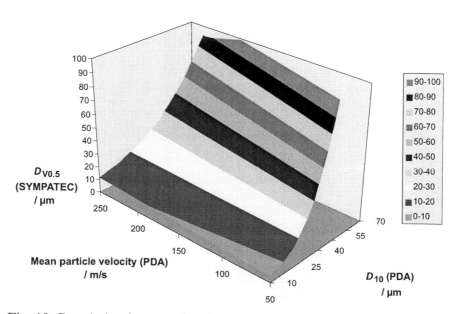

Fig. 19: Correlation between local PDA measurement of size and velocity and global powder size measured by the SYMPATEC system

ameter and the mean particle velocities measured by PDA and the mean diameter of the inserted powder fraction measured with the SYMPATEC system is shown in Fig. 19. The relationship between the three variables can be approximated by

$$D_{V0.5,\ global} = C_0 \exp\left(\frac{D_{10,\ PDA} - \left(C_1 U_{PDA} + C_2\right)}{C_3\ U_{PDA} + C_4}\right) \qquad (1)$$

with $C_0 = 1$ μm, $C_1 = -0.28$ μm/(m/s), $C_2 = 1.36$ μm, $C_3 = 0.10$ μm/(m/s), and $C_4 = 8.14$ μm.

4.4 Global characterization of the spray by a local PDA measurement

Using this correlation, the local mean particle sizes measured on-line by may be related to the global mean powder size measured by the SYMPATEC system. In Fig. 20, the volume medians $D_{V0.5,\ global}$ calculated by means of Eq. (1) from the PDA measured D_{10} and mean velocity are plotted versus the off-line characterization results ($D_{V0.5}$ SYMPATEC).

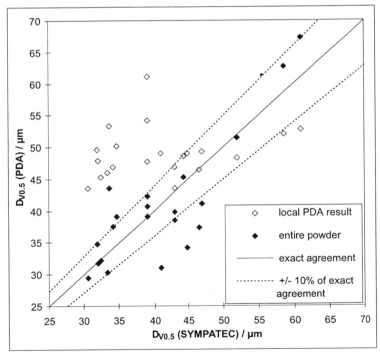

Fig. 20: Characterization of the IGA sprays (copper, copper-tin and silver melts)

540

The open symbols represent the "raw" PDA measurement data, the filled symbols represent the values calculated from Eq. (1). In general, the agreement is satisfactory and indicates that the local PDA measurement can characterize the global powder size with an uncertainty of only ±10% in most of the experiments. It should be noted that Fig. 20 includes measurements with melt tube diameters between 1.9 and 4 mm, gas pressures between 25 and 70 bar, and three different metals.

4.5   Definition of a closed control loop

A control circuit was composed, based on the global mean sizes, calculated from the PDA measured mean size and velocity according to Eq. (1). The control loop optimizes the atomization pressure of the compressed gas in IGA in order to reach a desired mean powder size. The PDA measurement and the process data recording are performed simultaneously, triggered by the control software on a Master PC.

The interval between the measurements can be shorter than 10 s if the data rate, detected by the PDA, is high enough to ensure a safe determination of the mean diameters by several thousands of single particle measurements per second. Compared with the overall run time of the industrial processes of several hours, the ruling time of around 5 minutes may be considered as negligible. In Fig. 21 the result of three on-line control loop runs are depicted in terms of PDA-measured global mean particle diameters as functions of process time.

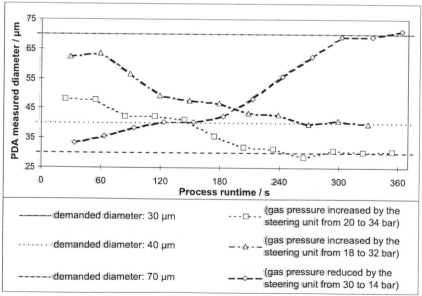

Fig. 21: Results of three controlled IGA processes

5 CHARACTERIZATION OF THE ATOMIZATION PROCESS BY LDA MEASUREMENTS

It was shown above that the velocity of the particles in the spray is an important parameter for the spray characterization. As an alternative to the method described in Sections 4.3 and 4.4, the pure velocity information of the particles at a given location in the flow field can be used for the characterization of the overall mean particle size of the spray in IGA. This method is based on the aerodynamic behavior of the powder particles in the gas flow field.

In Fig. 22 the axial velocity of the particles produced by IGA processes is plotted versus the particle diameter of the produced powders determined off-line. The single-component LDA measurements were carried out 12 cm downstream from the nozzle, at a radial distance of 4 mm from the spray center. The diagram represents results for different inner diameters of the melt tube, but constant outer diameter of the tube and constant geometry of the gas nozzles. At inner diameters of the melt tube of 2.3 mm or less, it was found that a unique correlation between the locally measured velocity of the particles and the mean particle diameter $D_{V0.5}$ measured by SYMPATEC exists. Basically, the mean velocity decreases with

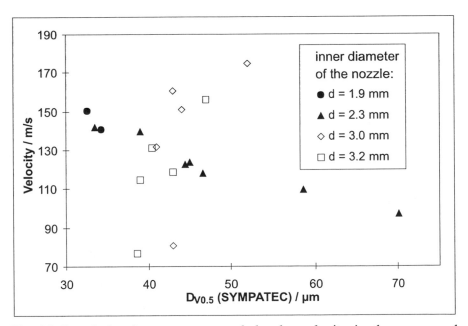

Fig. 22 Correlation between measured droplet velocity in the spray and the IGA powder mean sizes measured by SYMPATEC (Copper melts)

increasing mean diameter, as already shown above. At melt tube inner diameters of more than 2.3 mm the situation changes completely, probably due to a change in the pressure field at the melt tube outlet. For these tubes, no useful correlation was found.

Using the correlation between velocity and global mean particle size of the powder shown in Fig. 22, the prediction of the global mean powder particle size is feasible by calibrating the correlation function with on-line LDA measurements in a set of calibration processes. With this calibration, the prediction of the global mean diameter of the powder as measured by SYMPATEC is possible with a deviation of not more than ±10% (Fig. 23).

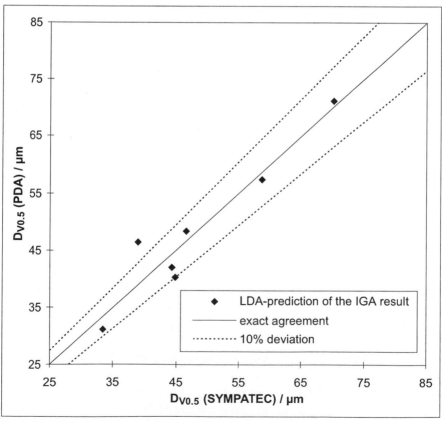

Fig. 23: LDA characterization of the IGA sprays (copper melts)

6 SUMMARY AND OUTLOOK

A deeper understanding of the IGA process of molten metals was obtained through detailed off-line and on-line particle size and velocity measurements in the metal sprays and powders using the PDA technique. A correlation between the results of local PDA measurements and off-line particle size measurements characterizing the entire powder yield was found for a specific atomizer geometry. Thus the characterization of the whole spray by a single PDA measurement is possible with an uncertainty of ±10% in the volume median particle diameter. In addition, it could be shown that, for well defined conditions, a single LDA measurement can be sufficient to characterize the whole spray with the same degree of uncertainty. This method is based on a sufficient stability of the two-phase flowfield in the atomization chamber.

In the future, detailed optimization tests of the close-coupled nozzles applied in IGA must be performed. This optimization must be based on detailed investigations of the (gas) flow field near the nozzle with experimental and numerical methods. Furthermore it is necessary to develop a technique for measuring the melt flow rate on-line.

ACKNOWLEDGEMENTS

The financial support of the present work from the EU in the frame of the BRITE/EURAM project 8209 entitled "On-line process control of liquid gas and water atomization by extended phase-Doppler anemometry" is gratefully acknowledged.

REFERENCES

Anderson, I. E., Figliola, R. S. & Morton, H. 1991, Flow mechanisms in high pressure gas atomization. *Materials Science and Engineering* A **148**, 101-114.

Bauckhage, K. & Schreckenberg, P. 1989, Control of Powder-Metal Production: A New Application of Phase-Doppler-Anemometry. *Proc. of the Int. Conf. on Mech. of Two-Phase Flows*, Taipeh, Taiwan, 267-272.

Capus, J. M. 1996, METAL POWDERS: A Global Survey of Production, Applications and Markets 1992-2001. Elsevier Science Ltd, Oxford, UK.

Domnick, J., Raimann, J. & Wolf, G. 1996, Application of Extended Phase-Doppler Anemometry in Liquid Gas Atomization. *Proc. of the 8th Int. Symp. on Appl. of Laser Techniques to Fluid Mechanics*, Lisbon, Portugal, 20.4.1-20.4.8.

Domnick, J., Raimann, J., Wolf, G., Schubert, E. & Bergmann, H. W. 1995, On-Line Characterization of Particle Jets During the Atomization of Molten Metal by High Velocity Cryogenic Liquid Gas Jets. *Proc. ILASS-95*, Nürnberg, Germany, 137-146.

Göbel, G., Doicu, A., Wriedt, T. & Bauckhage, K. 1997, Influence of surface roughness of conducting spheres on the response of a phase-Doppler anemometer. *Part. & Part. Syst. Char.* **14**, No. 6, 283-289.

Kuntz, D. W. & Payne, J. L. 1995, Simulation of powder metal fabrication with high pressure gas atomization. *Advances in Powder Metallurgy and Particulate Materials* **1**, No. 1, 63-77.

Miller, S. A. & Miller, R. S. 1992, Real Time Visualization of Close-Coupled Gas Atomization. *Advances in Powder Metallurgy and Particulate Materials* **1**, No. 1, 113-125.

Raimann, J., Brenn, G., Brito Correia, J., Domnick, J. & Shohoji, N. 1997, On-line measurement of powder size and velocity in the water atomization process. *Proc. of the European Conference on Advances in Structural PM Component Production*, Munich, Germany, 415-422.

Ridder, D., Osella, S. A., Espina, P. I. & Biancaniello, F. S. 1992, Intelligent Control of Particle Size Distribution During Gas Atomization. *Int. J. Powder Metallurgy* **28**, No. 2, 133-147.

Vielhaber, U. 1989, Genauigkeit der Partikelgrößenmessung mit Laserbeugungsspektrometern. *Sonderdruck Verfahrenstechnik, Interkarma-Report*.

White, D. G. 1997, Challenges for the 21st Century. *Int. J. Powder Metallurgy* **33**, No. 5, 45-54.

VI.2. Effect of Variable Liquid Properties on the Flow Structure within Shear-Driven Wall Films

A. Elsässer[1], W. Samenfink[2], J. Ebner[3], K. Dullenkopf[3], and S. Wittig[3]

[1] KNECHT Filterwerke GmbH, D-70376 Stuttgart, Germany
[2] Robert Bosch GmbH, D-70049 Stuttgart, Germany
[3] Lehrstuhl und Institut für Thermische Strömungsmaschinen, Universität Karlsruhe (TH), D-76128 Karlsruhe, Germany

Abstract. Liquid wall films driven by air flow occur in fuel preparation processes of advanced prefilming gas turbine combustor nozzles or in intake manifolds of SI-engines. There is a need to develop new and more accurate models to describe the heat transfer from the wall to the film. To obtain data necessary for the modeling a new measurement device in combination with an enhanced data processing technique has been developed. The device combines two novel measurement systems. It consists of a film structure measurement system and a specially adapted LDV-system with a miniaturized probe volume. This combination allows simultaneous measurements of film thickness and flow velocity within the film. For the first time detailed information of the internal flow structure are provided under consideration of the effect of variable liquid properties on the film hydrodynamics.

Keywords. film hydrodynamics, LDV

1. Introduction

Shear-driven liquid wall films play an important role in the fuel preparation process of advanced prefilming gas turbine combustors and intake manifolds of SI-engines (Roßkamp et al. (1997), Rottenkolber et al. (1998)). Advanced two-phase flow CFD-codes have been proven as a valuable tool in design and optimization of these components. Nevertheless, there is still a strong need to develop new and more accurate models for the fluid motion as well as for the heat transfer from the wall to the film. These advanced models have to consider the waviness and the internal flow structure of the wall film. However, for a more realistic simulation of the film flow detailed and accurate experimental data is crucial for an improved physical understanding as well as for the verification of fundamental assumptions used in the numerical code. The major problem governing the experimental investigation of shear-driven liquid films is the very small mean film height $\overline{h}_F$. For technical relevant film flows it is only $100 - 150 \ \mu m$ for technical relevant film flows. Conventional probe techniques to determine the internal flow structure for such films are not applicable.

Due to the multitude difficulties in the experimental access there is a lack of information about the internal hydrodynamics of shear-driven liquid films. To solve these problems, a new measurement and data processing technique was developed. The two main components of this novel measurement technique had been presented at the Lisbon Conference 1996 (Wittig et al. (1996) and Samenfink et al. (1996)). It consists of a laser-based film structure measurement system and a novel LDV-setup with a miniaturized probe volume. Their key feature is that they allow simultaneous measurements of the film thickness and the flow velocities within the film. This provides detailed information of the internal flow structure and allows for the first time to consider the effect of variable liquid properties on the flow structure and the film transport mechanisms respectively.

2. Measurement Techniques

In order to gain the above mentioned comprehensive insight into the physics of liquid transport phenomena, detailed information of the instantaneous film height, wave shape and velocity profiles within the film are required. To perform the measurements a special test facility for film flow investigations was designed and built. Because commercial measurement systems do not provide the capability necessary for these investigations. The development of new suitable equipment was necessary. The general working principles of these instruments will briefly be explained in the following sections. A more detailed description can be found in our earlier papers (Wittig et al. (1996) and Samenfink et al. (1996)).

2.1 Test Rig for Film Flow Investigations

The main item of the test rig used for the experiments was a transparent duct made of acrylic glass as shown in **Figure 1**. It has a rectangular cross section $(30 \times 196 \ mm)$ and a length of $1000 \ mm$. To establish the air flow in the duct it was connected with the suction side of a blower. The air flow (left to right) generates a shear stress τ at the film surface transporting the liquid downstream. The fluid was supplied by a row of small holes through the channel floor thus forming a continuous wavy film at the bottom wall. For the LDV measurements presented here the film fluid was seeded with starch particles. Free stream velocity of the air flow and wall film conditions including temperature, pressure and volume flux were controlled continuously.

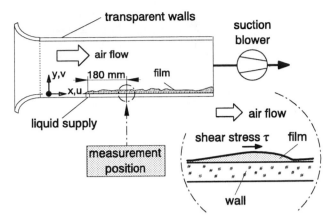

Figure 1: Test section

2.2 Film Thickness and Film Structure

In numerous studies different electrical approaches had been applied by various authors to determine film thickness (capacitive or inductive sensors as well as electrical resistance probes were used). However, the spatial resolution of these methods was generally limited by the sensor dimensions (about 3 to 5 mm in diameter), so individual waves usually could not be detected. Alternative optical methods based on the analysis of a fluorescent dye added to the film or the absorption of light passing a dyed liquid layer have major drawbacks especially in evaporating films. In general the dyes have different evaporating properties and therefore accumulate in the remaining liquid layer leading to false results. During various studies of prefilming atomizers at the Institut für Thermische Strömungsmaschinen at the University of Karlsruhe (ITS), different measurement techniques had been studied and a comprehensive overview was given by Himmelsbach (1992).

An improved approach avoiding the above mentioned problems in the exact determination of film thickness was presented by Samenfink et al. in 1996. An advanced technique based on light absorption without dyes was described, applicable in water, water/glycerine mixtures or alcohols. It uses the effect of absorption in wavelengths of the near infrared range.

The main principle of the system is schematically drawn in **Figure 2**. A laser beam of the intensity I_0 passes the liquid sheet from the bottom wall. Depending on the thickness of the sheet and the absorption coefficient k' the intensity I_0 is reduced to the value I_M according to the Lambert-Beer law (see **Figure 2**). In the technical relevant thickness range of $h_F \leq 1\ mm$ a value of $k' \approx 1\ mm^{-1}$ is necessary to achieve a sufficient absorption. In the presented system a wavelength of $\lambda = 1480\ nm$ is used for water and water/glycerine measurements (water: $k' = 3.14\ mm^{-1}$). For example a water layer thickness of only 1 mm leads to an intensity reduction of more

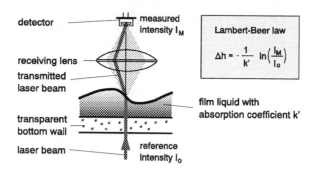

Figure 2: Light absorption in the liquid film

than 95%. With a large size receiving lens above the film the remaining intensity is focussed on the detector to determine the intensity I_M. In order to enhance the accuracy of the film thickness measurements, the intensity losses through reflections at the film surface or optical components (e.g. fiber coupling) were compensated by superposing a second laser beam with a wavelength of $\lambda = 830\ nm$. The secondary beam was therefore not absorbed by the film liquids used in the tests. The intensity losses were exclusively caused by effects not depending on the film thickness, so this information could be used to correct the intensity I_M of the primary beam mainly absorbed in the film. By means of this technique the intensity absorption through the liquid film could be determined with an excellent level of accuracy. The spatial resolution of this system is about $\Delta h_F \simeq 1\ \mu m$.

However, this measurement technique does not give direct information on the spatial surface structure, but on the film thickness as a function of time.

In general shear-driven fuel films in technical applications show a three-dimensional wave pattern on the surface. These waves play a decisive role the internal flow in the film. Therefore, it is necessary to record the surface structure and the film thickness simultaneously. The surface structure is represented by the wave angle α. It can be calculated from the deflection of the transmitted laser beam as shown in **Figure 3**. After the beam passed the bottom wall and the liquid film it enters the air space with an angle δ at the phase boundary. This is caused by the surface angle and the change in the index of refraction $n_2 \to n_1$. If the angle β can be determined experimentally, α can be calculated by means of the Snellius law and some trigonometric relations as demonstrated in **Eqn. 1**.

$$\alpha = \arcsin\left(\frac{n_1}{n_2}\sin\delta\right)$$

$$\delta = \beta + \alpha$$

$$\Rightarrow \quad \alpha = \arctan\left(\frac{\sin\beta}{n_2/n_1 - \cos\beta}\right) \tag{1}$$

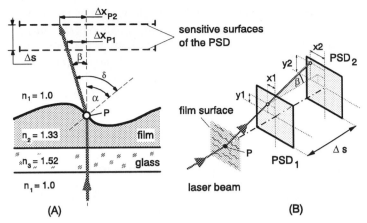

Figure 3: Beam deflection at the film waves and principle of surface angle measurement

However, the main problem remaining is that a simple measurement setup can only detect the displacement Δx with respect to the optical axis. As a consequence the major source of errors is the variation in the height of point P on the film surface. This effect is caused by the wavy film and is directly affecting the beam displacement Δx. Therefore, a more sophisticated technique to determine the correct exit angle δ had to be applied.

As demonstrated in **Figure 3A** and **Eqn. 2** two independent position sensitive detectors (PSD) separated by the known distance Δs are necessary. Because the wavy surfaces of the films under investigation were generally three-dimensional, beam deflection occurs in cross stream direction also. The determination of the exit angle β is therefore based on a two-dimensional measurement to detect the components β_x and β_y in main and cross stream direction respectively. In the design used for the later describes measurements, the detector consists of a complex system of beam splitters and PSD devices, as described in detail by Samenfink et al. (1996).

$$\beta \;=\; \arctan\left(\frac{\sqrt{(x_2 - x_1)^2 + (y_2 - y_1)^2}}{\Delta s}\right) \tag{2}$$

For the exact assignment of the velocity data inside the film derived by means of the LDV system (as described in the following chapter) to the position within the wave, the temporal thickness traces had to be transferred to spatial coordinates. The precise spatial reconstruction of the film surface structure was possible by combining the simultaneous film thickness and surface angle data. Calculating the derivative of the time-dependent film thickness the variation of film thickness $\dot{h}_F(t)$ as a function of time is obtained. The function $\dot{h}_F(t)$ can be splitted into a time dependent term and a spatial dependent term.

$$\dot{h}_F(t) = \frac{dh_F}{dt} = \frac{dh_F}{dx} \cdot \frac{dx}{dt} \tag{3}$$

The spatial term represents the surface angle variation and can be directly expressed as a function of the measured wave angle $\alpha_F(t) = dh_F/dx$. Assuming that the mean transport velocity of the wave $\overline{u}_W = dx/dt$ is constant, $\overline{u}_W$ can be determined by the deviation of the temporal film thickness and the corresponding surface angle measurements. It allows to express the film thickness as a function of the main flow coordinate x with high spatial resolution. The calculated surface structure represents the shape of the waves passing the probe volume with a high level of accuracy.

2.3 Film Velocity Profile Determination

Due to the lack of commercially available systems for the measurement of velocity profiles in thin, wavy liquid films a novel LDV system had to be developed, especially adapted to the required conditions. The crucial feature was an excellent resolution in film height direction coupled with the possibility of avoiding photomultiplier overload through reflections at the persistent moving film surface.

2.3.1 Transmitting Optics

As basic system for the transmitter a two-component fiber optic LDV system was used, looking vertically up through the transparent bottom wall of the test section (see **Figure 4**). In this way excellent optical access to the film flow was achieved. In order to reduce the length of the probe volume a beam expander and a short focal length ($f = 80\ mm$) was utilized. They reduce the transmitter probe volume to a length of $\Delta y_F \simeq 260\ \mu m$ and a diameter

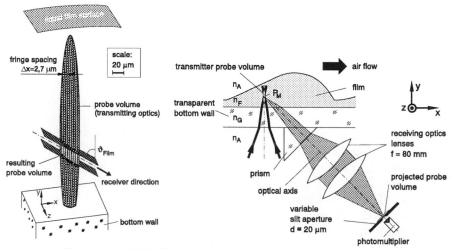

Figure 4: LDV-System with probe volume size reduction

of $\Delta x_F \simeq \Delta z_F \simeq 25~\mu m$ at the $1/e^2$ intensity point as shown in **Figure 4**. However, the size Δy_F is still one order of magnitude larger than the desired spatial resolution of about $25 - 30~\mu m$ in film height direction. Therefore, the probe volume length has to be reduced further by the receiver design and orientation.

2.3.2 Receiving Optics

As previously pointed out a 180° backscattering setup would not provide the necessary small probe volume size and furthermore the reflections at the film surface would cause serious problems considering the signal to noise ratio in a wide range of surface angles. To solve these problems the detector was shifted to an off-axis position, thus cutting a narrow section of the probe volume as shown in **Figure 4**. Simultaneously, this off-axis angle helps to solve the reflection problem at the air/liquid interface. An angle of 45° to the optical axis was chosen, because our previous investigations had shown that reflections in this range occurs relatively rare under conditions used in the tests. To collect data only in a well defined section of the probe volume in the y-axis direction, a slit aperture was positioned in front of the color splitter in the receiver. The optical signals were detected by a two-component photo-multiplier setup in combination with counter processors and a self developed software for data acquisition and analysis.

To reduce astigmatism due to the off-axis angle a prism was fixed perpendicular to the optical axis in order to obtain equal refraction conditions for the scattered light. Different angles of incidence appear only at the boundary between glass and liquid. However, this interface had a small variation in the index of refraction and the optical pathes in the film to the probe volume were very short due to the small film thickness. By means of a prism a minimal abberation could be achieved.

As mentioned earlier the problem of strong reflections at the moving film surface seriously disturbing the measurements had to be solved. The problem and its solution built in the LDV receiving optics is outlined in **Figure 5**. Because the photomultiplier tubes had to be protected from the intense light directly reflected at the liquid/air interface. In order to avoid overload and erroneous Doppler bursts, a Bragg-cell was implemented to switch off the laser beams for reflecting conditions. To detect reflecting conditions reliably an additional cw-Laser (pilot laser) with ($\lambda_{Pilot} = 673~nm$) was coupled into the receiver fiber of the transmitter probe (**Figure 5A**), thus forming a focus by means of the front lens, too. To separate the different wavelengths in the receiving optics a red reflective mirror was used, light of shorter wavelength can pass onto the photomultipliers. The time span necessary for the electronic and the Bragg-cell to switch off the main beams was gained by a specific optical arrangement in the receiver. The front lens of the receiver collects scattered light in an angular range of 20° and this complete field is

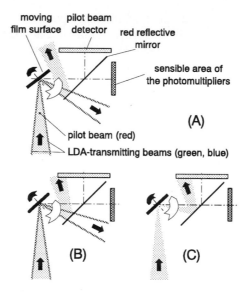

Figure 5: Principle of photomultiplier protection using an additional pilot beam

observed by the red light detector. The angular range of the green and the blue photomultipliers is reduced by an aperture to 5°. Therefore, red light is reflected into the pilot beam detector first (cf. **Figure 5B**). The main beams can now be switched off prior to full illumination of the photomultipliers by reflection, as shown in **Figure 5C**. Moving further, the reflected light leaves the pilot detector and the measurement can be continued by switching the main beams back on.

2.4 Simultaneous Film Velocity and Structure Measurements

The possibility of simultaneous velocity and film thickness respectively surface structure measurements would provide a major step in the understanding the internal transport mechanisms of shear-driven films. Combining the two instruments to perform time resolved measurements created a valuable tool with high spatial and temporal resolution.

The final layout of the complete optical system is demonstrated in **Fig. 6**. The system was mounted on a high precision 3D-traversing device. An independent spatial movement of the receiving optics relative to the transmitting optics was necessary to scan the height profile. For this a second extremely accurate 2D-traverse unit was applied with a step size of 0.5 μm. By using the same transmitting optics for the LDV and the film thickness measurement system an identical focus location for both devices was guaranteed. This is

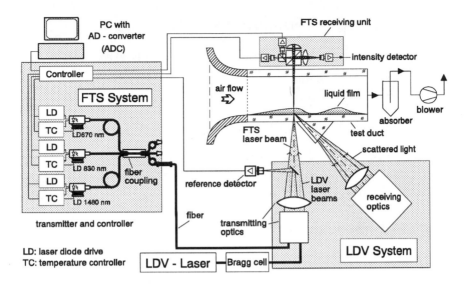

Figure 6: Setup for simultaneous measurement of film velocity and structure

an important characteristic in the light of measurement accuracy considering the small wave dimensions.

The electronic coupling, triggering, and control was done under computer control. Film structure information was stored for the passage time of long waves prior and after each LDV event. This variable recording time, depending on the boundary conditions, was necessary to ensure that the whole area of interest (e.g. wave structure) was recorded. Data processing and analysis was done offline after the measurements as described in sections 3.2 and 3.3.

2.5 Mean Shear Stress Measurement

A crucial parameter for each analytical or experimental study of liquid films on a surface is the driving shear stress at the liquid/gas interface. To determine the corresponding shear stress at the measurement location the air velocity profiles in the duct above the film have to be measured. This was done by a standard LDV-system through the transparent side walls, as shown in **Fig. 7**. For the measurements the air flow was seeded with 1 μm water droplets from an ultrasonic atomizer. The air sees the wavy film surface as an additional roughness influencing the flow profile close to the prefilmed wall strongly. The resulting shear stress could be qualified by matching the air flow data to the linear range of the logarithmic law of the wall (Roßkamp, Willmann and Wittig (1997)).

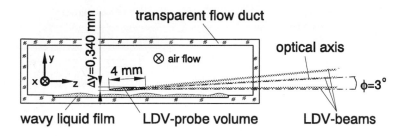

Figure 7: Air speed measurements above the film

3. Data Processing

3.1 Film Flow Characteristics

The result of a typical temporal record from the optical film thickness measurement system is shown in **Figure 8** together with characteristic parameters of wavy film flow. An individual wave is characterized by a steep gradient at the front (left side) and a moderate gradient at the wave back facing the air flow. The film height between the waves does not fall below a typical value. Generally a thin sublayer is continuously present, in this case a minimum thickness of $h_{F,min} \approx 0.040\ mm$. The diagram on the right of **Figure 8** shows the time mean probability of film liquid existing at a certain height h_F determined from the measurements. Close to the wall a continuous sublayer exists leading to a probability $p = 1$ for $h_F \leq h_{F,min}$. As expected, the probability decreases constantly by moving higher in the film because only a few very large waves are reaching up to this height.

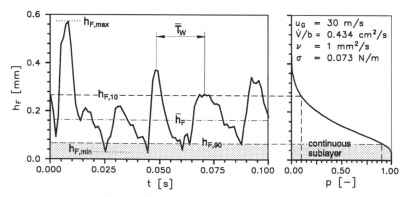

Figure 8: Characteristics of a wavy film flow

In order to characterize the stochastic nature of the film thickness distribution some statistical parameters are given in the diagram. One is the time mean film thickness $\overline{h}_F$, average from measured data with constant sampling rate.

From the probability distribution two other quantities can be derived. The thickness of the continuous liquid sublayer is described by $h_{F,90}$, a thickness which is exceeded for 90% of the measured time span. The corresponding value $h_{F,10}$ is characteristic for the height of the waves and is exceeded for only 10% of the total time. Together with the mean wave velocity $\overline{u}_W$, described in section 2.2, a typical wave length can be derived from $\overline{\lambda} = \overline{u}_W \overline{T}_W$.

3.2 Characterization of Typical Waves

To reach a comprehensive and correct insight to the flow phenomena in shear-driven liquid films, postprocessing of the measured data from the film velocity analyzer and the film structure analyzer was necessary. The individual LDV-measurements at a certain distance to the wall were stochastically distributed in waves of different height, length, and shape, because LDV data could only be taken if a particle passes the probe volume. To link the LDV results to their position inside representative wave types a differentiation and classification of wave forms occuring was necessary.

The conditional averaging procedure was based on a selection of typical waves from the data of the wave structure analysis, according to each flow condition. In order to get reliable statistical values about the hydrodynamics of liquid films an extensive set of information from the simultaneous measurements was processed and demonstrated in the following section. As an example the pure water film data already shown in **Figure 8** was used.

In a first step the measured thickness traces were examined visually or by computer based analysis tools to find and select a frequently appearing wave form. A sample of typical wave forms for this water film is shown in **Figure 9**, where the chosen basic form is marked with the thick black line. In this case it was a large main wave of 0.430 *mm* maximum height and a short wave back. However, each wave has distinct individual characteristics with humps, peaks, and dents. These items disturbed the allocation of the selected wave shape to the measured waves containing the velocity data. Therefore, a reliable smoothing technique had to be developed.

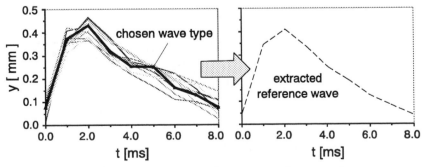

Figure 9: Determination of representative wave forms

In the second step some wave structures were selected from the stored data sets which were similar to the basic ones. The comparison was done by means of the 'least square fit' method. Particulary due to the varying individual wave shape different weighting factors along the wave length had to be applied. The wave front had a large factor, the back of the main wave a moderate value, and the remaining part a small one. With this procedure the major shape characteristics could be preserved. This is demonstrated on the left side of **Figure 9** where some of the selected waves are plotted together with the one chosen as basic wave form in the first iteration. Finally the extracted reference wave is plotted on the right as the result of the averaging procedure. Using the reference wave instead of the initial basic one, the assignment rate of the wave form to the measured film structures could be improved significantly in a second iteration.

3.3 Velocity Data Allocation

The extracted reference wave was then used to allocate the velocity data location inside the waves of similar surface structure (see **Figure 10**). By shifting the reference wave over all the measured film structure elements centered around the LDV-measurement locations (position 1 to position 10 in **Figure 10**) the position of maximum similarity was searched. The examined wave shape was accepted if the similarity level was high enough. For the processing the same general procedure and weighting factors as described in the previous section were used.

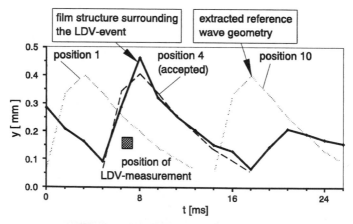

Figure 10: LDV data allocation

The processed velocity value was then assigned to a position in a two dimensional table of discrete positions inside the reference wave form. Considering the numerous variations in wave shape, several thousand data sets had to be acquired and checked for a reliable statistic of the velocity distribution.

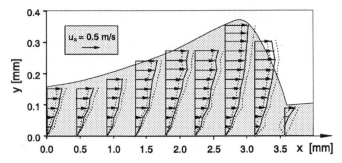

Figure 11: Mean velocity profiles inside the wave

Figure 11 shows an exemplary result of the evaluated velocity distribution for a typical wave structure. Due to the conversion from a time dependent abscissa to spatial coordinates the wave movement is from left to right, so the steep wave front is now located on the right side. At each location inside the wave the mean velocities in main flow direction are indicated. An indication of the velocity deviations is given by the dotted lines around the mean values representing the $\overline{u} \pm RMS$ range. Due to the averaging process, the velocity deviations can result from velocity fluctuations as well as from a slightly different flow structure of the individual waves.

For pure water films the profiles in the wave center and at the wave back showed no significant differences. Particulary the velocity gradients near the wall are very similar, so the conclusion can be drawn that the wall shear stress τ_W is similar all over this region (cf. section 4.2). However, at the wave front the mean velocity u_x decreases to zero and the deviation is centered around zero. A detailed analysis shows, that at the wave front many LDV-measurements had negative u_x velocities up to a height of about $h_{F,90}$. This leads to the conclusion that a rolling mechanism is responsible for the liquid transport in the wave, as indicated in **Figure 12**. The fluid bulk is driven by shear forces at the film surface and penetrates the continuous sublayer in the wave front region. This effect forces compensational flows which are oriented opposite to the main flow direction.

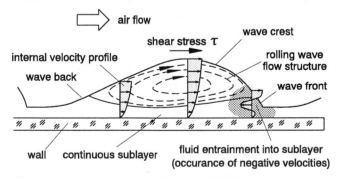

Figure 12: Flow structure of rolling waves

4. Film Hydrodynamics at various Liquid Properties

Even from simple visualization experiments it is well known, that film transport is influenced significantly by boundary conditions like shear stress and liquid flux. The influence of these parameters to the internal film flow characteristics had been discussed previously by Elsäßer et al. (1997). However, the properties of the liquid particulary the viscosity have significant influence on the resulting film hydrodynamics, too. The capability of the measurement and data processing systems allowed for the first time a comprehensive and detailed investigation of film characteristics with different liquid properties. The results of this study are given in the following.

4.1 Wave Hydrodynamics

As demonstrated in **Figure 13** a variation in the kinematic viscosity ν of the liquid leads to a significant change in film surface structure. The wavy surface on films with low viscosity is three-dimensional. The single wave trains are sickle-shaped and relatively short in cross flow direction. At higher viscosity the waves becomes more linear with a smooth surface in the spacing in between.

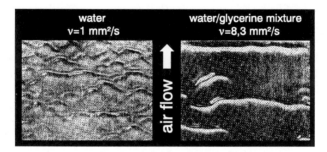

Figure 13: Surface structure visualization

To vary the liquid properties in the tests water and water/glycerine mixtures had been used. By means of water/glycerine mixtures the viscosity could be varied in a wide range in this work ($\nu = 1$, 2.2 and 8.3 mm^2/s) but the surface tension was kept almost identical.

The effect of an increase in kinematic viscosity on the structure of the film surface is demonstrated in **Figure 14** showing the plots of typical temporal thickness traces. The air velocity was kept constant at $u_G = 30$ m/s and the mass flux normalized by the width was $\dot{V}/b = 0.434$ cm^2/s. The film surface of pure water ($\nu = 1$ mm^2/s) shows an ensemble of waves with various size and length. Smaller subwaves occur in the spacings between the larger main waves. With increasing viscosity the surface becomes more regular and the subwaves disappear. A decreasing number of large main waves of growing

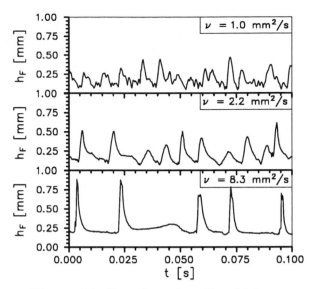

Figure 14: Time dependent film thickness

maximum height dominate the film structure. Additionally, the wave front
and the wave back gets steeper. However, exact statements about the length
of the waves are not possible in this time dependent plot because of the different wave travelling velocities $\overline{u}_W$.

The mechanism of rolling waves travelling on top of a thin but continuously
present sublayer is responsible for the major fraction of the liquid transport
in shear-driven films. Recent investigations (Elsäßer et al. (1997)) indicated,
that approximately 70% of the liquid mass is transported in the waves for
pure water films. This clearly shows the importance of a detailed investigation of the wave hydrodynamics under different liquid properties. The
individual wave characteristics for the three cases becomes even clearer by
comparing the extracted reference waves. Their shapes represent the structure of typical average waves under the specific liquid parameters (**Figure
15**). The maximum wave height was about $1.1 \cdot h_{F,10}$ for the boundary conditions mentioned before. The dimensions of the different structures can be
discussed easier transforming the temporal to spatial x-axes by means of $\overline{u}_W$
and $\overline{T}_W$.

Figure 15 demonstrates that the wave length and particulary the height of
the wave tips increase with viscosity. Also the wave back seems to thicken.
Furthermore, from the non-dimensionized plot $y/h_{F,max}$ versus x/λ it can
be depicted, that the fraction of the wave crest region significantly decreases
with growing viscosity compared to the mean wavelength $\overline{\lambda}$.

For a more detailed analysis of the internal flow field the distributions of
the velocity component u_x are given as contour plots in **Figure 16**. The
grey scale indicates the level of mean velocity for each position in the wave

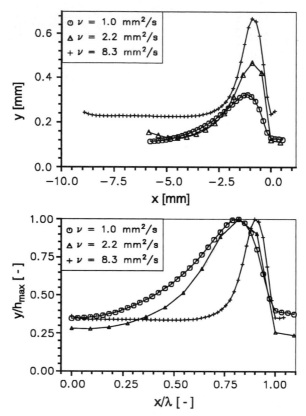

Figure 15: Reference wave forms for different liquid properties

giving a good survey of the flow field characteristics. In the water film with $\nu = 1 \ mm^2/s$ the isolines of the velocity are almost parallel to the bottom wall. This indicates from the bottom to the wave tip a nearly uniform velocity gradient exists over the wave length. In pure water films the internal velocity profiles differ only with absolute height, but not in the overall shape as also could be seen in **Figure 11** already. A similar behavior can be observed in films with slightly increased viscosity ($\nu = 2.2 \ mm^2/s$), but with smaller velocities near the wall. A deviation from this uniform distribution only occurs at the wave fronts where high momentum fluid from the rolling waves entrains the sublayer. This can be observed at the wave back ($x/\lambda \leq 0.1$) also, where interaction with the subsequent wave takes place.

In contrast to these evenly distributed velocities the high viscosity fluid ($\nu = 8.3 \ mm^2/s$) shows steeper gradients in the region below the wave crest compared to the long back of the wave. In the wave back mean velocities decrease to nearly $u_x = 0 \ m/s$ even to a height of 0.200 mm. This effect leads to a negligible liquid transport in the gap between the waves, so the transport of fluid takes place mainly inside the wave crests.

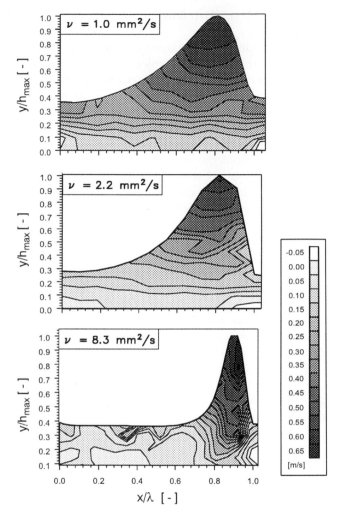

Figure 16: Distribution of axial velocity inside the waves

With respect to modelling approaches the fact that almost the whole mass is transported in the waves has to be taken into account to get a realistic simulation. Additionally not only the transport characteristic is influenced but also the heat transfer perpendicular to the film. It depends on the internal mixing caused by the different wave shapes and frequencies. Taking into account these results a new and more realistic model could be developed by Roßkamp et al. (1998). It leads to an enhancement of the Nusselt number up to factor of nearly 3.5 compared to a laminar film model with smooth surface.

4.2 Shear Stress Distribution

The before mentioned transport mechanisms influence the modeling of shear-driven liquid films for numerical flow calculation significantly. Due to the steeper velocity gradient in the wave crest region the wall shear stress calculated by the gradients at the wall from $\tau_W = \mu \cdot \Delta u / \Delta y$ increases remarkable (τ_W represents the forces from the air flow driving the film). It comprises the Newtonian shear forces from roughness effects to the air flow as well as pressure drag components due to the shape of the wave crest. By means of the results from the simultaneous measurements of film structure and velocity the shear stress distribution along the surface of a typical wave could be determined with a high level of high accuracy.

In **Figure 17** these values are compared to the time mean shear stresses $\bar{\tau}$ derived from the velocity profile measurements in the air flow (section 2.5). For a film formed of a fluid with low viscosity the agreement is excellent, except for the wave front region. There shear stress changes the direction due to the local negative velocity near the wall. In the other region wall shear stress is uniform due to the homogeneous velocity distribution. As already indicated in **Figure 16** the fluid with the high viscosity shows a significant increase of τ_W in the wave crest. This can lead to a strong misjudgement of the mass transport if $\bar{\tau}$ is used for film flow simulations, also resulting in erroneous film thickness and film roughness prediction. However, our investigation has clearly shown that for low viscosity fluids up to $\nu \leq 2 \ mm^2/s$ the simplification of constant mean shear stress is valid and can be used in predictions of global film transport. If ν exceeds $2 \ mm^2/s$ improved models have to be applied, taking into account the strong variation of $\bar{\tau}$ as a function of the wave length.

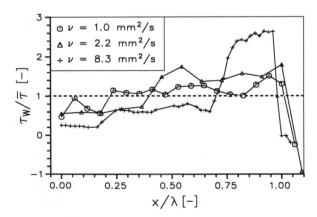

Figure 17: Normalized wall shear stress versus wave length

5. Conclusions

To improve the knowledge about the transport mechanisms in wavy shear-driven liquid films, important in many technical applications, a novel LDV system was designed. It allows accurate velocity profile measurements from the wall up to the wave tips in thin wavy films. The instrument was combined with an optical film thickness measurement system to determine the structure of the wavy film flow accurately. An extensive data processing procedure was developed to allocate the measured film velocity data to wave forms with typical shape and size under the actual flow conditions.

With this comprehensive set of information it was possible to determine internal wave quantities with a high level of accuracy, for example the effective shear stress distribution over the wave length. This enabled a more detailed insight in the flow hydrodynamics of the liquid film at various properties. The comprehensive results give fresh stimulus to improve the film transport and heat transfer modelling significantly due to the better understanding of the internal film flow mechanisms.

Acknowledgements

Special thanks is done to our students Jens Läuger for assistance in the experimental part of this work and Stephan Hentz for his excellent contribution in preparing the analysis software.

References

ELSÄSSER, A., SAMENFINK, W., EBNER, J., DULLENKOPF, K. & WITTIG, S., 1997 Dynamics of Shear-driven Liquid Films. *Proceedings of the 7th International Conference on Laser Anemometrie - Advances and Applications, Karlsruhe, Germany, September 8 – 11.*

ELSÄSSER, A., SAMENFINK, W., EBNER, J., DULLENKOPF, K. & WITTIG, S., 1998 Effect of Variable Liquid Properties on the Flow Structure within Shear-driven Wall Films. *Proceedings of the Ninth International Symposium on Applications of Laser Techniques to Fluid Mechanics, Lisbon, Portugal, July 13 – 16.*

HIMMELSBACH, J., 1992 *Zweiphasenströmungen mit schubspannungsgetriebenen welligen Flüssigkeitsfilmen in turbulenter Heißluftströmung - Meßtechnische Erfassung und numerische Beschreibung.* Dissertation, Institut für Thermische Strömungsmaschinen, Universität Karlsruhe (T.H.).

564

Rosskamp, H., Elsässer, A., Samenfink, W., Meisl, J., Willmann, M. & Wittig, S., 1998 An Enhanced Model for Predicting the Heat Transfer to Wavy Shear-driven Liquid Wall Films. *3rd International Conference on Multiphase Flow, Lyon, France, June 8 – 12*.

Rosskamp, H., Willmann, M. & Wittig, S., 1997a Computation of Two-Phase Flows in Low-NO_x Combustor Premix Ducts Utilizing Fuel Film Evaporation. *ASME-Paper 97-GT-226* .

Rosskamp, H., Willmann, W. & Wittig, S., 1997b Heat up and Evaporation of Shear Driven Liquid Wall Films in Hot Turbulent Air Flow. *Proceedings of the 2nd International Symposion on Turbulence, Heat and Mass Transfer, Delft, Netherlands, June 9 – 12*.

Rottenkolber, G., Dullenkopf, K. & Wittig, S., 1998 Visualization and PIV-Measurements inside the Intake Port of an IC-Engine. *Proceedings of the Ninth International Symposium on Applications of Laser Techniques to Fluid Mechanics, Lisbon, Portugal, July 13 – 16*.

Samenfink, W., Elsässer, A., Wittig, S. & Dullenkopf, K., 1996 Internal Transport Mechanisms of Shear-driven Liquid Films. *Proceedings of the Eighth International Symposium on Applications of Laser Techniques to Fluid Mechanics, Lisbon, Portugal, July 8 – 11*.

Wittig, S., Elsässer, A., Samenfink, W., Ebner, J. & Dullenkopf, K., 1996 Velocity Profiles in Shear-driven Liquid Films: LDV-Measurements. *Proceedings of the Eighth International Symposium on Applications of Laser Techniques to Fluid Mechanics, Lisbon, Portugal, July 8 – 11*.

VI.3. Some Aspects of Electrohydrodynamic Spraying in Cone-Jet Mode

K. Ohba, P.H. Son, and Y. Dakemoto

Department of Mechanical and Systems Engineering, Kansai University,
1-3-35 Yamate-cho, Suita, Osaka 564-8680, Japan

Abstract. Herein we report an experimental study on the electrohydrodynamic (EHD) spraying in cone-jet mode. The effects of electric field strength, surrounding ambient fluid, surface tension of the working fluid, pressure head in the working fluid supply line and the electrodes configuration of the apparatus for spraying were investigated. The working fluid, ethanol or water, of different surface tension was issued out into the surrounding fluid, air or CO_2 gas, by using a needle-plate apparatus for EHD spraying. The formation of the cone-jet at the needle tip was monitored by using a hypervideo microscope. The droplet size and velocity were measured by using a Phase Doppler Particle Analyzer. The results indicate that a stable cone formed in CO_2 gas in a certain range of the applied voltage for either ethanol or water, while it takes place for only ethanol sprayed in air. The droplet size decreased with increasing applied voltage and is of order of 2-4 μm for both ethanol and water when produced in stable cone-jet mode.

Keywords. Electrohydrodynamic (EHD) spraying, Cone-jet mode, PDPA.

1. Introduction

The electrohydrodynamic (EHD) atomization is a process of producing a droplets spray by imposing a strong electric field or electric charge on the liquid. This atomization method has a great potential in producing droplets of small size and provides additional controlling parameters such as the applied voltage and the electric charge on the droplet. The EHD spraying is employed in a number of technologies, among which are ink-jet printing, crop spraying, paint spraying surface coating and powder production. Despite the extensive practical applications of the method, the droplet formation process still remains elusive. There is a number of modes of droplet formation depending on the electric field strength and its configuration, the physical properties of the working fluid and its ambient medium. A comprehensive review on the EHD spraying modes is given by Cloupeau and Prunet-Foch (1994). Among those modes, the EHD spraying in cone-jet mode is one of the most important, since it can produce a monodispersed spray of very fine droplets of micrometer order. However the conditions for the formation of a steady cone-jet are not clarified. This work does not seek to answer fundamental questions about the mechanism of the cone-jet formation, but rather to investigate the effects of electric field strength, ambient fluid, surface tension of working fluid and hydrostatic pressure of liquid supply line on the cone-jet formation. In the present experiments, the characteristic parameters of the spray such as the droplet size and its velocity were acquired by using a Phase Doppler Particle Analyzer (PDPA). Also, the process of cone-jet formation at the capillary tip was monitored. The experimental droplet size then is compared to the theoretical one, which was calculated based on our theoretical and experimental work (1998).

2. Droplet Formation in Cone-Jet Mode

A striking feature of producing droplet spray in cone-jet mode is that a monodispersed spray of very fine droplets of some micrometers can be obtained. In the present experiments, the liquid is gravitationally supplied from the reservoir to the needle at a specific pressure head. In the absence of an electric field, a pendant meniscus of pear shape forms at the needle tip. The meniscus grows in size with time until it detaches and falls from the needle tip under action of the gravitational force. The application of a DC voltage to the needle and its increase cause the reduction in the droplet size and the increase in droplet emission frequency. In general, the diameter of droplets formed in this so-called dripping mode remains greater than the needle diameter. With further increasing the applied voltage, the meniscus attains a conical form. At a certain voltage, a very small jet compared to the capillary diameter emanates from the vertex of the conical meniscus with subsequent disintegration to form a spray of fine droplets.

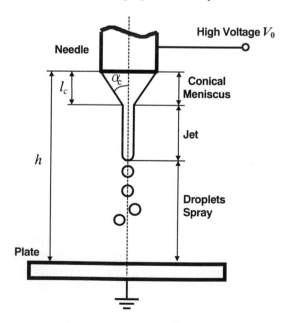

Figure 1. Sketch of an EHD spray in cone-jet mode.

It would be a difficult task to conduct an analytical description for the complete EHD spray. Therefore, an EHD spray in cone-jet mode can be split into three interacting regions with quite distinct behavior (see figure 1):

1) Conical meniscus, which forms at the needle tip;
2) Liquid jet, which breaks up into droplets by the wellknown Rayleigh's capillary instability;
3) Droplets spray, where the information on the droplet size and velocity, as well as the droplet charge is important.

The effects of the surface tension, pressure head in the liquid supply line and electric field strength can be estimated in the light of the normal pressure balance at the meniscus surface, which is expressed as

$$\sigma\left(\frac{1}{R_1}+\frac{1}{R_2}\right)-\rho_1 gH-\frac{1}{2}\varepsilon_0 E_n^2 = 0 \tag{1}$$

where σ is surface tension of liquid in contact with air, R_1 and R_2 are principle radii of surface curvature, ρ_1 is liquid density, H is height level between liquid surface in reservoir and needle tip, ε_0 is permittivity of surrounding air and E_n is normal electric field strength at the liquid surface.

The breakup of jets is thought to occur due to the Rayleigh's capillary instability. We have carried out the linear analysis on the instability of an electrically charged liquid jet. By using the basic equations of fluid mechanics and electromagnetism, Son and Ohba (1998) have derived the following equation for the dimensionless growth rate of the disturbances developed on the surface of the jet.

$$\omega^{*2} = -\frac{k^{*2}\rho^* IK}{(I-\rho^* K)^2}We + \frac{k^*}{I-\rho^* K}\left[\left(1-m^2-k^{*2}\right)-\alpha^2 Bo_E\left(1+k^*\frac{1}{K}\right)\right] \tag{2}$$

where

$\rho^* = \dfrac{\rho_2}{\rho_1}$: ρ_1 and ρ_2 are density of continuous and dispersed phase, respectively;

$I = \dfrac{I_m(k^*)}{I'_m(k^*)}$ and $K = \dfrac{K_m(k^*)}{K'_m(k^*)}$: $I_m(k^*)$ and $K_m(k^*)$ are modified Bessel functions of the 1st and 2nd kind;

$k^* = ka$: Dimensionless wavenumber, where a denotes the radius of undisturbed liquid jet;

$We = (U_1 - U_2)^2 \dfrac{\rho_1 a}{\sigma}$: Weber number, U_1 and U_2 are velocities of continuous and dispersed phase;

$Bo_E = \dfrac{\varepsilon_0 E_n^2 a}{\sigma}$: Electrical Bond number; α is a constant and equal to 0.707 in our work.

The calculation results of this equation have revealed that the varicose mode of instability takes place at low electric field strength and the kink mode of instability

568

is enhanced at high electric field strength. The kink instability causes the lateral motions of the jet, while the varicose instability causes the breakup of the jet. By assuming the jet breaks up with the wave of maximum growth rate, we can calculate the droplet diameter according to the following formula.

$$d = (12\pi)^{1/3}(\lambda_{cr}^{*})^{1/3} a \tag{3}$$

where $\lambda_{cr}^{*}(=1/k_{cr}^{*})$ is dimensionless critical wavelength. The equation (3) is used for theoretically predicting the droplet size.

3. Experimental Apparatus and Procedure

In the present experiments, an electrohydrodynamic spraying was produced by establishing a high electric field between the tip of a needle, from which liquid is emanated, and a washer-plate. The experimental apparatus is shown schematically in figure 2. The main components of the system included a liquid delivery system, an electrical charging system, a CO_2 gas supply system, a laser measurement system and a visualization system.

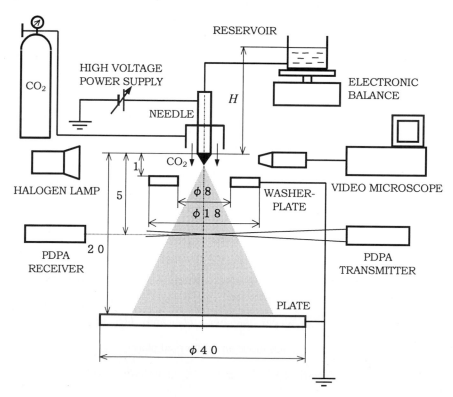

Figure 2. Schematic diagram of experimental apparatus.

The fluid delivery system consisted of a liquid reservoir, where the liquid surface level was maintained at the heights H of 15 and 20 mm above the surface of the needle tip. The working fluids were distilled water and ethanol with surface tensions of 0.0728 and 0.0228 N/m, respectively. Liquid was fed to the needle under gravity. The mass flow rate was measured by defining the mass reduction of liquid in one minute with the help of an electronic balance. The needle is made of a stainless steel with 260 μm I.D. and 510 μm O.D. Two kinds of needle were used: one with flat tip and another with conically tapered end with semivertex angle 30°. A Glassman High Voltage Power Supply with variable voltage of 0-10 kV was used to apply to the needle a high DC voltage of positive polarity with respect to an electrically grounded plate of washer type. This aluminum washer-plate with outer diameter of 18 mm and hole diameter of 8 mm was placed perpendicularly below the needle at the distance of 1 mm from the needle tip. Another electrically grounded plate of aluminum with diameter of 40 mm was placed further downstream at 20 mm from the needle tip to collect the droplets. Coaxial with the needle, a flow of CO_2 gas with 3 l/min flow rate isolated the needle tip from the ambient air in order to suppress the onset of corona discharge. The droplet size and its axial component of velocity at 5 mm far from the needle tip on the needle centerline were measured by using an Aerometrics Phase Doppler Particle Analyzer (PDPA). The PDPA was configured to receive the scattered light in the 30° off-axis forward scattering mode. A Keyence Hypervideo Microscope with exposure time of 20 μsec was used to monitor the shape and behavior of the meniscus formed at the needle tip with the aid of a halogen lamp.

In a typical run, the applied voltage was increased by 10 V step starting from zero and finishing at corona discharge occurrence. At each value of applied voltage, the droplet size and its axial component of velocity, mass low rate were measured and the video recording was taken.

4. Experimental Results and Discussion

4.1. Cone-Jet Configuration

Taylor was the first to make a theoretical prediction of the cone angle. Considering the case of an isolated drop located in a strong electric field, Taylor has showed that the half-angle at the apex of the cone is equal to 49.3°. However our experiments showed stable conical menisci formed at the needle tip for a certain range of applied voltage and hydrostatic pressure. Moreover, the value of the cone angle and its length is variable. The generatrix of the cone may be either convex or concave.

Ethanol. Since ethanol has a low surface tension, the pendant drop forms at the needle tip for H of both 15 and 20 mm. These specific heights were selected based on the initial observations identifying those at which the formation of the cone-jet mode was best characterized for both ethanol and water. The photographs showing the shape of the meniscus at various applied voltages are given in figure 3(a). At 2.6 kV, the meniscus attains a convex conical form and a jet begins emitting from the apex of the cone. With increasing applied voltage, the cone length shortens, the cone angle becomes larger, cone generatrix becomes concave and the jet diameter decreases. At about 3.7 kV, a second jet forms as a result of abundant accumulation of charge at the meniscus surface. The number of jets formed at the rim of the needle tip in this so-called multi-jet mode increases with further increasing in applied voltage. The process of cone-jet formation is almost similar for spraying of ethanol in either CO_2 gas or air.

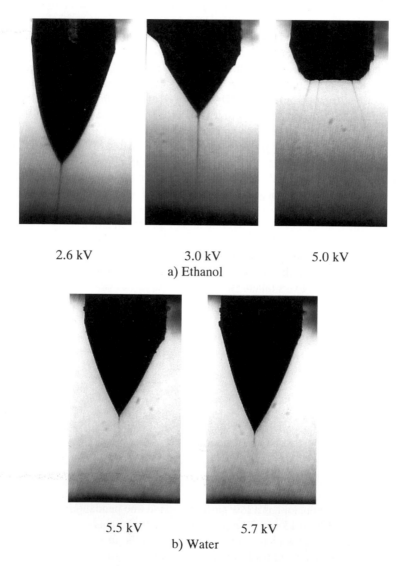

| 2.6 kV | 3.0 kV | 5.0 kV |

a) Ethanol

| 5.5 kV | 5.7 kV |

b) Water

Figure 3. Photographs of conical menisci at the needle tip at various applied voltages.

Water. No spray of water in the air in cone-jet mode could be obtained. Since the surface tension of water is relatively high, the corona discharge occurs before the electric field becomes high enough to create the electrostatic pressure that exceeds the pressure due to surface tension. The spraying in cone-jet mode was obtained only in CO_2 gas. For needle with flat tip, the cone-like meniscus forms with strong lateral oscillation. By contrast, a stable cone occurs for needle with cone ending, as shown in figure 3(b). Although the cone length shows a little change compared to that for ethanol at various applied voltages, it increases with increasing applied voltage. We could not define whether the jet formed by the video microscope used in the present experiments. There are two possible mechanisms of droplet emission from the apex of the cone. In one mechanism, the droplets are emitted directly from the cone apex due to strong electric field present there by the growth of unstable surface deformations. In another mechanism, the liquid jet is ejected and subsequently breaks up into droplets.

4.2. Droplet Size

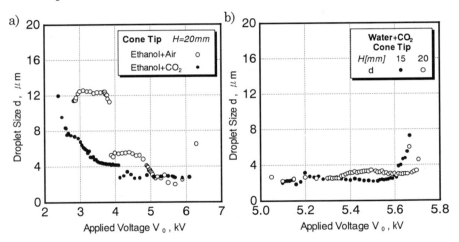

Figure 4. Droplet size variation with applied voltage.

For ethanol, the increase in applied voltage reduces the jet diameter and its breakup wavelength, and thus, the droplet size as well. For the case of ethanol issued in CO_2 gas, droplet size decreases with increasing applied voltage and stays unchanged at about 3 μm at applied voltage above 4 kV for needle with cone tip, as shown in figure 4(a). For ethanol issued in air, with increasing applied voltage, droplet size stays constant at 12 μm until 3.7 kV, then sharply decreases in size down to 5 μm. It is constant at about 3 μm from 5 kV until corona occurs. For the flat tip, the droplet size is a little larger and about 4 μm at applied voltage above 5 kV. Reducing the height level between liquid surface in reservoir and the needle tip from 20 mm to 15 mm resulted in higher voltage range where the droplet size remains unchanged and minimum. The comparison between the experimental droplet size and calculated one according to formula (3) for ethanol is shown in figure 5. A relatively good agreement is achieved.

For water, droplet diameter is almost constant at 2-3 μm for the entire range of applied voltage where the cone is stable, as shown in figure 4(b). But this range of voltage is more narrow compared with the case of ethanol and is from 5.1 to 5.7 kV. Droplet size is of the same magnitude for two different pressure heads (H=15 and 20 mm) in the present experiments.

4.3. Axial Component of Droplet Velocity

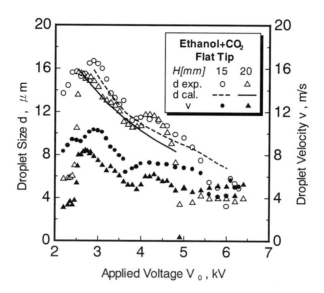

Figure 5. Variation of droplet size and its axial component of velocity.

The variation of the axial component of droplet velocity has a similar tendency of droplet size variation with increasing applied voltage. The value of velocity changes within the interval 2-10 m/s depending on the applied voltage and droplet size, as shown in figure 5 for the case of ethanol sprayed in CO_2 gas. The larger the droplet, the higher its momentum. Besides, the larger droplets are thought to carry a larger amount of charge, which results in a greater electrostatic force acting on the droplet in the direction of electric field and attracting the droplet downward to the grounded plate.

4.4. Flow Rate

In the present experiments, instead of keeping the flow rate constant we kept the hydrostatic pressure in the liquid supply line constant. For needle with cone tip, flow rate is small and almost constant at nearly 0.01 ml/min (see figure 6). However, for needle with flat tip, the flow rate increases with increasing applied voltage and reaches the maximum of about 0.16 ml/min at 6 kV where the corona discharge occurs.

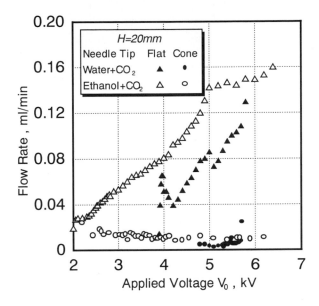

Figure 6. Variation of volumetric flow rate with applied voltage.

5. Concluding Remarks

The experimental study on the EHD spraying in cone-jet mode has showed that:

- The surface tension plays an important role in the formation of cone-jet. The cone-jet mode was established in a large region of applied voltage for liquids with low surface tension such as ethanol. For water, the cone-jet mode took place only in CO_2 gas and within a narrow range of applied voltage.

- In cone-jet mode, very fine droplets of 2-4 μm were obtained for both ethanol and water.

- A comparison between the experimental droplet size and that calculated by our theoretical result of instability analysis of charged liquid jets shows a relatively good agreement.

REFERENCES

CLOUPEAU, M. & PRUNET-FOCH, B. 1994 Electrohydrodynamic spraying functioning modes: A critical review. *J. Aerosol Sci.* **25**, 1021-1036.

SON, P.H. & OHBA, K. 1997 Instability of a perfectly conducting liquid jet in electrohydrodynamic spraying: Perturbation analysis and experimental verification. *J. Phys. Soc. Japan* **67**, 825-832.

VI.4. The Velocity Field in an Air-Blasted Liquid Sheet

A. Lozana[1], I. García Pallacín[2], F. Barreras[2], and C. Dopazo[1,2]

[1] LITEC/CSIC, Maria de Luna, 3, 50015-Zaragoza, Spain
[2] Centro Politécnico Superior de Ingenieros, Area de Mecánica de Fluidos
Universidad de Zaragoza, Maria de Luna, 3, 50015-Zaragoza, Spain

Abstract. The instability growth that causes a liquid sheet break up when it is subjected to high velocity parallel air streams has been analyzed in some recent experimental studies. Most of them have been based on visual observations from instantaneous spray images or oscillation frequency measurements. The purpose of this work is the study of the air field near the sheet interfaces to improve the understanding of the air/liquid interaction. To this end, flow visualization and particle image velocimetry (PIV) have been used to ascertain the flow structure and to obtain two-component velocity maps of planar sections of the air streams.

Keywords. Atomization, PIV, Instabilities

1 Introduction

Two-dimensional liquid sheets are becoming increasingly popular as a test flow to study instability growth and atomization processes. Large aspect ratio configurations facilitate the study of transverse perturbations in regions close to the nozzle exit, and edge effects are minimized. Results can, thus, be extrapolated to axisymmetric geometries that are more common in industrial applications. Experiments and simulations regarding large aspect ratio sheets have their earliest precedent in the work of Hagerty and Shea (1955), where perturbations were introduced by oscillating the liquid nozzle. In that analysis the water sheet was exiting in a quiescent atmosphere. In the present work, atomization is produced with the aid of a high speed air coflow, in a configuration similar to those of Rizk and Lefevbre (1980) or Arai and Hashimoto (1985).

It might be interesting to add that although initial interest in the liquid sheet problem has been mostly academic, recent papers describe applications where the specific two-dimensional characteristics of this flow are especially advantageous (e.g. Mansour et al. 1998).

2 Description of the Experiment

The design of the experimental set up used in the present study is similar to that described in Mansour and Chigier (1990) and Lozano *et al.* (1996). It differs, however, in that both water and air nozzle profiles have been contoured fitting 6th order polynomials to ensure uniform and parallel air and water velocity profiles at the exit. Water injected at the top of the nozzle head exits vertically through a 0.35 mm wide slit. The nozzle has a 23:1 contraction ratio. The span of the sheet is 80 mm, yielding an aspect ratio of 230. Air is also introduced from the top following a settling chamber with two honeycombs and a wire mesh screen to smooth the flow. The air channels located at both sides of the liquid nozzle have contraction ratios of 15:1 and exit widths of 3.45 mm.

This geometry might not provide the most efficient atomization, as air impinging at an angle with respect to the liquid sheet would probably produce a faster break up. Its simplicity, however, is an advantage in order to be compared with numerical simulations and to identify basic break up mechanisms. For the conditions under study, water velocities have ranged from 0.6 to 6 m/s, while air velocities have been varied between 15 and 35 m/s.

To visualize the air field and measure the velocity using particle image velocimetry (PIV), one of the air channels has been seeded with glycerin/water droplets from a commercial seeder. To illuminate, a double cavity Quantel Nd:YAG laser has been used, capable of producing 6 ns, 125 mJ pulses at 532 nm. The temporal spacing between the pulses has been adjusted according to the velocity of the flow, being on the order of 30 μs. Images have been obtained either with a slow-scan 16 bit Princeton Instruments CCD camera, or with a cross-correlation 1000x1000 pixels TSI PIVCAM 10-30 camera with Kodak array. With the appropriate TSI synchronizer, this second camera is capable of working in "frame straddling" mode, with a minimum temporal delay of 25 μs between each image in a pair.

3 Visualization Results

Images have been acquired for longitudinal sections of the flow, perpendicular to the exit nozzle. Special attention has been devoted to the near field region, close to the nozzle, to study the triggering mechanisms of the instabilities. Several fields of view have been imaged: 4.3 x 4.3 cm with a resolution of 43 μm/pixel, 2 x 2.6 cm with a resolution of 70 μm/pixel, and 1 x 2 cm with resolution of 20 μm/pixel.

Initial test images without water flow have been registered to ensure proper seeding levels and air flow characteristics. The experimental conditions substantially change when water is added. The large density difference between air and water produces the growth of longitudinal perturbations. The air/water interface interaction results in a Kelvin-Helmholtz instability which quickly causes the sheet break-up. Increasing the air velocity with respect to that of water, the sheet breaking point recedes to the nozzle exit, hindering the study of the instability waves. For this reason, only moderate air velocities have been considered.

The presence of the liquid sheet also modifies the image recording process due to the different intensities between the seeded glycerin droplets and the spray droplets from the atomized sheet. Trying to increase the registered signal from the air can cause a total image saturation if water droplets are present because they have much larger diameters.

It has to be noted that although it is possible to seed both air channels, the illuminating sheet becomes so distorted after crossing the liquid curtain that only the air stream closer to the laser source can be analyzed.

Instantaneous images for a fixed air velocity of 15 m/s and decreasing water velocities (2.1, 1.5 and 0.9 m/s) are depicted in Figs.1 a), b) and c). Figure 2 shows similar images for the same water velocity values, but for an air velocity of 25 m/s. In the images, the water flows from left to right although in the experiment it was moving downwards.

As has been reported in other works, until the air/water velocity ratio exceeds a limit, no atomization occurs. As air velocity increases, the sheet starts oscillating, initially with a dominant dilatational or varicose mode. At some point, the dominant mode becomes antisymmetric, and when this happens, the growth rate is highly enhanced resulting in a quick break-up at a distance of a few wavelengths from the nozzle exit. The oscillation frequency increases linearly with the air velocity, with a much weaker dependence on water velocity. The spray angle, however, presents a relatively sharp maximum when varying the water velocity for a fixed air flow.

Comparing Figs. 1 and 2 it can be noticed how the higher air velocity causes the sheet break up to occur closer to the nozzle exit. The quick atomization in Fig. 2 originates a cloud of water droplets impeding the visualization of the glycerin-air markers. For higher air speeds, not shown in the images, atomization is almost immediate and more efficient, resulting in a spray formed by smaller droplets.

The seeded air is clearly visible in Figs. 1 a) and b). The mixing layer between the air stream and the quiescent room air results in vortical structures with scales smaller than the sheet oscillation wavelength. In the air/water interfaces no vortical structures are clearly discernible. The air gets decelerated by the liquid sheet, and the possible vortices that could be shed from the splitter plate seem to be washed out by the water. This is an important point to clarify the origin of the instability waves, and the mechanisms that could contribute to their growth..

The water displacements are followed by the air streams. As the liquid sheet oscillation amplitude increases, the transverse component of the movement is also increased. Although the water is flowing downwards, under the effect of the surface tension it moves like a pendulum, and the sinusoidal wave degenerates in

578

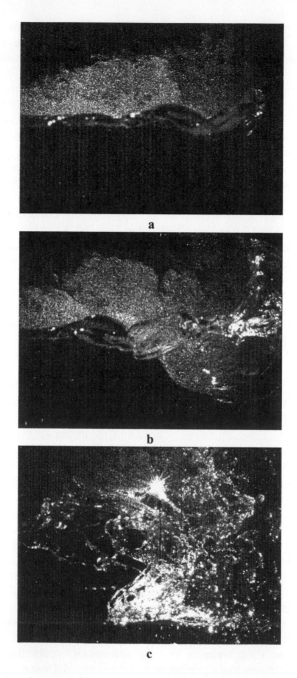

Fig. 1: Liquid sheet images. Air velocity: 15 m/s. Water velocities: a) 2.1 m/s, b) 1.5 m/s, c) 0.9 m/s. Field of view: 4.3 x 3.2 cm. In the experiment, the sheet was flowing downwards.

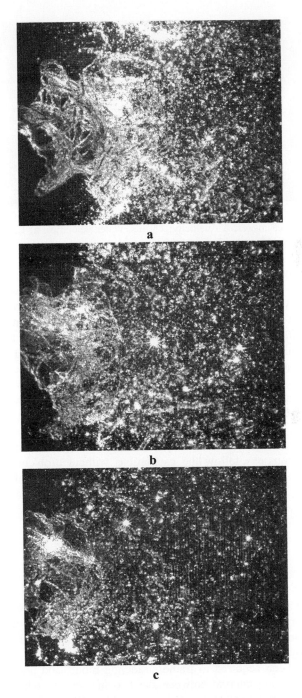

Fig. 2: Liquid sheet images. Air velocity: 25 m/s. Water velocities: a) 2.1 m/s, b) 1.5 m/s, c) 0.9 m/s. Field of view 4.3 x 3.2 cm (as in Fig. 1).

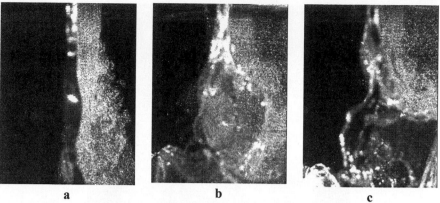

Fig. 3: Liquid sheet images. Air velocity: 15 m/s. Water velocities: a) 1.5 m/s, b) 1.2 m/s, c) 1.0 m/s. Field of view 2.6 x 2 cm.

a zigzag shape with pointy edges that are curved upwards. The air following this movement is also turned upwards and separates creating large recirculation rollers, and occasionally small vortices detached from the sharp vertices of the wave (see Lozano et al. 1997). When this situation takes place close to the nozzle head, the recirculation effect is enhanced by the presence of the plate where the exit slits are located, that partly blocks the entrance of room air. The air deflected upwards collides with the plate and is reingested by the stream. This recirculation zone could contribute to the fast amplitude growth of the instability waves and to the steep increase in the spray angle for determinate values of air and water velocities.

This process can be clearly seen in Fig. 3 where again, a sequence of frames with decreasing water velocities (1.5 m/s, 1.2 m/s and 1.0 m/s) for a fixed air speed (15 m/s) is presented. For the higher water velocity, the wave amplitude is small, and the air stream can follow the oscillations without problems. For 1.2 m/s, although the amplitude is larger, the wave maintains its sinusoidal shape, and the air is capable of flowing along the curve maximum. When the velocity reduces to 1.0 m/s, the zigzag shape is apparent, although the curve vertex is out of the image. The air stream turns upwards, generating the recirculation zone.

4 Velocity Measurements

The data set acquired with higher resolution has been used to study in more detail a region extending 2 cm from the nozzle, that yields a resolution of 20 μm/pixel. Only a low air velocity, 15 m/s, has been considered, to increase the length of the intact sheet.

Images have been acquired in frame straddling mode, and analyzed by cross-correlation. The time interval between the two images in each pair was set to 30 μs. For an air velocity of 15 m/s, this time yields a displacement of 450 μm,

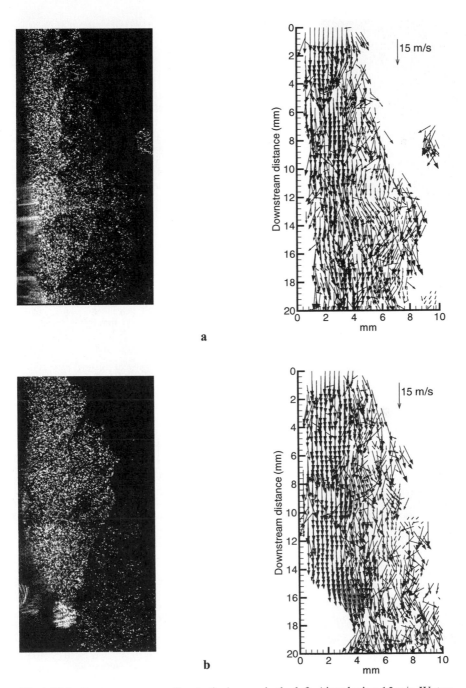

Fig 4: Velocity maps corresponding to the images in the left. Air velocity: 15 m/s. Water velocity a)1.8 m/s, b) 1.5 m/s.

582

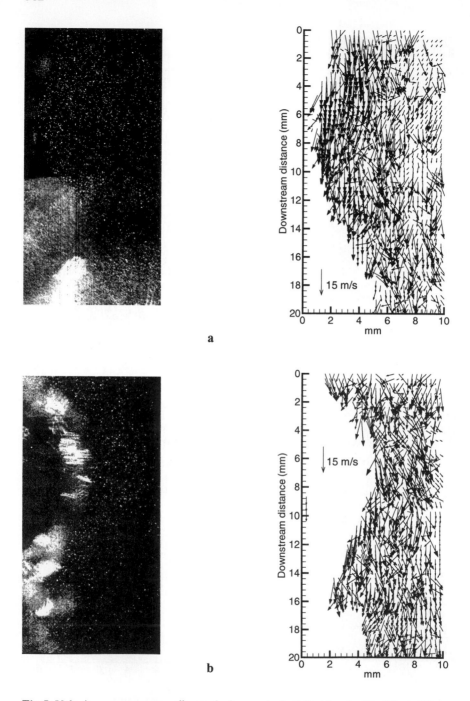

Fig 5: Velocity maps corresponding to the images in the left. Air velocity: 15 m/s. Water velocity a)1.2 m/s, b) 1.15 m/s.

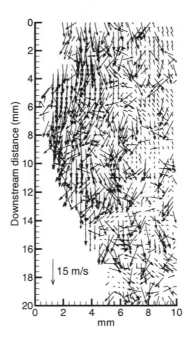

Fig. 6: Velocity map in Fig. 5 a), with mean velocity of the recirculation structure subracted

which for the given resolution corresponds to 22.5 pixels. A window 32 pixels wide has been used to calculate the cross-correlation, with an offset of 15 pixels.

Some results are shown in Figs. 4 and 5 for a fixed air velocity (15 m/s) and the following decreasing water exit velocities: 1.8 m/s (Fig. 4 a), 1.5 m/s (Fig.4 b), 1.2 m/s (Fig. 5 a) and 1.15 m/s (Fig. 5 b). Velocity maps in the right hand side of the figures correspond to the images in the left. No vectors have been interpolated in these maps. Axis values correspond to mm, according to the resolution of 20 µm/pixel. A reference vector has been included in each one of the images for the air exit velocity.

For the higher water velocities, when the sheet is not oscillating in sinusoidal mode, the air near the liquid interface closely follows the water direction, independently of the small vortical structures formed in the air/air interface in the outer side of the coflow. In Fig. 4 a) the external mixing layer can easily be distinguished from the unmixed air stream region near the sheet surface. Possible influences by these small vortical structures do not seem to alter the wavelength or amplitude of the sheet movement although from the lower regions of the images it can be seen how the entrained room air can occasionally get directly in contact with the sheet The influence of this entrainment could be analyzed by varying the exit width of the air conducts.

584

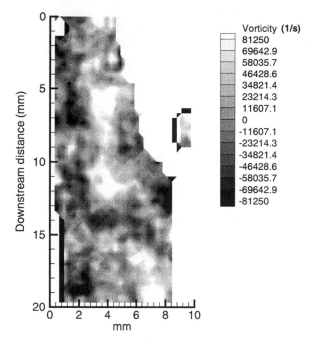

Fig. 7: Vorticity map corresponding to the velocities depicted in Fig. 4 a).

For lower water velocities, when the sheet is flapping, the width of the air streams increases (compare Figs. 4 a) and 5 a)) because quiescent room air is set in motion by the sheet displacements. In these cases, (Fig. 5) the separation between the unmixed air core and the external air mixing layers is not so clearly discernible.

The large recirculation zone visible in Fig. 3 appears also in Fig. 5. The presence of recirculation structures is sometimes masked by the flow mean velocity. In order to identify them more clearly, the velocity of the vortex center can be subtracted. After subtraction, the velocity map in Fig. 5 a) transforms into the one presented in Fig. 6, where the recirculation is more apparent.

Looking at the velocity maps, apparently, no clear vortical structures can be identified in the air/water interface, probably due to their small size. Nevertheless, recent experimental results suggest that the sheet oscillation is actually imposed by air vortex shedding (Barreras, 1998). If the perpendicular component of the vorticity field is calculated, some small vortices are revealed rolling over the liquid sheet. Figure 7 is the vorticity map calculated from the velocity values in Fig. 4 a). The referred vortices are the four dark structures at the left hand side of the image. These vortices, detached from the nozzle, are convected downstream, but for these flow conditions, their influence in the water oscillation amplitude is almost negligible.

5 Conclusions

The behavior of the air streams surrounding a liquid sheet has been studied seeding them with glycerin droplets and illuminating the flow with a laser sheet. The longitudinal waves formed in the water sheet by Kelvin-Helmholtz instability have been visualized simultaneously to the coflowing air. Fast acquisition of image pairs by frame straddling has enabled the calculation of the 2-D air velocity field by PIV cross correlation.

Small vortical structures are observed in the air/air interface between the coflowing air streams and the quiescent room atmosphere. The water/air interfaces present different characteristics. The liquid sheet which flows with slower velocity than the air does not curl in rollers due to its much higher density. Air vortices near the water surface are not clearly identified, although calculations of the perpendicular vorticity component reveal their presence. It is their shedding from the nozzle lip what appears to force the sheet instability. The resulting sinusoidal oscillations are subjected to very high amplification ratios that cause the film break up into small droplets.

The air streams follow the displacements of the liquid sheet. For large oscillation amplitudes, which ultimately result in the atomization, the liquid surface is curved upwards, and the air is "pushed" by the flapping sheet. In these cases, separation between the two layers can occur. It has also been observed in this study that when large amplitude oscillations take place close to the nozzle head, the concave curvature of the sheet forces the air into a recirculation bubble, that might in turn contribute to the fast growth of the longitudinal instability waves.

As a final consideration, it has to be pointed out that to completely explain the phenomena that produce the final atomization of the liquid sheet, the three dimensionality of the flow has to be taken into account, as transverse waves also develop in the interfaces.

Acknowledgments

This project has been partially supported by the Dirección General de Estudios Superiores of the Spanish Government under contracts PB96-0739-C03-03 and PB96-0719

References

ARAI, T. & HASHIMOTO, H., 1985, Disintegration of a thin liquid sheet in a concurrent gas stream. Proc. ICLASS-85, London.

BARRERAS, F., 1998, Experimental Study of the Break up and Atomization of a Liquid Sheet, Ph. D. Thesis Report, University of Zaragoza (in Spanish).

HAGERTY, W.W. & SHEA, J.F.,1955 A study of the stability of plane fluid sheets, *J. Appl. Mech.* December, 509-514.

LOZANO, A., CALL, C.J., DOPAZO & C. GARCÍA-OLIVARES, A., 1996, An Experimental and Numerical Study of the Atomization of a Planar Liquid Sheet, *Atomization and Sprays*, 6, 77-94.

LOZANO, A., BARRERAS, F., YATES, A., GARCÍA, I. & ANDRÉS, N., 1997, Experimental study of the near field of a breaking liquid sheet, Proc. ICLASS-97, Aug. 18-22, Seoul, Korea

MANSOUR, A. & CHIGIER, N., 1990, Disintegration of liquid sheets. *Phys. Fluids A*, 2,(5), 706-719.

MANSOUR, A., CHIGIER, N., SHIH, T. & KOZAREK, R.L., 1998, The Effects of the Hartman Cavity on the Performance of the USGA Nozzle Used for Aluminum Spray Forming, *Atomization and Sprays* 8, (1), 1-24.

RIZK, N.K., & LEFEBVRE, A.H., 1980, The Influence of Liquid Film Thickness on Airblast Atomization, *J. Eng. Power* 102, 706-710.

VI.5. Measurements of Film Characteristics Along a Stationary Taylor Bubble Using Laser Induzed Fluorescence

J.P. Kockx, R. Delfos, and R.V.A. Oliemans

J.M. Burgers Centre, Laboratory for Aero and Hydrodynamics,
Delft University of Technology, 2628 AL, Delft, The Netherlands

Abstract. In this study we measure the characteristics of the falling film along a stationary Taylor bubble. A Taylor bubble is a large gas bubble in a vertical tube that spans the tube diameter and rises with a high speed through the liquid. The film characteristics are used to verify the model proposed by Delfos (1996) for the gas flux out of the Taylor bubble, the so-called entrainment flux. We developed a laser induced fluorescence (LIF) technique to measure the instantaneous film thickness.

The film along the Taylor bubble agrees with the theoretical description of a free falling laminar film with friction on the wall until a Taylor bubble length of 0.7 m. After this length the film thickness increases till it reaches a constant value of 2.8 mm, which agrees with measurements on a fully developed turbulent film (Karapantsios, 1995). The film surface changes from smooth to rough at a Taylor bubble length of 0.5 m. The disturbances are found to be related to the turbulent flow present in the film. The area of the film surface covered by disturbances grows with the bubble length. This is described by the intermittency which increases from 0 at a bubble length of 0.5 m to 1 at a length of 1 m. The model for the entrainment flux appeared only to be in qualitative agreement with the incoming gas flux.

Keywords. LIF, Taylor bubble, film thickness, surface waves, entrainment

1 Introduction

Liquid-gas flow in a vertical pipe is one of the fundamental flow categories in the field of two-phase flow. Depending on the flow rates of the gas and liquid, the flow patterns in this case can differ considerably. One of these flow patterns is slug flow.

A slug flow is characterised by a liquid flow in combination with a series of large axi-symmetric bullet-shaped gas bubbles, so-called Taylor bubbles. These bubbles occupy most of the cross-section of the pipe and move upward much faster than the liquid. Between the Taylor bubble and the tube wall the liquid flows downwards as a thin free-falling film. The Taylor bubbles are separated by regions of continuous liquid phase that contain small gas bubbles. These regions are denoted as liquid slugs. A schematic picture of a vertical slug flow is given in Fig. 1.

588

Vertical slug flow is encountered in many industrial two-phase flow applications. For instance it occurs during oil production and in various process equipment. The strongly instationary flow conditions during slug flow may have a large mechanical impact on the equipment in which this flow occurs. For industrial applications it is therefore important to determine the conditions for which slug flow can appear. An important parameter is the void fraction, the fraction of gas, in the liquid slug. To predict the void fraction various models have been proposed. For a review see Fabre & Liné (1994).

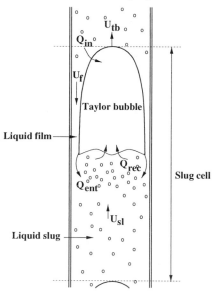

A fully developed slug flow can be considered as as train of so-called slug-cells, each consisting of one Taylor bubble together with the surrounding liquid film and one successive liquid slug. Such a slug-cell is indicated in Fig. 1. In their model, based on a description in terms of a slug cell, Fernandes *et al.* (1983) predict the void fraction in the liquid slug by considering the gas fluxes into and out of the Taylor bubble. When the liquid flows around the nose of the Taylor bubble it deforms into a film too thin to accommodate the bubbles carried within the liquid slug. As a result bubbles coalesce with the nose of the

Figure 1: A schematic picture of slug flow in a vertical tube. U_{tb}, U_f, U_{sl} are respectively the Taylor bubble velocity, the film velocity and the superficial liquid velocity.

Taylor bubble, which results in a gas flow rate into the Taylor bubble, Q_{in}. At the bottom of the Taylor bubble small bubbles are torn off due to the falling film which plunges into the liquid slug. This gas flow rate out of the Taylor bubble is indicated as the entrainment flux, Q_{ent}. A part of these entrained gas bubbles re-coalesces back into the Taylor bubble at its trailing edge, which results in a re-coalescence flux, Q_{rec}. For a stationary Taylor bubble these gas fluxes are in balance

$$Q_{in} = Q_{ent} - Q_{rec} \qquad (1)$$

The detailed mechanisms behind entrainment and re-coalescence are not yet fully understood (Dukler & Fabre, 1992). In particular the mechanism by which small bubbles are torn off from the Taylor bubble, i.e. the entrainment process, remains unclear. Delfos (1996) has suggested that the entrainment occurs due to the fact that the liquid interface at the bottom of the Taylor bubble is not able to follow the variations in thickness of the downward falling liquid film. Based on this hypothesis he has proposed a model for the entrainment flux. Given the uncertainties in the entrainment process and the role of the falling liquid film in this process, our objective here is to measure the characteristics of this falling film. In particular the mean film thickness, film thickness fluctuations and intermittency. For this purpose we developed a method to measure the film thickness based on a Laser Induced Fluorescence (LIF) technique.

In the present paper some theory of the falling film is explained in section 2. The experimental set-up as well as the measurement method are described in section 3. Finally results are presented and conclusions are drawn.

2 Theory

In preparation for the analysis of our experimental data, we concentrate in this section on some theoretical aspects of a Taylor bubble. Using potential flow, Dumitrescu (1943) has calculated the shape of a Taylor bubble. He finds for the radius of curvature R_n at the nose of the bubble

$$R_n = 0.36D \tag{2}$$

where D is the diameter of the pipe in which the Taylor bubble rises. The liquid layer alongside the Taylor bubble far from its nose is sufficiently thin to be described as a falling film. The shape of the Taylor bubble is completed by connecting the tangents of the bubble nose expression and the falling film equation. Then he finds for the rise velocity of a single Taylor bubble in a stagnant liquid

$$U_{tb} = 0.35\sqrt{gD} \tag{3}$$

2.1 The falling film

Let us consider the falling film surrounding the Taylor bubble in a frame where the Taylor bubble is fixed. Within this reference frame the film flows downward and it behaves as a free-falling film, i.e. the shear across the interface between film and bubble is assumed to be negligible. For a thin film ($h_f \ll D$) we can approximate this case as a two-dimensional film flowing along a flat vertical plate. In the simplest case of a free-falling, laminar film without friction on the wall, the velocity profile will be uniform. As a result of conservation of mass, the downward liquid flow rate must be constant at each position along the vertical pipe. This implies that the liquid flow rate above the fixed Taylor bubble must be equal to the liquid flow rate in the falling film:

$$\frac{\pi}{4}D^2 U_{tb} = \frac{\pi}{4}\left(D^2 - [D - 2h_f(z)]^2\right)U_f(z) \tag{4}$$

$$h_f(z) = \frac{D}{2}\left(1 - \sqrt{1 - \frac{U_{tb}}{U_f(z)}}\right) \tag{5}$$

where $h_f(z)$ is the mean film thickness and $U_f(z)$ the mean film velocity. The z denotes the axial position (taken positive downwards) along the pipe where the origin of z is chosen at the top of the Taylor bubble.

The velocity of a free-falling film in our reference frame reads

$$U_f(z) = U_{tb} + \sqrt{2gz}. \tag{6}$$

If we take friction with the wall into account, a laminar boundary layer will develop on the wall. The boundary-layer thickness for the case of a two-dimensional accelerating laminar flow reads (Delfos 1996, Appendix A):

$$\delta(z) = 3.41\sqrt{\nu\sqrt{\frac{z}{2g}}}$$

with ν the kinematic viscosity of the liquid. The accelerating boundary layer causes a non-uniform velocity profile within the film. As a consequence we have to correct the free-falling film thickness (5) by adding to h_f the displacement thickness δ^* due to the boundary layer (Delfos 1996, Appendix A), given by

$$\delta^* \approx 0.252\,\delta \tag{7}$$

A final point to consider is the transition from laminar to turbulent flow inside the liquid film. When the turbulent flow is fully developed across the whole film, the film should reach a constant film thickness. Takahama (1980) and Karapantsios (1995) did measurements on a falling film till a Reynolds number of $11 \cdot 10^3$. They found an experimental correlation, which gives a mean film thickness of respectively 2.7 and 2.8 mm for our Reynolds number of $3.5 \cdot 10^3$.

2.2 The entrainment flux

Delfos (1996) suggests that the entrainment flux, the gas flow out of the Taylor bubble into the liquid slug, occurs due to disturbances on the surface of the falling film along the Taylor bubble.

The proposed mechanism for entrainment is now the following: The falling film along the Taylor bubble becomes turbulent at a certain Taylor bubble length, L_{onset}. The turbulent eddies in the film then deform the film surface. Due to these surface disturbances, advected by the film flow, the free surface at the bottom of the Taylor bubble sees, so to speak, a transversely oscillating wall. The transverse acceleration of this oscillation is proportional with the rate $\frac{h_w}{\lambda^2}$, with h_w the amplitude of the wave and λ the wave length of the wave. When this acceleration a_t becomes much larger then the gravitational acceleration g, the pool surface is not able to follow these oscillations (disturbances) of the incoming film flow (Delfos, 1996). As a result an air gap is generated between the vertically falling film and the liquid slug at the bottom of the Taylor bubble. This gap may be closed by the crest of the next disturbance and as a result air is trapped inside the liquid. The entrained gas is then approximately equal to the volume of air enclosed between the crests of the disturbances.

To come to an expression for the entrainment flux we have to make some assumptions. First we assume that the disturbances are sinusoidal waves. Further more we suppose that the velocity of the waves is approximately equal to the bulk velocity. Finally we assume that the disturbances on the surface occur intermittently with an intermittency factor, I, which can be interpreted as the percentage of Taylor bubble circumference on which disturbances occur. The model for the entrainment flux then becomes

$$Q_{ent}(z) = I\pi(D_i - 2h_f(z))U_f(z)h_w(z) \tag{8}$$

To calculate the entrainment flux we thus need to measure the mean film thickness, the intermittency, the film velocity and the amplitude of the waves on the surface of the liquid film.

3 Experiment

3.1 Experimental set-up

The details of the experimental set-up that we have used are described by Delfos (1996). Here we will summarise only the main characteristics. A schematic layout of the experimental set-up is illustrated in Fig. 2. It consists of a vertical, cylindrical pipe made out of perspex with an inner diameter D_i = 100 mm. In this tube, a Taylor bubble is generated by an injection of air through a small tube with diameter D_t. At the end of this tube a spherical cap is attached. The purpose of this cap is to stabilise the bubble in the centre of the main vertical pipe. The radius of curvature of this cap is the same as that of the nose of a theoretical Taylor bubble as given in (2).

The bubble is kept at a fixed position by a water flow moving downward with a velocity equal to U_{tb} given by (3). Above the Taylor bubble a flow straightener, made out of honeycomb material, is placed to obtain a uniform velocity profile. Such a profile is representative for flow conditions encountered by a rising Taylor bubble in a stagnant liquid. Apart from creation of a flat velocity profile, the flow straightener also decreases the turbulence intensity in the flow in front of the Taylor bubble.

The part of the pipe where the Taylor bubble can be observed, is called the test section. The test section is enclosed by a rectangular, water-filled, box to provide a good optical access.

3.2 The LIF-technique to measure film thickness

The objective of our research is to measure the characteristics of the thin film which is present between the tube wall and the Taylor bubble. To measure the instantaneous thickness at a given location we need a one-point measurement method, which is able to measure the film thickness continuously with a high sample frequency. Several methods are available to perform such measurements, with as example conductive or needle contact probes. The

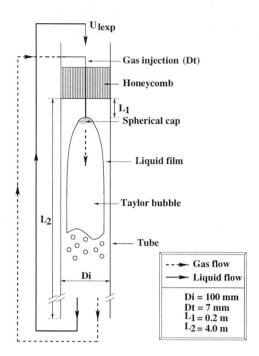

Figure 2: A schematic drawing of the experimental set-up. D_t is the external diameter of the gas injection tube.

disadvantage of these techniques is that the probes disturb the flow and the needle point probe measurement is relatively slow. Disturbance of the film flow should be avoided, because it can change the characteristics of the film. For instance, a disturbance can lead to early transition from laminar to turbulent flow. Furthermore the film thickness in our case has to be measured at various axial positions along the Taylor bubble. This is difficult to carry out with probes, because many holes in the wall are needed. Another method is to use the light absorption method to measure film thickness. However, this requires a detector inside the Taylor bubble, which moreover can transverse in axial direction. This is in practice very complicated to realise.

As an alternative to the methods discussed above, we have chosen to measure the film thickness by means of a Laser Induced Fluorescence technique (LIF) proposed by Hewitt (1964). The advantage of this technique is that it is non-intrusive and thus it does not disturb the flow, which is an important requirement in our case as discussed above. In the LIF-method that we have applied, we can measure the film thickness with the laser and the detector both at one side of the tube. Our method allows instantaneous observation of the film thickness at a given position as a function of time. From these measurements we aim to obtain statistical information, such as the mean and the variance of the film thickness.

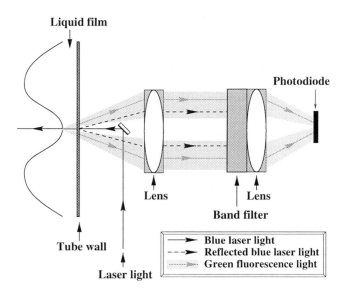

Figure 3: Schematic figure of film thickness measurement with the Laser Induced Fluorescence technique.

The principle of this LIF-technique is illustrated in Fig. 3. Blue light (λ_L = 488 nm) from an Argon-ion laser illuminates via a small circular mirror, D_m = 3 mm, the liquid film, which contains a constant concentration of a fluorescent dye, Uranine AP, $[C_{20}H_{14}Na_2O_5]$. When the dye is excited by the incident beam, a green fluorescence (λ_F = 515 nm) results. This green light is observed by an optical system which is positioned at the same side of the pipe as the incoming blue laser light. After passing a collimating lens, the fluorescent light is first separated from the reflected blue light by a bandpass filter (XM-535 Corion). Subsequently, it is focused on a photodiode by means of a second lens. The signal of photo-diode is proportional to the intensity of the fluorescent light. The proportional constant , k_{ph}, depends on the detector properties and the fraction of light captured by the detector.

The intensity of the fluorescent light depends on the intensity of the incoming blue light I_0, the concentration of the dye C, the film thickness h_f (i.e. the light path through the fluorescence solution) and the properties of the detector (Guilbault, 1973). The relation for the photo-diode signal reads

$$V_{ph} = k_{ph}I_f = k_{ph}\Phi I_0(1 - e^{-\epsilon C h_f}) \tag{9}$$

where Φ is the quantum efficiency (0.92 for Uranine AP) and ϵ is the extinction coefficient. During the experiment the concentration of the fluorescent dye and the intensity of the incoming laser light are kept constant. The measured intensity of fluorescence can thus directly be interpreted as a measure for the film thickness.

594

At intermediate concentrations, the fluorescence light is not evenly distributed along the light beam, due to the exponential term in (9). Its explanation is that the portion of fluorescence solution nearest to the light source absorbs radiation so that less is available for the rest of the solution. For sufficiently dilute solutions and small film thicknesses this effect gets small and (9) can be simplified to the following relationship

$$V_{ph} = k_{ph}I_f \approx k_{ph}\Phi I_0 \epsilon C h_f - \frac{1}{2}k_{ph}\Phi I_0 \epsilon^2 C^2 h_f^2 \qquad (10)$$

where the non-linear second term is called the inner-cell effect (Guilbault, 1973). Its explanation is that the portion of fluorescence solution nearest to the light source absorbs radiation, so that less is available for the rest of the solution. The dye concentration is $C = 1.10^{-3}$ gl^{-1}. It is chosen such that the fluorescence intensity is sufficiently high for the photo diode to detect, but sufficiently low to avoid the inner-cell effect.

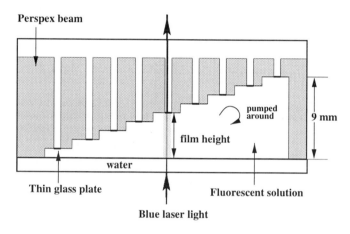

Figure 4: Schematic side-view of the calibration set-up.

The relationship (10) must follow from a separate calibration procedure in which the intensity of fluorescence is determined for a set of given film thicknesses. For this we have designed a special calibration set-up, which is shown schematically in Fig. 4. It consists of a rectangular container, made out of perspex, which is filled with the fluorescent solution. In this container a ladder-shaped perspex beam with holes is placed on the bottom. The holes are closed on the bottom side with thin glass plates ($d_g = 0.2$ mm). These thin glass plates replace the free surface (water-air) in the experiment. During calibration we correct for the fact that the water-glass-air surface reflects about 2 % more than the water-air interface. Between these thin glass plates and the bottom of the container a range of liquid layers with well-defined thicknesses are present. The bottom of the container consist of two perspex walls with a water layer in between, to imitate the construction of the test section in the experiment. The distance between the laser and the liquid layers is taken the same as in the experiment. During the calibration, the fluorescent solution is circulated by a pump to prevent photo decomposition (Guilbault, 1973) of the dye. The calibration procedure now consist of measuring the thickness of the liquid layers by means of the LIF-technique. The voltage of the photo-diode (which measures the fluorescence intensity) is found to fluctuate during the calibration measurement. These fluctuations occur since the laser itself fluctuates with 50Hz and its harmonics. The r.m.s. intensity of this ripple is 2 % of the total measured intensity at a given film height. These Nx50Hz fluctuations are digitally filtered out. The error in the thickness of the defined liquid layer is approximately 0.02 mm and the error in the calibration curve (based on the function given in (10) is 0.5 %. Each calibration is performed 3 times, including a complete new out-line of the calibration set-up. Based on the 4 calibrations the total error in the calibration becomes 3 %. For a typical film thickness of 3 mm this leads to an error of about 0.1 mm. An example of a calibration curve is shown in Fig. 5. The curve is almost linear which confirms that the inner-cell effect is small.

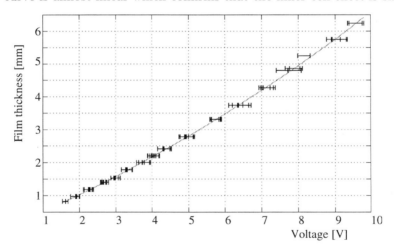

Figure 5: The calibration curve, which gives the relation between the film thickness and the voltage of the photo-diode (i.e. the fluorescence intensity). Measurement points of 3 successive calibrations.

After calibration the laser and detector are mounted on a traversing system along the test-section, so that we could measure the film thickness at different points along the Taylor bubble. At every measuring point the intensity of fluorescence is sampled during 15 seconds with at a frequency of 4000 Hz.

3.3 Determination of intermittency and wave height

To determine the intermittency and wave height, we have to look at the continuous film thickness signal. In Fig. 6 and Fig. 7 we show examples of the film signal for at a Taylor bubble length of 0.8 m.

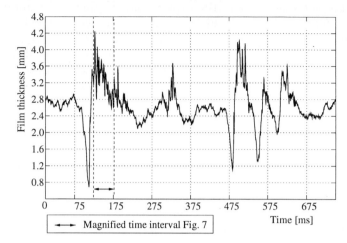

Figure 6: An example of the continuous film thickness signal at a Taylor bubble length of 0.8 m. Undulations are present on the surface of the film.

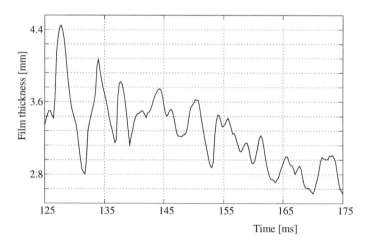

Figure 7: A magnification of the above time series. The small waves on top of the large waves can be seen.

In Fig. 6 we can see two kind of waves. Very large waves with a time period of 50-100 ms and an amplitude, h_w, of ≈ 1.5 mm and small waves on top of the large waves. In Fig. 7 we show an magnification of a part of the first figure. Here we can see clearly the small waves with time period of about 3-10 ms and a $h_w \approx 0.5$ mm. In the power spectrum we find the large waves at a frequency of about 10 Hz and the small waves at about 200 Hz. If we assume that the velocity of the waves is equal to the liquid film velocity, in our case about 3 ms^{-1}, we get a wave length, λ, of about 0.3 m an 15 mm for the large and small waves respectively.

The criterion for onset of entrainment, $\frac{a_t}{g} \gg 1$, can in our case only be satisfied if the rate $\frac{h_w}{\lambda^2} \gg 0.03$ m^{-1}. The large waves do not conform to this requirement. Thus, to determine the relevant wave height and intermittency for the entrainment, the film signal is filtered by a high pass filter with a cut off frequency of 30 Hz.

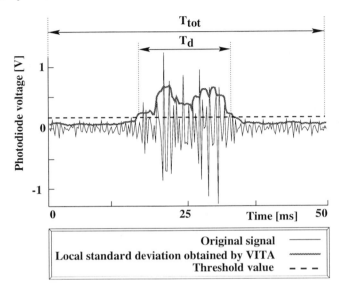

Figure 8: Definition of the intermittency with the VITA-technique. T_d is the time that disturbances on the film surface are present and T_{tot} is the total time interval.

The intermittency, which in our case is defined as the percentage of time that waves occur on the film surface, is computed by applying the VITA-technique (Johansson, 1983). In the VITA technique, shown in Fig. 8, the (local) standard deviation is calculated for small time intervals of the signal (in our case 10 msec). The time period that this local standard deviation exceeds a certain threshold value, is then defined as the disturbed time T_d. This time interval is the period of time that waves are present on the surface of the film. We define the threshold value as the standard deviation of a film without waves on the surface. This value is found by taking the mean local standard deviation of the film thickness of a small Taylor bubble ($L_{tb} < 0.4$ m). In our experiment the threshold value is 0.09 V, which is 0.1 mm at a mean film thickness of 3 mm. The intermittency is then obtained as the ratio between the disturbed time T_d and the total time interval, T_{tot}.

The amplitude of the waves is calculated with the standard deviation of the film thickness signal. This standard deviation of the waves on the surface, σ_w, is obtained from the standard deviation of the filtered film thickness signal σ_{tot} corrected for intermittency and the noise σ_0. This leads to the following expression

$$\sigma_w^2 = \frac{\sigma_{tot}^2 - (1 - I)\sigma_0^2}{I} \qquad (11)$$

If we assume that the disturbances are sinusoidal waves, the wave height becomes

$$h_w \approx \sqrt{2}\,\sigma_w. \qquad (12)$$

4 Results

4.1 The film thickness

We first give a general impression of the surface of the film at different Taylor bubble lengths. In Fig. 9 we show three time series of Taylor bubble lengths of 0.35 m, 0.77 m and 1.05 m respectively. The instantaneous film thickness is normalised by the mean film thickness of each time series. It can be seen that at short Taylor bubbles only some long waves occur with small wave heights. At a length of 0.77 m long waves with large wave heights occur and on top of the long waves short waves exist. At long Taylor bubble of 1.05 m the amount of large and small waves increases rapidly. This figure gives a visualisation of the development of waves on the surface of the falling film.

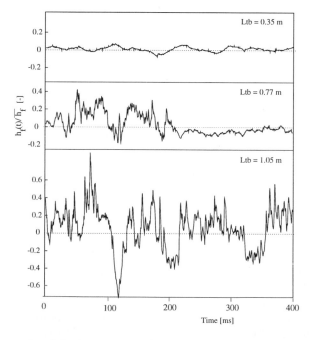

Figure 9: Time series of the instantaneous film thickness at three different Taylor bubble lengths. The instantaneous film thickness is normalised by the mean film thickness of the time series.

We will now focus on the mean film thickness of the film at different Taylor bubble lengths. In Fig. 10 we give the measured mean film thickness as function of the distance from the Taylor-bubble front. These data have been obtained with a gas influx $Q_{co} = 0.36$ ls^{-1} and a downward liquid velocity of $U_{tb} = 0.35$ ms^{-1}. The film thickness profile is only given for this particular case of the gas influx, because the profile is independent of the incoming gas flux.

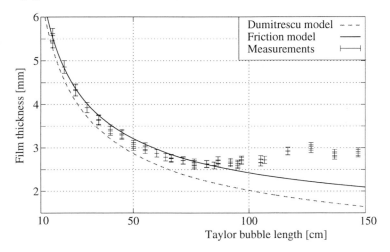

Figure 10: The measured film thickness at $Q_g = 0.36$ ls^{-1} and $U_{tb} = 0.35$ ms^{-1} compared with the models for a free-falling laminar liquid film with and without friction.

In Fig. 10 we also show the theoretical models for the Taylor bubble shape as discussed in section 2.1. We see that the measured film thickness agrees rather well with the theoretical film thickness for the case of a laminar free-falling film with friction until a bubble length of about 0.7 m. The difference between the measurements and the friction model after this point is probably caused by the development of turbulence. The turbulent friction at the wall causes a decrease of the mean velocity in the film which results in thicker film. Eventually the film thickness seems to reach a constant value of ± 2.8 mm, which agrees with the thickness of a fully-developed turbulent film determined from the empirical correlation by Karapantsios (1995).

4.2 Intermittency distribution and wave height

The intermittency distribution along the Taylor bubble, determined as described in section 3.3, is shown in Fig. 11. From our measurements we see that the intermittency is very low until the bubble reaches a length of about 0.5 m. This point is called 'the onset of entrainment' and we interpret this as the transition point. Beyond this point the intermittency grows rapidly to 1 at about 1 m.

In the same figure the intermittency of the surface of the falling film

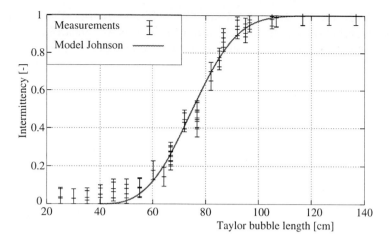

Figure 11: Measured intermittency distribution along the Taylor bubble at $U_{tb} = 0.35$ ms^{-1} compared with the empirical model by Johnson & Fashifar (1994) for intermittency distribution in a boundary layer on a flat plate.

is compared with a model for the intermittency distribution in a boundary layer during transition. In this latter case the intermittency is equal to the percentage of time that turbulent spots occur in the boundary layer (Johnson & Fashifar, 1994). Their empirical model agrees well with our measurements, which is consistent with our hypothesis that the disturbances are caused by turbulence.

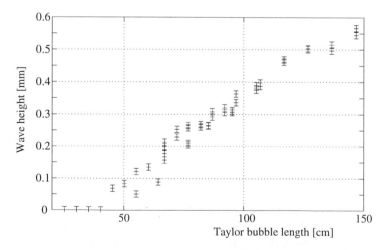

Figure 12: The wave height of the waves on the surface of the falling liquid film.

The wave height of the disturbances on the falling-film surface, computed according to (11) and (12) is shown in Fig. 12. The wave height increases

from 'the onset of entrainment point' at about 0.5 m till a height of about
0.55 mm at a Taylor bubble length of 1.4 m. The trend in the data shown in
this figure, suggests that the wave height will increase even more after this
length.

4.3 Derivation of the entrainment flux

We finally calculate the entrainment flux. For this we need an estimate for all
the parameters in (8) which are respectively the mean film velocity U_f, the
mean film height h_f, the intermittency I and the height of the waves on the
falling film surface h_w. Apart from the mean film velocity, these parameters
can all be obtained directly from our measurements. The remaining parameter, i.e. the mean film velocity, is calculated with help of the measured film
thickness by means of conservation of mass given by (5).

In Fig. 13 we give the entrainment flux, Q_{ent} obtained from (8) as a
function of the bubble length. The entrainment flux is zero till a Taylor
bubble length of 0.5 m. This agrees with the hypothesis of Delfos (1996)
that the entrainment process starts as soon as waves appear on the surface
of the film. The the entrainment flux increases gradually from 0 to 0.5 ls^{-1}
between 0.5 and 1.5 m Taylor bubble length.

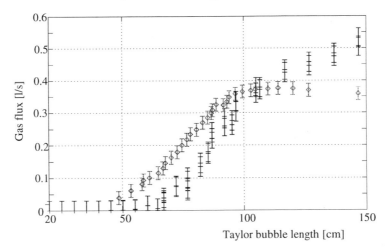

Figure 13: The modelled entrainment flux,Q_{ent}, compared with the gas influx at $U_{tb}=$
0.35 ms^{-1}

In the same figure we show the gas influx at different Taylor bubble
lengths. The gas in-flux is equal to the entrainment flux if we neglect the
re-coalescence flux, according to equation (1). It can be seen that the entrainment flux does only qualitatively compares with the gas influx. Further
research on the onset criterion of entrainment and the re-coalescence flux is
therefore warranted to come to a better comparison between the entrainment
flux and the gas in-flux.

5   Conclusions

Based on the results presented in this study we can draw the following conclusions

- We have applied a technique to measure the film thickness based on Laser Induced Fluorescence and found the results to be reliable. The film thickness can be measured within an accuracy of 0.1 mm.

- The measured mean film thickness along a stationary Taylor bubble agrees with the model for a laminar free-falling film with friction on the wall until the Taylor bubble reaches a length of 0.7 m. After this bubble length the increase in film thickness indicates that there is a transition from laminar to turbulent flow in the liquid film. The film thickness reaches a constant value of 2.8 mm, that agrees with the empirical correlation for the film thickness of a fully-developed turbulent film measured by Karapantsios (1995).

- The film surface changes from smooth to rough at a Taylor bubble length of 0.5 m. As soon as the film surface becomes rough entrainment is observed, which is in agreement with the hypothesis of Delfos (1996) that disturbances on the film surface are responsible for entrainment. These disturbances are found to be related to the turbulent flow present in the film. The area of the film surface covered by disturbances grows with the bubble length. This is described by the intermittency which increases from 0 at a bubble length of 0.5 m to 1 at a length of 1 m. The intermittency distribution compares with the intermittency distribution of turbulent spots in a boundary layer in transition (Johnson, 1994).

- The effective wave height of the disturbances increases from 0 at the onset of entrainment point ($L_{tb} = 0.5$ m) to 0.5 mm at a Taylor bubble length of 1.5 m.

- The simple model for the entrainment flux (Delfos, 1996) is only in qualitative agreement with the gas in flux. Further study on the onset of entrainment criterion and the re-coalescence flux is therefore warranted.

Acknowledgements
The authors thank T.D. Tujeehut and J.L.J. Zijl for their work on the experimental part of this research. We also thank Prof. dr. ir. F.T.M. Nieuwstadt for his useful discussions.

References

DELFOS, R. 1996. *Experiments on air entrainment from a stationary slug bubble in a vertical tube.* Ph.D. thesis, Delft University of Technology.

DUKLER A.E., FABRE J. Gas-liquid slug flow; knots and loose ends. *Third int. workshop on Two-Phase fundamentals, ICL, june 1992.*

DUMITRESCU, D.T. 1943. Strömung an einer Luftblase im senkrechten Rohr. *Z. angew. Math. Mech.*, **23**(3), 139–149.

FABRE J., LINÉ A. 1994. Advancements in two-phase slug flow modelling. *Proc. of the Univ. of Tylsa Cent. SPE Symp, Tulsa.*

FERNANDES R.C., SEMIAT R., DUKLER A.E. 1983. Hydrodynamic model for gas-liquid slug flow in vertical pipes. *AIChE Journal*, **29**, 981–89.

GUILBAULT, G.G. 1973. *Practical Fluorescence.* Marcel Dekker Inc., New York.

HEWITT G.F., LOVEGROVE P.C., NICHOLLS B. 1964. Film thickness measurement using fluorescence technique. *Atomic Energy Research, Harwell*, **AERE - R4478**.

JOHANSSON A.V., ALFREDSSON P.H. 1983. Effects of imperfect spatial resolution on measurements of wall-bounded turbulent shear flows. *J. Fluid Mech.*, **137**, 409.

JOHNSON M.W., FASHIFAR A. 1994. Statistical properties of turbulent bursts in transitional boundary layers. *Int. J. Heat Fluid Flow*, **15**, 283–290.

KARAPANTSIOS T.D., KARABALAS, A.J. 1995. Longitudinal characteristics of wavy falling films. *Int. J. Multiphase Flow*, **21**(1), 119–127.

TAKAHAMA H., KATO S. 1980. Longitudinal flow characteristics of vertically falling liquid films without concurrent gas flow. *Int. J. Multiphase Flow*, **6**, 203–215.

VI.6. Experimental Analysis of a Confined Bluff Body Flow Laden with Solid Particles

T. Ishima[1,2], J. Borée[1], P. Fanouillère[3], and I. Flour[3]

[1] Institut de Mécanique des Fluides , UMR CNRS/INPT-UPS 5502
Allée du Professeur Camille Soula , 31400 TOULOUSE, France
[2] Gunma University , Tenjin, KIRYU 376-8515, Japan
[3] EDF – DER – LNH , 6, Quai Watier , 76400 CHATOU, France

Abstract. An experimental study of the two-phase flow downstream a confined bluff body is presented in this paper. This situation is particularly interesting for the development of gas-solid flow turbulence closures. Measurements are made with a phase Doppler anemometer. Glass beads are added to the inner jet. The initial particle size distribution covers a wide range of size classes from 20 μm to 110 μm with a mean value of 60 μm. The single phase configuration is chosen to establish a stagnation point in the recirculating region. This configuration emphasises the role of the inertia of the particles. This study shows that no local equilibrium with the fluid turbulence can be assumed and that one has to write and solve transport equations for the fluctuating velocity of the particles. In particular, production and transport effects are clearly observed. Contrasted mass-loading ratio of the inner jet ranging from 22 % to 110 % are selected in order to show the two-way coupling between continuous and discrete phases. We show that it has a noticeable effect on the structure of the flow downstream the bluff body and on the spatial distribution of the particulate phase.

Keywords. Gas-solid two-phase flow, PDA, Bluff body, Polydispersed particles

1 Introduction

Particle-laden gas flows are common in the industry and receive presently a lot of attention in the laboratories. Significant efforts are developed to improve numerical predictions of engineering complex situations with Lagrangian (Berlemont, Desjonqueres, and Gouesbet 1990) or Eulerian (Elgobashi and Abou-Arab 1983; Simonin 1991) predictions. Owing to the large number of new dimensionless parameters involved when one adds particles in a flow, development of turbulence models require a large amount of accurate experimental data in basic homogeneous or inhomogeneous flows.

Laser based measurements now provide insight into dispersed two-phase flows. A large number of basic experiments are reported in channel flows (Kulick, Fessler, and Eaton 1994), boundary layer flows (Rogers and Eaton 1991) and in simple free shear flows as shear layers (Hishida, Ando, and Maeda 1992; Ishima,

Hishida, and Maeda 1993) and jets (Hardalupas, Taylor, and Whitelaw 1989; Modaress, Tan, and Elgobashi 1984; Prévost, Borée, Nuglisch, and Charnay 1996). The role of the Stokes number of the particles, which compares the particle Stokesian time scale to the time scale of the large eddy motion seen by the particles is largely discussed in the shear flows where a typical turbulent time scale is easy to define. Modifications of the turbulence for moderate mass-loading (defined as the ratio of particles mass fluxes to air mass fluxes) is clearly demonstrated in these works. Fluid particle correlations are measured in (Prévost *et al.* 1996) and used to analyse the contrasted evolution of the particles Reynolds stresses along the jet.

These basic situations present essential test cases for numerical predictions. For instance, jet flow studies have clearly shown that the influence of ejection conditions and the local production by the mean particle velocity gradients are dominant mechanisms for the dispersed phase fluctuating motion (Hardalupas *et al.* 1989) that have to be taken into account in continum models (Simonin 1991). The bluff body flow presented in this paper was designed as a severe test case for the modelling. This configuration is typical of an industrial application where the objective is to control the mixing of a fuel (pulverised coal) with surrounding air flow. The outlet velocity of the inner jet is chosen low enough in order to obtain two stagnation points (see figure 1) in single phase. Polydispersed glass particles are added to the inner tube flow only. The stagnation point configuration obtained here is therefore interesting as inertia properties of the particles and fluid/particle coupling in the inner jet are expected to play a dominant role. The initial particle size distribution covers a wide range of size classes from 20 µm to 110 µm with a mean value of 60 µm. Measurements obtained with a Phase Doppler anemometer will be presented. A size class analysis will illustrate the contrasted evolution of the fluid and particle turbulence.

The experimental configuration and the measurement method are first described. The evolution of the fluid and particles mean and fluctuating velocity field for two typical size classes is presented in section 3. The air flow in the presence of particles obtained for two different mass-loading ratio of the inner jet $M_j = 22\%$ and $M_j = 110\%$ is then compared with the single-phase flow data. The consequences as spatial distribution of the particulate phase is concerned are finally discussed.

2 Experimental Set-Up

2.1 Flow Configuration

The experiment presented here has been performed in the flow loop Hercule of EdF-LNH. The vertical axisymmetric air flow downstream the bluff body is shown in figure 1. The outer and inner radius of the annular outer region are respectively $[R_2 = 150\text{mm}; R_1 = 75\text{mm}]$. The maximum velocity of the annular flow is 6 m/s and the Reynolds number of the flow is $R_e = U(R_2 - R_1)/\nu \approx 30000$.

This relatively low value of the external velocity was chosen in order to use a reasonable external volume flux and to overcome seeding problems. The external volume flux is kept constant at $Q_e = 780$ Nm³/h for a relative static pressure at x = 0 of $\Delta p = 10$mbar. The length of the straight annular section upstream the test section is $L_e = 2$m. Several honeycomb structures are added in order to remove any swirl motion. An inner tube jet [$R_j = 10$mm; $U_j(r = 0) = 4$m/s] is generated by compressed air flows on the axis of the recirculation as shown in figure 1. With $L_e/D_j = 100$, the tube flow is established. The volume flux of the jet is $Q_j = 3.8$ Nm³/h at the relative static pressure $\Delta p = 10$mbar. This corresponds to a mean velocity $\overline{U}_j = 3.4$m/s. The ratio of the jet volume flux to the annular flow volume flux is very low (0.5%). The maximum velocity $U_j = 4$m/s was chosen in order to get a single phase flow with stagnation points in the recirculation. The Reynolds number of the pipe flow is Re = $\overline{U}_j D_j / v \approx 4500$, the flow is turbulent with $\overline{U}_j / U_j (r = 0) = 0.85$. Moreover, we could check that the velocity profile at the exit is well fitted by a power law of exponent approximately (1/7).

A particular care was devoted to seed both outer and inner single phase flows. Two water injectors designed for spreading water mist as tracer particle are located upstream the establishment region and are used to seed the annular flow. Smoke tracers generated by a smoking machine (produced by Dantec) are added to the central jet flow.

The test section is 1500 mm in length. For further descriptions, The origin is set on the edge of the bluff body and at the center of the inner jet. The measurements on the axis were made at 3mm, 10mm, 20mm and after every 20mm up to 500mm. Flow characteristics were also measured in planes of 3, 80, 160, 200, 240, 320 and 400mm. The flow will be described henceforth using a cylindrical coordinate system (x,r,θ) to indicate the axial (downward), radial and azimuthal directions. The components of the mean and fluctuating velocity field are denoted respectively by (U,V,W) and (u,v,w) where V is the radial component and W is the azimuthal component. No swirling motion was detected to within our measurement precision. Subscripts "f" and "p" indicate respectively fluid and particles properties. The symbol $< >_f$ and $< >_p$ indicates averaging operators associated respectively to fluid and particle phases. The expressions u' and v' stand respectively for longitudinal and radial standard deviation.

2.2 Glass Particles

Glass particles are released in the conducting pipe of the inner jet by a particle feeder. All flow conditions and particle mass loading are monitored on a work station. The mass loading is controlled accurately in closed loop by weighting continuously the particle feeder. Two mass flux of particles of 1 kg/h (resp. 5 kg/h) corresponding to contrasted inner jet mass loading ratio M_j of $M_j = 22\%$ (resp. $M_j = 110\%$) have been selected in order to show the coupling between continuous and discrete phases. Note that the mass loading ratio of the global flow is in each case very weak: $M_t = 0.1\%$ (resp. $M_t = 0.5\%$).

The material density of the glass particle is $\rho_p = 2470$ kg/m^3. The initial particle size distribution covers a wide range of size classes from 20 µm to 110 µm with a mean value of 60 µm. The Stokes number of a particle size class is defined as the ratio of the particle aerodynamic time constant τ_p to an appropriate turbulent time scale τ_j. The Stokesian particle relaxation time was chosen for $\rho_p >> \rho_f$: $\tau_p = \rho_p d_p^2/18\mu$ where d_p is the median diameter of the particle size class, ρ_p is the particle density and μ is the fluid viscosity. While the assumption of Stokes flow around an isolated particle is not satisfied for many cases in our experimental conditions, this characteristic time calculation remains interesting for particle Reynolds number up to order of unity. The particle time scale varies significantly. For size classes 20 µm, 60 µm and 110 µm, its value is respectively $\tau_p = 3.0$ms , 27.5ms and 92.0ms.

2.3 PDA Settings

For making measurement with a Laser Doppler system, 15 optical windows consisting of thin plastics sheets with a thickness of 0.3 mm are located along the test section. A two components phase Doppler anemometer produced by Dantec (particle Dynamics Analyser: PDA) was used. For accuracy and convenience of the displacement, optical fibber system was preferred. Total inter section angle of incident beam was 1.5 degree. The receiving optics was settled at 67 degrees of off-axis angle from the incident beam to minimise the contribution of the reflected light.

The optical windows were put only on the incident beam side. The actual off-axis angle was therefore changed from the 67 degree at the outer side of the measuring range because of the refraction effect on the cylindrical wall of the test section. The off-axis angle error by this reason was checked before making diameter measurement. It is less than 3% and is thus not significant for diameters measurements.

The used refractive index was set as 1.51 which is for normal glass particle. In this experiment, water droplet and smoke were used as the tracer particles. Their refractive indexes are not equal to that of glass particle. However, the result from the single phase flow measurement with only seeding particle indicates that the mean diameter of the water droplets is less than 5µm and that of smoke is 2µm, where the values were obtained with the refractive index of the glass particle. In addition, the results from two-phase flow without tracers show that very few glass particles with less than 10 µm in diameter exist in the experimental region. Therefore, all the particles with less than 5 µm in diameter were treated as tracer particles in the present experiment. We have verified that all the particles in this size-class can be considered as good tracers of the continuous phase. Measurements are carried out for the continuous and dispersed phases simultaneously. Unless specified, we perform statistical averaging of at least one thousand independent samples in each size-class. Consequently, estimated statistical absolute errors for mean values (resp. relative errors for standard deviation values) are respectively $\Delta U \approx 0.06u$' (resp. $E_\sigma \approx 5\%$) with a 95% confidence level.

3 Two-Phase Flow at Moderate Mass Loading $M_j = 22\%$

3.1 Qualitative Description of the Flow

The sketch of the confined bluff body flow presented in figure 1 was derived from systematic axial and lateral mean velocity profiles in single phase configuration (Ishima and Boree 1998). The flow is schematically decomposed into three longitudinal regions. Region A ranges from the jet nozzle to the first stagnation point S_1 with $x_1 \approx 120$mm or $x_1/D_j \approx 6$. The jet is therefore rapidly stopped by the recirculating flow. Note also that the positive static pressure gradient due to the section increase should play an important role. In A, the central jet is surrounded by a recirculation upward flow which "feeds" both the initial entrainment in the jet and the annular shear layer developing at the edge of the bluff body. Region B ranges from S_1 to S_2 and represents the recirculation region. An intense upward flow on the axis is detected (see profiles of U_f in fig. 4a). The second stagnation point S_2 is located approximately at $x_2 \approx 220$mm or $x_2/D_j \approx 11$. Profiles measured further downstream at x = 200mm, 240mm, 320mm and 400mm show that region C downstream S_2 corresponds to the development of a wake flow.

For small particles much denser than the surrounding fluid, a first step is to compare the particle stokesian time τ_p with representative time scales of the turbulent flow.

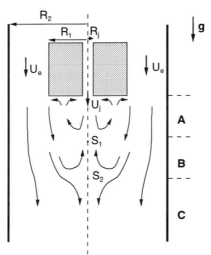

Figure 1: Sketch of the confined bluff body flow

We can characterise qualitatively important features of this configuration :

- With $\tau_j = D_0/U_0 \approx 5ms$, most particles are only partly responsive to the initial inner jet flow.
- The time scale of the strong longitudinal velocity decrease about the stagnation point S_1 represent the deceleration imposed to the particles. $\tau_{dec} = \left(\partial U_f / \partial x\right)^{-1} \approx 15ms$ can be obtained from figure 3a. Only the smallest particles can therefore recirculate.
- The annular shear layer developing at the edge of the bluff body or the downstream wake are characterised by $\tau_s = \left(\partial U_f / \partial r\right)^{-1}$ (Hishida *et al.* 1992). τ_s is evaluated from measurements (Ishima and Boree 1998) resulting in τ_s = 4ms at x = 80mm, τ_s = 7ms at x = 160mm and τ_s =20ms at x = 400mm. Most particles are therefore not responsive and dispersion of the largest particles is expected to occur later in the far wake.

3.2 Exit Velocity Profiles of the Two-Phase Flow

The fluid and particle mean velocity profiles at x=3mm (nozzle exit) are displayed in figure 2. The 60μm particles differs slightly from the continuous phase. They should experience rebounds on walls which have an obvious signature on mean longitudinal velocity profile.

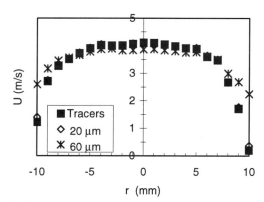

Figure 2: Fluid and particle mean velocity profiles at nozzle exit

3.3 Axial Evolution of the Two-Phase Flow

Figures 3a, b and c compare respectively the axial evolution of the mean axial velocity, rms axial velocity and rms radial velocity for the air and for the two contrasted size classes $d_p \in$ [15μm, 25μm] and $d_p \in$ [55μm, 65μm] noted for simplicity d_p = 20μm and d_p = 60μm. Figure 3a shows that particles lag the fluid behaviour so that their axial velocity decreases more slowly. The difference induced by particle inertia is particularly striking just downstream the fluid

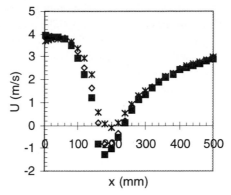

Fig. 3a : Mean longitudinal velocity

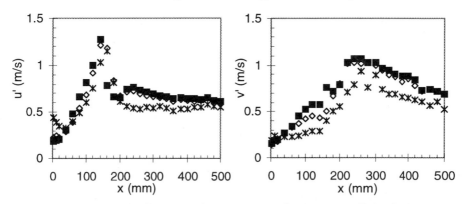

Fig. 3b : rms longitudinal velocity   Fig. 3c: rms radial velocity

Figure 3: Longitudinal evolution on the axis.
■, Tracers ; ◊, d_n = 20μm ; ✗, d_n = 60μm

stagnation point where the two size classes exhibit mean velocities of opposite signs. This trend increases with the size class and no mean negative velocity is detected in the downstream evolution of particles larger than 60 μm not displayed here.

Quite high relative Reynolds number based on the mean velocities differences are reached (The effect of a drift velocity (Simonin 1991) is not considered) in the recirculation corresponding to max(R_{ep}) = 1 for d_p = 20μm and max(R_{ep}) = 6 for d_p = 60μm.

A clear maximum of the fluid and particle axial fluctuating velocity is detected. Let's focus first on the fluid behaviour. The particular configuration with two stagnation points is interesting for turbulence studies. S_1 and S_2 have very different natures: At S_1, $\dfrac{\partial U}{\partial x} \approx -75s^{-1} < 0$ and at S_2, $\dfrac{\partial U}{\partial x} \approx +29s^{-1} > 0$. On the axis of an axisymmetrical flow, mean continuity equation tells us that $\dfrac{\partial V}{\partial r} = -\dfrac{1}{2}\dfrac{\partial U}{\partial x}$. At S_1, turbulence is therefore submitted to an axial compression and radial strain while axial strain and radial compression occur at S_2. The axial evolution of longitudinal and radial fluctuating components (see profiles of u'_f and v'_f in figures 3b and 3c) show clearly that turbulence react strongly to these solicitations. For an axisymmetrical turbulent flow, productions terms for the longitudinal, radial and azimuthal fluctuating components are respectively :

$$Prod_{uf} = -2\left[\left\langle u'^2_f\right\rangle_f \frac{\partial U_f}{\partial x} + \left\langle u'_f\, v'_f\right\rangle_f \frac{\partial U_f}{\partial r}\right] \quad (1)$$

$$Prod_{vf} = -2\left[\left\langle u'_f\, v'_f\right\rangle_f \frac{\partial V_f}{\partial x} + \left\langle v'^2_f\right\rangle_f \frac{\partial V_f}{\partial r}\right] \quad (2)$$

On the axis, using mean continuity equation, they reduce to :

$$Prod_{uf}(r=0) = -2\left\langle u'^2_f\right\rangle_f \frac{\partial U_f}{\partial x} \quad (3)$$

$$Prod_{vf}(r=0) = \left\langle v'^2_f\right\rangle_f \frac{\partial U_f}{\partial x} \quad (4)$$

S_1 is associated with an intense production of u'_f and destruction of v'_f which explains the sharp peak and the very strong anisotropy $\max(u'_f / v'_f) \approx 2.2$ at $x_1 \approx 140$mm. S_2 is associated with a moderate destruction of u'_f and production of v'_f. Moreover, mean and turbulent transport terms are known to be important contributors to v'_f balance on the axis. Downstream S_2, v'_f exceeds u'_f with a quasi constant value of $u'_f / v'_f \approx 0.7$ which increases slowly downstream.

A clear maximum of the axial fluctuating velocity of the particle is also detected in figure 3b. The maximum corresponds to the location of maximum production of u'_p by mean longitudinal velocity gradients (see (Simonin *et al.* 1995) for the corresponding equations) and is shifted downstream of u'_f maximum. The evolution of the radial fluctuating velocity of the particles (fig. 3c) is significantly different. The initial increase of v'_p is slow, particularly for the largest particles. The anisotropy of the bead fluctuating motion is very high in the recirculation zone with $\max(u'_{p20} / v'_{p20}) \approx 2.8$ and $\max(u'_{p60} / v'_{p60}) \approx 3.6$ at $x_1 \approx 140$mm. v'_p increases sharply at the downstream end of the recirculation zone about point S_2 and exceeds the longitudinal fluctuating motion. Qualitatively, this high level of v'_p is believed to be associated with a global unsteadiness of the end of the recirculation bubble. Any lateral or azimuthal flapping of this region will indeed result in a high level of radial rms velocity on the axis. A slow decrease of the radial rms velocity is measured in the early wake flow (x > 240 mm). The memory of the recirculation zone is of course important. Noticeably, the anisotropy u'_p / v'_p of the bead fluctuating motion is very similar to the fluid one. These results show clearly that any assumptions of local dragging of the particulate phase by the turbulence in such flow put aside the dominant mechanisms.

3.4 Radial Evolution of the Two-Phase Flow

To emphasise the previous discussion concerning the influence of inertia, the whole set of radial evolution at x=160 mm is drawn in figure 4. The role of the inertia of the particles is particularly clear as mean longitudinal velocities are concerned (fig. 4a). A local maximum is detected on the axis for U_{p20} and U_{p60} at r=0 is still positive while U_f is typical for a recirculating flow. Note that, in this highly intermittent region, the downward movement of particles is also probably associated with downward moving fluid pockets. It would be very interesting to draw profiles of the mean fluid velocity seen by the particles which may also exhibit a jet-like behaviour near the axis. More work is presently devoted to this point using the method of (Prévost *et al.* 1996). The mean movement of the large particles (fig. 4a, b) differs from the mean movement of fluid and 20 μm particles. In the outer region, their mean longitudinal velocities is significantly lower than the fluid velocity while their mean radial velocity shows an outward motion. Two reasons can contribute to this observation. First, the estimated turbulent time scale in the shear layer is $\tau_t = \left(\partial U_f / \partial r \right)^{-1} \approx 7$ ms which is larger than $\tau_{p20} = 3$ms but smaller than $\tau_{p60} = 27.5$ms. Due to their inertia, large particles can cross the shear layer (for x < 80 mm) and are convected downstream by the external flow. A second explanation comes when one looks at the radial evolution of the pdf of particle size measured at x = 160 mm. The probability of finding a 55/65 μm particle decreases sharply from r = 60 mm to r = 90 mm. The presence of large particles is therefore associated only with outward fluid motion of low longitudinal velocity which also contributes to the results. Note also that the statistical convergence associated with large particles in the outer region (r > 80 mm) is not so good as less than two hundred samples were used to limit duration of data acquisition in this region.

The radial evolution of the fluctuating velocities deserve a lot of comments. The longitudinal fluctuating velocities u'_{p20} and u'_{p60} are much greater than u'_f on the jet axis with $u'_{p20}/u'_f \approx u'_{p60}/u'_f \approx 1.7$. Production of the particle longitudinal fluctuation by the longitudinal mean particle velocity gradient is maximum on the axis and explains the important peak. The lateral evolution of u'_f is typical for a recirculating flow with an obvious signature of the shear layer. Local production terms (relations (1) and (2)) were inferred from the data. The maximum value of Prod_{uf} correspond closely to the region of maximum shear rate (max(Prod_{uf}) = 150m^{2}s^{-3} at $r \approx 50$mm) and to the peak detected in figure 4c. Contrary to what is observed classically in quasi parallel shear flows, a significant maximum of v'_f production (max(Prod_{vf}) = 110m^{2}s^{-3} $\approx$ 0.75max(Prod_{uf})) is located at r = 40 mm. This location corresponds to the radial edge of the recirculation zone (fig. 4a), to the maximum value of $\left(\partial V_f / \partial r \right)$ and to the peak value of v'_f.

Peak values of u'_p and v'_p are detected at these radial locations also and are believed to be associated with production mechanisms. An interesting result can be obtained if one draws the radial evolution of the correlation coefficient $R_{uv} =$

614

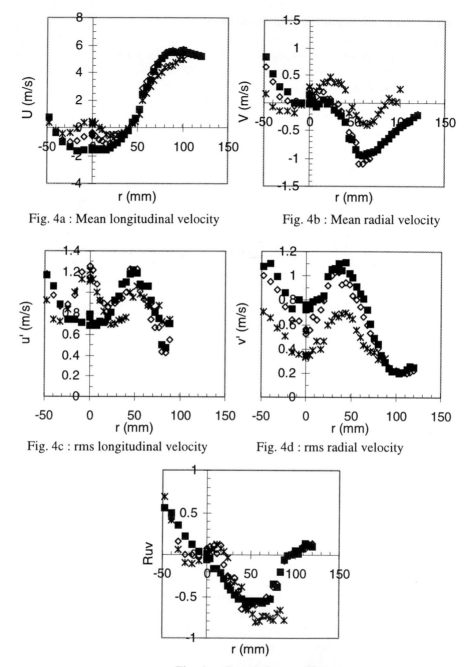

Fig. 4a : Mean longitudinal velocity

Fig. 4b : Mean radial velocity

Fig. 4c : rms longitudinal velocity

Fig. 4d : rms radial velocity

Fig. 4e : Correlation coefficient

Figure 4: Radial evolution at x=160 mm.
■, Tracers ; ◇, d_p = 20μm; ✕, d_p = 60μm

$\overline{uv}/u'v'$ (fig. 4e). R_{uvp} and R_{uvf} have opposite signs as expected near the flow axis. The large scale time scale of the shear layer is $\tau_t \approx 7$ms and the 20 μm particles follow closely the evolution of the fluid phase. On the contrary, one sees that the fluctuating velocities of the 60 μm particles are highly correlated and that R_{uv60} is quasi constant throughout the whole shear layer. This value corresponds to recent theoretical results by Fevrier and Simonin (1998).

4 Influence of the Mass Loading of the Inner Jet

The value of the mass loading provides information about the proportion of the momentum flow rate carried by the particles to that carried by the gaseous phase in the inner jet. Particles have a mean slip velocity that causes momentum transfer from the dispersed phase to the fluid. In two phase jets, (Hardalupas *et al.* 1989; Modaress *et al.* 1984) have shown that this exchange and the particle/turbulence interaction depend strongly on the particle Stokes number that separates fully responsive, partly responsive and unresponsive particles. Our polydispersion ranges from responsive to unresponsive glass beads and a strong two-way coupling is therefore expected.

Note that the central jet in region A (fig. 1) is the only dense region of the two-phase flow while the other parts of the flow are very dilute. M_j has however here a strong effect on the whole organisation of the mean flow. The axial evolution of the gas phase velocity in presence of particles is plotted in figure 5. Figure 5a shows that the axial evolution of the air mean axial velocities for $M_j = 0$, 22% and 110 % differ significantly. One notes in particular that the region of strong velocity decrease is shifted downstream when M_j increases and that no mean stagnation point is found for $M_j = 110\%$. The initial strong longitudinal increase of u'_f and v'_f seems to be delayed after 4 jet diameters at higher mass loading. This is not surprising as the relative motion between the fluctuating velocities of the particles and the fluid provides an additional mechanism for turbulent energy dissipation. Moreover, an estimation of the inter-particle collision time τ_{coll} (Lavieville et al 1997) for $M_j = 110\%$ leads to $\tau_{coll} \approx 15$ms which is comparable with τ_p. Inter-particle collisions should therefore be important near the jet outlet at the highest loading. Elsewhere in the flow, the local mass loading becomes very weak as the two stream mixes. At first order, the modification of the fluid mean flow in presence of the glass beads will thus be responsible for turbulence modifications. As an example, figure 5b shows that the location of the peak of axial rms velocity is shifted downstream together with the shift of maximum longitudinal gradient of the mean axial velocity (fig. 5a). Radial profiles not shown here would lead to similar observations.

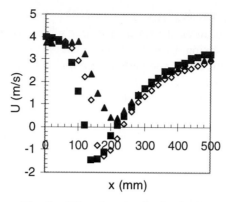

Fig. 5a : Mean longitudinal velocity

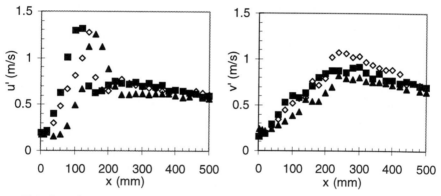

Fig. 5b : Rms longitudinal velocity   Fig. 5c : Rms radial velocity

Figure 5: Longitudinal evolution on the axis.   ■, Single phase flow;
◇, Air at M_j = 22%; ▲, Air at M_j = 110%

These changes in the flow evolution have important consequences on the distribution of the particulate phase in the flow. Figure 6 presents three number distributions of particle size measured on the axis of the flow for M_j = 22% at respectively x=10 mm, x=200 mm and x=320 mm. The probability of finding a particle belonging to a given size class at the measurement point displayed in figure 6 was estimated by dividing the number of detected particles of the given size class by the total number of particles detected. This method supposes that the proportion of valid data is the same in each size class. This statement is not true because it is well known that the probability of detection depends on the size class. The presented results are therefore biased toward large particles. No correction is attempted here (Sommerfeld and Qiu 1995) and results are only qualitative.

x=200 mm is located in the recirculation zone. We see clearly that the probability of detection of particles larger than 60 μm is greater. This is natural if one realises that the particle of largest class should not experience any return flow on the axis. The small particles are therefore dispersed while big ones go straight forward. The situation (not shown here) is very different at $M_j = 110\%$ (Ishima and Boree 1998). The probability of finding larger classes increases regularly when one moves downstream. This is due to the absence of any recirculation on the axis and to radial dispersion of the small classes. Predictions of these distributions by turbulent two-phase flow models are particularly important in combustion chambers or streams of pulverised solid fuels to ultimately predict accurately the combustion phase.

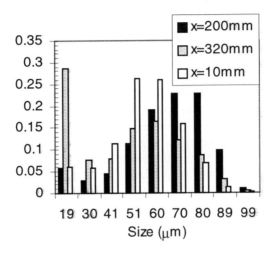

Figure 6: Qualitative number distribution of particle size measured along the axis

5 Conclusion

An experimental study of two-phase flow behind a confined bluff body has been presented in this paper. This configuration has direct links with typical industrial applications in coal combustion. Such situation is also very interesting for the development of two phase flow turbulent closures.

The initial particle size distribution covers a wide range of size classes from 20 μm to 110 μm with a mean value of 60 μm. Dealing with a polydispersion is possible when phase Doppler measurements are made. A size class analysis was presented. This analysis is necessary for such a wide range of particles time scale. Moreover, contrasted inner jet mass loading ratio M_j ranging from 22 % to 110 % have been selected in order to show the coupling between continuous and discrete phases.

For the chosen ejection velocity of the inner jet, we have shown that three different longitudinal zones can be distinguished in the flow characterised on the axis by respectively a jet flow, a recirculating flow and the establishment of a wake flow. Two stagnation points separate these regions. These stagnation points

618

have different nature described by the sign of the longitudinal rate of deformation. The reaction of the turbulent field is clear : a very high anisotropy is measured and explained on the flow axis.

This study shows particularly well that no local equilibrium with the fluid turbulence can be assumed and that one has to write and solve transport equations for the fluctuating velocity of the particles. Production effects due to either mean axial strain near the stagnation points or more classically to high shear in the initial jet, in the annular shear layer and in the "wake like" flow is clearly observed. The locations of these maxima can be highly dependant of the diameter of the particles when the influence of the inertia is dominant and the polydispersion has to be taken into account. This is expected to be a major difficulty for models, particularly when two-way coupling is significant.

The effect of mass loading was discussed and is clear when one looks at the axial evolution of the mean longitudinal velocities of the fluid in the presence of particles. The region of strong velocity decrease is shifted downstream when M_j increases and no mean stagnation point is found for $M_j = 110\%$. Moreover, the significant initial increase of rms velocities is delayed at $M_j = 110\%$. The mass loading has therefore a global effect on the flow organisation in the wake of the bluff body. By changing the flow organisation, it also modifies the region of turbulence production and the spatial distribution of the particulate phase. A bad prediction of the effects of the mass loading will thus result in a wrong prediction of the whole two-phase flow and eventually of the resulting combustion phase.

This situation is believed to be a severe test case for turbulent two-fluid models. Measurements of fluid/particle correlated motion in relevant part of the flow and comparison with model predictions are presently on their way.

Acknowledgements
We want to express a special acknowledgement for O. Simonin who initiated this research and to G. Balzer who accepted to keep on the long way necessary to obtain reliable experimental results.

References
BERLEMONT, A., DESJONQUERES, P. & GOUESBET, G. 1990 Particle lagrangian simulation in turbulent flows. *Int. J. Multiphase Flow* **16**, 19-34.

ELGOBASHI, S.E. & ABOU-ARAB, T.W. 1983 A two-equation turbulence model for two-phase flows. *Phys. Fluids* **26**, 931-938.

FEVRIER, P. & SIMONIN, O. 1998 Constitutive relations for fluid-particle velocity correlations in gas-solid turbulent flows. In *Third International Conference on Multiphase Flows, ICMF'98*, Lyon, June 8-12.

HARDALUPAS, Y., TAYLOR, A.K.M.P. & WHITELAW, J.H. 1989 Velocity and particle-flux characteristics of turbulent particle-laden jets. *Proc. R. Soc. Lond. A* **426**, 31-78.

HISHIDA, K., ANDO, A. & MAEDA, M. 1992 Experiments on particle dispersion in a turbulent mixing layer. *Int. J. Multiphase Flow* **18** (2), 181-194.

Isʜɪᴍᴀ, T. & Boʀéᴇ, J. 1998 *Presentation of a two-phase flow data base obtained on the flow loop Hercule.* , IMFT /EEC- 68.

Isʜɪᴍᴀ, T., Hɪsʜɪᴅᴀ, K. & Mᴀᴇᴅᴀ, M. 1993 Effects of particle residence time on particle dispersion in a plane mixing layer. *J. Fluids Eng.* **115**, 751-759.

Kᴜʟɪᴄᴋ, J.D., Fᴇssʟᴇʀ, J.R. & Eᴀᴛᴏɴ, J.K. 1994 Particle response and turbulence modification in fully developed channel flow. *J. Fluid Mech.* **277**, 109-134.

Lᴀᴠɪᴇᴠɪʟʟᴇ, J., Sɪᴍᴏɴɪɴ, O., Bᴇʀʟᴇᴍᴏɴᴛ, A. & Cʜᴀɴɢ, Z. 1997 Validation of inter-particle collision models based on large eddy simulation in gas-solid turbulent homogeneous shear flow. In *ASME FEDSM' 97 , June 22-26.*

Mᴏᴅᴀʀᴇss, D., Tᴀɴ, H. & Eʟɢᴏʙᴀsʜɪ, S.E. 1984 Two-component LDA measurement in a two-phase turbulent jet. *AIAA J.* **22 (5)**, 624-630.

Pʀéᴠᴏsᴛ, F., Boʀéᴇ, J., Nᴜɢʟɪsᴄʜ, H.J. & Cʜᴀʀɴᴀʏ, G. 1996 Measurements of Fluid/particle correlated motion in the far field of an axisymmetric jet. *Int. J. Multiphase Flow* **vol. 22, n° 4**, 685-703.

Rᴏɢᴇʀs, C.B. & Eᴀᴛᴏɴ, J.K. 1991 The effect of small particles on fluid turbulence in a flat-plate, turbulent boundary layer in air. *Phys. Fluids A* **3(5)**, 928-937.

Sɪᴍᴏɴɪɴ, O. 1991 Prediction of the dispersed phase turbulence in particle-laden jets. In *Proc. 4th Int. Symp. on gas solid flows, ASME FED*, vol. 121, 197-206.

Sɪᴍᴏɴɪɴ, O., Dᴇᴜᴛᴄʜ, E. & Boɪᴠɪɴ, M. 1995 Large eddy simulation and second moment closure model of particle fluctuating motion in two–phase turbulent shear flows. In *Selected Papers from the Ninth Int. Symp. on Turbulent Shear Flows, Springer-Verlag,* 85-115.

Sᴏᴍᴍᴇʀғᴇʟᴅ, M. & Qɪᴜ, H.H. 1995 Particle concentration measurements by phase-doppler anemometry in complex dispersed two-phase flows. *Experiments in fluids* **18**, 187-198.

VI.7. 3-D Measurements of Bubble Motion and Wake Structure in Two-Phase Flows Using 3-D Scanning Particle-Image-Velocimetry (3-D SPIV) and Stereo-Imaging

C. Brücker

Aerodynamisches Institut der RWTH Aachen, Wüllnerstr. zw. 5 u. 7, D-52062 Aachen, Germany

Abstract. Scanning-Particle-Image-Velocimetry is used in combination with a stereoscopic mirror system to record the three-dimensional wake structure and the bubble coordinates simultaneously over time in a bubbly two-phase flow. Pulsed planar swarms of 8 bubbles with a mean diameter of 0.8 cm were released simultaneously in a water tank and where retained in the same vertical position during the observed period by an imposed counter water stream. For the bubble size studied we found a regular shedding of hairpin vortices building a chain of vortex loops in the wake downstream. The results demonstrate the important role of the wake for the bubble interaction. Only when trailing bubbles are captured within a wake of leading ones, bubble collision and possibly coalescence happens. A video-animation of the temporal evolution of the flow demonstrates this dynamic process which is difficult to express in single pictures.

Keywords. scanning, 3-D PIV, stereo, wake, collision, hairpin vortices

1. Introduction

Bubble interaction in swarms is understood as a main factor in amplification of turbulent kinetic energy production in bubbly two-phase flows. The interaction is mainly caused by the individual wake structure, the wake capture process and the interaction of several bubbles in clusters or "chimneys" (see Stewart 1995). Shedding of vortices with ordered structure from the bubble wake plays an important role in these mechanisms.

Most of the knowledge about the wake structure of bubbles has been achieved so far from flow visualization studies of single ascending bubbles (Fan 1990) or two bubbles in line (Komasawa et al. 1980). However, for a more detailed knowledge of bubble dynamics, three-dimensional and time-recording measurement techniques are required which are able to reconstruct the vortical wake structures generated within the flow. For this reason we utilized a three-dimensional extension of classical Particle-Image-Velocimetry termed as Scanning-Particle-Image-Velocimetry (SPIV) which is based on tomographic scanning of the flow volume of interest using a scanning light-sheet technique (Brücker 1995, Brücker 1996a). The scanning light-sheet samples the flow volume rapidly in depth within a set of staggered planes in each of which the flow field can be analyzed with advanced three-dimensional PIV algorithms. This technique offers a better resolution compared to 3-D Particle-tracking techniques, however it requires high-speed imaging. It already has been successfully applied to the wake of a solid spherical cap as an approximation of a spherical cup bubble (Brücker 1996b) and is here utilized to resolve the time-dependent three-dimensional flow field in the wake of a bubble cluster rising in quiescent fluid.

Under the difficult conditions of instantaneous flow measurements in multiphase flows containing swarms of larger bubbles with random location, the multi-planar method of SPIV has the advantage to provide information in the scanned volume at each location. Therefore, important dynamical features driven by the wake structures are not overseen as possibly in case of a stationary sheet used in conventional PIV. Similar important as the three-dimensional aspect is the fact, that the time-history of the flow is recorded with sufficient accuracy. Otherwise, the dynamical behavior of both phases cannot be related to each other within a swarm of bubbles.

This paper presents the results of time-recording measurements of bubble motion and wake structures via SPIV and digital high-speed imaging in a pulsed swarm of relative few bubbles. The major aim of the study is the understanding of bubble interactions in swarms with bubbles corresponding in size to the typical bubble size as used in real chemical reactors (of order of 1 cm and more). For these conditions, the maximum void fraction of the dispersed phase at which PIV measurements could reveal the wake structure is limited to about 5%. Otherwise strong optical distortion at the phase boundaries prohibit any quantitative correct measurements over larger fields

2. Experimental setup

2.1 Bubble generator and water channel

In the present study we focus on large bubbles with a diameter of the order of 1 cm as typical for real chemical reactors. A bubble release mechanism similar as the one described in Stewart (1995) was used. This bubble generator enables the controlled generation of bubbles with near equal volume and a simultaneous planar release. A rotating plate with 8 holes is used which is placed between two fixed plates from which the top is the release plate and the bottom the filling plate, respectively. In the filling position, each chamber is filled with a defined gas volume using a proportioning screw with 8 syringes in parallel, connected with small tubes to the filling plate. In the release position, the chambers open at the top and bottom and the pressure difference drives the bubbles out of the generator.

In the first version, a plate with 8 holes with a diameter of 6 mm over an area of $4 \times 4 cm^2$ was used. The diameter of the thereby generated bubbles was on average 0.8 cm with a scatter of less than 0.1 cm. The resulting Eötvös-number is about 10 and the Reynolds-number is approximately 2000, based on the bubble diameter. Therefore, the bubbles can be expected to be of elliptical shape which would ascend in irregular rocking motion in case of a single bubble. Within a cluster of bubbles, the interaction of the bubbles and the wakes lead to more complex behavior.

The described bubble generator is placed at the bottom of a vertical water column with a cross-section of 10 cm by 10 cm in square. All side walls of the 2 meter long test section are made of transparent acrylic glass to provide optical access from 360°. The test section is fixed to a large water reservoir at the top in which the small tracer particles (Vestosint, $\varnothing \approx 30\mu m$) can be homoheneously mixed to the fluid. Either experiments with bubbles rising in quiescent fluid or within a counter stream can be realized with the vertical water channel. For the latter case, a single swam is generated and – after the swarm reaches the center of the test section – a valve at the bottom left side wall is opened to a certain position such that the swarm of bubbles remains nearly in the same position

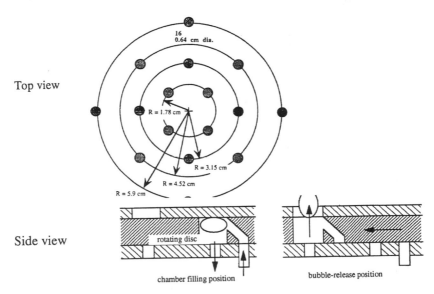

Fig. 1: Bubble generator (from Stewart 1995)

2.2 Optical setup with scanning device and stereoscopic mirror system

For application of SPIV in the bubbly flow, we used a digital high-speed video camera (*Speedcam+512*, 512 × 512 pixel resolution) with a recording rate of 1000 frames/sec and a high-speed rotating drum scanner for synchronized sequential scanning of the flow volume. The principle of the drum scanner can be found in the paper of Delo & Smits (1993). Fig. 2 depicts a sketch of the optical arrangement for the tomographic recording using the high-speed camera. The scanning light-sheet is oriented in a vertical plane and is scanned in direction parallel to the viewing direction within the flow volume. The total scanning width was 2.4 cm with a number of 5 planes, each 0.6 cm spaced apart. From successive images in each plane (temporal separation $\Delta t = 5$ ms), the 2-D velocity field within the light-sheet can be determined by cross-correlation technique.

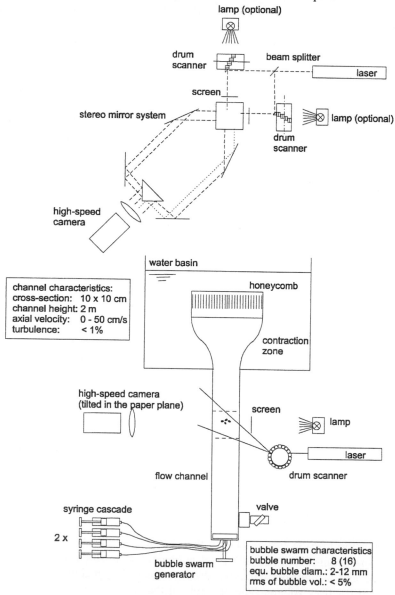

Fig. 2: Sketch of the optical arrangement for Scanning PIV in a vertical bubble column using two simultaneous scanning light-sheets and a stereo mirror system

To obtain also the bubble motion in addition to the flow field, we applied a shadow-method simultaneous to the illumination in a light-sheet and used a stereoscopic viewing arrangement. The simultaneous recording was realized here with a stereoscopic mirror arrangement and the split-screen technique, see Fig. 2. The viewing directions have an angle of 90° perpendicular to each other which offers the determination of the 3-D position of the bubbles in the volume of interest. The incoming light from both viewing directions is deflected in the mirror system such that the imaging sensor area is illuminated in two equal rectangular segments which corresponds to both views (split-screen technique). Note, that for this reason the effective horizontal resolution of each view is reduced by a factor of two. Thus, the viewing area is 3cm x 6cm with a resolution of 256x512 pixel (magnification of 1/10).

An alternative arrangement uses the shadow-technique with background illumination from two sides if only the bubble dynamics is of interest. In other experiments we also used two drum scanners scanning the flow simultaneous in two perpendicular light-sheets to obtain all three velocity components in the scanned volume with the stereoscopic recording system. As we see later, images of the bubbles can be obtained even without any background illumination only by the shadow of the light scattered from the particles in the light-sheet. Though in this case the image quality of the bubbles is bad, it is often still good enough for the three-dimensional reconstruction of the bubble position.

In order to avoid direct light from the scanning laser beam on the video chip, the rotating drum unit was tilted such that the direct beam crosses the flow under a certain offset angle and passes above the drum unit. If two laser sheets are used simultaneously, we used the both lines of the Ar^+ laser with 488nm and 514nm separate in the multi-line mode and placed color filters in the path of the non-corresponding viewing direction such that each of both segments of the sensor area only receives the scattered light from the corresponding light-sheet.

2.3 Image processing

From the images of the bubble in both views the three-dimensional bubble shape and location can be determined via back-projection procedure. In addition, the particle images within the fluid are used to determine the flow field. The necessary initial image processing steps for phase discrimination are shown in the following figure. The left half of the recorded image shows the light scattered by the particles in the light-sheet as typical for PIV. On the right-hand side, one can see the shadow-images of the bubbles within the volume, illuminated from the back by a lamp. In this image configuration, the projection direction is parallel to the light-sheet orientation which means, that the bubbles are seen from perpendicular views, one view from the scattered light originating from the light-sheet and one view from the direct projected light of the lamp. The position of the light-sheet is well seen on the right as a bright vertical bar. By means of the shadow of the laser behind the bubbles one can also find the images of the bubbles on the left and right side which correspond to one bubble in the illuminated volume (those, which are at the same vertical coordinate and which shade the light-sheet). A calibration grid was recorded prior to the experiments to find out the correct relation between object coordinates and image coordinates.

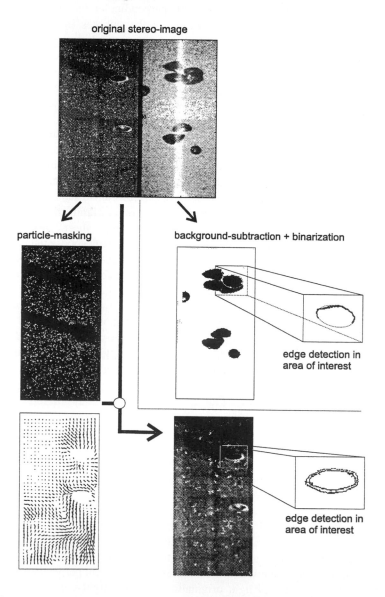

Fig. 3: Processing steps on the left-hand and right-hand sided image segment for a stereo-image in an experiment with combined light-sheet method (left view) and shadow-technique (right view). The processing steps include phase discrimination, bubble contour detection and velocity field determination

The particle images are separated from the original image by means of a particle masking technique. Therefore, the original image is first binarized (subtraction of a background image and thresholding) and those regions with bright areas of 2-10 pixels are taken as candidates of particle images. After a final dilatation procedure we obtain a binary mask which is used for masking of the original image. With a logical AND and OR procedure of the original image, one obtains the image of solely the bubbles and the image of solely the particles.

The bubble contours in both views are determined by edge-detection algorithms. Out of the bubble contours, the three-dimensional position and orientation of the bubble is estimated. Therefore a rotational symmetric revolution of the bubbles with an elliptical cross-section, i.e. a spheroidal shape of the bubbles is assumed as a first approximation. The data of the ellipsoidal contours in each view are obtained by a least-square fitting method from the determined contour coordinates. From the data of the ellipse in both views, we obtain the orientation of the bubble to the vertical (the angle of attack of the bubble).

The flow field is determined by an improved DPIV method using the method of cross-correlation with adaptive window shifting and iterative window size reduction (see Willert 1997). This allows - compared to previous methods - a higher spatial resolution and a higher accuracy of the estimated velocity vector. In addition, the signal to noise ratio increases (the bias error is eliminated due to the shifting method) and due to the window refinement, the cross-correlation can be applied also close to the surface of the bubbles without lost of accuracy. Due to the prior binary masking of the particle images, all image information of the bubbles is removed and no additional falsification due to the dispersed phase exists. The cross-correlation procedure runs over the complete image first. At the end, all vectors are removed at those locations, where the average number of particles is less than 3 in the interrogation window. At these locations, the local fluid vector is estimated by the average of the neighboring vectors. At the blank regions – the area of the dispersed phase - the cross-correlation fails and no information is obtained. This procedure yields at the end the 2-D velocity field within the light-sheet plane. In case of dual-scanning with two perpendicular light-sheets, all three velocity components can be determined in the scanned volume. For the dispersed phase, the temporal information of the center of the ellipsoids gives the trajectories of the bubbles.

3. Results of measurements

With the described bubble generator, first experiments of the bubble wake structures were carried out for bubbles ascending in quiescent fluid. In this study we used a dual-scanning method, scanning two light-sheet perpendicular to each other with two drum scanners through the volume of interest. This has a cross-section of 3.5 cm × 3.5 cm in square and a vertical extension of 7 cm. In the experiments the bubbles were released and as soon as they enter the observation volume, the camera memory loop is triggered and the next 1000 images are stored. The following result represents an experiment out of numerous others, where the wake capturing process is well seen. Fig. 4 gives the reconstructed bubble trajectories in a perspective view and a view from top together with an indication of the bubble sizes in the lower right corner (the bubble geometry could be determined from the shadows of the light scattered from the particles in the illuminated planes without any additional background illumination).

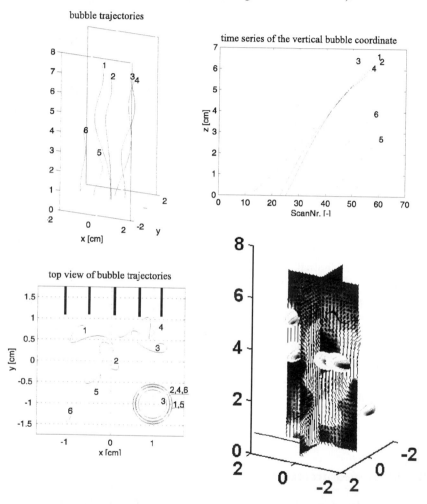

Fig. 4: Three-dimensional trajectories of bubbles in a swarm with 8 bubbles ascending in quiescent flow and corresponding wake structure to a characteristic moment of the bubble cluster (Re ≈ 2500, the background color in the vector plot indicates the amount of vertical upwards flow)

All bubbles have more or less the same diameter of the projected area of order of 1 cm. The Reynolds-number based on this diameter amounts ca. Re=250. The figure displays in addition at the right the vertical coordinate of the bubbles over the recording time. A characteristic picture of the induced flow field is shown in a vector plot, where the background color in a rainbow spectrum represents the amount of upwards velocity (red color indicates a region with high upwards motion, blue a region of low vertical motion). The flow fields in the two perpendicular light-sheet planes are shown together to illuminate the three-dimensional nature of the wake. In the present experiment, two bubbles enter first the observation volume (no. 2 and no. 3) which are followed by two other bubbles (no. 1 and no. 4), which show a much increased vertical velocity and collide later with the leading bubbles. This is well documented in the plot showing the vertical bubble motion where the trailing bubbles show an increased ascending velocity due to the suction effect.

The top view demonstrates, that the leading bubbles move more or less in a helical path overlayed with a linear sidewards motion, whereas the trailing bubbles have no characteristic trajectory and are mainly driven by the wake of the leading bubbles. To later moments they collide with the leading bubbles and drift laterally away. The colored region shown in the vector plot indicate that wake fluid with increased vertical velocity is left in a helical structure behind the bubble.

Another experiment demonstrates the interaction of bubbles over longer periods which was carried out in counter flow to keep the bubbles in the observation region. The mean downwards flow velocity was abrupt adjusted from rest to 20 cm/s after the bubbles entered the frame. Here, only a single light-sheet was used for scanning and the other view was used for shadow-imaging of the bubbles. The graphical presentation is equal to that of the last figure. In the vector plot the downwards velocity component by the counter flow is subtracted. Here, the color represents the amount of the vorticity component normal to the light-sheet plane. It was calculated by means of central difference schemes from the measured velocity field and was smoothed to remove erroneous small scale regions of high vorticity and to clarify the large-scale vortices.

A two-fold capturing event of a small satellite bubble no. 2 (this bubble was generated with a reduced gas volume in one of the chambers of the bubble generator) into the wake of a larger leading bubble no. 3 could be observed in the recorded period (scan no. 50-70 and scan no. 100-135 in Fig. 5). Note, that in all experiments collisions were observed exclusively after wake capturing (this agrees with the findings from Stewart 1995, and Fan 1990) as also seen in the presented results where prior to collision the small bubble is captured by the wake of the larger one. The small bubble first reaches to a certain moment (scan no. 50 in the plot of the vertical bubble coordinates over time in Fig. 5) a critical minimum distance to the large leading bubble and is then trapped into the wake rapidly. It is then accelerated and approaches the large bubble, however, it happens that it is repelled again (scan no. 60 in Fig. 5). To a later moment, the small bubble is captured to the second time into the wake of the large bubble (scan no. 120) and then collides with it (scan no. 135).

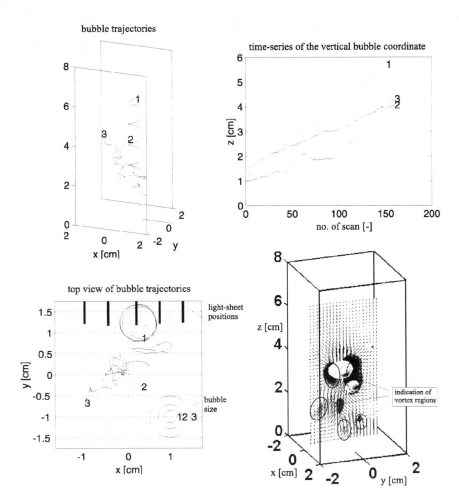

Fig. 5: Three-dimensional trajectories of bubbles in a swarm with 8 bubbles in a counter flow and corresponding vortical structure in the wake to a characteristic moment of the bubble cluster (Re $\approx$ 2500, the background color in the vector plot indicates the amount of vorticity)

The vorticity distribution clarifies the ring-like vortical structure in the attached wake of the large bubble. Farther downstream, the remainder of a loop-like vortical structure can be seen. This structure was found to be shed periodically from the larger bubble similar as observed in the wake of a spherical cap (see Brücker 1996). A computer animation of the dynamical process can be downloaded at http://wwwold.aia.rwth-aachen.de/.public_html/christoph/home as an avi-file (also published on the CD of the Proc. 3rd Int. Conf. Multiphase Flows, Brücker 1998).

A more clarifying presentation of the wake structure is shown in Fig. 6 by means of a spatio-temporal reconstruction of the vorticity field. First, from each scan the plane with the closest location to the center of the leading bubble was chosen (in most cases plane no. 4). From the derived vorticity field in this plane, a horizontally line out of the field at a fixed location 2 cm downstream of the centroid of the bubble was taken and the vorticity information there was staggered scan-by-scan in a data array. The resulting data matrix displayed as a contour plot gives the spatio-temporal evolution of the vorticity in this location of the wake of the leading bubble over the complete period of recorded bubble motion. Because the shedding of the vortices was found to be nearly periodic in time with an approximate constant convection velocity, the vertical coordinate in Fig. 6 can be interpreted either as the time coordinate or alternatively as the vertical coordinate in a frozen picture of the wake. Contour lines of constant vorticity are shown where the vorticity exceeds a certain cut-off level.

The spatio-temporal reconstruction demonstrates the nature of the wake as a chain of alternately tilted ring-like vortex structures with a regular zig-zag orientation in this plane. The neighboring positive and negative contour lines represent therein slices through ring-like vortical structures. The orientation of these structures are marked in the graph by ellipses, which also clarify the correspondence of the positive and negative contour lines. From inspection of the instantaneous vorticity field in the other planes of the scans, we could see that these structures do not represent closed vortex rings but the heads of downstream oriented hairpin-like vortices which are arranged in a chain of vortex loops as explained in the sketch on the right-hand side beneath the vorticity reconstruction.

These results clearly demonstrate to the first time quantitatively that the wake of bubbles at this range of Reynolds-number consists of a chain of vortex loops similar as observed in the wake of solid axisymmetric bluff bodies like prolate spheroids, spheres or spherical caps (Fan 1990, Brücker 1997). It is also obvious from the measurements that the wake is not a helical vortex structure as quoted by Lindt 1972. Our measurements confirm earlier results using qualitative visualizations of the wake of spherical cap bubbles made by Yabe and Kunii (1978) and recent flow visualization studies of the wake of ellipsoidal polystyrol particles rising in water (Lunde and Perkins 1997).

Further interesting to note is that the capture process of the small bubble into the wake of the leading bubble is triggered by the shedding of the vortex loops. The moment of wake capturing initiation in the second capturing cycle (see scan no. 100-135 in Fig. 5) is marked in the graph in Fig. 6 at point "A". The smaller bubble at this moment is located at the right-hand side of the large bubble at the region where the arrow is located in the sketch. The arrow marks the direction of the flow field induced by the loop ahead to this moment. This induced flow is responsible that the smaller bubble is driven into the wake. Both events, on one hand that the bubbles reach a minimum distance to each other to that moment of their individual paths, and on the other hand that the small bubble is then at the right place with respect to the induced flow field (and to the right time within the shedding cycle) are the conditions for the wake capturing to take place.

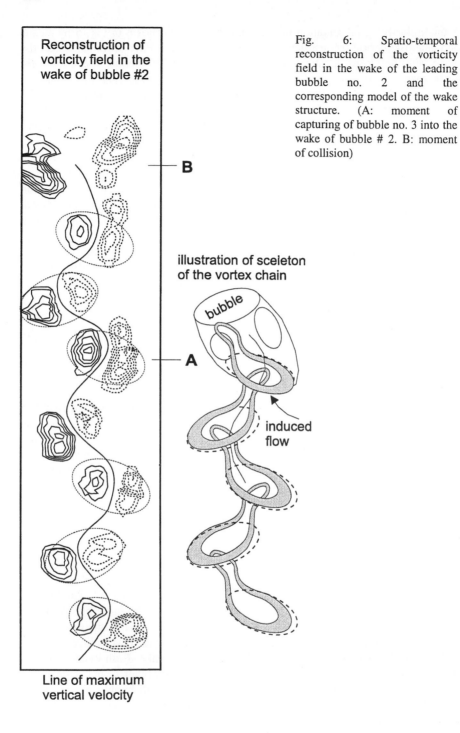

Reconstruction of vorticity field in the wake of bubble #2

B

Line of maximum vertical velocity

illustration of sceleton of the vortex chain

bubble

A

induced flow

Fig. 6: Spatio-temporal reconstruction of the vorticity field in the wake of the leading bubble no. 2 and the corresponding model of the wake structure. (A: moment of capturing of bubble no. 3 into the wake of bubble # 2. B: moment of collision)

It needs a further complete shedding cycle until the captured bubble collides with the leading bubble which is marked in Fig. 6 as "B". During this period the bubble motion is not continuous - see the plot of the vertical bubble coordinate over time in Fig. 5 - but the bubble is alternately accelerated (scan no. 100), decelerated (scan no. 115) and accelerated again (scan no. 120). Careful inspection of the shedding cycle revealed that the acceleration phase of the small bubble is coupled in time to the passing by of a loop with the head oriented to the right. Vice versa, the following deceleration phase is coupled by the passing by of the next loop with the head oriented to the left. This is due to the change of the induced flow field at the right-hand side of the vortex chain from where the small bubble approaches the large one. With the period of the alternate loop orientation and zig-zag arrangement the induced velocity at the considered location switches alternately from a positive to a negative vertical component. While in the first wake capture cycle (scan no. 50-70) the deceleration was strong enough for the bubble to be repelled again from the leading large bubble, in the here described second wake capturing cycle the deceleration is not sufficient to reject the bubble again out of the wake so that the bubbles finally collide.

Overall, the results demonstrate that the wake capturing process is strongly coupled to the periodical shedding. Therefore, the shedding of vortices in the wake of bubbles trigger the probability of wake capturing and therefore the probability of collision and coalescence. This could possibly help to develop a more accurate statistical model of bubble interaction in multiphase flows.

Finally, the obtained bubble trajectories are discussed in context with the observed bubble wakes. A first striking results is that the smallest bubble no. 1 with a diameter of 3 mm has the maximum rising velocity and a near perfect helical path (see Fig. 5). It has also the highest orientation to the vertical with an angle of attack of about 40° and a large ratio of the helix diameter to the bubble diameter of about 3. An explanation therefore can be found in the theory put forward by Saffman (1956) which is the subject of a following paper. No indications of a periodic shedding could be found from the obtained flow field which gives the hint, that the wake consists of a pair of attached streamwise vortices as also found at lower Reynolds-numbers of order of 200 in the wake of solid axisymmetric bluff bodies (Brücker 1997). This has to be proven with PIV experiments where the light-sheet is oriented horizontally which is subject of our future work. Bubble no. 2 initially shows a zig-zag-like motion which is coupled with the shedding of vortex structures similar as in case of the large bubble no. 3, see Fig. 6. The orientation of the bubble is about 25°-30° to the vertical and the ratio of oscillation amplitude to the bubble diameter is about 2 (see Fig. 5). As soon as the bubble is captured in the wake of bubble no. 3, the motion is more or less determined by the pressure field in the wake of the large bubble which can be seen in a large increase of the angle of attack of the bubble. Bubble no. 3 shows initially a rocking motion which persist until short before the collision. It moves in an irregular helical path with a ratio of the oscillation amplitude to the bubble diameter of less than 0.5. The path oscillation frequency is definitely coupled to the vortex shedding frequency similar as found for bubble 2.

4. Conclusion

A scanning PIV method has been developed for application in multiphase flows, where simultaneously the bubble trajectories and shape as well as the flow field can be obtained in the three-dimensional flow volume. A stereoscopic mirror arrangement and the split-screen technique have been used to store two perpendicular views of the flow field on a digital high-speed video-camera. A maximum of 1000 images could be recorded with a frame frequency of 1000 Hz which was sufficient to follow the bubbles motion and flow evolution in detail. Cross-correlation technique was used to reconstruct the velocity field within the light-sheet planes. Using special phase discrimination algorithms we could also detect the bubbles and track their position during time of recording within the scanned volume.

The results demonstrate that the wake of the bubbles at Reynolds-numbers of order of 1000 consists of a regular chain of hairpin-like vortices with their heads oriented downstream and arranged in a zig-zag manner (the heads point alternately to the left and right-hand side). From previous SPIV measurements of the wake structure behind a solid spherical cap it can be said that the basic wake structure is the same. In case of the solid spherical cap it was found that the hairpin-like structures result from the instability of the wake against helical waves with a wave number of 1 in azimuthal direction which is similar for all axisymmetric solid bluff bodies and seems to hold also for the wake of spherical or ellipsoidal bubble. The loops are generated by two counter-rotating azimuthal waves of the same amplitude and phase velocity.

The observation of an incomplete and a following complete capturing processes of a smaller bubble into the wake of a leading large bubble during the recorded period could be directly related to the vortex shedding in the wake of the leading on. Hence, the shedding of vortices in the wake of bubbles trigger the probability of wake capturing and therefore the probability of collision and coalescence. This demonstrates that the wake nature is a dominant feature in multiphase flows which has a large influence on bubble-bubble interaction and cluster formation. In addition, it was found that the bubble path oscillations are strongly related to their wake nature. In general, the results demonstrate the necessity of time- and whole-volume measurement techniques to describe the flow in the wake of the bubbles.

Acknowledgements
The author thanks the DFG for sponsoring the work in the Focus-Research-Program "Analyse, Modellierung und Berechnung mehrphasiger Strömungen", Kr 387/35

References

BRÜCKER, CH. 1995 Digital-Particle-Image-Velocimetry (DPIV) in a scanning light-sheet: 3-D starting flow around a short cylinder, *Exp. in Fluids* **19**, 255-263

BRÜCKER, CH. 1996a 3-D Scanning Particle-Image-Velocimetry (3-D SPIV), *VKI Lecture Series on Particle–Image-Velocimetry* LS-1996-03, Von Karman Institute, Rhode-St-Genese, Belgium

BRÜCKER, CH. 1996b 3-D Scanning-Particle-Image-Velocimetry: Technique and Application to a spherical Cap Wake Flow, *Applied Sci. Res.* **56**(2-3), 157-179

BRÜCKER, CH. 1997 Untersuchung des Nachlaufs einer Kugelschale mit der dreidimensionalen Particle-Image-Velocimetry, *Abhandlungen aus dem Aerodynamischen Institut* **32**, 58-67.

BRÜCKER, CH. 1998 Bubble interaction in swarms: a study of the wake structures with tomographic Particle-Image-Velocimetry, *Proc. 3rd Int. Conf. Multiphase Flows*, June 8-12, Lyon, published on a CD

DELO, C. & SMITS A. J. 1993 Visualization of the three-dimensional, time evolving scalar concentration field in a low Reynolds-number turbulent boundary layer. In *Near Wall Turbulent Flows*, Launder, B.E. Eds, 573-582

FAN, L.S.F. 1990 Gas-Liquid-Solid Fluidization Engineering", *Butterworths Ser. in Chem. Eng.*, Butterworths, Boston, London

FAN L.S. & TSUCHIYA, K. 1990 Bubble wake dynamics in liquid and liquid-solid suspensions, *Butterworths Ser. in Chem. Eng.*, Butterworths, Boston, London

KOMASAWA, I., OTAKE, T. & KAMOJIMA M. 1980 Wake behavior and its effect on interaction between spherical-cap bubbles, *J. Chem. Eng.* **13**(2), 103-109.

LINDT, J.T. 1972 On the periodic nature of the drag on a rising bubble, *Chem. Eng. Sci.* **26**, 1776-1777

LUNDE K & PERKINS, R.J. 1997 Observations on wakes behind spheroidal bubbles and particles. Paper no. FEDSM97-3530, ASME-FED Summer meeting, Vancouver, Canada

SAFFMAN, P.G. 1956 On the rise of small air bubbles in water, *J. Fluid Mech.* **1**, 2.

STEWART, C. W. 1995 Bubble interaction in low viscosity liquids, *Int. J Multiphase Flows* **21**(6), 1037-1046.

WILLERT C. 1997 Stereoscopic Digital-Particle-Image-Velocimetry for wind tunnel flows, *Meas. Sci. Techn.* **8**, 1465-1479.

YABE, K. & KUNII, D. 1987 Dispersion of molecules diffusing from a gas bubble into a liquid, *Int. Chem. Eng.* **18**, 666-671

Authors' Index

Springer
and the
environment

At Springer we firmly believe that an international science publisher has a special obligation to the environment, and our corporate policies consistently reflect this conviction.
We also expect our business partners – paper mills, printers, packaging manufacturers, etc. – to commit themselves to using materials and production processes that do not harm the environment. The paper in this book is made from low- or no-chlorine pulp and is acid free, in conformance with international standards for paper permanency.

Springer

Printing: Saladruck, Berlin
Binding: H. Stürtz AG, Würzburg